ADVANCES IN MOLECULAR GENETICS OF PLANT-MICROBE INTERACTIONS

Current Plant Science and Biotechnology in Agriculture

VOLUME 10

Aims and Scope
The book series is intended for readers ranging from advanced students to senior research scientists and corporate directors interested in acquiring in-depth, state-of-the-art knowledge about research findings and techniques related to plant science and biotechnology. While the subject matter will relate more particularly to agricultural applications, timely topics in basic science and biotechnology will also be explored. Some volumes will report progress in rapidly advancing disciplines through proceedings of symposia and workshops while others will detail fundamental information of an enduring nature that will be referenced repeatedly.

The titles published in this series are listed at the end of this volume.

Advances in Molecular Genetics of Plant-Microbe Interactions

Vol. 1

*Proceedings of the 5th International Symposium
on the Molecular Genetics of Plant-Microbe Interactions,
Interlaken, Switzerland, September 9–14, 1990*

edited by

HAUKE HENNECKE

ETH, Zürich, Switzerland

and

DESH PAL S. VERMA

The Ohio State University, Columbus, Ohio, U.S.A.

KLUWER ACADEMIC PUBLISHERS

DORDRECHT / BOSTON / LONDON

Library of Congress Cataloging-in-Publication Data

International Symposium on the Molecular Genetics of Plant-Microbe
   Interactions (5th : 1990 : Interlaken, Switzerland)
     Advances in molecular genetics of plant-microbe interactions :
   proceedings of the 5th International Symposium on the Molecular
   Genetics of Plant-Microbe Interactions, Interlaken, Switzerland,
   September 9-14, 1990 / edited by Hauke Hennecke and Desh Pal S.
   Verma.
        p.   cm. -- (Current plant science and biotechnology in
   agriculture ; v. 10-)
     Includes indexes.
     ISBN 0-7923-1082-9 (acid-free paper)
     1. Plant-microbe relationships--Genetic aspects--Congresses.
   2. Plant-microbe relationships--Molecular aspects--Congresses.
   I. Hennecke, Hauke. II. Verma, D. P. S. (Desh Pal S.), 1944-   .
   III. Title. IV. Series: Current plant science and biotechnology in
   agriculture ; 10.
   QR351.I53   1990
   581.2'3--dc20   90-25677

Published by Kluwer Academic Publishers,
P.O. Box 17, 3300 AA Dordrecht, The Netherlands.

Kluwer Academic Publishers incorporates
the publishing programmes of
D. Reidel, Martinus Nijhoff, Dr W. Junk and MTP Press.

Sold and distributed in the U.S.A. and Canada
by Kluwer Academic Publishers,
101 Philip Drive, Norwell, MA 02061, U.S.A.

In all other countries, sold and distributed
by Kluwer Academic Publishers Group,
P.O. Box 322, 3300 AH Dordrecht, The Netherlands.

Printed on acid-free paper

Printed in the Netherlands

PREFACE

Research on the interaction between plants and microbes has attracted considerable attention in recent years. The use of modern genetic techniques has now made possible a detailed analysis both of plant and of microbial genes involved in phytopathogenic and beneficial interactions. At the biochemical level, signal molecules and their receptors, either of plant or of microbial origins, have been detected which act in signal transduction pathways or as co-regulators of gene expression. We begin to understand the molecular basis of classical concepts such as gene-for-gene relationships, hypersensitive response, induced resistance, to name just a few. We realize, and will soon exploit, the tremendous potential of the results of this research for practical application, in particular to protect crop plants against diseases and to increase crop yield and quality.

This exciting field of research, which is also of truly interdisciplinary nature, is expanding rapidly. A Symposium series has been devoted to it which began in 1982. Recently, the 5th International Symposium on the Molecular Genetics of Plant-Microbe Interactions was held in Interlaken, Switzerland. It brought together 640 scientists from almost 30 different countries who reported their latest research progress in 47 lectures, 10 short oral presentations, and on over 400 high-quality posters. This book presents a collection of papers that comprehensively reflect the major areas under study, explain novel experimental approaches currently in use, highlight significant advances made over the last one or two years but also emphasize the obstacles still ahead of us. Thus, the book gives a good overall impression of the state-of-the-art of research on molecular plant-microbe interactions, and will hopefully serve as a useful reference source for the coming years.

The Interlaken Symposium would not have been possible without the generous support by various foundations, research organizations, private corporations as well as the assistance from the host institution, the Swiss Federal Institute of Technology Zürich (see overleaf). My special thanks go to Hélène Paul for her excellent secretarial work during the preparation of this book.

Zürich   Hauke Hennecke
October 1990

The following organizations and industrial companies are gratefully acknowledged for their generous support:

- Eidgenössische Technische Hochschule, Zürich, Switzerland
- Ciba-Geigy AG, Basel, Switzerland
- Hoffmann-La Roche AG and Maag AG, Basel and Dielsdorf, Switzerland
- Sandoz AG, Basel, Switzerland
- Schweizerischer Nationalfonds, Bern, Switzerland
- Schweizerische Akademie der Naturwissenschaften, Bern, Switzerland
- Nestlé AG, Vevey, Switzerland
- Schering AG, Berlin, FRG
- Schweizerische Gesellschaft für Mikrobiologie
- International Society for Plant Molecular Biology
- Bayer AG, Leverkusen, FRG
- Hoechst AG, Frankfurt, FRG
- E.I. du Pont de Nemours & Co., Wilmington DE, USA
- Northrup King, Stanton MN, USA
- Rahn & Co., Zürich, Switzerland
- The American Phytopathological Society, USA
- Rhône-Poulenc, Lyon, France
- MBR Bio Reactor AG, Wetzikon, Switzerland
- Technomara AG, Wallisellen, Switzerland
- Lucerna Chem AG, Luzern, Switzerland
- Medipack AG, Trüllikon, Switzerland
- Du Pont de Nemours (Deutschland) GmbH, Bad Homburg, FRG
- Digitana AG, Horgen, Switzerland
- Suchema AG, Kaltenbach, Switzerland
- Bender & Hobein AG, Zürich, Switzerland
- Gibco/BRL AG, Basel, Switzerland

Table of Contents

SECTION I: BACTERIA-PLANT INTERACTIONS (PATHOGENIC)

SECTION II: BACTERIA-PLANT INTERACTIONS (SYMBIOTIC)

SECTION III: PLANT-FUNGUS INTERACTIONS

SECTION IV: HOST PLANT RESPONSE TO MICROBIAL INFECTION

SECTION V: BIOCONTROL AND RHIZOSPHERE ASSOCIATIONS

Section I

BACTERIA-PLANT INTERACTIONS (PATHOGENIC)

MOLECULAR STRATEGIES IN THE INTERACTION BETWEEN *AGROBACTERIUM* AND ITS HOSTS

Eugene W. Nester, Ph.D. and Milton P. Gordon, Ph.D*#
* Department of Microbiology; *# Department of Biochemistry
University of Washington, Seattle, WA 98185

Crown Gall Tumor Formation By *Agrobacterium*. *Agrobacterium* induces plant tumors by inserting a fragment (T-DNA) of its tumor inducing (Ti) plasmid into the plant chromosome. Some understanding of the functioning of genes required for this process is being elucidated. The *vir* genes are activated by two classes of plant signal molecules, plant phenolic compounds and a variety of plant sugars. These molecules interact with the VirA protein which then activates the VirG protein, which in turn transcriptionally activates all of the *vir* genes following binding to their promoter regions. The VirA and VirG proteins, members of a two component regulatory system undergo the phosphorylation and phosphate transfer demonstrated by other members of the group. The *virD* operon is concerned with early stages in the processing of T-DNA. The VirD1 protein has topoisomerase activity for the Ti plasmid. The *virD2* locus codes for a site specific endonuclease which nicks at the right and left borders of the T-DNA. The VirC1 protein interacts with the overdrive sequence and in conjunction with the VirD1 and VirD2 proteins promotes T-strand formation and the efficiency of transfer of the T-DNA. The *virE* operon codes for a single stranded DNA binding protein which associates with the T-DNA. The *virB* operon has now been sequenced and many of its protein products have been shown to be associated with the cytoplasmic membrane. Presumably these protein components provide the pore through the bacterial envelope for the exit of the T-DNA. The *virB11* gene product has ATP'ase activity and autophosphorylates. Its DNA sequence is similar to that of a gene in *B. subtilis* required for development of competence. The VirB10 protein is oligomeric and anchored in the inner membrane. Open reading frames 9, 10 and 11 all code for gene products which are essential for tumor formation.

The genes of the T-DNA code for enzymes of auxin and cytokinin synthesis and presumably one or more genes code for the alteration of cytokinin and perhaps auxin activity. Another has been reported to code for transport of opines out of the plant cell.

Overall Features of Crown Gall Tumor Formation

Agrobacterium tumefaciens induces a disease, crown gall, in a wide variety of dicotyledonous plants by transferring a piece of its tumor-inducing (Ti-) plasmid into the plant cell where it

3

H. Hennecke and D. P. S. Verma (eds.),
Advances in Molecular Genetics of Plant-Microbe Interactions, Vol. 1, 3–9.
© 1991 *Kluwer Academic Publishers. Printed in the Netherlands.*

becomes integrated and functions in the plant. The overall features of this disease are illustrated in Figure 1.

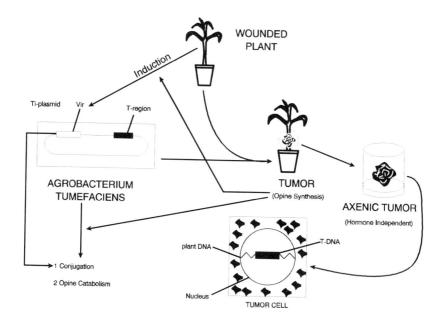

Figure 1. Overall features of crown gall tumor formation by *Agrobacterium*. It is possible to delete the T-DNA and insert useful genes under the control of plant promoters. The expression of these genes integrated into the plant DNA confers desired properties on the plant. This technology forms the basis of genetic engineering of plants using *Agrobacterium*. Note that an arrow from the vir genes enhance the conjugal transfer of the Ti plasmid to Agrobacterium (Gelvin and Habeck(1990)), (Steck and Kado(1990)). Further, a recent report presents evidence that certain opines promote T-DNA transfer into plant cells (Veluthambi(1989)).

The transferred and integrated DNA (T-DNA) codes for the synthesis of the two growth regulators, auxin and cytokinin, as well as for a group of amino acid derivatives termed opines. The expression of the genes for phytohormone synthesis, gives rise to the symptoms of crown gall tumor formation. The transfer of the T-DNA requires the expression of a variety of other genes on the Ti-plasmid, the virulence (*vir*) genes. These genes, which are involved in the processing and transfer of the T-DNA, are not expressed when *Agrobacterium* grows in the absence of plant cells, but are activated by plant cell metabolites synthesized by the wounded plant. As far as we are aware, the *Agrobacterium*-plant interaction is unique in that the end result is the transfer and integration of a piece of bacterial DNA into the plant chromosome. However, the interaction of this organism with the plant can serve as a model system for many other bacterial-plant interactions, since it involves such features as the attachment of the bacteria to their host cells, the activation of genes required for pathogenicity by plant signal molecules, and the production of phytohormones, auxin and cytokinin, as the virulence factors which result in the gall-like symptoms of the disease. These features are common to other bacterial-plant interactions.

Early Events in the Transfer of T-DNA into Plant Cells

ATTACHMENT OF BACTERIA TO PLANT CELLS

Considerable evidence exists that *Agrobacterium* must bind to plant cells in order to cause crown gall tumors. However, the molecular basis of this attachment process remains elusive. A number of mutants of *Agrobacterium* have been identified which map to three loci, all of which map to the chromosome and are termed *Chv*. None of these mutants is capable of attaching, and all are avirulent. Further, all of these mutants are involved with the synthesis or transport of a low molecular weight polysaccharide, β-1,2-glucan).

The *Chv*A codes for a protein that is necessary for the transport of the β-1,2-glucan into the periplasm; the *Chv*B gene codes for a 235kd membrane-associated protein which converts glucose into the cyclic β-1,2-glucan. The *exo*C locus codes for an enzyme which converts glucose 6phosphate to glucose 1phophate, a step required for the synthesis of cellulose as well as β-1,2-glucan. How β-1,2-glucan functions in attachment is not clear. It has not been possible to demonstrate that the addition of concentrated cell supernatants from β-1,2-glucan synthesizing cells to β-1,2-glucan negative cells can complement the negative cells in attachment (Cangelosi, unpublished observation). Significantly, the β-1,2-glucan negative cells are pleiotrophic and some property other than β-1,2-glucan synthesis may be responsible for the cells' inability to attach.

TI-PLASMID

Although a number of genes necessary for tumor formation have been identified which map to the chromosome, it appears that most of the genes required for tumor formation are located on the large 180 kilobase Ti-plasmid). The two sets of genes required for tumor formation include the T-DNA and the *vir* genes.

DESIGNATION OF VIR GENES

The virulence genes are comprised of approximately 35 kilobases of DNA and are essential for tumor formation, although they are not transferred into the plant. They can be divided into six highly studied transcription units, or operons. These include *vir*A, *vir*G, *vir*B, *vir*C, *vir*D and *vir*E. Two additional *vir* operons have been identified in the octopine Ti *vir* regions (*vir*H and *vir*F), but they have not been studied extensively. The most relevant data on the *vir* genes are summarized in Figure 2.

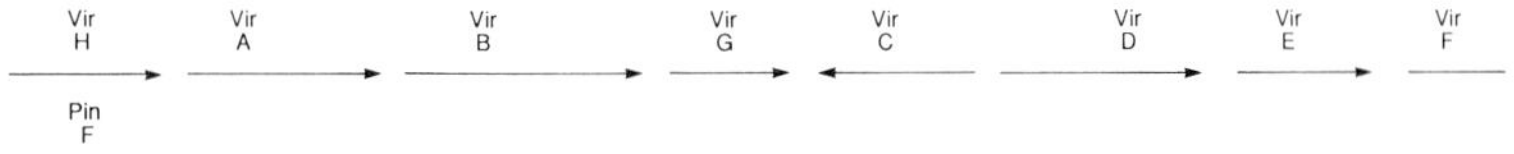

| Vir H | | Vir A | | Vir B | | Vir G | | Vir C | | Vir D | | Vir E | | Vir F |

Pin F

Vir	Inducibility	Size (Kb)	ORF's	Function
A	+	2.8	1	Plant Signal Sensor
G	+	1.0	1	Transcriptional Activator
B	+	9.5	11	Pore?
D	+	4.5	4	Processing of T-DNA; Topoisomerase and Endonuclease
C	+	1.5	2	Promotes Transfer of T-DNA; Binds O.D.
E	+	2.2	2	Single-Strand DNA Binding Protein
H	+	3.4	2	Cytochrome P450 Enzyme
F	?	?	?	?

Figure 2. *Vir* gene order and relevant information.

It seems likely that additional *vir* genes may be identified in the future. These probably will be recognized as genes which are required for tumor formation on some, but not other, plants, and/or as genes which are required for optimal tumor formation. Different strains of *Agrobacterium* have different *vir* genes. For example, strains of *Agrobacterium* that induce tumors that synthesize the opine octopine have the *vir*H region, whereas strains which induce nopaline synthesizing tumors do not. However, the nopaline-inducing strains have a vir gene coding for the synthesis of isopentenyl adenosine monophosphate at the same relative map position. There are other less well-identifiable differences between these two types of Ti-plasmids.

CHEMICAL INDUCER OF VIR GENES

The induction of *vir* genes depends upon the synthesis of plant signal molecules synthesized by the wounded plant. One such molecule which has been identified is acetosyringone (3,5-dimethoxy4hydroxy acetophenone) (Stachel et al(1985)). This compound, which seems to be a derivative of a precursor of lignin biosynthesis, has been found in a wide variety of plants following wounding. A number of other compounds chemically related to acetosyringone also have inducing activity (Spencer et al(1988)).

It is now clear that in addition to acetosyringone, a variety of sugars are vir gene inducers. The sugars, all of which are components of the cell wall of dicotyledonous plants, have their greatest effect when acetosyringone is limiting (Cangelosi, Ankenbauer and Nester (in press)). Mutants which lack the periplasmic protein which binds these sugars can induce tumors on some but not on other plants. These mutants also are defective in chemotaxis to these monosaccharides and also grow poorly on them as a carbon and energy source. Recent experiments have shown that the monosaccharides interact with the periplasmic binding protein which then interacts with the periplasmic portion of virA or with the membrane proteins associated with chemotaxis and sugar uptake (Cangelosi, Ankenbauer and Nester (in press)).

ACTIVATION OF PVRA PROTEIN

Two of the *vir* genes, *vir*A and *vir*G are required for the expression of all *vir* genes, since mutations in either locus eliminate the expression of all other *vir* genes (Winans et al(1986)). Protein homology searches have shown that the VirA and VirG proteins are homologous to a large number of two-component regulatory systems including EnvZ/OmpR, NtrB/NtrC and CheA/CheY (Winans et al(1986)). The first protein of each pair detects a particular environmental signal, and then transfers this information to a second component which in turn, in most cases, activates the expression of a series of genes whose gene products respond to the environment. The VirA protein is a transmembrane protein (Winans, et al(1989)). Two hydrophobic regions anchor the N-terminal portion of the protein to the cytoplasmic membrane with a region protruding into the periplasmic space and the C-terminal domain remaining in the cytoplasm. Presumably in some undefined way, the VirA protein interacts with the plant signal molecule, thereby becoming activated, and in turn activates the VirG protein. The mechanism of VirA protein activation is now being clarified. It has been shown that the VirA protein has an autophosphorylating activity which is the most likely mechanism for VirA protein activation (Jin et al(1990a)). The role that acetosyringone plays in autophosphorylation is unclear, since the addition of g-labeled ATP with purified VirA protein but without acetosyringone still results in the labeling of the VirA protein.

A histidine residue located in the highly conserved block of amino acids which is found in all of the VirA homologues is the amino acid phosphorylated. If this histidine is mutated to a glutamine residue which cannot be phosphorylated, then the vir genes are not inducible nor is the strain capable of inducing crown gall tumors. This demonstrates that phosphorylation of the VirA protein is essential for its proper functioning.

INTERACTION OF VIRA AND VIRG

The phosphate residue of the phosphorylated histidine in the VirA protein is transferred directly to the VirG protein (Jin, unpublished observation). Thus, when the phosphorylated VirA protein was mixed with unphosphorylated VirG protein and samples withdrawn at various times, with increasing time of incubation, the phospho-VirA signal got weaker while the phospho-VirG signal got stronger. The phosphate transfer occured very rapidly since five seconds after the VirA and VirG proteins were mixed together, transfer could be observed.

The phosphate bond in the VirG protein is highly unstable to both acid and base, which suggests that either an aspartic acid or glutamic acid is being phosphorylated. By appropriate techniques, the unstable phosphate-amino acid bond was stabilized and the protein was then cleaved and the cleavage products sequenced. From these studies, we conclude that a specific aspartic acid residue is phosphorylated in the VirG protein (Jin et al, unpublished observation). When this particular aspartic acid was mutated *in vitro* to asparagine by site-directed mutagenesis, the mutant was no longer able to induce *vir* genes nor was it able to induce tumors on kalanchoe leaves. This indicates that phosphorylation of the VirG protein is essential for its biological function.

INTERACTION OF THE VIRG PROTEIN WITH THE "VIR BOX"

The VirG protein recognizes a 12 base pair conserved sequence called the "*vir* box" which is located upstream of each of the *vir* genes (Winans et al(1987)). Footprinting analysis of each of the promoter regions indicates that the VirG protein covers the "*vir* box" on both strands of the DNA (Jin et al(1990b)). However, phosphorylation does not seem to be required for the binding of the VirG protein to the "*vir* box" region. The VirG protein was isolated from *E. coli* cells into

8

which the gene had been cloned, and alkaline phosphatase was added to remove any phosphate from the protein. This preparation of the VirG protein still bound specifically to a region of the DNA that included the "*vir* box". However, these data do not rule out the possibility that the phosphorylated VirG protein may have a higher affinity for this region than the nonphosphorylated form. Alternately, the action of the phosphorylated VirG protein may be beyond the binding step, such as the activation of transcription by RNA polymerase.

The overall features of the role of the VirA and VirG interaction with one another and the interaction of the VirG protein with the "*vir* box" are shown in Figure 3.

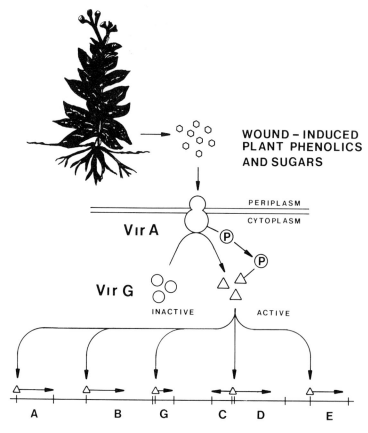

Figure 3. Model for the interaction of the VirA and VirG proteins with each other and with the "*vir* box."

Acknowledgments

We would like to thank the members of our research groups both past and present for providing much of the data presented in this paper. This work was supported by NIH grant 5RO1GM32618, NSF #DMB8704292, and USDA #88-37234-3618.

References

Gelvin, S. and Habeck, L. (1990) '*vir* genes influence conjugal transfer of the Ti plasmid of *Agrobacterium tumefaciens*', J. Bacteriol. 172, 1600-1808.

Jin, S. et al. (1990a) 'The VirA protein of *Agrobacterium tumefaciens* is autophosphorylated and is essential for *vir* gene regulation', J. Bacteriol. 172, 525-530.

Jin, S. et al. (1990b) 'The regulatory VirG protein specifically binds to a cis-acting regulatory sequence involved in transcriptional activation of *Agrobacterium tumefaciens* virulence genes', J. Bacteriol. 172, 531-537.

Spencer, P. and Towers, G. H. N. (1988) 'Specificity of signal compounds detected by *Agrobacterium tumefaciens*', Phytochemistry 27, 2781-2785.

Stachel, S., et al. (1985) 'Identification of the signal molecules produced by wounded plant cells that activate T-DNA transfer in *Agrobacterium tumefaciens*', Nature 318, 624-629.

Steck, T. R. and Kado, C. I. (1990) 'Virulence genes promote conjugative transfer of the Ti plasmid between *Agrobacterium* strains', J. Bacteriol. 172, 2191-2193.

Veluthambi, K. et al. (1989) 'Opines stimulate induction of the *vir* genes of the *Agrobacterium tumefaciens* Ti plasmid', J. Bacteriol. 171, 3696-3703.

Winans, S. et al. (1986) 'A gene essential for *Agrobacterium* virulence is homologous to a family of positive regulatory loci', Proc. Natl. Acad. Sci. USA 83, 8278-8282.

Winans, S. et al. (1987) 'The role of virulence regulatory loci in determining *Agrobacterium* host range', in D. Von Wettstein and N. Chua (eds), Plant Molecular Biology, Plenum Publishing Corp., New York, pp. 573-582.

Winans, S. et al. (1989) 'A protein required for transcriptional regulation of *Agrobacterium* virulence genes spans the cytoplasmic membrane', J. Bacteriol. 171, 1616-1622.

Young, C. and Nester E. (1988) 'Association of the $VirD_2$ protein with the 5' end of T strands in *Agrobacterium tumefaciens*', J. Bacteriol. 170, 3367-3374.

SIGNAL TRANSDUCTION VIA VIR A AND VIR G IN *AGROBACTERIUM*

P.J.J. Hooykaas, L.S. Melchers, A.J.G. Regensburg-Tuïnk, H. den Dulk-Ras, C.W. Rodenburg and S.Turk.

Clusius Laboratory
Dept. of Plant Molecular Biology
Wassenaarseweg 64
2333 AL Leiden
The Netherlands

ABSTRACT. The virulence genes of *Agrobacterium tumefaciens* are inducible by certain phenolic compounds. Regulation is mediated by the VirA and VirG gene products, which together form a two component regulatory system. Results indicate that the VirA protein forms a sensor for phenolic compounds and that the VirG protein becomes activated via the VirA protein.

The VirA protein was found to be located in the inner membrane of the bacterium. Its topology was studied in more detail using fusions of *virA* with the *phoA* coding sequence. In this way it was found that the VirA protein has most of its N-terminal domain in the periplasm connected via a transmembrane spanning segment (TM2) with a large cytoplasmic C-terminal domain. In contrast to expectation the domain which acts as a receptor for phenolic compounds turned out not to be located within the periplasmic domain of VirA, but probably is present in the TM2 membrane spanning segment.

REGULATION OF THE VIRULENCE GENES

The virulence genes of *Agrobacterium tumefaciens* are silent, but can be induced during growth of *Agrobacterium* in medium containing plant exudate (e.g. Stachel et al, 1986). The substances responsible for induction were isolated from tobacco tissue by Stachel et al (1985) and shown to be the phenolic compounds acetosyringone and hydroxy-acetosyringone. By screening a large number of aromatic compounds it was found that many phenolic compounds are capable of inducing *vir*-genes albeit generally with lower efficiency than acetosyringone

10

H. Hennecke and D. P. S. Verma (eds.),
Advances in Molecular Genetics of Plant-Microbe Interactions, Vol. 1, 10–18.
© 1991 *Kluwer Academic Publishers. Printed in the Netherlands.*

(Spencer and Towers, 1988; Melchers et al, 1989b). Amongst these *vir*-inducing compounds are well-known plant substances such as the lignin-precursor coniferyl alcohol. The *vir*-inducer with the most simple chemical formula is guiacol, which is a derivative of phenol with a methoxy-substitution at the *ortho*-position. The presence of inducing compounds in the standard *Agrobacterium* growth media (pH7) does not lead to the induction of *vir*-genes. The system seems specifically adapted to conditions which are prevalent in the plant and in plant fluids. Thus a low pH (5-6) is required for induction to occur (Stachel et al, 1986) as well as a high sugar (sucrose) content (Alt-Moerbe et al, 1988; Melchers et al, 1989b). Although *Agrobacterium* has its optimum temperature for growth at about $32°C$, virulence gene induction is much reduced at this and higher temperatures (Alt-Moerbe et al, 1988; Melchers et al, 1989b). Therefore, inefficient *vir*-gene induction may be one of the factors which render *Agrobacterium* incapable of inducing plant tumours at temperatures above $30°C$ (Braun, 1943). Although the *vir*-genes of different types of Ti and Ri plasmids can be induced in similar inducer containing media, there are small differences in their requirement for optimal induction. Using a construct in which the octopine Ti *virB* promoter was fused to the *lacZ* gene (coding for ß-galactosidase) we found that while octopine and leucinopine (supervirulent) Ti strains required pH 5.3 for optimal induction, nopaline Ti and agropine Ri strains needed pH 5.8 for maximal *vir*-induction.

Certain *Agrobacterium* strains have a limited host range for tumour induction in contrast with the well-characterized wide host range strains which form tumours on a large variety of dicotyledonous plant species. Limited host range strains have been isolated from *Vitis vinifera* (e.g. Panagopoulos and Psallidas, 1973) and *Lippia canescens* (Unger et al, 1985). We studied *vir* gene induction in some of these strains in order to elucidate whether any peculiarities in the *vir* induction system might at least partially explain their limited host range for tumour induction. Lippia strains turned out te be inducible by acetosyringone in the same media as wide host range strains. However, limited host range *Vitis* strains (such as AG57) were not inducible in these media. We tested a large number of compounds to find one that could induce the Vir-system of strain AG57 in induction medium, but could not find any that did. Subsequently, we found that the physical requirements needed for induction in this type of strain were somewhat different from those of wide host range strains. Temperature had to be preferably below $28°C$, pH below 5.3 and inoculum size small (Turk et al, 1990). The fact that strains such as AG57 have more stringent requirements for *vir*-induction to occur may partially form an explanation for their limited host range for tumour induction.

THE REGULATORY SYSTEM

Two *vir*-operons (*virA* and *virG*) show expression in the absence of an inducer. In fact the VirA and VirG proteins that are determined by these operons form a two component regulatory system that is necessary for expression of the other *vir* operons (Stachel and Zambryski, 1986). The DNA sequences of the *virG* genes of the octopine Ti (Melchers et al, 1986; Winans et al, 1986), nopaline Ti (Powell et al, 1987) and agropine Ri plasmid (Aoyama et al, 1989) have been determined, as well as those of the *virA* genes of the octopine Ti (Leroux et al, 1987; Melchers et al, 1987) and nopaline Ti plasmid (Morel et al, 1989). The gene products determined by the *virG* and *virA* genes of these different types of plasmids were found to be very similar. This is in agreement with earlier findings that showed that *vir*-genes from different plasmid types were interchangeable (Hooykaas et al, 1984). It was found that the couple *virA/virG* forms one of a family of two-factor regulatory systems (Nixon et al, 1986) controlling the expression of genes responding to such diverse signals as osmolarity (*envZ/ompR*), nitrogen supply (*ntrB/ntrC*) and presence of oxygen (*fixL/fixJ*). The VirA protein as well as most of the related proteins from other two factor systems are predicted to have two (or more) membrane spanning α-helices. This component of the system therefore is probably located in the membrane, where it can act as a sensor for a particular signal. The other component of the system probably functions as a DNA-binding protein capable of stimulating the transcription of particular genes after having been activated via the sensor protein. Evidence in favour of this model was recently obtained for several two component systems. Results showed that the sensor proteins had autophosphorylating activity and could act as kinases of their accompanying activator proteins. While autophosphorylation occurred at a conserved histidine residue in the sensor component, phosphorylation of a conserved aspartate residue took place in the activator (Hess et al, 1988; Keener and Kustu, 1988; Weiss and Magasanik, 1988; Igo et al, 1989). Phosphorylation of the activator protein in turn led to altered DNA-binding properties and the enhanced transcription of genes controlled by the system (Aiba and Mizuno, 1990; Forst et al, 1989). Recently, autophosphorylating activity of the VirA protein was shown in *in vitro* experiments (Huang et al, 1990; Jin et al, 1990a), and also phosphorylation of VirG by VirA was reported (Jin et al, 1990b). It was also shown recently that the VirG protein is indeed a sequence specific DNA-binding protein capable of binding to *vir*-promoter regions (Jin et al, 1990c; Pazour and Das, 1990). Together these new results show that the VirA-VirG couple behaves similarly to other couples of this family of two component regulatory sequences.

TOPOLOGY OF THE VIRA PROTEIN

Since we were particularly interested in the
functioning of the VirA sensor protein, we performed
experiments to find out whether this was a membrane protein
indeed. To this end we overproduced portions of the VirA
protein in *E.coli* and used these to isolate VirA
antibodies. With these antibodies we could show that the
VirA protein was only present in inner membrane fractions
of *Agrobacterium* (Melchers et al, 1987). In order to
understand the topology of the VirA protein in more detail
we constructed *in vitro* fusions between different N-
terminal parts of the *virA* gene and a *phoA* gene that lacked
its signal peptide. The *phoA* gene encodes the enzyme
alkaline fosfatase, which is normally present in the
periplasm of the bacterium. In the absence of a signal
peptide the protein is retained in the cytoplasm and there
it does not show any enzymatic activity, not even if the
cells are disrupted. Apparently, transport across the inner
membrane is necessary to render the protein enzymatically
active (Manoil and Beckwith, 1985). When a signal peptide
or a membrane spanning domain of a heterologous protein is
coupled to the N-terminus of such a truncated PhoA protein,
transport and exposure to the periplasmic space occurs and
alkaline phosphatase activity can be detected. We applied
this system developed by Manoil and Beckwith to study the
VirA protein. Our data (Melchers et al, 1989a) and similar
data from Winans et al (1989) showed that in accordance
with the predictions from DNA sequence analysis the VirA
protein has in the N-terminal part a periplasmic domain
that is bordered by two membrane spanning domains
(TM1,TM2). The large domain following TM2 in the direction
of the C-terminus was found to be located in the cytoplasm
as predicted. On the basis of these findings we postulated
that the VirA protein would function in a similair way as
other chemoreceptor proteins and would use the periplasmic
domain to sense signals (phenolic compounds) and the
cytoplasmic domain to transmit these signals to VirG via
phosphorylation. In order to find out whether this is
indeed the case, we made hybrids between VirA and the
E.coli chemoreceptor protein Tar, which activates the
chemotaxis apparatus after sensing the presence of
aspartate or loaded maltose binding protein. The Tar-
protein has a basic structure similar to that of the VirA
protein, and it has been shown that the periplasmic domain
of Tar is the sensor domain, while the cytoplasmic domain
contains the signal transmitter region. When replacing the
periplasmic part of VirA by that of Tar, we expected that,
if the model postulated above would be correct, this hybrid
protein would now no longer respond to acetosyringone but
rather to aspartate or maltose. Unexpectedly, however, the

hybrids responded still to acetosyringone, but not to aspartate or maltose (Melchers et al, 1989a). Certain of these hybrid proteins were even more active in mediating the activation of *vir*-genes than the wild-type VirA protein, and rendered the system less sensitive to the negative influence of a relatively high pH or temperature. Significant induction took place even at pH8 and at temperatures as high as 33°C and 37°C. Thus the results obtained with VirA-Tar hybrids show that 1) the sensor domain is not located in the periplasmic region of VirA 2) the amino acid composition of the periplasmic domain is an element conferring pH-and temperature-sensitivity on the system. In line with this we found that the deletion of the periplasmic domain from the VirA protein did not render the protein inactive. In fact such deleted VirA proteins behaved like the wild-type VirA protein in acetosyringone-dependent signal transduction. Further work showed that the deletion of the second transmembrane domain (TM2) from VirA or the replacement of this domain by the corresponding domain of Tar led to proteins that no longer responded to plant phenolics, but rather mediated a low, signal-independent level of *vir*-gene induction. We conclude that the second transmembrane domain of VirA is an essential element of VirA, which possibly contains the sensor (signal receiver) function of this protein. That the location of the sensor domain of VirA is different from that in the more classically studied chemoreceptors does not have to be too surprising in light of the fact that the signals in the case of VirA are formed by relatively apolar phenolic compounds that probably have a relatively high affinity for membranes. Future research will hopefully shed more light on the exact molecular details of the signal communication between plant cells and agrobacteria as mediated by the VirA and VirG proteins.

REFERENCES

Aiba,H. and Mizuno,T. (1990) Phosphorylation of a bacterial activator protein, OmpR, by a protein kinase, EnvZ, stimulates the transcription of the ompF and ompC genes in Escherichia coli. FEBS Lett 261:19-11.

Alt-Moerbe,J., Neddermann,P., Van Lintig,J., Weiler,E.W., and Schröder,J. (1988). Temperature-sensitive step in Ti plasmid vir-region induction and correlation with cytokinin secretation by Agrobacteria. Mol.Gen.Genet. 213:1-8.

Aoyama,T., Hirayama,T., Tamamoto,S., and Oka,A. (1989). Putative start codon TTG for the regulatory protein VirG of the hairy-root-inducing plasmid pRiA4. Gene 78:173-178.

Braun,A.C. (1947). Thermal studies on the factors responsible for tumor initiation in crown gall. Am.J.Bot. 34:234-240.

Forst,S.A., Delgado,J., and Inouye,M. (1989). DNA-binding properties of the transcripton activator (OmpR) for the upstream sequences of ompF in Escherichia coli are altered by envZ mutations and medium osmolarity. J.Bacteriol. 171:2949-2955.

Hess,F.J., Oosawa,K., Kaplan,N., and Simon,M.I. (1988). Phosphorylation of three proteins in the signaling pathway of bacterial chemotaxis. Cell 53:79-87.

Hooykaas,P.J.J., Hofker,M., Den Dulk-Ras,H., and Schilperoort,R.A. (1984). A comparison of virulence determinants in an octopine Ti plasmid, a nopaline Ti plasmid, and an Ri plasmid by complementation analysis of Agrobacterium tumefaciens mutants. Plasmid 11:195-205.

Huang,Y., Morel,P., Powell,B., and Kado,C.L. (1990). VirA, a coregulator of Ti-specified virulence genes, is phosphorylated in vitro. J.Bacteriol. 172:1142-1144.

Igo,M.M., Ninfa,A.J., and Silhavy, T.J. (1989). A bacterial environmental sensor that functions as a protein kinase and stimulates transcriptional activation. Genes and Development 3:598-605.

Jin,S., Roitsch,T., Ankenbauer,R.G., Gordon,M.P., and Nester,E.W. (1990). The VirA protein of _Agrobacterium_ _tumefaciens_ is autophosphorylated and is essential for _vir_ gene regulation. J.Bacteriol. 172:525-530.

Jin,S., Prusti,R.K., Roitsch,T., Ankenbauer,R.G., and Nester,E.W.(1990). The VirG protein of _Agrobacterium_ _tumefaciens_ is phosphorylated by the autophosphorylated VirA protein and this is essential for its biological activity. J.Bacteriol. : in press.

Jin,S., Roitsch,T., Ankenbauer,R.G., Gordon,M.P., and Nester,E.W. (1990). The regulatory VirG protein specifically binds to a cis-acting regulatory sequence involved in transcriptional activation of _Agrobacterium_ _tumefaciens_ virulence genes. J.Bacteriol. 172:531-537.

Keener,J., and Kustu,S. (1988). Protein kinase and phosphoprotein phosphatase activities of nitrogen regulatory proteins NTRB and NTRC of enteric bacteria: roles of the conserved amino-terminal domain of NTRC. Proc.Natl.Acad.Sci.USA 85:4976-4980.

Leroux,B., Yanofsky,M.F., Winans,S.C., Ward,J.E., Ziegler,S.F., and Nester,E.W. (1987). Characterization of the _virA_ locus of _Agrobacterium_ _tumefaciens_ a transcriptional regulator and host range determinant. EMBO J. 6:849-856.

Manoil,C., and Beckwith,J. (1985). Tn _phoA_: a transposon probe for protein export signals. Proc.Nath.Acad.Sci.USA 82:8129-8133.

Melchers,L.S., Thompson,D.V., Idler,K.B., Schilperoort,R.A., and Hooykaas,P.J.J. (1986). Nucleotide sequence of the virulence gene _virG_ of the _Agrobacterium_ _tumefaciens_ octopine Ti plasmid: significant homology betweeen _virG_ and the regulatory genes _ompR_, _phoB_ and _dye_ of _E.coli_. Nucleic Acids Res. 14:9933-9942.

Melchers,L.S., Thompson,D.V., Idler,K.B., Neuteboom,S.T.C., De Maagd,R.A., Schilperoort,R.A., and Hooykaas,P.J.J. (1987). Molecular characterization of the virulence gene _virA_ of the _Agrobacterium_ _tumefaciens_ octopine Ti plasmid. Plant Mol.Biol. 9:635-645.

Melchers,L.S., Regensburg-Tuïnk,T.J.G., Bourret,R.B., Sedee.N.J.A., Schilperoort,R.A., and Hooykaas,P.J.J. (1989). Membrane topology and functional analysis of the sensory protein VirA of _Agrobacterium tumefaciens_. EMBO J. 8:1919-1925.

Melchers,L.S., Regensburg-Tuïnk,A.J.G., Schilperoort,R.A., and Hooykaas,P.J.J. (1989). Specificity of signal molecules in the activation of _Agrobacterium_ virulence gene expression. Mol. Microbiol. 3:969-977.

Morel,P., Powell,B.S., Rogowski,P.M., and Kado,C.I. (1989). Characterization of the _virA_ virulence gene of thenopalin plasmid, pTiC58, of _Agrobacterium tumefaciens_. Mol. Microbiol. 3:1237-1246.

Nixon,B.T., Ronson,C.W., and Ausubel,F.M. (1986). Two-component regulatory systems responsive to environmental stimuli share strongly conserved domains with the nitrogen assimilation regulatory genes _ntrB_ and _ntrC_. Proc.Natl.Acid.Sci.USA 83:7850-7854.

Panagopoulos,C.G.,and Psallidas,P.G. 1973. Characteristics of Creek isolates of _Agrobacterium tumefaciens_ (Smith and Townsend). J.Appl.Bact. 36:233-240.

Pazour,G.J., and Das,A. (1990). _VirG_, an Agrobacterium tumefaciens transcriptional activator, initiates translation at a UUG codon and is a sequence-specific DNA-binding protein. J.Bacteriol. 172:1241-1249.

Powell,B.S., Powell,G.K., Morris,R.O., Rogowsky.P.M., and Kado,C.I. (1987). Nucleotide sequence of the _virG_ locus of the _Agrobacterium tumefaciens_ plasmid pTiC58. Mol.Microbiol. 1:309-316.

Spencer,P.A., and Towers,G.H.N. (1988). Specificity of signal compounds detected by _Agrobacterium tumefaciens_. Phytochemistry 27:2781-2785.

Stachel,S.E., and Zambryski,P.C. (1986). _virA_ and _virG_ control the plant-induced activation of the T-DNA transfer process of _Agrobacterium tumefaciens_. Cell 46:325-333.

Stachel,S.E., Messens,E., Van Montagu,M., and Zambryski,P. (1985). Identification of the signal molecules produced by wounded plant cells that activate T-DNA transfer in _Agrobacterium tumefaciens_. Nature 318:624-629.

Stachel,S.E., Nester,E.W., and Zambryski,P. (1986). A plant cell factor induced <u>Agrobacterium tumefaciens vir</u> gene expression. Proc.Natl.AcadSci.USA 83:379-383.

Turk,S.C.H.J., Melchers,L.S., Den Dulk-Ras,H., Regensburg-Tuïnk,A.J.G., and Hooykaas,P.J.J. (1990). Virulence gene induction in different <u>Agrobacterium</u> strains is differentially affected by environmental conditions. Role of the VirA sensor protein.Plant Mol. Biol, submitted.

Unger,L., Ziegler,S.F., Huffman,G.A., Knauf,V.C., Peet.R., Moore,L.W., Gordon.M.P., and Nester,E.W. (1985). New class of limited-host range <u>Agrobacterium</u> mega-tumor-inducing plasmids lacking homology to the transferred DNA of a wide-host-range tumor-inducing plasmid. J.Bacteriol. 164:723-730.

Weiss,V., and Magasanik,B. (1988). Phosphorylation of nitrogen regulator I (NRI) of <u>Escherichia coli</u>. Proc.Natl.Acad.Sci, USA 85:8919-8923.

Winans,S.C., Ebert,P.R., Stachel,S.E., Gordon.M.P., and Nester,E.W. (1986). A gene for <u>Agrobacterium</u> virulence homologous to a family of positive regulatory loci. Proc.Natl.Acad.Sci.USA 83:8278-8282.

Winans,S.C., Kerstetter,R.A., Ward,J.E., and Nester,E.W. (1989). A protein required for transcriptional regulation of <u>Agrobacterium</u> virulence genes spans the cytoplasmic membrane. J.Bacteriol. 171:1616-1622.

THE T-DNA ON ITS WAY FROM *AGROBACTERIUM TUMEFACIENS* TO THE PLANT

B. Hohn[1], Z. Koukolíková-Nicola[1], F. Dürrenberger[1,2], G. Bakkeren[1,3] and C. Koncz[4]

[1]Friedrich Miescher-Institut, P.O. Box 2543, CH-4002 Basel, Switzerland; [2]Present Address: University of Geneva, Department of Molecular Biology, CH-1211 Geneva; [3]Present Address: Biotechnology Laboratory, Vancouver, B.C., Canada V6T 1W5; [4]MPI für Züchtungsforschung, D-5000 Köln 30, FRG.

ABSTRACT

Agrobacterium tumefaciens can, under suitable conditions, transfer the T-segment of its tumor inducing plasmid into plants. There it is usually detected integrated in nuclear DNA and expressing its natural or chimaeric genes. The virulence D2 protein, responsible for the endonucleolytic processing of T-DNA, was found covalently attached to the right end of free T-DNA molecules. Unintegrated T-DNA was isolated from plants using a plant viral replicon as T-DNA. Analysis of resulting viruses revealed that the right end of T-DNA was much better preserved than the left one, implying protection by a (virulence?) protein. Comparison of T-DNA integrated in plant DNA with respective preinsertion sites showed that integration invariably was accompanied by a small deletion at the insertion site. Short homologies between target DNA and segments at or close to T-DNA ends seem to have been instrumental in the integration process which may have been aided by virulence protein(s) at the right junction.

INTRODUCTION

Microorganisms of the genus *Agrobacterium* have developed a sophisticated mechanism of genetically and stably transforming their hosts. As a consequence the plant, under the government of the transferred DNA, deviates part of its metabolic resources to build products which only the inciting bacterium can catabolize.

The bacterium only transfers a specific segment of its large Ti (Tumor inducing) plasmid, the T-DNA (Tranfer DNA), into plant cells. Analysis of the transfer process has revealed sequences and functions that are required in *cis* and in *trans* to the T-DNA, respectively. The T-DNA is delimited by two almost perfect direct repeats of 25 bp called border sequences. The T-DNA transfer is mediated by products of the virulence region, located on the Ti plasmid. The bacterium does not carry the T-DNA in a constitutively transferable form but has to be induced by wounded plant cells to render its T-DNA

19

H. Hennecke and D. P. S. Verma (eds.),
Advances in Molecular Genetics of Plant-Microbe Interactions, Vol. 1, 19–27.
© 1991 *Kluwer Academic Publishers. Printed in the Netherlands.*

movable. The virulence region, also localized on the Ti plasmid, but separated from the T-DNA, is responsible for this DNA transfer. Signal molecules synthesized in wounded plant tissue induce the expression of virulence genes, the induction being mediated by the action of the VirA and VirG proteins [1-3]. Two virulence loci are directly involved in processing the T-DNA. Proteins D1 and the N-terminal part of D2 code for a DNA topoisomerase and borderspecific endonuclease, respectively [4; 5-8]. As a result a specific nick is introduced in the lower strand of the border repeat sequences. Besides these nicked molecules single stranded T-DNA molecules of lower strand polarity, the so called T-strands, are found as well as doublestranded linear T-DNA versions and a small amount of covalently closed circular molecules [1-3]. The virulence E locus codes only for auxiliary functions as in its absence virulence is not abolished but diminished. Virulence E2 protein is a sequence unspecific single stranded DNA binding protein [9-12].

In this communication we will describe T-DNA molecules detected in induced bacterial cells, speculate about the transfer process, describe molecules related to unintegrated T-DNA in plants and finally discuss comparisons of T-DNA inserts into plant DNA with respective preinsertion sequences.

THE T-DNA IN THE BACTERIUM

Upon induction of the virulence region specific nicks are introduced in the lower strand of border sequences (Fig.). Single stranded molecules of lower strand polarity, composed exclusively of T-DNA, are formed as a result, possibly by displacement by a newly synthesized bottomstrand [3] see Fig. structure 2. Other T-DNA molecules found in induced *Agrobacterium* cells consist of double stranded linear (structure 3) and a small amount of covalently closed circular molecules. Processing at the border thus explains the absolute requirement of border sequences for transformation, as found in genetic experiments. The endonucleolytic activity resides in the combination of the proteins VirD1 and the N-terminal half of virD2, as experiments in *Agrobacterium* and in *Escherichia coli* [4] have shown. Thereby virD1 exhibits topoisomerase activity whereas in VirD2 the sequence specificity seems to reside.

The VirD2 protein (filled circles in the Figure), after having cleaved the T-DNA between the third and fourth basepair of each border sequence, remains attached to the 5' end of the lower T-DNA strand [13-17]. This protein-DNA association was found to be resistant to SDS, mercaptoethanol, mild alkali, piperidine, and hydroxylamine, indicating that it involves a covalent linkage [16]. The attached protein rendered the right end of doublestranded T-DNA resistant to 5' → 3' exonucleolytic attack, at least *in vitro* [16]. Thus, this bond may prevent T-DNA degradation also *in vivo* and inhibit religation of T-DNA to the Ti plasmid. The major role(s) of attached VirD2 protein, however, may be in later stages such as T-DNA transfer and/or integration.

T-DNA TRANSFER

In what form the DNA travels to the plant is not known, although an involvement of single

THE T-DNA ON ITS WAY FROM
AGROBACTERIUM TUMEFACIENS TO THE PLANT

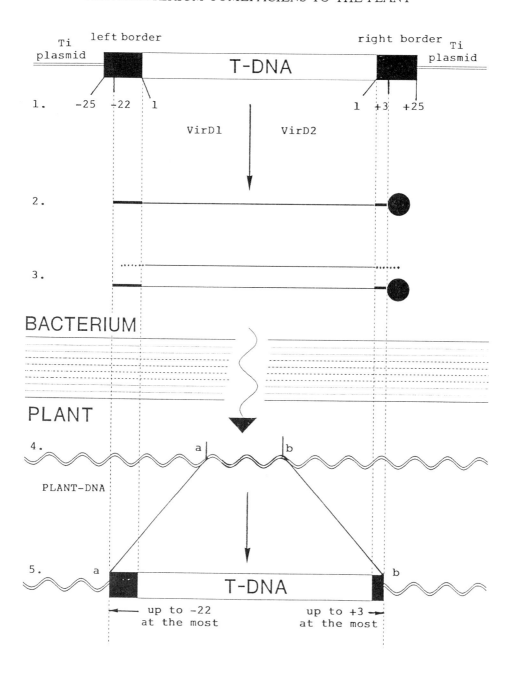

stranded T-DNA at some stage is very likely. It may be complemented to a double stranded form in the bacterium, in the plant, or only upon integration. Adaptation to preexisting bacterial mechanisms seems to be a possible route for the evolution of this unique interkingdom DNA transfer. Several principles may have been combined: a) perception and transduction of environmental stimuli, followed by activation of genes responding to the stimuli, c) induction of proviruses and d) conjugation. Accordingly, T-DNA may be mobilized like conjugating plasmid DNA or like a virus. The need for attachment is not an argument against a virus theory as only closely juxtaposed cells may represent the correct trigger for a virus-transmission. Tumor-inducing virus-particles have been looked for, but so far not found. The single stranded nature of a large part of processed T-DNA molecules may suggest their direct involvement in the transfer, along a relatively safe "mating bridge". Complete safety may not be guaranteed, however, since protecting virulence E2 protein molecules may have to be recruited. The extracellular complementation of *virE* mutants [18,11] can be taken as an argument in favour of involvement of this protein in the actual transfer.

Mobilization functions and origin of transfer of the mobilisable plasmid RSF 1010, in conjuction with Ti plasmid virulence genes, could also be demonstrated to transfer genes to plants [19]. This process may actually mimic an intermediate in evolution, between conjugation between bacteria and "modern" T-DNA transfer from bacterium to plant.

THE T-DNA IN THE PLANT

Once inside the plant-nucleus, by whatever route, T-DNA may not immediately integrate. Experiments to explore the fate of T-DNA inside the plant by chemical methods are difficult to interpret. A different approach to analyze T-DNA molecules in the plant, presumably before/independently of their insertion in the nuclear DNA, has been taken with the use of agroinfection [20]. In these experiments T-DNA was constructed to consist exclusively of a plant viral genome. Once inside a plant, the T-DNA's life as a plant virus can begin as soon as it is circularized. This genetic rescue of T-DNA molecules is plant specific, as opposed to any chemical rescue which would severely suffer from contamination by *Agrobacterium* specific material. Analysis of circle joints, conveniently cloned and amplified by the virus, revealed the following: sequences at the right border were relatively well conserved, with 40% of all rescued molecules including the third base of the border corresponding to the nicking site in the bacterium. This contrasts with a more ragged representation of sequences at the left border. However, even here the rule was followed that not a single base left of the left nick site is ever transferred. A left border mutant (deletion of a base in the conserved core sequence) diminished the efficiency of agroinfection altogether. In addition, it allowed transmission of sequences to the plant which actually were located beyond the nick site, outside of the T-DNA.

These results imply that the right T-DNA end is protected from degradation during transfer. Since the VirD2 protein is found covalently attached to the 5' end of the right T-DNA terminus in *Agrobacterium* (see above), these results suggest the conservation of this bond in the plant.

As discussed earlier [16], several functions can be imagined for the VirD2 protein covalently attached to T-DNA, possibly in association with other virulence proteins and/or plant proteins: protection of T-DNA during transfer, targeting to the plant nucleus, priming of DNA replication, and integration into the host genome.

T-DNA INTEGRATION

T-DNA integrates into nuclear DNA. *In situ* hybridisation and mapping using genetic or RFLP markers have established that no preferential target for T-DNA integration could be observed [reviewed in 3]. T-DNA insertion units consist of one to several copies of T-DNA. Independent locations of several inserts are also found frequently and have been shown to segregate in subsequent generations. Tandem arrangements of T-DNA have been found to be composed of direct and inverted repeats. Both of these have been found in single T-DNA arrays [3].

Several junctions of T-DNA element to T-DNA element and of T-DNA to plant DNA have been mapped and some of the joints have been sequenced [reviewed in 1,3]. Of the right border repeat maximally three bases have been found retained whereas 24 bases have been found preserved from the left border. Analysis of T-DNA molecules rescued as replicating units allowed a similar conclusion, based on many independently isolated viruses [20]. Since they did not have to go through an integration step, the similarity of the T-DNA ends in the two systems allows the conclusion that integration *per se* did not lead to major distortion of T-DNA ends.

Only two studies have been undertaken until recently, in which the junction sequences of integrated T-DNA was compared to the sequence of unoccupied target DNA. One study revealed a 158 bp duplication of target sequences, now bordering inserted DNA [21]. Some minor rearrangements of target sequences, a small deletion as well as the insertion of small stretches of DNA of unknown origin seem to have accompanied the integration event. The other study comparing pre- and postinsertion sequences revealed that a rather truncated T-DNA had integrated, possibly using small regions of homology at the junctions and again with concommitant creation of a small target deletion [22].

In a more extended analysis Mayerhofer et al [23] analysed the integration pattern of T-DNA into the genome of *Arabidopsis thaliana*. Analysis of seven transformants revealed:

1) no sequence homologies between target sequences and no special features of the target sequences. However, preferential integration into potentially transcribed areas of the analysed and other transformants has been noted, as transcriptional and translational fusions of a T-DNA gene with adjacent cellular sequences were found at similar frequencies in transformants of *Arabidopsis thaliana* and *Nicotiana tabacum* [24]. Since these two species have a markedly differing density of transcribed sequences these results cannot be interpreted as random insertions. Also in mammalian systems DNA in the process of being transcribed may more easily be accessible to invading (Adeno- or Retrovial [25,26]) DNA or, more specifically (but possibly related), may contain more nicks that can serve as entry

points for integrating DNA.

2) small homologies between the right T-DNA end and the target sequence (at b, see Figure). In cases where the right T-DNA end was "complete" (i.e. including up to three bp of border sequence), homologies were less apparent. This implies a special role of the right T-DNA end in accomplishing or helping assimilation of T-DNA into plant DNA and suggests involvement of special proteins such as virulence proteins, possibly in combination with others, in the integration process.

3) small homologies between the left T-DNA end and the target sequence (at a, see Figure).

4) Small rearrangements (direct and inverse duplications) of T-DNA or insert sequences in three cases at one T-DNA/plant DNA junction each.

5) small (29 to 73 bp) deletions (between a and b, Figure) of target DNA.

As suggested by these results, T-DNA (in single- or double stranded form, but in any case with an exposed left 3' end) actively or passively searches for entry into chromosomal DNA. A nick may be detected (or produced?) by the virulence protein armed right T-DNA end; ligation at this point may be followed or preceded by homology search in nearby target sequences. T-DNA may help in exposing target homologies or may exploit DNA replication dependent on temporary single strandedness of target DNA. This homology search must be followed by annealing, mismatch repair, repair synthesis, exo- and endonucleolytic "adaptations" of the recombining partners, and ligation.

Assimilation of right and left end must be independent of each other, as suggested by the independent entry points (a and b, see Figure). The close proximity of these, however, implies a transient link between the T-DNA ends. Such a link could be established by host encoded end to end joining proteins such as have been proposed for *Xenopus laevis* eggs [27], or by (a) virulence protein(s) attached to the right T-DNA end but having affinity to the left end, or a combination of the two. Alternatively, or in addition, the involvement of a 3' protruding upper T-DNA strand in the adherence of T-DNA ends could be invoked. Such T-DNA molecules (see Figure, structure 3) have indeed been detected in induced cells of *Agrobacterium*, but their significance in T-DNA transfer, if any, has remained unknown [Z. Koukolíková-Nicola et al,; 1, and unpublished].

Analysis of tandem T-DNA insertions is not as complete. Such arrangements of T-DNA were found to be composed of direct and inverted repeats. Interestingly, the breakpoints of each pair of T-DNA elements involved in an inverted repeat were, as far as analysed, identical [28]. This may most easily be explained by replication of a double stranded T-DNA molecule, with template strand switching of the replication machinery at the end of a particular element. Since such structures have never been detected in the bacterium they most likely arose in the plant. This would imply that the T-DNA contains a replication origin. A double stranded T-DNA molecule, the suggested template, either arrived from the bacterium as such or is a single stranded molecule converted to the double stranded form. Whether replication of T-DNA before integration is a prerequisite in general is not known. A different kind of mechanism has to be invoked in explaining the head to head link found in a plant transformed with two different T-DNA molecules originating from one *Agrobacterium* strain [29]: in this case ligation of two double stranded T-DNA elements must have occurred, since 5' ends of single stranded DNA molecules cannot ligate to each other. Thus it seems that several mechanisms may operate within plants to replicate, assemble and integrate T-DNA.

SUMMARY

The T-DNA of *Agrobacterium tumefaciens* is localized like a prophage on the Ti-plasmid. Upon induction of the virulence genes by a plant signal it excises and assembles into a kind of transfer complex. Then it is expelled into a plant cell *via* a bacterium-plant bridge, in a process mimicking certain aspects of bacterial conjugation. In the target cell the T-DNA "plasmid" or "virus" integrates, exploiting bacterial and plant functions.

These are certainly simplifications. Undoubtedly, a more detailed understanding of the mechanisms used in this unique example of interkingdom DNA transfer will be required.

REFERENCES

[1]   Koukolíková-Nicola, Z., Albright, L. and Hohn, B. (1987) 'The mechanism of T-DNA transfer from *Agrobacterium tumefaciens* to the plant cell', in Th. Hohn and J. Schell (eds.), Plant DNA Infectious Agents, New York Springer Verlag, pp. 109-148.

[2]   Ream, W. (1989) '*Agrobacterium tumefaciens* and interkingdom genetic exchange', Annu. Rev. Phytopathol. 27, 583-618.

[3]   Zambryski, P., Tempé, J. and Schell, J. (1989) 'Transfer and function of T-DNA genes from *Agrobacterium* Ti and Ri plasmids in plants', Cell 56, 193-201.

[4]   Ghai, J. and Das, A. (1989) 'The *virD* operon of *Agrobacterium tumefaciens* Ti plasmid encodes a DNA-relaxing enzyme', Proc. Natl. Acad. Sci. USA 86, 3109-3113.

[5]   Yanofsky, M.F., Porter, S.G., Young, C., Albright, L.M., Gordon, M.P. and Nester, E.W. (1986) 'The *virD* operon of *Agrobacterium tumefaciens* encodes a site-specific endonuclease', Cell 47, 471-477.

[6]   Alt-Moerbe, J., Rak, B. and Schröder, J. (1986) 'A 3,6-kbp segment from the *vir* region of Ti plasmids contains genes responsible for border sequence-directed production of T region circles in *E. coli*', EMBO J. 5, 1129-1135.

[7]   Yamamoto, A., Iwahashi, M., Yanofsky, M.F., Nester, E.W., Takebe, I. and Machida, Y. (1987) 'The promoter proximal region in the *vir*D locus of *Agrobacterium tumefaciens* is necessary for the plant-inducible circularization of T-DNA', Mol. Gen. Genet. 206, 174-177.

[8]   Jayaswal, R.K., Veluthambi, K., Gelvin, S.B. and Slightom, J.L. (1987) 'Double-stranded cleavage of T-DNA and generation of single stranded T-DNA molecules in *Escherichia coli* by a *virD*-encoded border-specific endonuclease from *Agrobacterium tumefaciens*', J. Bacteriol. 169, 5035-5045.

[9]   Gietl, C., Koukolíková-Nicola, Z. and Hohn, B. (1987) 'Mobilization of T-DNA from

26

Agrobacterium to plant cells involves a protein that binds single stranded DNA', Proc. Natl. Acad. Sci. USA 84, 9006-9010.

[10] Das, A. (1988) '*Agrobacterium tumefaciens virE* operon encodes a single stranded DNA-binding protein', Proc. Natl. Acad. Sci. USA 85, 2909-2913.

[11] Christie, P.J., Ward, J.E., Winands, S.C. and Nester, E.W. (1988) 'The *Agrobacterium tumefaciens virE2* gene product is a single stranded-DNA-binding protein that associates with T-DNA', J. Bacteriol. 170, 2659-2667.

[12] Citovsky, V., De Vos, G. and Zambryski, P. (1988) 'Single stranded DNA binding protein encoded by the *virE* locus of *Agrobacterium tumefaciens*', Science 240, 501-504.

[13] Herrera-Estrella, A., Chen, Z., Van Montagu, M. and Wang, K. (1988) 'VirD proteins of *Agrobacterium tumefaciens* are required for the formation of a covalent DNA - protein complex at the 5' terminus of T-strand molecules', EMBO J. 7, 4055-4062.

[14] Ward, E.R. and Barnes, W.M. (1988) 'VirD protein of *Agrobacterium tumefaciens* very tightly linked to the 5' end of T-strand DNA', Science 242, 927-930.

[15] Young, C. and Nester, E.W. (1988) 'Association of the *virD2* protein with the 5' end of T strands in *Agrobacterium tumefaciens*', J. Bacteriol. 170, 3367-3374.

[16] Dürrenberger, F., Crameri, A., Hohn, B. and Koukolíková-Nicola, Z. (1989) 'Covalently bound *virD2* protein of *Agrobacterium tumefaciens* protects the T-DNA from exonucleolytic degradation', Proc. Natl. Acad. Sci. USA 86, 9154-9158.

[17] Howard, E.A., Winsor, B.A., De Vos, G. and Zambryski, P. (1989) 'Activation of the T-DNA transfer process in *Agrobacterium* results in the generation of a T-strand-protein complex: Tight association of VirD2 with the 5' ends of T-strands', Proc. Natl. Acad. Sci. USA 86, 4017-4021.

[18] Otten, L., DeGreve, H., Leemans, J., Hain, R., Hooykaas, P.J.J. and Schell, J. (1984) 'Restoration of virulence of *vir* region mutants of *Agrobacterium tumefaciens* strain B653 by coinfection with normal and mutant *Agrobacterium* strains', Mol. Gen. Genet. 175, 159-163.

[19] Buchanan-Wollaston, V., Passiatore, J.E. and Cannon, F. (1987) 'The *mob* and *oriT* mobilization functions of a bacterial plasmid promote its transfer to plants', Nature 328, 172-174.

[20] Bakkeren, G., Koukolíková-Nicola, Z., Grimsley, N. and Hohn, B. (1989) 'Recovery of *Agrobacterium tumefaciens* T-DNA molecules from whole plants early after transfer', Cell 57, 847-857.

[21] Gheysen, G., Van Montagu, M. and Zambryski, P. (1987) 'Integration of *Agrobacterium tumefaciens* transfer DNA (T-DNA) involves rearrangements of target plant DNA sequences', Proc. Natl. Acad. Sci. USA 84, 6169-6173

[22] Matsumoto, S., Ito, Y., Hosoi, T., Takahashi, Y. and Machida, Y. (1990) 'Integration of *Agrobacterium* T-DNA into a tobacco chromosome: Possible involvement of DNA homology between T-DNA and plant DNA', Mol. Gen. Genet., in press.

[23]   Mayerhofer, R., Koncz-Kalman, Z., Nawrath, C., Bakkeren, G., Crameri, A., Angelis, K., Redei, G.P., Schell, J., Hohn, B. and Koncz, C. (1990) 'T-DNA integration: a mode of illegitimate recombination in plants', submitted.

[24]   Koncz, C., Martini, N., Mayerhofer, R., Koncz-Kalman, Z., Körber, H., Redei, G.P. and Schell, J. (1989) 'High frequency T-DNA-mediated gene tagging in plants', Proc. Natl. Acad. Sci. USA 86, 8467-8471.

[25]   Schulz, M., Freisem-Rabien, U., Jessberger, R. and Doerfler, W. (1987) 'Transcriptional activities of mammalian genomes at sites of recombination with foreign DNA', J. Virol. 61, 344-353.

[26]   Scherdin, U., Rhodes, K. and Breindl, M. (1990) 'Transcriptionally active genome regions are preferred targets for retrovirus integration', J. Virol. 64, 907-912.

[27]   Thode, S., Schäfer, A., Pfeiffer, P. and Vielmetter, W. (1990) 'A novel pathway of DNA end-to-end joining', Cell 60, 921-928.

[28]   Jorgensen, R., Snyder, C. and Jones, J.D.G. (1987) 'T-DNA is organized predominantly in inverted repeat structures in plants transformed with *Agrobacterium tumefaciens* C58 derivatives', Mol. Gen. Genet. 207, 471-477.

[29]   Jouanin, L., Bouchez, D., Drong, R.F., Tepfer, D. and Slightom, J.L. (1989) 'Analysis of TR-DNA/plant junctions in the genome of a *convolvulus arvensis* clone transformed by *Agrobacterium rhizogenes* strain A4', Plant Mol. Biol. 12, 75-85.

FUNCTIONAL ORGANIZATION OF THE REGIONS RESPONSIBLE FOR NOPALINE AND OCTOPINE CATABOLISM IN TI PLASMIDS OF *AGROBACTERIUM TUMEFACIENS*

J. SCHRÖDER, J. VON LINTIG, AND H. ZANKER
Universität Freiburg
Institut für Biologie II
Schänzlestr. 1
D-7800 Freiburg
F.R.G.

ABSTRACT. The catabolism of opines synthesized in transformed plant cells by Agrobacterium is an important aspect in the bacteria/plant interaction. We investigate Ti plasmid genes in the noc-region of pTiC58 and in the occ-region of pTiAch5 which are necessary for catabolism of nopaline and octopine, respectively. Results with a binary vector system indicate that both regions code for constitutively expressed proteins which are necessary and sufficient for activation in trans of the catabolic functions in presence of the opines. These proteins from the noc- and the occ-region reveal significant similarities to each other and to regulatory proteins characterized in other bacteria. The positions of the inducible promoters have been mapped to regions of less than 0.5 kbp. Apart from proteins directly involved in opine catabolism, both regions code for at least two proteins which are related to transport proteins in other bacteria. These and previous results indicate that several genes in the noc- and the occ-region are related, but adapted for function in nopaline and octopine utilization.

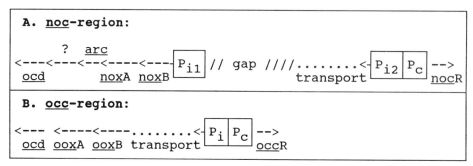

Fig. 1. Functional organization of the noc-region (pTiC58) and the occ-region (pTiAch5). P_i = opine-inducible and P_c = constitutive promoters. ocd = ornithine cyclodeaminase; ? = 40 kDa gene; arc = arginase; noxA, noxB, ooxA, ooxB = nopaline and octopine oxidase, respectively. nocR and occR = genes involved in the regulation of the inducible promoters.

H. Hennecke and D. P. S. Verma (eds.),
Advances in Molecular Genetics of Plant-Microbe Interactions, Vol. 1, 28–31.
© 1991 *Kluwer Academic Publishers. Printed in the Netherlands.*

1. INTRODUCTION

Plant tumors induced by Agrobacterium tumefaciens synthesize a group of substances (opines) which are metabolized by the bacteria. This is an essential part of the interaction and presumedly was the driving force in the evolution of these systems. The genes essential for catabolism are on the Ti plasmids; in the case of nopaline and octopine plasmids the regions are called noc- and occ-region, respectively.

Our previous work has shown that the noc-region encodes at least three enzyme functions (Fig. 1A): nopaline oxidase (requires the coaction of two proteins, noxA and noxB), arginase (arc), and ornithine cyclodeaminase (ocd) [1]. These enzymes are induced by nopaline, and together they serve to convert nopaline via L-arginine and L-ornithine into L-proline which is then further metabolized by enzymes encoded in the chromosome. Between these genes is a coding region for a 40 kDa polypeptide of unknown function. The genes arc and ocd have been sequenced and the enzyme properties were characterized [2,3]. A gene for OCD (induced by octopine) was also discovered in the occ-region. A detailed analysis showed that the ocd genes and proteins from the noc- and the occ-region are closely related, and that the proteins differ in the regulation of enzyme activity by L-arginine [4].

This report summarizes our recent studies on the regulation and function of genes in the noc- and in the occ-region.

2. THE NOC-REGION IN PTIC58

Our present work focusses on the promoters, regulatory genes, and transport functions. Studies with a binary vector system and a lacZ reporter gene construct in Agrobacterium defined at least three promoters (Fig. 1A). One is located directly in front of the catabolic genes, and it is responsible for expression of all of these genes. It is induced in presence of nopaline, and its activation requires the presence of a gene (nocR, at the right end of the noc-region) which is expressed from a constitutive promoter. Sequence analysis revealed significant homology with regulatory genes in other gram-negative bacteria, and this supports the proposal that the polypeptide is a key element in the regulation of the noc-region genes by nopaline (manuscript in preparation). The protein also activates transcription from a second inducible promoter just to the left of nocR. This controls the expression of at least two polypeptides with significant similarity to bacterial proteins involved in transport of low molecular weight substances. Functional studies of these proteins are in progress.

3. THE OCC-REGION IN PTIACH5

Recent studies mapped the position of octopine oxidase (Fig. 1B) which, like in nopaline plasmids, requires two polypeptides for function (ooxA and ooxB). Interestingly, no genes for arginase or a 40 kDa polypeptide were detected in the occ-region.

Two promoters have been discovered sofar. One is inducible in presence of octopine, and it controls the expression of an operon which includes the catabolic enzymes and at least two other polypeptides. The latter are most likely involved in transport, because the DNA sequence indicates a relation to bacterial polypeptides of similar functions and to the corresponding polypeptides in the noc-region. The second promoter is constitutive and directs the expression of a gene (occR) which is necessary for the activation of the inducible promoter. The DNA sequence predicts significant homology with nocR from the noc-region as well as with other regulatory genes in bacteria (manuscript in preparation). These results indicate that occR represents a key element in the regulation of occ-region genes by octopine. We are currently investigating the genes and proteins which are involved in transport functions.

4. COMPARISON OF THE NOC- AND OCC-REGIONS

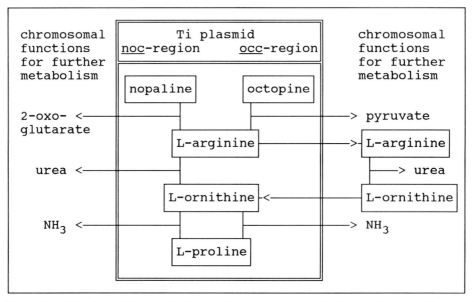

Fig. 2. Ti plasmid encoded reactions for catabolism of nopaline and octopine in Agrobacterium.

The available results suggest a basically similar organization of the noc- and the occ-regions: both contain constitutively expressed regulatory genes, functions for transport, and genes for catabolic conversion of the opines into substances which can be used in the general metabolism of Agrobacterium. The similarity extends to extensive homologies of the proteins. This has been shown previously for the ornithine cyclodeaminases [4], and our recent results indicate significant relationships of the regulatory proteins nocR and occR as well as of the transport functions.

There are, however, important differences in details of the organization and the functions. While the <u>noc</u>-region is split into two parts, the <u>occ</u>-region appears to represent a single block of genes. Also, the <u>occ</u>-region seems to lack the arginase and the 40 kDa protein gene which are present in the <u>noc</u>-region. The absence of arginase but presence of an octopine inducible ornithine cyclodeaminase in the <u>occ</u>-region is interesting, because it indicates that the pathway from octopine to L-proline requires an arginase which is encoded in the chromosome (Fig. 2). These genes are present in most <u>Agrobacterium</u> strains, and the same appears to be the case for additional copies of ornithine cyclodeaminase genes (see [3,4] for detailed discussions).

ACKNOWLEDGMENTS

Supported by Deutsche Forschungsgemeinschaft (SFB206) and Fonds der Chemischen Industrie.

REFERENCES

[1] Sans, N., G. Schröder, and J. Schröder. 1987. The *noc*-region of Ti plasmid C58 codes for arginase and ornithine cyclodeaminase. Eur. J. Biochem. 167:81-87.

[2] Sans, N., U. Schindler, and J. Schröder. 1988. Ornithine cyclodeaminase from Ti plasmid C58: DNA sequence, enzyme properties, and regulation of activity by arginine. Eur. J. Biochem. 173:123-130.

[3] Schrell, A., J. Alt-Mörbe, T. Lanz, and J. Schröder. 1989. Arginase of <u>Agrobacterium</u> Ti plasmid C58. DNA sequence, properties, and comparison with eucaryotic enzymes. Eur. J. Biochem. 184:635-641.

[4] Schindler, U., N. Sans, and J. Schröder. 1989. Ornithine cyclodeaminase from octopine Ti plasmid Ach5: identification, DNA sequence, enzyme properties, and comparison with gene and enzyme from nopaline Ti plasmid C58. J. Bacteriol. 171:847-854.

GENE-FOR-GENE RELATIONSHIPS SPECIFYING DISEASE RESISTANCE IN PLANT-BACTERIAL INTERACTIONS

A. Bent, F. Carland, D. Dahlbeck, R. Innes, B. Kearney, P. Ronald,
M. Roy, J. Salmeron, M. Whalen, and B. Staskawicz.
Department of Plant Pathology
University of California, Berkeley, CA 94720. U.S.A.

ABSTRACT.

We are currently developing three model plant-bacterial interactions that will allow us to answer specific questions concerning the molecular basis of disease resistance and genetic mechanisms involved in pathogen mutation to overcome host resistance. We are studying the interaction of *Pseudomonas syringae* pv. *tomato* (*Pst*) with *Arabidopsis thaliana* and *Lycopersicon esculentum* (tomato) with the goal of cloning corresponding avirulence-resistance gene pairs that specify disease resistance. We have demonstrated that certain strains of *Pst* can cause disease on both hosts while others are only capable of causing disease on specific ecotypes of *A. thaliana* or cultivars of tomato. Two avirulence genes have been cloned and characterized from *Pst* that correspond to two genetically characterized resistance genes in both tomato and *A. thaliana*.. *Pst* strains carrying the avirulence gene *avrPto* are avirulent on tomato cultivars carrying the resistance gene *Pto*, while *Pst* strains harboring the *avrRpt2* gene are avirulent on *A. thaliana* ecotypes containing the corresponding gene *Rpt2*. Finally, we have demonstrated that the avirulence gene *avrBs2* from *Xanthomonas campestris* pv. *vesicatoria* is involved in the fitness of the pathogen and may explain why the corresponding pepper resistance gene *Bs2* seems to stable under natural field conditions.

INTRODUCTION.

Genetic analyses of plant-bacterial interactions have demonstrated that race-specific disease resistance is specified by the presence of single avirulence genes in the bacterium that correspond to single resistance genes in the host (1,2,3,4,5,6,8,9,10,11,12,13,14, 16 17,18,21,22,23) The major goal of our laboratory is to clone and characterize corresponding avirulence-resistance gene pairs that control the expression of disease resistance. We have chosen to study the interactions between *Pseudomonas syringae* pv. *tomato* (*Pst*) and *Arabidopsis thaliana* and *Lycopersicon esculentum* (tomato). These interactions serve as excellent model systems because we are able to manipulate the host and the pathogen employing the tools of both classical and molecular genetics (16,21)

As mentioned previously, we have also discovered that the avirulence gene *avrBs2* from *X.c. vesicatoria* is involved in both the induction of a race-specific defense reaction and also has a role in pathogen fitness(7). This is the first example we have examined that has demonstrated this phenomenon.

H. Hennecke and D. P. S. Verma (eds.),
Advances in Molecular Genetics of Plant-Microbe Interactions, Vol. 1, 32–36.
© 1991 *Kluwer Academic Publishers. Printed in the Netherlands.*

RESULTS

We have recently cloned two avirulence genes (*avrPto, avrRpt2*) from *Pseudomonas syringae* pv. *tomato* strain JL1065 that are involved in specifying disease resistance on either *Arabidopsis thaliana* or *Lycopersicon esculentum*. These genes were identified from a JL1065 cosmid library by inoculating DC3000 and T1 exconjugants on the appropriate host lines. For instance, the *avrPto* avirulence gene when introduced into the normally virulent *P.s. tomato* strain T1 is recognized by the tomato cultivar 76R that contains the resistance gene *Pto*, whereas the *avrRpt2* avirulence gene is recognized by the *Arabidopsis thaliana* ecotype Col-O.

The following table depicts the disease interactions between *Pseudomonas syringae* pv. *tomato* and *Arabidopsis thaliana* and *Lycopersicon esculentum* (Table 1.)

P.s. tomato strains	*Arabidopsis thaliana* Col-O	Hs-O	Po-1	*Lycopersicon esculentum* 76R(*Pto*)	76S(pto)
DC3000	S	S	S	R	S
JL1065	R	R	R	R	S
T1	R	R	R	S	S
T1(avrPto)	R	R	R	R	S
DC3000(avrRpt2)	R	S	S	R	S

The resistance gene, *Pto*, was originally identified from the sexually compatible wild species *Lycopersicon pimpinellifolium* and has been genetically mapped to a single locus on chromosome five (14). As mentioned previously, we have identified and characterized the avirulence gene *avrPto* (Ronald et. al. 1991 manuscript in preparation) by the inoculation of T1(*avrPto*) exconjugants on the isogenic *Pto* containing cultivar 76R and the conversion of T1 to avirulence on 76R. In our laboratory we are also attempting to clone the *Pto* resistance gene employing the strategy of transposon tagging. We have successfully introduced the maize transposable element *Ac* into *Pto* containing lines and have demonstrated that this element is capable of moving from the original T-DNA insertion to new locations in the genome. Thus we will be able to detect an insertion into the *Pto* gene by the conversion of a resistant phenotype to a susceptible phenotype. The ability to screen for *Pto* mutants is further facilitated by the availability of isogenic bacterial strains that differ in a single gene that contains *avrPto* activity. The screening of potential *Pto* mutants employing a naturally occurring avirulent race of *Pst* may not uncover a mutation due to the possibility of additional avirulence genes in that strain.

As can be observed from Table 1., we have also identified the avirulence gene *avrRpt2* from *Pst* that is differentially recognized by *A. thaliana* ecotypes, Col-0. Hs-0, and Po-1. These data will now allow us to determine if the ability of Col-0 to be resistant to DC3000(*avrRpt2*) is due to a single genetic locus. We are currently scoring segregating populations between Col-0 x Hs-0 and Col-0 x Po-1. Again, it should be stressed, that the construction of isogenic bacterial strains differing in a single avirulence gene will greatly facilitate the genetic characterization of the corresponding resistance gene.

A major effort in our laboratory is currently underway to isolate plant mutants that have been converted from resistance to susceptibility. We are concentrating our efforts on both tomato and *Arabidopsis*. We are currently screening M2 mutagenized seed of the *Arabidopsis* ecotype Col-0 by inoculating with DC3000 (*avrRpt2*). The seed has been mutagenized with the mutagen diepoxybutane (DEB), which has been shown to cause deletions in other organisms. This is especially significant as it is now possible to clone genes from *Arabidopsis thaliana* using the protocol of genomic subtraction (19). Along these lines we have also developed a rapid and efficient screen to identify mutations. Basically, this procedure involves dipping whole plants into a bacterial suspension of $5x10^8$ cfu/ml of *Pseudomonas sryingae* pv. *tomato* in a solution of .01% of Silwet L-77. Silwet L-77 is a silicon based copolymer that acts as a surface tension depressant and facilitates the movement of bacteria through the stomata into the intercellular space. Thus we are able to screen many potential mutants in a relatively small amount of time.

Finally we are studying the molecular basis of durable disease resistance in the bacterial spot disease of pepper. We have recently suggested that the Bs2 gene of pepper is stable under natural field conditions for at least two important reasons(7) Firstly, all strains of *Xanthomonas campestris* pv. *vesicatoria* contain the corresponding avirulence gene avrBs2. This is different from other *Xanthomonas campestris* pv. *vesicatoria* avirulence genes in that only certain strains contain the avirulence genes *avrBs1* or *avrBs3*.(1,14,20) But more importantly, it seems that spontaneous mutants that overcome the *Bs2* resistance gene are less fit and do not grow to as high a level as the wild type strain in pepper plants that do not contain the *Bs2* gene. These data suggest that *avrBs2* not only is involved in controlling the induction of a defense reaction on *Bs2* pepper plants , but also must have a role in virulence. Southern blot analysis of *Xanthomonas campestris* pv. *vesicatoria* spontaneous mutants that have over come *Bs2* resistance seem to have no major DNA alterations as revealed by restriction enzyme analysis. Thus, we are in the process of sequencing the mutants alleles in attempts to localize which nucleotides have been altered and to determine if we can separate the two functions of virulence and avirulence.

REFERENCES

1. Bonas U., R.E. Stall and B. Staskawicz. 1988. Genetic and structural characterization of the avirulence gene avrBs3 from *Xanthomonas campestris* pv. *vesicatoria*. MGG 218:127-136.

2. Ellingboe, A. 1982. Genetical aspects of active defense. p. 179-192, In Active defense mechanisms in plants. Ed. by R.K.S. Wood. Plenum Press, New York.

3. Flor, H. 1971. Current status of the gene-for-gene concept. Ann. Rev. Phytopath. 9:275-296.

4. Gabriel, D.W., A. Burges and G.R. Lazo 1986. Gene-for-gene recognition of five cloned avirulence genes from *Xanthomonas campestris* pv. *malvacearum* by specific resistance genes in cotton. P.N.A.S. 83:6415-6419.

5. Hitchin, F.E., C. Jenner, S. Harper, J. Mansfield, C. Barber and M. Daniels 1989. Determinant of cultivar specific avirulence cloned from *Pseudomonas syringae* pv. *phaseolicola* race 3. Physiol. Mol. Plant Path. 34:309-322.

6. Huynh, T., D. Dahlbeck and B. Staskawicz. 1989. Bacterial blight of soybeans: Regulation of a pathogen gene determining host cultivar specificity. Science 245:1374-1377.

7. Kearney, B. and b. Staskawicz 1990. Widespread distribution and fitness contribution of *Xanthomonas campestris* avirulence gene avrBs2. Nature 346:385-386.

8. Keen, N. and B. Staskawicz 1988. Host range determinants in plant pathogens and symbionts. Ann Rev. Micro. 42:421-440.

9. Keen, N., S. Tamaki, D. Kobayashi, D. Gerhold, M. Stayton, H. Shen, S. Gold, J. Lorang, H. Thordal-Christensen, D. Dahlbeck, and B. Staskawicz. 1990. Bacteria expressing avirulence gene D produce a specific elicitor of the soybean hypersensitive reaction. MPMI (in press).

10. Kelemu, S. and J. Leach 1990. Cloning and characterization of an avirulence gene from *Xanthomonas campestris* pv. *oryzae*. MPMI 3:59-65.

11. Kobayashi, D., S. Tamaki, and N. Keen 1989. Cloned avirulence gene from the tomato pathogen *Pseudomonas syringae* pv. *tomato* confer cultivar specificity on soybean. PNAS 86:157-161.

12. Minsavage, G., D. Dahlbeck, M. Whalen, B. Kearney, U. Bonas, B. Staskawicz and R. Stall. 1990. Gene-for-gene relationships specifying disease resistance in Xanthomonas campestris pv. vesicatoria - pepper interactions. MPMI (in press).

13. Napoli, C., B. Staskawicz. 1987. Molecular characterization and nucleic acid sequence of an avirulence gene from Race 6 of *Pseudomonas syringae* pv. *glycinea*. J. Bacteriol. 169:572-578.

14. Pitblado, R.E. and B. MacNeil and E. Kerr. 1984. Chromosomal identity and linkage relationships of Pto, a gene for resistance to Pseudomonas syringae pv. tomato in tomato. Can. J. Plant Path. 5:251-255.

15. Ronald P. C., B. J. Staskawicz. 1988. The avirulence gene avrBs[1] from *Xanthomonas campestris* pv. *vesicatoria* encodes a 50 kd protein. Molec. Plant-Microbe Interact. Vol. 1, No. 5:191-198.

16. Somerville, C. 1989. Arabidopsis blooms. The Plant Cell 1:1131-1135.

17. Staskawicz, B. J., D. Dahlbeck, N. Keen, C. Napoli. 1987. Molecular characterization of cloned avirulence genes from race 0 and race 1 of *Pseudomonas syringae* pv. *glycinea*. J. Bact. 169:5789-5794.

18. Staskawicz, B., D. Dahlbeck and N. Keen 1984. Cloned avirulence gene of Pseudomonas syringae pv. glycinea determines race-specific incompatibility on Glycine max (L.) Merr.

PNAS 81:6024-6028.

19. Straus, D. and F. Ausubel 1990. Genomic subtraction for cloning DNA corresponding to deletion mutations. PNAS 87:1889-1893.

20. Swanson, J., B. Kearney, D. Dahlbeck, B. J. Staskawicz. 1988. Cloned avirulence gene of *Xanthomonas campestris* pv. *vesicatoria* complements spontaneous race change mutant. Molecular Plant-Microbe Interactions 1:5-9.

21. Tsuji, J. and Somerville, S.C. 1988. Xanthomonas campestris pv. campestris-induced chlorosis in Arabidopsis. Arabidopsis Inf. Serv. 26:1-8.

22. Vivian, A., G. Atherton, J. Bevan, I. Crute, L. Mur, and J. Taylor. 1989. Isolation and characterization of cloned DNA conferring specific avirulence in *Pseudomonas syringae* pv. *pisi* to pea (Pisum sativum) cultivars, which possess the resistance allele, R2. Physiol. Mol. Plant Path. 34:335-344.

23. Whalen, M. C., R. E. Stall, B. J. Staskawicz. 1988. Characterization of a gene from a tomato pathogen determining hypersensitive resistance in a non-host species and genetic analysis of this resistance in bean. P.N.A.S. 85:6743-6747.

AVIRULENCE GENE *D* FROM *PSEUDOMONAS SYRINGAE* PV. *TOMATO* AND ITS INTERACTION WITH RESISTANCE GENE *Rpg4* IN SOYBEAN

N. KEEN, D. KOBAYASHI*, S. TAMAKI*, H. SHEN, M. STAYTON*, D. LAWRENCE*, A. SHARMA, S. MIDLAND, M. SMITH** AND J. SIMS

Department of Plant Pathology, University of California, Riverside, CA 92521, USA

*Department of Molecular Biology, University of Wyoming, Laramie, WY 82071, USA

**Natural Products and Instrumentation Branch, Food and Drug Administration, Washington, DC, 20204, USA

ABSTRACT. Certain pathogen avirulence genes have been associated with production by the pathogen of specific elicitors, chemicals that initiate hypersensitive defense responses (HR) only in those plant cultivars which carry complementary disease resistance genes. We have cloned and characterized one of these avirulence genes, *avrD*, from *Pseudomonas syringae* pv. *tomato*. Various Gram-negative bacteria containing this cloned avirulence gene produce a low molecular weight compound(s) that causes the HR exclusively in soybean cultivars carrying the disease resistance gene, *Rpg4*. We isolated two major *avrD* elicitor-active molecules and have partially characterized them as homologous C13 and C15 hydrocarbons that are heavily substituted with oxygen atoms and cyclized at one end. The physiologic role of the *avrD* elicitor molecules in *P.s.* pv. *tomato* remains unclear since mutants deficient in *avrD* retained virulence in tomato plants. However, the fact that expression of *avrD* was greatly stimulated when *P.s.* pv. *tomato* was grown on plant leaves or in cell suspension cultures suggests that the gene may be important for the survival of the bacteria on or in plant leaves. The purified elicitor-active molecules specifically induced PAL and CHS mRNAs in the soybean cultivar Norchief, which carries the *Rpg4* resistance gene, but not in Acme which lacks this gene. Thus, gene-for-gene specificity mediated by the elicitor was reflected at the level of defense gene expression. The *avrD* specific elicitor preparations also elicited necrosis and phytoalexin production in soybean callus and cotyledons. The evidence therefore indicates that the *avrD* elicitor molecules are the signal which elicits the HR in the *avrD-Rpg4* gene-for-gene interaction.

H. Hennecke and D. P. S. Verma (eds.),
Advances in Molecular Genetics of Plant-Microbe Interactions, Vol. 1, 37–44.
© 1991 *Kluwer Academic Publishers. Printed in the Netherlands.*

1. INTRODUCTION

Invocation of the defensive hypersensitive response (HR) in plants frequently involves the interaction of single genetic loci in the plant (disease resistance genes) as well as in an attacking pathogen (avirulence genes). Only recently, however, have we begun to understand their function. Indirect evidence suggests that plant disease resistance gene products may function as receptors for signal molecules produced by the pathogen, called elicitors. This is unproven, largely because resistance genes have not yet been molecularly cloned. However, avirulence genes have been cloned and characterized from several pathogens and some of them have been associated with the production of specific elicitors. Among these are the coat protein gene and its primary gene product, the capsid protein, of tobacco mosaic virus (Culver and Dawson, 1989), a peptide elicitor produced by *Cladosporium fulvum* isolates carrying the *ACf9* avirulence gene (De Wit, 1990) and a low molecular weight elicitor produced by bacteria expressing *avrD* from *Pseudomonas syringae* pv. *tomato* (Keen et al., 1990). In this paper, we will discuss recent progress on identification of the *avrD* elicitor, its biochemical effects on soybean tissues and the regulation of *avrD* gene expression in *P. syringae* pathovars.

2. CHARACTERIZATION OF THE *avrD* GENE

Three classes of cosmid clones were isolated from a genomic library of *P. syringae* pv. *tomato* PT23 DNA which, when introduced into race 4 of *P.s.* pv. *glycinea*, caused this bacterium to elicit a hypersensitive defense reaction on certain soybean cultivars (Kobayashi et al., 1989). One of these classes was subsequently shown to harbor a single avirulence gene, called *avrD* (Kobayashi et al., 1990a). A second class of cosmid clones contained a gene with very high homology to *avrA*, previously cloned from *P. syringae* pv. *glycinea* race 6 (Staskawicz et al., 1984). A single cosmid clone was also recovered from the *P.s.* pv. *tomato* library that yielded a weak hypersensitive reaction on 9 of the 10 standard soybean differential cultivars, but it proved difficult to subclone the phenotype to a small DNA fragment. These facts raised the possibility that this clone might contain all or part of the *P.s.* pv. *tomato* hrp gene cluster. These genes are essential for the pathogenicity of several phytopathogenic bacteria (Lindgren et al., 1988) and, more particularly, have been shown to cause hypersensitive reactions when introduced into certain heterologous pathogens (Beer et al., 1989; Huang et al., 1988). Our recent results, obtained collaboratively with C. Boucher and B. Staskawicz, have shown that the avirulence phenotype of the third *P.s.* pv. *tomato* clone is indeed conferred by a ca. 11 kb DNA fragment which lies within or immediately adjacent to the *hrp* cluster of *P.s.* pv. *tomato* (Lorang et al., 1990).

Kobayashi et al. (1990a) sequenced the *avrD* gene from *P.s.* pv. *tomato* and showed that it was the first of five tandem open reading frames which appear to be organized as an operon. The *avrD* gene showed no significant homology to previously sequenced genes and encoded a 34 kDa protein that appeared to remain in the cytoplasm of Gram-negative bacterial hosts. Kobayashi et al (1990b) observed that *P.s.* pv. *glycinea* race 4 also contained a gene with high homology to *avrD*, but it did not function as an avirulence gene. However, Keen and Buzzell (1990) showed that soybean cultivars which reacted hypersensitively to *P.s. glycinea* race 4 cells carrying the cloned *avrD* gene from *P.s.* pv. *tomato* contained a single dominant disease resistance gene, called *Rpg4*. Thus, the interaction of *avrD* and *Rpg4* represents a classical gene-for-gene interaction. Unlike other bacterial avirulence genes thus far cloned, *E. coli* cells expressing *avrD* caused a hypersensitive reaction in soybean cultivars containing *Rpg4*. Further, it was observed that *E. coli* as well as several other Gram-negative bacteria harboring the cloned *avrD* gene produced a low molecular weight factor in the culture fluids which elicited the soybean HR (Keen et al., 1990). This is of considerable importance since it illuminates the biochemical basis of signalling that occurs between pathogens expressing *avrD* and plants harboring *Rpg4*. Since much of the early work with *avrD* has already been published, we will concentrate on more recent results in this paper.

3. The *avrD* ELICITOR

The *avrD* elicitor has been produced in relatively large amounts from M9-glucose culture fluids of *E. coli* DH5α, *P. syringae* pv. *tomato* or *P.s.* pv. *glycinea* cells containing high expression plasmid constructs of *avrD* (Keen et al., 1990 and unpublished). *P. syringae* pv. *tomato* normally makes elicitor activity due to presence of the native *avrD* gene, but *P.s.* pv. *glycinea* only makes exceedingly low levels and *E. coli* does not produce detectable activity in the absence of the cloned *avrD* gene. The elicitor activity has proved to be somewhat unstable during purification, but crystalline preparations have recently been obtained that elicit the HR only in leaves of soybean cultivars carrying *Rpg4* (minimal concentration, ca. 10 ng in 10 μl water per injection site for production of a visible HR on *Rpg4* primary soybean leaves). M9 culture fluids of bacteria over-expressing *avrD* were processed in two different ways: first, the fluids were passed through XAD-7 columns and the adsorbed elicitor activity was eluted with ethanol. The resulting solution was dried, redissolved in water and passed through a Dowex 1 column and the elicitor was then eluted with 0.01 N acetic acid. In the second method, culture fluids were adjusted to pH 5.0 with 1 N HCl and directly extracted three times with ca. 0.2 volumes of ethyl acetate. The organic fraction, containing most of the elicitor activity was dried, redissolved in 95% ethanol and further fractionated by HPLC.

Phenyl and C18 reversed phase HPLC column chromatography disclosed the presence of two well separated and elicitor-active peaks from the preparations above that were detectable at 254 nm. *E. coli* as well as the two *P. syringae* pathovars overexpressing *avrD* produced peaks with identical retention times, although the relative heights of the two peaks were different for the pseudomonads as compared to *E. coli*. Proton and carbon NMR spectra of the two HPLC peaks in $CDCl_3$ were similar, but certain peaks did not integrate properly and varied between preparations, leading to the suspicion that the preparations were impure or that decomposition occurred after isolation. Upon standing in $CHCl_3$ at high concentration, colorless crystals formed from both HPLC peaks 1 and 2. These compounds proved to be elicitor-active in soybean leaves and yielded reproducible NMR and mass spectral data. Both crystalline compounds have saturated alkyl chains (C_5H_{11} in the case of HPLC peak 1 and C_7H_{15} for HPLC peak 2) attached to a common nucleus ($C_8H_9O_6$), the structure of which has not been fully determined. This part of the molecules appears to contain a gamma-lactone ring and two free hydroxyl groups. Such a highly oxygenated structure is rare in bacteria and it is not clear if the elicitor molecules are produced from the conventional malonate fatty acid pathway or the acetate polyketide pathway. The function of these molecules in *P. syringae* pv. *tomato* is also unclear, particularly in view of the fact that *avrD* mutants appear to retain normal virulence in tomato plants (H. Shen, unpublished data). The structures of the elicitors indicate that they may possess surfactant properties, but this possibility is only now being tested.

4. EFFECTS OF THE *avrD* SPECIFIC ELICITOR ON CELLS OF SOYBEAN AND OTHER PLANTS

Since the *avrD* elicitor caused hypersensitive reactions on leaves of only those soybean cultivars which were resistant to *P.s. glycinea* race 4 carrying the cloned *avrD* gene, it appeared that the elicitor was the biochemical signal perceived by resistant plants to elicit the HR. To further test this hypothesis, segregating progeny of a cross of soybean cultivars Flambeau (resistant to *P.s. glycinea* race 4 carrying *avrD* and sensitive to the *avrD* elicitor) x Merit (susceptible to *P.s. glycinea* race 4 carrying *avrD* and insensitive to the elicitor) were examined (Keen and Buzzell, 1990). In the segregating F_2 and F_3 progeny examined to date, all of the 875 segregants resistant to the bacteria were also sensitive to the *avrD* elicitor. All of the 281 segregants that were susceptible to the bacteria were also insensitive to the elicitor. No plants were obtained in which this linkage was broken. These data establish that a single dominant gene (for which we propose the designation *Rpg4*) occurs in soybean that complements avirulence gene D. The data also strongly argue that the *avrD* elicitor interacts with the *Rpg4* gene product and that the elicitor is the biochemical signal from infecting bacteria that is recognized by *Rpg4* soybean plants to elicit the HR.

In addition to soybean cultivars carrying the dominant *Rpg4* resistance gene, the *avrD* elicitor has been found to elicit necrotic leaf reactions in some but not all cultivars of chrysanthemum and also in the petunia cultivar Mitchell, but not other cultivars thus far tested

(Stayton and Keen, unpublished data). Representatives of several other plant species tested did not react when the *avrD* elicitor was infiltrated into leaves. These results raise the possibility that genes functionally identical to the soybean *Rpg4* disease resistance gene occur in chrysanthemum, petunia and possibly other plant taxa.

The *avrD* specific elicitor was investigated for its effects on soybean gene expression, phytoalexin elicitation and its ability to cause cell necrosis in soybean. Glyceollins are antibiotic compounds produced by soybean plants in response to pathogen infection that appear to be involved in disease resistance (Keen and Yoshikawa, 1990). We have observed that both leaves and cotyledons of soybean cultivars carrying the *Rpg4* resistance gene produce these phytoalexins in response to the purified *avrD* elicitor (A. Sharma, unpublished data). Phenylalanine ammonia lyase (PAL) is the first enzyme in the general phenylpropanoid pathway, while chalcone synthase (CHS) is the first enzyme specific to the flavonoid-isoflavonoid pathway utilized by soybean for glyceollin biosynthesis. Both PAL and CHS mRNAs were rapidly induced in leaf tissue of the cultivar Norchief, which carries the *Rpg4* disease resistance gene, in response to infiltration of the specific elicitor or inoculation with *P.s.* pv *glycinea* race 4 expressing the cloned *avrD* gene (D. Lawrence and M. Stayton, unpublished data). The PAL and CHS mRNAs were not induced by either treatment in the cultivar Acme that lacks the *Rpg4* gene. The toxic effect of the specific elicitor on soybean leaves was also investigated by measuring protein synthesis as the incorporation of labelled methionine. The specific elicitor inhibited protein synthesis in cv. Norchief leaf tissue, but had no effect in Acme (S. Tamaki, D. Lawrence and M. Stayton, unpublished data). Therefore, the toxicity of the specific elicitor in soybean mirrors the situation seen with certain host selective toxins of fungal pathogens (Nishimura and Kohmoto, 1983). The major distinction is that the *avrD* elicitor does not function as a toxin important in pathogenicity but instead appears clearly to be involved in the initiation of the hypersensitive defense reaction in *Rpg4* soybeans.

5. THE B PLASMID

The *avrD* gene and associated cistrons occur on an indigenous ca. 75 kb plasmid in *P. syringae* pv. *tomato* (Kobayashi et al., 1990a). This plasmid was originally detected by Bender and Cooksey (1986) and designated the 'B' plasmid, but no functions have yet been assigned to it. The plasmid has been detected in all *P.s. tomato* isolates thus far examined. This, coupled with the fact that the B plasmid proved difficult to cure from *P.s. tomato* PT23 using temperature shifts and classical curing agents, led us to suspect that it might play an important role in the bacterium. Recently, the plasmid has been cured by insertion of a *kan-sac* cartridge into it by marker exchange mutagenesis and selection for survivors on high sucrose media (H. Shen, unpublished data). The resultant cured mutant strain, however, appears to grow normally and produces normal lesions on tomato plants. A mutant bacterial strain carrying a deletion of ca. 15 kb of DNA in the B plasmid including *avrD* also grew normally and gave normal lesions on tomato plants. However, deletion of only ORFs 3, 4 and 5 of the *avrD* operon but not *avrD* itself resulted in *P.s. tomato* cells that grew more slowly than the wild-type and were not pathogenic on tomato plants.

Over-expression of the *avrD* gene in *P.s. tomato* by a single *E. coli lac* promoter increased production of the *avrD* elicitor (Keen et al., 1990), but did not affect growth rates or virulence in tomato plants. On the other hand, higher level expression of *avrD* in *P.s. tomato* using the triple *lac* UV5 promoters of pAVRD12 (Keen et al., 1990), led to cells that grew less rapidly in culture and were less virulent in tomato plants. These data indicate that the *avrD* protein product or the *avrD* elicitor may be toxic to *P.s.* pv. *tomato* cells at high *avrD* expression levels.

6. EXPRESSION OF *avrD*

Using the *Vibrio fischeri lux* reporter gene system, we observed that the expression of avrD was much higher in *P.s.*pv. *tomato* or *P.s.* pv. *glycinea* cells inoculated into soybean, tomato or tobacco leaves than when the cells were grown on several minimal or rich culture media (Shen and Keen, 1989). We have also recently found that the same bacteria express high *avrD* promoter activity when added to tobacco suspension culture cells. However, induction was not observed when bacteria were added to the unused suspension culture medium or to medium from which the tobacco cells had been removed. While the significance of these results has not been fully established, they may indicate the necessity of cell-cell contact for *avrD* induction.

7. FUTURE DIRECTIONS

When the structure of the *avrD* elicitor is completely known, it will be of interest to construct suitable derivatives in order to search for the putative receptor occurring in *Rpg4* soybean plants. This would in turn permit an approach to isolation of the *Rpg4* gene. A major objective is also to determine the role of *avrD* and the *avrD* elicitor in the biology of *P. syringae* pv. *tomato*. Indications that the *avrD* elicitor-active molecules possess surfactant properties may provide an important clue because such molecules are known to be important in the ecology of many bacteria (Cooper and Zajic, 1980). We will accordingly examine the possible role of *avrD* in growth of the bacteria on leaf surfaces, since they can exist in a resident phase on leaf surfaces without causing disease (Schneider and Grogan, 1977). We are also interested in other functions located on the B plasmid of these bacteria which might be important for their survival.

8. REFERENCES

Beer, S.V., Zumoff, C.H., Bauer, D.W., Sneath, B.J., and Laby, R.J. (1989) 'The hypersensitive response is elicited by *Escherichia coli* containing a cluster of pathogenicity genes from *Erwinia amylovora*' Abstr. 169, American Phytopatholgical Society meeting, Richmond, Virginia.

Bender, C.L. and Cooksey, D.A. (1986) 'Indigenous plasmids in *Pseudomonas syringae* pv. *tomato*: conjugative transfer and role in copper resistance' J. Bacteriol. 165, 534-541.

Cooper, D.G. and Zajic, J.E. (1980) 'Bacterial speck of tomato: sources of inoculum and establishment of a resident population' Adv. Appl. Microbiol. 26, 229-253.

Culver, J.N. and Dawson, W.O. (1989) 'Tobacco mosaic virus coat protein: an elictor of the hypersensitive reaction but not required for the development of mosaic symptoms in *Nicotiana sylvestris*' Virology 173,755-758.

DeWit, P.J.G.M. (1990) 'Functional models to explain gene-for-gene relationships in plant-pathogen interactions' In W. Boller and F. Meins (eds.) Genes involved in plant defense, Plant Gene Research, vol. 8, Springer-Verlag, Vienna (in press).

Huang, H-C., Schuurink,R., Denny, T.P., Atkinson, M.M., Baker, C.J., Yucel, I., Hutcheson, S.W., and Collmer, A. (1988). 'Molecular cloning of a *Pseudomonas syringae* pv. *syringae* gene cluster that enables *Pseudomonas fluorescens* to elicit the hypersensitive response in tobacco plants' J. Bacteriol. 170,4748-4756.

Keen, N.T. and Buzzell, R.I (1990) 'New disease resistance genes in soybean against *Pseudomonas syringae* pv. *glycinea*: evidence that one of them interacts with a bacterial elicitor' Theor. Appl. Genet. (in press).

Keen, N.T. and Yoshikawa, M. (1990) 'The expression of resistance in soya beans to *Phytophthora megasperma* f.sp. *glycinea*' In Biological Control of Soil-borne plant pathogens, p. 329-344. Ed. by D. Hornby et al., CAB International, Wallingford, Oxon, U.K.

Keen, N.T., Tamaki, S. Kobayashi, D., Gerhold, D., Stayton, M., Shen, H., Gold, S.,. Lorang, J., Thordal-Christensen, H., Dahlbeck D., and Staskawicz, B. (1990) 'Bacteria expressing avirulence gene D produce a specific elicitor of the soybean hypersensitive reaction' Molec. Plant-Microbe Inter. 3,122-121.

Kobayashi, D.Y., Tamaki, S.J., and Keen, N.T. (1989) `Cloned avirulence genes from the tomato pathogen *Pseudomonas syringae* pv. *tomato* confer cultivar specificity on soybean' Proc. Natl. Acad. Sci., USA 86,157-161.

Kobayashi, D.Y., Tamaki, S.J., and Keen, N.T. (1990a) `Molecular characterization of avirulence gene D from *Pseudomonas syringae* pv. *tomato*' Molec. Plant-Microbe Inter. 3,94-102.

Kobayashi, D.Y., Tamaki, S.J., Trollinger, D., Gold, S. and Keen, N.T (1990b) `A gene from *Pseudomonas syringae* pv. *glycinea* with homology to avirulence gene D from *P.s.* pv. *tomato* but devoid of the avirulence phenotype' Molec. Plant-Microbe Inter. 3, 103-111.

Lindgren, P.B., Panopoulos, N.J., Staskawicz, B.J., and Dahlbeck, D. (1988) `Genes required for pathogenicity and hypersensitivity are conseved and interchangeable among pathovars of *Pseudomonas syringae*' Molec. Gen. Genet. 211,499-506.

Lorang, J.M., Boucher, C.A., Dahlbeck, D., Staskawicz, B. and Keen, N.T. (1990) `An avirulence function from *Pseudomonas syringae* pv. *tomato* is located within a *hrp* cluster' Abstract 28, American Phytopathological Society meeting, Grand Rapids, Michigan

Nishimura, S and Kohmoto, K. (1983) `Host-specific toxins and chemical structures from *Alternaria* species' Annu. Rev. Phytopathol 21, 86-116.

Schneider, R.W. and Grogan, R.G. (1977) `Bacterial speck of tomato: sources of inoculum and establishment of a resident population' Phytopathology 67, 388-394.

Shen, H. and Keen, N.T. (1989) `Regulation of avirulence gene D (*avrD*) from *Pseudomonas syringae* pv. *tomato* studied using a novel Tn*7-lux* system' Abstract 349, American Phytopathological Society meeting, Richmond, Virginia.

Staskawicz, B.J., Dahlbeck, D., and Keen, N.T. (1984) `Cloned avirulence gene of *Pseudomonas syringae* pv. *glycinea* determines race-specific incompatibility on *Glycine max* (L.) Merr.' Proc. Natl. Acad. Sci., USA 81,6024-6028.

GENES AND SIGNALS CONTROLLING THE *PSEUDOMONAS SYRINGAE* PV. *PHASEOLICOLA*-PLANT INTERACTION

R. FELLAY, L. G. RAHME, M. N. MINDRINOS[1], R. D. FREDERICK[2],
A. PISI[3], AND N. J. PANOPOULOS
Department of Plant Pathology, University of California,
Berkeley, CA 94729, USA.
Current addresses of some authors:
[1]Department of Molecular Biology, Massachusetts General
Hospital, Boston, MA 02114, USA;
[2]Department of Plant Pathology, Ohio State University,
Columbus, OH 43210-1087;
[3]Institute of Plant Pathology, University of Bologna,
Bologna, Italy.

ABSTRACT. The hrp cluster of Pseudomonas syringae pv.
phaseolicola spans nearly 22 kb and comprises seven
complementation groups, some of which represent multici-
stronic operons. The transcriptional organization of this
region was established by quantifying the expression levels
of chromosomal hrp::inaZ fusions in planta and in vitro and
corroborates sequence data available for several genes/
operons. Two genes, hrpS and hrpL, are involved in the
regulation of several hrp operons, whose expression additionally requires the function of ntrA, the structural gene
for sigma-54. Analysis of site-directed substitutions of a
glycine residue to asparagine in the putative ATP binding
motif of the HrpS protein and of the putative product of hrpR
indicated that HrpS shares common functional features with
other similar activators and that the hrpR product, which is
not in itself sufficient for the activation of hrp gene
expression, may modulate the function of other components
involved in regulation. The entire regulon is controlled by
medium osmolarity as well as medium composition. The activation of the hrpSR operon and the hrpL gene requires an
unidentified plant signal(s).

1. Introduction

The hrp ("harp") genes of Pseudomonas syringae pv.
phaseolicola were first identified as a functionally
significant group for the development of the disease symptoms
and elicitation of the hypersensitive response in our
laboratory in 1986 [8]. Phenotypically similar genes and hrp
gene homologues are also present in other members of P.
syringae and in other gram-negative plant pathogenic bacteria
[rev. in 2, 13]. It is now apparent that the majority of

45

H. Hennecke and D. P. S. Verma (eds.),
Advances in Molecular Genetics of Plant-Microbe Interactions, Vol. 1, 45–52.
© 1991 *Kluwer Academic Publishers. Printed in the Netherlands.*

these phytopathogens share a set of "core" pathogencity functions and that hrp genes constitute an important part of this set. Therefore, understanding the role of hrp genes is essential for building a complete picture of the disease process. This report summarizes our ongoing structural, functional and regulatory analysis of the hrp genes in P. s. phaseolicola.

2. Materials and methods

The Tn3-Spice transposon [9] was used extensively in our studies, both as a mutagen and as a sensitive reporter of target gene expression. The growth conditions and inoculation methods for in vitro and in planta expression studies are described in ref. [12] and in an upcoming publication (Rahme, Mindrinos and Panopoulos, in preparation). Complementation analysis was carried out as described elsewhere [12]. The ntrA gene of P. s. phaseolicola NPS3121 was identified in a genomic library by screening for homology to an ntrA probe from P. aeruginosa (Fellay and Panopoulos, in preparation). The methods used for site directed mutagenesis of HrpS and HrpR and for the cloning of the ntrA gene will be described elsewhere (Fellay and Panopoulos, in preparation).

3. Results and discussion

3.1. GENETIC ORGANIZATION.

The majority of hrp genes of P. s. phaseolicola are clustered in a ca. 22 kb region (the "hrp cluster") located on the bacterial chromosome, while a single additional locus (hrpM) is found outside this cluster. The hrp cluster comprises seven complementation groups that are organized and transcribed as shown in Fig. 1 [10, 12]. The hrpM locus contains two open reading frames (M1 and M2 in Fig. 1) [4] and is not tightly linked to the hrp cluster.

3.2. REGULATION OF hrp GENES.

3.2.1. Inductive and repressive signals. Expression of hrp genes is regulated by several different mechanisms, namely, osmolarity, medium composition, and a plant signal(s). Studies with hrp::inaZ fusions in several operons showed that hrp genes are expressed in a low osmolarity medium (M9) and this expression is prevented if the osmolarity is raised by addition of several different solutes (Rahme, Mindrinos and Panopoulos, in preparation). In vitro expression levels are also influenced by medium composition in a manner that is suggestive of catabolite repression. The hrpS (Fig. 2) and hrpL genes behave distinctly different from the other hrp

operons in that they are induced only to a very low degree
in vitro and only when the genes themselves are functional.

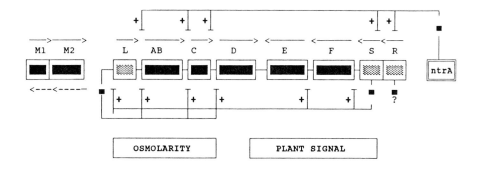

Fig. 1. The genetic organization and transcription of hrp
genes in P. s. phaseolicola. A + sign indicates known
positive regulatory interactions.

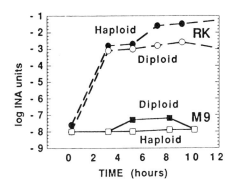

Fig. 2. Expression
of a chromosomal
hrpS::inaZ fusion
in haploid (hrpS⁻)
and diploid (hrpS⁺)
strain in M9 medium
and in Red Kidney
bean leaves.

Expression of the hrp genes in planta differs in one
important respect from the pattern observed in vitro.
Specifically, hrpL and the hrpSR operon are expressed to very
high levels, and the other operons in the hrp cluster are
maximally expressed. The behavior of the hrpS (Fig. 2) and
hrpL genes provides definitive evidence for the involvement
of a plant signal in the transcriptional activation of the
hrp operons during the bacterium-plant interaction.
   The nature of this signal(s) is not presently known. The
hrpL and hrpS genes are involved in the signal transduction
pathway. We have indications that the expression of these
genes is regulated differently, depending on whether the host
is resistant or susceptible to the pathogen.

3.2.2. Regulatory genes. The specific functions of most hrp

genes remain unknown at present, except for two of them which regulate the expression of others (Fig. 1). Thus, six of the seven operons present in the hrp cluster are regulated at the transcriptional level by the product of the hrpS gene and at least three of these operons are also regulated by hrpL [5, 10, 12]. The hrpS product is a 34 kD protein that resembles several other transcriptional regulators (NtrC, DctD, NifA, PgtA, and XylR) of enteric bacteria, Rhizobium spp. and Pseudomonas [6]. The product of the hrpR gene resembles in its amino acid sequence HrpS and the other proteins mentioned above (Grimm, Panopoulos, Dahlbeck, and Staskawicz, in preparation). However, this protein plays no direct role in the activation of hrp operons (Mindrinos, Rahme and Panopoulos, in preparation).

The nitrogen regulatory protein NtrC and other similar transcriptional activators require ATP to promote the formation of "open" (transcriptionally active) complexes [11]. Both HrpS and HrpR posses the consensus ATP binding motif that is characteristic of NtrC-like proteins. A conserved glycine residue in this motif of HrpS and HrpR was replaced with asparagine by site directed mutagenesis (Fellay and Panopoulos, in preparation). Plasmids that carried the entire hrpRS operon but had the G -> N mutation in hrpS, hrpR or in both genes were introduced into a strain that lacked the entire operon and carried a hrpD::inaZ reporter fusion. The ability of these plasmids to activate hrpD expression was compared with that of a plasmid carrying the wild type hrpRS operon. The hrpS mutant allele was completely non functional in hrpD activation. Furthermore, the presence of the hrpR mutant allele resulted in significant decrease of hrpD activation. These findings confirm that the ATP binding motif of HrpS is important for the protein's activator function, and suggest that hrpR, although not required for hrpD expression, may affect some other step(s) in the mechanism by which hrpD is regulated.

3.2.3 The ntrA gene is required for hrp gene expression.
Transcription of bacterial genes involved in specialized metabolic functions or environmental adaptations often require special sigma factors, in addition to specific activator proteins. Since genes that are controlled by NtrC and other similar proteins require sigma-54, the product of the ntrA gene we investigated whether hrp gene expression had a similar requirement. Cosmids carrying the putative ntrA gene of P. s. phaseolicola were identified in a genomic library by hybridization with an ntrA probe from P. aeruginosa and were able to functionally complement an ntrA mutant of this bacterium (Fellay and Panopoulos, in preparation). The ntrA gene was localized by subcloning, insertional mutagenesis and partial sequencing, and ntrA

mutant of P. s. phaseolicola was constructed by marker exchange mutagenesis. This mutant required glutamine for normal growth and was neither pathogenic on bean nor able to elicit the hypersensitive response on tobacco. Furthermore, it showed strongly diminished expression of hrp operons, produced phaseolotoxin and was had normal motility. We conclude that genes such as hrp, that are required for primary symptom development in bean halo blight and for the hypersensitive response by P. s. phaseolicola are under ntrA control, while genes such as tox, that encode secondary symptom determinants (phaseolotoxin production) in this pathogen are ntrA independent.

3.3. THE "HARP" BOX.

Inspection of sequence data allowed us to identify a conserved sequence motif ("harp box" Fig. 3) which is found in the vicinity of the promoter regions of four hrp operons (AB, D, E, F) as well as upstream of the hrpS coding region (Fellay and Panopoulos, in preparation):

Fig. 3. Consensus sequence of the putative "harp box"

This motif is not present in the region upstream of the hrpM promoter, which is not osmoregulated. Interestingly, the hrp box shares similarity with the "vir boxes" found upstream of the vir operons of Ti plasmids an with the osmoregulated ompF/ompC system of E. coli.

3.4. HOST CULTIVAR SPECIFIC INTERACTIONS MAY BE MEDIATED BY hrp GENES.

Certain hrp genes appear to function as host specificity determinants at the host cultivar level. Specifically, mutations in hrpC and some in hrpF impair the ability of the bacterium to multiply and/or to form lesions to a different degree depending upon the host cultivar on which it is inoculated. Genetic crosses between bean cultivars that react differently to these mutants reveal single plant genes to be involved, with resistance segregating as recessive character. Furthermore, the interaction between these mutants and the host shows greater or lesser sensitivity to environmental variables such as humidity and/or photoperiod. This suggests that one role of the products of hrpC and hrpF may render the plant-bacterium interaction less sensitive to environmental variables.

4. Implications

Our studies provide clear evidence for the involvement of a plant signal(s) in the activation of hrp operons in P. s. phaseolicola. The osmotic repression of hrp genes is the first example of its kind for bacterial plant pathogens. The opposite phenomenon, namely osmotic induction of pathogencity related genes (for alginate biosynthesis) has been previously shown for the human pathogen P. aeruginosa and is thought to be an adaptation to a key physicochemical parameter in the infection court, namely the elevated level of salt in human lung afflicted by cystic fibrosis [1, 3]. In the case of pathogens such as P. s. phaseolicola, that infect aerial parts of plants, low osmolarity makes good biological sense as an environmental sensory signal. It's potential significance can be appreciated from three different perspectives. First, expression of hrp genes under conditions of low osmotic potential is the appropriate response to a key environmental factor favoring bacterial disease epidemics, namely rain. Second, regulation by osmotic factors as well as by a plant signal may be important aspect of the pathogen's cellular economy, enabling it to initiate infection without maintaining maximal constitutive expression of the hrp operons until a clear signal indicates the presence of the host and without risk that the presence of the host will go undetected. Thus, the plant apoplast, the pathogen's infection court, normally contains low concentrations of osmolytes, therefore it is a permissive environment for the expression of hrp genes up to a level necessary to initiate the bacterium-plant communication. The hrpS and hrpL operons become maximally expressed only in response to the plant signal and in turn activate the remaining hrp operons to the full level necessary to sustain infection.

A third scenario is that osmotic repression may be part of the mechanism by which pathogen growth is inhibited in the plant following hypersensitive necrosis or at late stages of lesion development. For example, massive electrolyte leakage occurs from plant cells that are damaged by pathogen infection at the early stages of the resistance response (hypersensitive necrosis) as well as at late stages of infection when disease lesions no longer expand. These osmolytes may increase the osmotic potential of the apoplast into which they are released sufficiently to dampen the level of expression of hrp genes, whose continued function is necessary for the development of disease symptoms. The validity of these hypotheses remain to be tested.

ACKNOWLEDGEMENTS. Work reported here was supported in part by grants from the National Science Foundation (DMB-8706129), the UC Biotechnology Program and the McKnight Foundation.

REFERENCES

1. Berry, A., De Vault, J. D. and Chakrabarty, A. M. (1989). High osmolarity is a signal for enhanced algD transcription in mucoid and non mucoid strains of Pseudomonas aeruginosa. J. Bacteriol. 171:2312-2317.

2. Daniels, M. J., Dow, J. M., and Osburn, A. E. (1988). Molecular genetics of pathogenicity in phytopathogenic bacteria. Annu. Rev. Phytopathol. 26:285-312.

3. De Vault, J. D., Berry, A., Misra, T. K., Darzins, A. and Chakrabarty, A. M. (1989). Environmental sensory signals and microbial pathogenesis: Pseudomonas aeruginosa infection in cystic fibrosis. Bio/Technology 7:352-357.

4. Frederick, R., D. (1990). Molecular characterization and nucleotide sequence analysis of two hrp loci from Pseudomonas syringae pv. phaseolicola. Ph. D. Thesis, University of California, Berkeley.

5. Grimm, C. G., Panopoulos, N. J. (1989). The predicted protein product of a pathogenicity locus from Pseudomonas syringae pv. phaseolicola is homologous to a highly conserved domain of several procaryotic regulatory proteins. J. Bacteriol. 171:5031-5038.

6. Gross, R., Arico, B. and Rapuoli, R. (1989). Families of signal-transducing proteins. Mol. Microb. 3:1611-1667.

7. Helman. J. D. and Chamberlin, M. J. (1988). Structure and function of bacterial sigma factors. Ann. Rev. Biochem. 57:839-72.

8. Lindgren, P. B., Peet, C. R. and Panopoulos, N. J. (1986). Gene cluster of Pseudomonas syringae pv. phaseolicola controls pathogencity on bean and hypersensitivity on non-host plants. J. Bacteriol. 168:512-522.

9. Lindgren, P. B., Frederick, R. D., Govindarajan, A. G., Panopoulos, N. J., Staskawicz, B. J. and Lindow, S. E. (1989). An ice nucleation reporter gene system: identification of inducible pathogenicity genes in Pseudomonas syringae pv. phaseolicola. EMBO J. 8:2990-3001.

10. Mindrinos, M. N., Rahme, L. G., Frederick, R. D., Hatziloukas, E., Grimm, C., and Panopoulos, N. J., (1989). Structure, function, regulation and evolution of genes involved in the expression of pathogenicity, the hypersensitive response and phaseolotoxin immunity in the bean halo blight pathogen. In S. Silver, A. M. Chakrabarty, B. Iglewski, and S. Kaplan (eds.), "Pseudomonas-89:Biotransformations, Pathogenesis and Evolving Biotechnology". American Society for Microbiology, Washington, D.C.. p. 74-81.

11. Popham, D. L., Szeto, D., Keener, J. and Kustu, C.(1989). Function of a bacterial activator protein that binds to transcriptional enhancers. Science 243:629-635.

12. Rahme, L. G., Mindrinos, M. N. and Panopoulos, N. J. (1990). The genetic and transcriptional organization of the hrp gene cluster of Pseudomonas syringae pv. phaseolicola. J. Bacteriol.: in press.

13. Willis, D. K., Rich, J. J., and Hrabak, E. M. (1990). hrp genes of phytopathogenic bacteria. Molec. Plant-Microbe Interact. 3: in press.

THE *HRP* GENE CLUSTER OF *ERWINIA AMYLOVORA*

S. V. BEER, D. W. BAUER, X. H. JIANG, R. J. LABY, B. J. SNEATH,
Z.-M. WEI, D. A. WILCOX, and C. H. ZUMOFF
Department of Plant Pathology
Cornell University
Ithaca, NY 14853 U.S.A.

ABSTRACT. From a library of *E. amylovora* DNA, a clone, pCPP430, was identified that contains the entire *hrp* gene cluster. The cosmid complemented all (18) original Hrp⁻ transposon-induced mutants of *E. amylovora* and bestows on *Escherichia coli,* and all other members of the *Enterobacteriaceae* tested, the ability to elicit the hypersensitive response in tobacco and other plants. Mutagenesis of pCPP430 revealed that the cluster spans ca. 40kb. Only the rightmost ca. 25 kb of the cluster is necessary to elicit the K^+ efflux / H^+ influx exchange reaction in tobacco cell suspension culture. This region hybridizes with a portion of the *hrp* cluster of *Pseudomonas syringae*, the *wts* (watersoaking) region of DNA of *E. stewartii*, and with genomic DNA of several other phytopathogenic bacteria. Studies of the regulation of *hrp* genes in *E. amylovora,* using the ß-glucuronidase reporter gene and through isolation of mRNA showed that the *hrp* genes are regulated in response to the nutritional status of the bacterium, rather than to compounds produced by plant cells or osmotic concentration. The predicted protein of one *hrp* locus is homologous to the conserved domain of several prokaryotic regulatory proteins including HrpS from *P. syringae*, NifA from *Klebsiella pneumoniae* and *Rhizobium* and NtrC from several bacteria.

1. Introduction

Erwinia amylovora was the first bacterium shown to cause disease in plants. It causes the often devastating disease known as fire blight. The bacterium affects pear, apple, quince, and several important ornamental plant species including *Crataegus, Cotoneaster, Pyracantha* and *Sorbus*. Certain host-specific strains infect only *Rubus* species (Aldwinckle and Beer, 1979).

On nonhosts, *E. amylovora* elicits the hypersensitive response (HR) (Klement, 1982). This response is characterized by the rapid (within 24 hours) collapse of leaf tissue following infiltration of intercellular spaces with suspensions containing ca. 10^7 colony forming units/ml or more. Experimentally, tobacco is used most commonly, but other nonhosts also undergo the HR when leaves are infiltrated with suspensions of *E. amylovora*, e.g. bean, sunflower, tomato, and pepper.

The genes that are required for elicitation of the HR are required also for pathogenicity to host plants, so called *hrp* genes (Lindgren et al., 1986). In *E. amylovora* all *hrp* genes are situated in a cluster (Beer et al., 1989). This paper deals with the *hrp* gene cluster and some

53

H. Hennecke and D. P. S. Verma (eds.),
Advances in Molecular Genetics of Plant-Microbe Interactions, Vol. 1, 53–60.

of its characteristics.

Several years ago, our laboratory began molecular genetic work on *Erwinia amylovora*. The work was undertaken as a new approach to understanding the genetics and mechanism of pathogenesis by the fire blight pathogen. The first step was an attempt to identify genes required for pathogenicity. Determination of the function of the genes should provide clues as to how, on the molecular level, *E. amylovora* causes fire blight.

2. Early Molecular Genetic Studies

Initially, many strains of *E. amylovora* were tested for efficiency of transformation and conjugation. A French strain (CFNB1367) was selected, which we refer to as Ea321 (Bauer, 1989). This strain was used for most of our molecular genetic studies, and it was the strain from which the entire *hrp* cluster of *E. amylovora* was cloned. Several low copy number mobilizible plasmids and high-capacity cosmid vectors were constructed (Bauer et al., 1990). These are particularly stable in *E. amylovora* and thus are suitable for introducing DNA into the fire blight pathogen and maintaining it without selection pressure. This property is needed to test pathogenicity functions in plant tissue.

Transposon mutagenesis systems were adapted for use in E. *amylovora* and transposon-insertion mutants altered in pathogenenic function were created using Tn5 (Steinberger and Beer, 1987). Some mutants proved to be Hrp⁻; they lack pathogenicity to pear and failed to elicit the HR. Other mutants were nonpathogenic to pear, but still were capable of eliciting the HR (Steinberger and Beer, 1988).

To identify *hrp* genes, the Tn5-containing fragment from an Hrp⁻ mutant was cloned. It was used as a probe in colony-hybridization procedures of a plasmid library of Ea321 DNA constructed in pBR325. Clones that hybridized to the probe were selected and the plasmids isolated. These were introduced into the Hrp⁻ mutants by transformation. The transformants were tested for pathogenicity and ability to elicit the hypersensitive response. Some mutants were restored to pathogenicity, although the reaction was weak relative to the wild-type symptoms and signs. Analysis of the bacteria present in infected tissue indicated that a high proportion of the cells had lost the complementing plasmid. When the insert was recloned in pCPP8, a vector constructed for stability in *E. amylovora* (Bauer et al. 1990), pathogenicity to pear was restored to near wild-type levels. Tests of the transformants for elicitation of the HR in tobacco were successful (Bauer and Beer, 1987).

In a second complementation procedure, members of a cosmid library of Ea321 DNA were mobilized into an Hrp⁻ mutant of *E. amylovora*. The transconjugants were inoculated to immature pear fruit slices, by touching the freshly cut slices to a plate containing well-separated colonies. Some slices exhibited symptoms and bacterial ooze following incubation for two to four days. The ooze was streaked onto agar plates, the resulting colonies purified and the cosmids transferred to RecA⁻ strains of *E. coli*. All cosmids that had the ability to complement one Hrp⁻ mutant for pathogenicity and HR eliciting ability were tested for similar ability in other Hrp⁻ mutants. Some, but not all Hrp⁻ mutants, were complemented. Preliminary evidence for the existance of a cluster of *hrp* genes was derived from studies of the effect of subclones on the complementation of specific mutants (Bauer and Beer, 1990).

3. Cloning of the Entire Intact *hrp* Gene Cluster of *Erwinia amylovora*

Mutagenesis of Ea321 with a lambda-vectored Tn*10* derivative using the techniques of

Steinberger and Beer (1988) resulted in a total of 18 hrp⁻ mutants. Only five of these had been complemented for the Hrp phenotype by cosmid and plasmid clones identified through 1988. Attempts to complement other Hrp⁻ mutants to identify other *hrp* genes led to the identification of pCPP430. This cosmid was found to complement all Hrp⁻ mutants of *E. amylovora*. In addition, it bestowed on *E. coli* and other members of the *Enterobacteriaceae* including *Klebsiella, Salmonella, Enterobacter,* and six species of *Erwinia,* the ability to elicit the HR when introduced into leaves of tobacco, sunflower, tomato, pepper and other plants. Based on these data and evidence that pCPP430 is colinear with Ea321 genomic DNA, we surmised that pCPP430 contained the entire intact *hrp* cluster of *E. amylovora* (Beer et al., 1989).

When strains of *E. coli* containing pCPP430 were infiltrated into tobacco leaf panels, the hypersensitive response developed very rapidly (Table 1). Two other cosmids, pCPP440 and pCPP450, were identified at about the same time as pCPP430, based on their hybridization to subclones of two plasmids that initially complemented Hrp⁻ mutants for pathogenicity and HR. In *E. coli* DH5, pCPP440 and pCPP450 also elicited the HR at a rate comparable to that elicited by *E. amylovora* strain Ea321 (Table 1). While cosmid pCPP430 complemented all Hrp⁻ mutants, pCPP440 and pCPP450 failed to complement some whose transposon insertions are located near the left end of the *hrp* cluster.

4. Extent of the *hrp* Cluster of *E. amylovora*

The *hrp* gene cluster of *E. amylovora* was delineated by several approaches. The cluster contained in pCPP430 was mapped with restriction endonucleases (Laby et al., 1989), and the locations of transposon insertions in Hrp⁻ mutants were determined. Portions of the *E. amylovora* insert DNA in pCPP430 were subcloned and the subclones tested for their ability to complement various Hrp⁻ mutants. In addition, pCPP430 was mutagenized with several transposons including Tn5PhoA, to locate possible membrane associated proteins and Tn5-*gus,* for use in regulation studies. Based on mutagenesis of pCPP430 and complementation of Hrp⁻ mutants, the *hrp* cluster spans between 40 and 42 kb of DNA. The region of the *hrp* cluster that is needed for elicitation of the XR is about 25 kb of the right end part of the *hrp* cluster. A rudimentary map of the *hrp* gene cluster is presented in Figure 1.

5. Relationships of the *hrp* Cluster of *E. amylovora* to DNA of Other Bacteria

To determine the possible existence in other bacteria of DNA homologous to that of the *hrp* gene cluster of *E. amylovora,* hybridization studies were carried out. Seven *Eco*RI fragments encompassing the portions of pCPP430 in which transposon insertions resulted in the Hrp⁻ phenotype; were used as a composite probe for genomic DNA of various bacteria and for two cosmids involved in plant pathogenicity which had been identified by other molecular plant pathologists. Genomic DNA of 24 strains of *E. amylovora* of diverse host and geographic origin hybridized with the composite probe. The patterns of hybridization were identical for all strains of *E. amylovora* that originated from pomaceous hosts. Strains of *E. amylovora* that originated from *Rubus* species also hybridized to the composite probe, but these (6) strains exhibited some polymorphism relative to the pCPP430 control and the pomaceous strains. Overall, there was a remarkable similarity in the patterns of hybridization between DNA of the *hrp* cluster of strain Ea321 and the 30 strains tested. These hybridization studies indicate a remarkable degree of *hrp* sequence conservation among strains of *E. amylovora*.

Hybridization of the composite probe from pCPP430 occurred also with genomic DNA of seven species of *Erwinia* and several species of *Pseudomonas*. When hybridizations and washes were conducted under low stringency conditions, one to six hybridizing bands were detected.

Hybridization of the composite probe also occurred with pHIR11, a cosmid containing the *hrp* cluster of *P. syringae* strain 61 identified by Huang et al. (1989). Further hybridization studies involving subclones of pHIR11 and pCPP430 indicated that a central region of the cluster of *E. amylovora*, encompassing about 15 kb, hybridized in a colinear manner with a region encompassing about 15 kb of DNA of pHIR11. These results suggest that these two, not so closely related bacteria, contain regions of conserved DNA that encode functions required for plant pathogenicity.

The *hrp* gene cluster of *E. amylovora* also hybridized with pES1044, a cosmid containing the *wts* region of *E. stewartii* (Frederick and Coplin, 1989). Subclones of pES1044 were tested for their ability to complement specifc transposon-induced Hrp⁻ mutants of *E. amylovora*. Subclones of pES1044 that hybridized to portions of pCPP430 restored the Hrp phenotype to mutants containing transposon insertions in the region of homology. *Erwinia amylovora* mutants with insertions lying outside the area of hybridization were not complemented. These studies indicate that *E. stewartii*, which does not under typical laboratory conditions elicit the HR, nevertheless, contains *hrp*-like genes that hybridize with DNA of *E. amylovora* and functionally complement *E. amylovora* Hrp⁻ mutants (Beer et al., 1990; Laby and Beer, 1990). Complementation for pathogenicity to maize of Wts⁻ mutants of *E. stewartii* recently has been achieved with DNA from the *hrp* gene cluster of *E. amylovora* (D. L. Coplin, personal communication).

6. *Escherichia coli* **has genes that functionally complement a portion of the *hrp* cluster of *Erwinia amylovora*.**

Based on the fact that pCPP440 and pCPP450 bestow on strains of *E. coli* the ability to elict HR (Figure 1), but are unable to complement Hrp⁻ mutants of *E. amylovora* whose transposons insertions are situated in the left end of the *hrp* gene cluster, it is possibile that *E. coli* contains some genes that can functionally complement the left end of the cluster. Evidence in support of this possibility was found (Wei et al., 1990). pCPP430 was mutagenized with transposons and the derivative cosmids tested for their ability to elicit the HR in *E. amylovora* and *E. coli*. Insertions in the left-most 12.8 kb portion of pCPP430 (which were marker-exchanged into wild-type Ea321) reduced the ability of *E. amylovora* strains to elicit the HR and abolish pathogenicity to immature pear fruits. In contrast, the same pCPP430 derivatives were able to elicit the HR in *E. coli* strains. Several transposon insertions in the left most 12.8 kb portion of pCPP430 were marker exchanged into *E. coli*. When the pCPP430 derivatives containing the insertion or pCPP440 or pCPP450 were tested for HR eliciting ability in the mutant *E. coli*, no HR developed. However, when the 12.8 kb *Eco*RI fragment from the left end of pCPP430 was introduced into the marker-exchanged mutants of *E. coli*, the same cosmids bestowed the ability to elicit the HR. These data provide strong evidence that *E. coli* contains genes that can functionally complement the *hrp* cluster of *E. amylovora*. Further complementation studies indicated that a 2.9 kb *Hind*III fragment was sufficient to restore that function, and furthermore was responsible for the rate of development of the HR in *E. amylovora* (Wei et al., 1990). This region likely regulates expression of the *hrp* genes.

7.   Regulation of the *hrp* gene cluster of *E. amylovora*

Regulation studies have been aided by the use of the ß-glucuronidase (GUS) gene. Cosmid pCPP430 was mutagenized with a Tn5-*gusA* transposon on a lambda vector (Sharma and Signer, 1989). Following mutagenesis, the cosmids were tested for elicitation of the HR. Eight GUS insertions situated throughout the *hrp* cluster were found to abolish the ability of the cosmid to elicit the HR in *E. coli*. The eight GUS insertions were marker-exchanged into Ea321 and expression of *hrp* genes were tested for expression of GUS activity in tobacco leaf tissue and in various media. Greater GUS expression occurred in tobacco leaf tissue than in media. The GUS insertions were repressed in rich media (Luria broth), but were expressed at higher levels in two minimal media. These data indicated that the complex nitrogenous compounds present in Luria broth might be responsible for at least part of the repression of the *hrp* genes. Analysis of the mRNA expressed in media also indicated enhanced expression of *hrp* genes in minimal media verses Luria broth (Wilcox and Beer, data not shown).

Greatest expression of *hrp* genes, as revealed by GUS activity, occurred in a modification of the tobacco cell suspending buffer used for the XR developed by Atkinson, et al. (1985). This medium contains mannitol, ammonium sulfate, and inorganic salts. When ammonium sulfate was added at concentrations greater than 50 mM, *hrp* genes were repressed. Expression of *hrp* genes is greater at pH 5.5 than at pH 7. Other components that repress *hrp* gene expression include nicotinic acid and histidine. The addition of sodium sulfate and sodium chloride did not significantly reduce expression, while ammonium chloride repressed the *hrp* genes (Wei et al., 1990). Based on these studies, the *hrp* genes of *E. amylovora* appear to be influenced by ammonium concentration, complex nitrogen sources and pH; their expression is independent of osmotic concentration. These results are in contrast to those of Panopoulos (personal communication) who has evidence for repression of *hrp* genes of *P. syringae* by high osmotic concentrations.

8.   Products of the *hrp* Gene Cluster of *Erwinia amylovora*.

The *hrp* gene cluster of *E. amylovora* is extensive, involving perhaps 40 kb of DNA. One 2.7 kb fragment corresponding to a *Bam*HI/*Hind*III fragment that complemented Billings strain P66 for the Hrp phenotype has been sequenced (Sneath et al., 1990). The predicted protein product bears striking similarity to the HrpS sequence of Grimm and Panopoulos (1989). The predicted protein product has a 43% amino acid similarity to HrpS of *P. syringae* and lower similarity to the NifA proteins of *Rhizobium*, *Bradyrhizobium*, and *Klebsiella* and the NtrC protein of *Bradyrhizobium* species. These proteins all belong to a superfamily of prokaryotic regulatory proteins.

Comparison of the proteins from *E. coli* strains with and without pCPP430 has revealed the presence of proteins likely to be involved in the Hrp phenotype. Proteins expressed by *E. coli* containing pCPP430 included a prominent spot on two-dimensional polyacrylamide gels that was not present in comparable preparations from *E. coli* containing only the vector of pCPP430. This spot corresponds to a protein of approximately 45 kd with a pI of 7.2.

Analysis of the proteins expressed by pCPP430 derivatives containing transposons that inactivate the Hrp phenotype has delineated the region encoding this protein to a 3.5 kb region of the *hrp* cluster. When the wild-type subclone corresponding to the location of the Tn insertion was included with the mutant version of the pCPP430, the protein again was expressed. The nature and function of the protein remains to be determined.

9. Summary of Characteristics of the *hrp* Gene Cluster of *Erwinia amylovora*.

1. The entire *hrp* cluster of *E. amylovora* has been cloned on a high-capacity cosmid, pCPP9.
2. The cluster complements all Hrp⁻ mutants of *E. amylovora* for pathogenicity to host plants and elicitation of the HR in nonhost plant.
3. When transferred to other bacteria, including *E. coli*, the *hrp* cluster bestows on them the ability to elicit the HR.
4. The cluster encompasses about 42 kb of chromosomal DNA. Only about 25 kb is needed for elicitation of the XR.
5. Based on analysis of GUS insertions and complementation of mutants with cloned DNA, seven transcriptional units have been identified. Based on complementation of mutants with cloned DNA, additional transcriptional units are likely.
6. Expression of *hrp* genes is controlled by pH and the concentration of nitrogenous compounds.
7. One transcriptional unit at left end of the cluster seems to influence the speed of development of the HR.
8. Two transcriptional units encode membrane-associated proteins, based on studies of TnPhoA insertions.
9. One transcriptional unit encodes a regulatory protein similar to HrpS, NifA and NtrC.
10. Sequences of the *hrp* cluster are conserved among many phytopathogenic bacteria, including *E. stewartii* and *P. syringae*.
11. Interspecific-complementiaton for pathogenicity has been achieved between pathogenicity genes and pathogenicity mutants of *E. amylovora* and *E. stewartii*.
12. *Escherichia coli* has functional homology with a portion of the *hrp* cluster of *E. amylovora*.

10. References

Aldwinckle, H. S. and S. V. Beer. 1979. Fire blight and its control. Pages 423-474 *in* J. Janick, ed. Horticultural Reviews, Vol. 1. Avi Publishing. Westport, CT.

Atkinson, M. M., J. S. Huang, and J. A. Knopp. 1985. The hypersensitive reaction of tobacco to *Pseudomonas syringae* pv *pisi*: activation of a plasmalemma K^+/H^+ exchange mechanism. Plant Physiol. 79:843-847.

Bauer, D. W. and S. V. Beer. 1987. Cloning of a gene from *Erwinia amylovora* involved in induction of hypersensitivity and pathogenicity. Pages 425-429 in: Proc. VI Int. Conf. Plant Pathogenic Bacteria, College Park, MD, June, 1985.

Bauer, D. W. 1989. Molecular genetics of pathogenicity of *Erwinia amylovora*: techniques, tools and their application. Ph.D. thesis. Cornell University, Ithaca, NY.

Bauer, D. W., A. B. Sprenkle, and S. V. Beer. 1990. Construction of stable, mobilizable plasmid and cosmid vectors, and their use in *Erwinia amylovora*. Gene (Submitted).

Bauer, D. W., and S. V. Beer. 1990. Identification of an *hrp* gene cluster of *Erwinia amylovora*. Mol. Plant-Microbe Interact. (Submitted).

Beer, S. V., C. H. Zumoff, D. W. Bauer, B. J. Sneath, and R. J. Laby. 1989. The hypersensitive response is elicited by *Escherichia coli* containing a cluster of pathogenicity genes from *Erwinia amylovora*. Phytopathology 79:1156.

Beer, S. V., R. J. Laby, and D. L. Coplin. 1990. Complementation of Hrp mutants of *Erwinia amylovora* with DNA of *Erwinia stewartii*. Phytopathology 80:(in press).

Grimm, C., and N. J. Panopoulos. 1989. The predicted protein product of a pathogenicity locus from *Pseudomonas syringae* pv. *phaseolicola* is Is homologous to a highly conserved domain of several procaryotic regulatory proteins. J. Bacteriol. 171:5031-5038.

Huang, H.-C., Schuurink R. Denny, T. P., Atkinson, M. M., Baker, C. J., Yucel, I., Hutcheson, S. W., and Collmer, A. 1988. Molecular cloning of a *Pseudomonas syringae* pv. *syringae* gene cluster that enables *Pseudomonas fluorescens* to elicit the hypersensitive response in tobacco plants. J. Bacteriol. 170:4748-4756.

Klement, Z. 1982. Hypersensitivity. Pages 149-177 in: Phytopathogenic Procaryotes, Vol. 2. M.S. Mount and G. H. Lacy, eds. Academic Press, New York.

Laby, R. J. and S. V. Beer. 1990. The *hrp* gene cluster of *Erwinia amylovora* shares DNA homology with other bacteria. Phytopathology 70:(in press).

Laby, R. J., C. H. Zumoff, B. J. Sneath, D. W. Bauer, and S. V. Beer. 1989. Cloning and preliminary characterization of an *hrp* gene cluster from *Erwinia amylovora*. Phytopathology 79:1211.

Lindgren, P. B., Peet, R. C, and Panopoulos, N. J. 1986. Gene cluster of *Pseudomonas syringae* pv. *phaseolicola* control pathogenicity on bean plant and hypersensitivity on non-host plants. J. Bacteriol. 168:512-522.

Sharma, S. B., and E. R. Signer. 1990. Temporal and spatial regulation of the symbiotic genes of *Rhizobium meliloti* in planta revealed by transposon *Tn5-gus*A. Genes and Development 4:344-356.

Sneath, B. J., X.-H. Jiang, H. M. Howson, and S. V. Beer. 1990. A pathogenicity gene from *Erwinia amylovora* encoded a predicted protein product homologous to a family of procaryotic response regulators. Phytopathology 80:(in press).

Steinberger, E. M. and S. V. Beer. 1987. Mutants of *Erwinia amylovora* altered in pathogenicity by transposon mutagenesis. Acta Horticulturae 217:167-168.

Steinberger, E. M. and S. V. Beer. 1988. Creation and complementation of pathogenicity mutants of *Erwinia amylovora*. Mol. Plant-Microbe Interact. 1:135-144.

Wei, Z-.M., and S. V. Beer. 1990. Functional homology between a locus of *Escherichia coli* and the *hrp* gene cluster of *Erwinia amylovora*. Phytopathology 80:(in press).

Wei, Z-.M., B. J. Sneath and S. V. Beer. 1990. Expression of the *hrp* genes of *Erwinia amylovora* under defined conditions. Abstracts: 5th Int. Symp. Mol. Genetics Plant-Microbe Interact. Interlaken, Switzerland. September, 1990.

Figure 1. The *hrp* gene cluster of *Erwinia amylovora*. Cosmid pCPP430 contains 46.1 kb of DNA of *E. amylovora* strain Ea321. It was identified initially based on its ability to restore pathogenicity to pear of one Hrp⁻ mutant. Triangles represent sites of Tn*5* or Tn*10* insertions in the chromosome of Ea321 that abolish the Hrp phenotype. pCPP430 complements all Hrp⁻ mutants and bestows on *Escherichia coli* the ability to elicit the hypersensitive response. pCPP440 and pCPP450 were identified based on hybridization to both probes "A" and "B", derived from plasmids identifed by Bauer and Beer (1987). They also bestow on *E. coli* the abtitiy to elicit the HR, but do not complement all Hrp⁻ mutants of *E. amylovora*. Based on the activity of transposon insertions, DNA required for the Hrp phenotype in *E. amylovora* and to elicit XR are indicated by arrows.

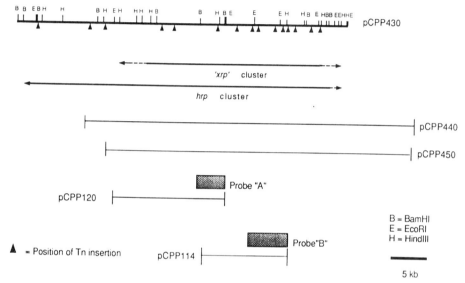

Table 1. Development of the hypersensitive response in tobacco leaf panels.

Strain	Reaction of Infiltrated Leaf Panel[1]		
	0-6 hr	6-9 hr	9-18 hr
E. coli DH5(pCPP9)	No reaction	No reaction	No reaction
DH5(pCPP430)	Flaccid	Collapse	Collapse
DH5(pCPP440)	No reaction	Flaccid	Collapse
DH5(pCPP450)	No reaction	Flaccid	Collapse
E. amylovora Ea321	No reaction	Flaccid	Collapse
Ea321(pCPP430)	Flaccid	Collapse	Collapse

[1]Bacteria were grown in minimal medium, centrifuged, resuspended at 10^8 cfu/ml and infiltrated into the intercellular spaces of tobacco leaves. Panels were examined periodically until 18 hours following infiltration.

CHARACTERIZATION OF GENES FROM *XANTHOMAS CAMPESTRIS* PV. *VESICATORIA* THAT DETERMINE AVIRULENCE AND PATHOGENICITY ON PEPPER AND TOMATO

RALF SCHULTE, KARIN HERBERS, STEFAN FENSELAU, ILSE BALBO, ROBERT E. STALL* AND ULLA BONAS
*Institut für Genbiologische Forschung Berlin GmbH, Ihnestr. 63, 1000 Berlin 33, Federal Republic of Germany; *Department of Plant Pathology, University of Florida, 1453 Fifield Hall, Gainesville, Florida 32611, U.S.A.*

Introduction

Xanthomonas campestris pathovar (pv.) *vesicatoria* (*Xcv*), a gram negative bacterium, is the causal agent of bacterial spot disease of pepper and tomato. After infection of a plant with *Xcv* two different types of reactions can be observed. If the plant is susceptible, the infection gives rise to watersoaked lesions (compatible interaction). In a resistant plant a hypersensitive response (incompatible interaction) is induced. The hypersensitive response (HR) is a local defense reaction accompanied by a rapid necrosis of the infected tissue (Klement and Goodman 1967). Such an incompatible interaction requires the presence of a resistance locus in the particular cultivar of pepper or tomato and a corresponding avirulence locus in the particular race of the pathogen (Minsavage et al. 1990).

We are interested in the molecular mechanism underlying the interaction between *Xcv* and the plant. To study the incompatible interaction we have chosen the avirulence gene *avrBs3* and a homologue thereof. The analysis of *avrBs3* (Bonas et al. 1989) indicated that it is necessary but not sufficient for the incompatible interaction with the resistant plant. Recently genes that control pathogenicity and the induction of an HR on pepper and tomato were isolated from *Xcv* (Bonas et al. 1990).

Results and Discussion

EXPRESSION OF AVRBS3

AvrBs3 encodes a 122 kDa protein. *Xcv* race 1 strains containing *avrBs3* specifically induce an HR on plants of pepper cultivar ECW-30R that carry the resistance locus *Bs3* (Minsavage et al. 1990). The nucleotide sequence of a 4.3 kb subclone of the *avrBs3* gene (Bonas et al. 1989) revealed the presence of several open reading frames (ORF). To determine promoter activity of the gene we fused DNA fragments containing putative promoter regions of the different ORFs to β-glucuronidase (GUS) as a reporter gene. The constructs were transferred into *Xcv* race 2 that does not contain *avrBs3* homologous sequences and is virulent on ECW-30R. GUS-Activity of the cells was measured after growth under different conditions: in broth or *in planta*. Only for ORF1 a promoter region could be defined and showed constitutive activity. Since mutations affecting ORF1 resulted in loss of avirulence activity and no expression of ORF2 or ORF3 could be found

61

H. Hennecke and D. P. S. Verma (eds.),
Advances in Molecular Genetics of Plant-Microbe Interactions, Vol. 1, 61–64.
© 1991 *Kluwer Academic Publishers. Printed in the Netherlands.*

we assume that the AvrBs3 protein is expressed from ORF1. Constitutive expression of ORF1 was confirmed by RNA and protein data. An affinity-purified polyclonal antibody specifically detects a 122 kDa protein in total extracts of *Xcv* race 1 cells. This protein is only present in race 1 and in race 2 transconjugants harboring an active *avrBs3* clone and not in other races (Bonas et al., in preparation). Biochemical fractionation experiments of *Xcv* race 1 cells showed that AvrBs3 is soluble. So far there is no indication for secretion of the protein. Hence the induction of the HR might be mediated by a different molecule.

AvrBs3 is a repetitive protein. The internal portion of AvrBs3 contains a 34 amino acid motif that is repeated in direct orientation 17.5 times thus representing 50% of the molecular weight of the protein. The amino acid sequence of the repeats is nearly identical. Are these repeats important for function? We addressed this question by randomly introducing additional repeats or by deleting repeat units within the repetitive region of the gene. Constructs of different lengths were transferred into *Xcv* race 2 to test for avrBs3 function. In all of the 31 derivatives tested a protein of the respective size was expressed, however, in 22 cases gene function was lost, i.e. the transconjugants were virulent on ECW-30R. Interestingly, the other deletion derivatives now induced an HR on the susceptible cultivar ECW and/or on a different pepper species both being susceptible to wild type race 2. The fact that internal deletions resulted in loss of function or in new specificities shows that the repeats are important for protein function.

ANALYSIS OF AN *AVRBS3*-HOMOLOGOUS GENE

Southern hybridizations of total genomic DNA of different strains of *Xcv* race 1, probed with an internal fragment of the *avrBs3* gene, revealed the presence of additional homologous sequences in a number of strains, e.g. strain 82-8 (Bonas et al. 1989). Like *avrBs3* the homologous sequences were plasmid-borne. Since mutants of strain 82-8 carrying marker gene exchanges in the *avrBs3* gene no longer induced an HR on pepper cultivar ECW-30R the homologous region obviously did not contain an active copy of *avrBs3*. We were interested in the structure of this "inactive" gene which was designated *avrBs3-i*. The gene was cloned using pLAFR3 and mobilized into Xcv race 2 (Bonas et al., in preparation). When protein extracts of *Xcv* race 2 transconjugants harboring *avrBs3-i* were tested by Western blot analysis using the AvrBs3-antibody a protein slightly smaller than AvrBs3 was recognized. Even more interesting was the finding that *avrBs3-i* induces HR on tomato cultivars (Canteros et al., in preparation). Preliminary nucleotide sequence data show that *avrBs3-i* shares high homology with *avrBs3*. There seem to be no sequence alterations in the N-terminal part of the ORF. However, the organization and sequence of the internal repeats is slightly different. This result is another indication that the repetitive region is important for function.

HRP GENES FROM *XCV*

Isolation and identification of a DNA region involved in pathogenicity of Xcv. Wild type cells of *Xcv* strain 82-8 were mutagenized with NTG and tested on pepper cultivar ECW for loss of pathogenicity. Six mutants of *Xcv* 82-8 (out of 1000 tested) were isolated. These mutant bacteria not only failed to grow in the plant and to induce disease on the susceptible plant but also to induce an HR on the resistant pepper cultivar ECW-30R and on tobacco. The expression of *avrBs3* in non-pathogenic mutants of strain 82-8 was unaltered as determined by Western blot analysis. The mutants were not restored by sequences from *Xanthomonas campestris* pv. *campestris* present in the plasmid clones pIJ3000, pIJ3020 (Daniels et al. 1984) and pIJ3070 (Tang et al. 1987). Complementation was successful with clones from a genomic library from wild type strain 75-3 of *Xcv* (Bonas et al. 1990). Different cosmid clones with overlapping inserts restored the ability to cause watersoaked lesions on leaves of ECW and an HR on ECW-30R. Two clones from *Xcv* that complemented most of the mutants, designated pXV2 and pXV9, were chosen for further analysis.

Insertion mutagenesis of the pathogenicity region and identification of complementation groups. The plasmids pXV2 and pXV9, covering a genomic region of approximately 30 kb, were mutagenized with the transposon Tn*3-gus*. Tn*3-gus* is a Tn*3*HoHo derivative containing a promoterless β-glucuronidase gene (Staskawicz, pers. communication). 32 different Tn*3-gus* insertions were introduced into the genome of *Xcv* strain 85-10 by marker gene exchange and the mutants scored for their reaction on pepper plants. Insertions near the borders of this 30-kb genomic region had no effect on pathogenicity. However, insertions distributed over a region of ca. 25 kb eliminated both pathogenicity and the ability to induce the HR on resistant pepper plants and the non-host plant tobacco. Several insertions mapping to a region of 4 kb within the 25-kb segment had no phenotypic effect. Since both pathogenicity and the ability to induce the HR were affected by the mutations the corresponding genomic region was designated *hrp* (Lindgren et al. 1986). To determine the genetic organization of the *hrp* region a detailed complementation analysis of the Tn*3-gus* mutants was performed. Non-pathogenic *Xcv*::Tn*3-gus* mutants were conjugated with different *E. coli* harboring pXV2::Tn*3-gus* or pXV9::Tn*3-gus* in which the transposon insertions were in a region flanking the site of insertion in the recipient. The transconjugants were tested on pepper cv. ECW and ECW-10R for symptom production. The results showed that the *hrp* region contains at least six different complementation groups.

The hrp genes from Xcv are plant-inducible. The GUS-activity of *Xcv* race 2 transconjugants harboring pXV2::Tn*3-gus* or pXV9::Tn*3-gus* was determined after growth of the bacteria in minimal or rich medium or in pepper leaves. In no case GUS-activity was measured after growth of the cells in medium. However, after growing *in planta* in a number of strains GUS-activity was induced. For each complementation group at least one insertion with inducible GUS-activity was found. The inducibility of *hrp*-gene expression was confirmed by RNA data (Schulte et al., in preparation). We are currently trying to identify the nature of the inducing compound.

Homology of the hrp region of Xcv to DNA of other Xanthomonas pathovars. In Southern hybridization experiments total genomic DNA of different races of *Xcv* and of different pathovars of *X. campestris,* digested with *Eco*RI, was hybridized with a fragment containing almost the entire *hrp* region. *Xcv* strains pathogenic on pepper or tomato showed identical hybridization patterns. Homologous DNA sequences were detected in pathovars *alfalfae, armoraciae, begoniae, campestris, citri, dieffenbachiae, glycines, malvacearum, oryzae, phaseoli, translucens* and *vignicola.* DNA of a non-pathogenic strain of *X. campestris,* T55, did not hybridize at all. We believe that strain T55 lacks a large portion of the pathogenicity region. DNA of *Pseudomonas solanacearum* showed only very weak homology to the *Xcv hrp* sequences although homology between *hrp* sequences from *P. solanacearum* and sequences from *Xcv* was reported (Boucher et al. 1987). DNA of strains of *Pseudomonas syringae* pathovars *glycinea, phaseolicola*, and *tomato* as well as of strains of *Rhizobium* and *Agrobacterium tumefaciens* did not hybridize to the *hrp* region of *Xcv*. The lack of homology between the *hrp* region of *Xcv* and sequences from different pathovars of *P. syringae* could indicate that the *hrp* genes of these pathogens have evolved divergently. Interestingly, the *hrp* genes from *P. s.* pv. *phaseolicola* also failed to hybridize to DNA from saprophytic or other pathogenic bacteria including *Xanthomonas* (Lindgren et al. 1988). Nevertheless, functional homologies cannot be ruled out.

Acknowledgements

This research was supported by grant # 322-4003-0316300A from the Bundesministerium für Forschung und Technologie to U. B.

References

Bonas, U., Stall, R.E., and Staskawicz, B. 1989. Genetic and structural characterization of the avirulence gene *avrBs3* from *Xanthomonas campestris* pv. *vesicatoria*. Mol. Gen. Genet. 218:127-136.

Bonas, U., Schulte, R., Fenselau, S., Minsavage, G.V., Staskawicz, B.J., and Stall, R.E. 1990. Isolation of gene cluster from *Xanthomonas campestris* pv. *vesicatoria* that determines pathogenicity and the hypersensitive response on pepper and tomato. Mol. Plant-Microbe Interact. (in press)

Boucher, C., Van Gijsegem, F., Barberis, P., Arlat, M., and Zischek, C. 1987. *Pseudomonas solanacearum* genes controlling both pathogenicity on tomato and hypersensitivity on tobacco are clustered. J. Bacteriol. 169:5626-5632.

Daniels, M. J., Barber, C. E., Turner, P. C., Sawczyc, M. K., Byrde, R. J. W., and Fielding, A. H. 1984. Cloning of genes involved in pathogenicity of *Xanthomonas campestris* pv. *campestris* using the broad host range cosmid pLAFR1. EMBO J. 3:3323-3328.

Klement, Z., and Goodman, R. N. 1967. The hypersensitive reaction to infection by bacterial plant pathogens. Annu. Rev. Phytopathol. 5:17-44.

Lindgren, P. B., Peet, R., and Panopoulos, N. J. 1986. Gene cluster of *Pseudomonas syringae* pv. "*phaseolicola*" controls pathogenicity on bean plants and hypersensivity on nonhost plants. J. Bacteriol. 168:512-522.

Lindgren, P. B., Panopoulos, N. J., Staskawicz, B. J., and Dahlbeck, D. 1988. Genes required for pathogenicity and hypersensivity are conserved and interchangeable among pathovars of *Pseudomonas syringae*. Mol. Gen. Genet. 211:499-506.

Minsavage, G. V., Dahlbeck, D., Whalen, M. C., Kearney, B., Bonas, U., Staskawicz, B. J., and Stall, R. E. 1990. Gene-for-gene relationships specifying disease resistance in *Xanthomonas campestris* pv. *vesicatoria*-pepper interactions. Mol. Plant Microbe Interact. 3:41-47.

Tang, J.L., Gough, C.L., Barber, E.E., Dow, J.M., and Daniels, M.J. 1987. Molecular cloning of protease gene(s) from *Xanthomonas campestris* pv. *campestris*: expression in *Escherichia coli* and role in pathogenicity. Mol. Gen. Genet. 210:443-448.

PECTIC ENZYME PRODUCTION AND BACTERIAL PLANT PATHOGENICITY

A. COLLMER[1], D. W. BAUER[1], S. Y. HE[1], M. LINDEBERG[1], S. KELEMU[1], P. RODRIGUEZ-PALENZUELA[1], T. J. BURR[2], AND A. K. CHATTERJEE[3]
Departments of Plant Pathology
[1]Cornell University, Ithaca, NY 14853
[2]New York State Agric. Experiment Station, Geneva, NY 14456
[3]University of Missouri, Columbia, MO 65211 USA

ABSTRACT. Aspects of pectic enzyme production were explored in three dissimilar plant pathogens. *Agrobacterium tumefaciens* biovar 3 produces polygalacturonase and causes both crown gall and root decay of grape. A polygalacturonase-deficient Tn5 mutant was isolated to permit analysis of the role of the enzyme in pathogenicity. *Pseudomonas syringae* pv. *lachrymans* causes angular leaf spot and fruit rot of cucumber. A pectate lyase-encoding *pel* gene was cloned from this strain by constructing a DNA library in a newly developed cosmid vector. A *pel*⁺ *P.s. syringae* transconjugant was then analyzed for ability to cause cucumber fruit rot. A mutant derivative of the soft-rot pathogen *Erwinia chrysanthemi*, containing site-directed mutations in genes encoding all of the known extracellular pectic enzymes but still able to macerate plant tissues, was found to produce several novel pectate lyases *in planta*. A gene encoding one of these enzymes was cloned from this strain, and the effect of the enzyme on host tissues was analyzed. A functional set of *E. chrysanthemi out* genes complementing mutations affecting the secretion of pectic enzymes beyond the periplasm was also cloned and found to confer upon *Escherichia coli* the ability to efficiently secrete several *E. chrysanthemi* extracellular proteins.

1. INTRODUCTION

Agrobacterium tumefaciens biovar 3, *Pseudomonas syringae* pv. *lachrymans*, and *Erwinia chrysanthemi* provide three distinct models for exploration of the role of pectic enzymes in plant pathogenesis. Pectic enzymes may have potentially diverse effects on the development of plant-microbe interactions, including: (i) maceration of parenchymatous tissues, (ii) killing of plant cells, (iii) facilitation of pathogen penetration, (iv) enabling pathogen colonization of the vascular system, and (v) elicitation of defense reactions [5]. An understanding of the role of pectic enzymes in the overall biology of the pathogen requires characterization of the enzymes, their regulation, and their secretion.

    A. tumefaciens biovar 3 and *P.s. lachrymans* are weakly pectolytic in culture and appear to produce only one or two pectic enzymes, respectively, whereas *E. chrysanthemi* is highly pectolytic and secretes a battery of pectic enzymes. We have addressed several questions regarding the production and activity of these enzymes in pathogenesis: Is polygalacturonase production necessary for *A. tumefaciens* biovar 3 to

65

H. Hennecke and D. P. S. Verma (eds.),
Advances in Molecular Genetics of Plant-Microbe Interactions, Vol. 1, 65–72.
© 1991 *Kluwer Academic Publishers. Printed in the Netherlands.*

cause root decay of grape seedlings and crown gall tumors on susceptible plants? Is pectate lyase production sufficient to enable a *P.s. syringae* strain which is a nonpectolytic pathogen of bean to colonize and rot cucumber fruit? What is the nature of the residual enzymes that enable an *E. chrysanthemi* strain with multiple mutations in known pectic enzyme genes to macerate plant tissues? Finally, what is the genetic basis for the ability of *E. chrysanthemi* to efficiently secrete multiple plant cell wall-degrading enzymes?

2. RESULTS AND DISCUSSION

2.1. The role of polygalacturonase in *A. tumefaciens* biovar 3 pathogenicity

Several *A. tumefaciens* biovar 3 strains had previously been reported to produce polygalacturonase, as indicated by their ability to cause cleared zones in thin agarose gels amended with polygalacturonic acid and stained with ruthenium red [2]. This observation was confirmed by analyzing culture supernatants from a variety of *Agrobacterium* strains with diagnostically buffered activity-stained isoelectric focusing gels [6]. Six biovar 3 strains from diverse geographical locations all produced a single polygalacturonase with a pI of 4.8 to 5.2 and a pH optimum of 4.5. The biovar 1 and 2 strains, C58 and R-3, produced no detectable pectic enzyme activity when grown on Kado 523 [10], the medium which supported the best polygalacturonase production by biovar 3 strains. Both tumorigenic and nontumorigenic biovar 3 strains produced polygalacturonase. The tumorigenic strain CG49 was chosen for further study. Polygalacturonase production was little affected by the carbon source in cultures of strain CG49, it was not induced by polygalacturonic acid, and it was higher in the rich Kado 523 medium than in basal media. In cultures grown for 36 h in this medium, 90% of the enzyme activity was extracellular.

To determine whether the enzyme was produced *in planta*, the roots of young Concord grape seedlings were inoculated with strain CG49 as described previously [3]. Necrotic lesions developed within 3 days and were excised and pooled from several seedlings along with an equivalent amount of tissue from the roots of mock inoculated seedlings. The plant material was crushed, and the extract clarified by centrifugation, concentrated, and then analyzed directly on an activity-stained isoelectric focusing gel. The necrotic lesions yielded a broad band of pectolytic activity that focused at the same pH as the polygalacturonase produced in culture by strain CG49, thus suggesting that *A. tumefaciens* biovar 3 does produce polygalacturonase during root decay.

To determine whether polygalacturonase production is necessary for *A. tumefaciens* biovar 3 to cause root decay or crown gall tumors, transposon mutagenesis was performed in strain CG49 using pSUP2021 [18] as a Tn5 delivery system. A polygalacturonase-deficient mutant was isolated by screening for loss of pectolytic activity on ultrathin polygalacturonate-agarose gels. The mutant was completely deficient in polygalacturonase production, both in extracellular and sonicated cell

extract fractions. Southern blot analysis indicated that the mutant
contained a single Tn5 insertion in a 2.0-kb *ClaI* fragment. The mutant
was unable to cause root decay in grape seedlings, but was
indistinguishable from the wild-type parent in sunflower, potato, and
carrot tumorigenicity assays. Further characterization of the Tn5
mutation and molecular cloning of the corresponding wild-type gene is
being done to establish the relationship between polygalacturonase
production and pathogenicity in *A. tumefaciens* biovar 3.

2.2. Construction of broad-host-range cosmid vector pCPP37 and cloning
of a gene encoding pectate lyase from *P. s. lachrymans*

The *P. syringae* pathovars differ in their ability to produce pectic
enzymes, as determined by their ability to cause pits in
polygalacturonate-agar plates at high and low pH [9]. For example, *P.s.
syringae* causes no pitting at any pH, whereas *P.s. lachrymans* causes
pitting at pH 4.6 and 8.5, indicating production of polygalacturonase
and pectate lyase, respectively. *P.s. lachrymans* causes angular leaf
spot of cucumber and can systemically invade and rot cucumber fruits
[15]. *P.s. syringae* B728A incites brown spot of bean and does not
produce disease symptoms in cucumber fruit. Both bacteria cause a
hypersensitive response in tobacco leaves. We were interested in
determining the role of pectic enzymes in various aspects of *P.s.
lachrymans* pathogenicity, especially the rotting of cucumber fruit, and
began by cloning a pectate lyase-encoding *pel* gene from *P.s. lachrymans*.
    Cloning of the *pel* gene was facilitated by the construction of
pCPP37, a new cosmid vector with several desirable characteristics. It
is derived from pMP92, a derivative of pTJS75, which is stably

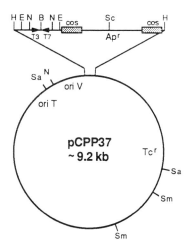

Fig. 1. Restriction map of cosmid vector pCPP37. The cloning site in
the cosmid vector construction cassette is not drawn to scale. *Bam*HI
(B), *Eco*RI (E), *Hind*III (H), *Not*I (N), *Sal*I (Sa), *Sca*I (Sc), *Sma*I (Sm).

maintained in *P. aeruginosa* and *P. putida* [19]. Its small size (9.2 kb) fosters the packaging of large DNA inserts. It possesses dual *cos* sites separated by a *bla* gene with a unique *Sca*I site. When pCPP37 cosmids are packaged, a 1.5-kb fragment consisting of *bla* and one *cos* sequence is eliminated, allowing inserts of 30-44 kb to be cloned. It is mobilizeable with helper plasmids like pRK2013. It contains a unique *Bam*HI cloning site flanked by T3 and T7 promoters and *Not*I sites for rapid mapping and characterization of inserts. Finally, as shown in Fig. 1, the cloning site and *cos* sequences are contained on a *Hind*III fragment, which can be transferred to other replicons for rapid construction of alternative cosmid cloning vectors.

Genomic libraries of *P.s. syringae* strain B728A (provided by D.K. Willis) and *P.s. lachrymans* strain 859 (provided by C. Leben) were constructed in pCPP34, a progenitor of pCPP37 containing a unique *Stu*I site in place of the *bla* fragment. Genomic DNA from the *P. syringae* strains was partially digested with *Sau*3A to generate fragments predominantly 33-50 kb. The fragments were then dephosphorylated and ligated to pCPP34 DNA that had been digested with *Stu*I and *Bam*HI. The ligations were performed in 5mM ATP to prevent blunt-end ligation so that vector concatamers could not form. The cosmids were packaged, transduced into *E. coli* ED8767, and Tc^r colonies were selected. Using this procedure, 1.3 x 10^6 cosmids/µg of insert DNA were obtained with *P.s. syringae* B728A DNA, and 2.3 x 10^6 cosmids/µg of insert DNA were obtained with *P.s. lachrymans* 859 DNA.

The average size of the inserts in these cosmids was 39 kb. In general, the cosmids were at least as stable as pLAFR5 in *P.s. syringae* and *P.s. lachrymans*. DNA inserts suffered no deletions when cosmids were propagated in the *recA E. coli* strain ED8767, but 25% of the cosmids showed deletions after repeated culturing in the Rec$^+$ *E. coli* strain LE392.

We screened the genomic library of *P.s. lachrymans* 859 DNA in *E. coli* ED8767 for pectolytic activity at pH 8.0. No pectolytic clones were found, suggesting that the strain 859 *pel* gene may be poorly expressed in *E. coli*. Consequently, the 859 library was conjugated *en masse* into the *recA P.s. syringae* strain BUVS1 (a derivative of B728A kindly provided by D.K. Willis). Eight transconjugants with the ability to pit pectate semi-solid agar were identified among 700 tested. Activity-stained isoelectric focusing gels showed that they all produced a single pectate lyase with a pI of approximately 9.4. Positive A$_{235}$ assays confirmed that the enzyme was a lyase.

One pectolytic *P.s. syringae* transconjugant was tested for its ability to elicit the hypersensitive response in tobacco and to cause fruit rot of cucumber. The production of pectate lyase did not interfere with elicitation of the hypersensitive response. Nor did it enable *P.s. syringae* to cause more than a slight, localized, softening in stab-inoculated cucumber fruit. It is noteworthy that production of a single pectate lyase enables the closely related pathogen, *P. viridiflava*, to rot cucumber and other market vegetables [13]. Although further analysis and manipulation of pectate lyase (and polygalacturonase) production by *P.s. lachrymans* is required, our results suggest that heterologous expression of the *P.s. lachrymans pel* gene is insufficient to extend the host range of *P.s. syringae* to cucumber fruit.

2.3. New pectate lyase isozymes produced by *E. chrysanthemi*

E. chrysanthemi EC16 is known to produce at least seven pectic enzymes: pectin methylesterase, exo-poly-α-D-galacturonosidase, exopolygalacturonate lyase, and four isozymes of pectate lyase [5]. These enzymes are typically induced by pectic compounds and all but exopolygalacturonate lyase are secreted to the medium by *E. chrysanthemi*. To genetically dissect this system, we have been using cloned DNA fragments and marker-exchange or exchange-eviction mutagenesis to direct deletions into the *E. chrysanthemi* pectic enzyme genes [8,17]. Mutant CUCPB5012 contains mutations affecting *pehX*, *pelX*, *pelA*, *pelB*, *pelC*, and *pelE*, and is correspondingly deficient in all of the above enzymes except pectin methylesterase. Nevertheless, mutant CUCPB5012, like the wild-type strain, causes macerated lesions in the leaves of chrysanthemum, its host of origin.

Several new pectic enzymes were discovered with the observation that culture fluids of *E. chrysanthemi* mutant CUCPB5012 could kill and macerate chrysanthemum leaf tissue, but only if the medium contained chrysanthemum tissue extracts. In contrast, the wild type produced macerating culture fluids when grown in media containing either chrysanthemum extracts or polygalacturonate. The best inducing condition for the mutant that we have found is growth in minimal medium containing plant cell walls. Analysis of macerating culture fluids from the mutant with activity-stained isoelectric focusing gels revealed at least five bands with pIs ranging from ca. 4.2 to 9.5. The new enzymes were also detectable in extracts of infected chrysanthemum cuttings. The plant-inducible pectic enzymes appear to be lyases based on their high pH optimum in assays with diagnostically buffered polygalacturonate-agarose overlays.

A genomic library of CUCPB5012 DNA was constructed in *E. coli* DH5, using previously described procedures [7] and the expression vector pTTQ18 [20]. Ten clones out of 9000 revealed pectolytic activity on pectate semi-solid agar medium supplemented with IPTG. One of the clones with the highest activity in A_{235} assays was chosen for further analysis. The plasmid was isolated and introduced into *E. chrysanthemi* CUCPB5012. The characteristics of the pectate lyase produced by the cloned gene in *E. coli* and in *E. chrysanthemi* were then compared. The *E. chrysanthemi* cells were grown in Kings medium B [12] lacking plant extracts but supplemented with IPTG for selective expression of the cloned *pel* gene. The pectate lyase from these recombinant *E. chrysanthemi* cultures yielded a band on activity-stained isoelectric focusing gels that had the same pI as the most alkaline of the plant-inducible enzymes. The enzyme preparation also macerated chrysanthemum leaves. In contrast, equivalent amounts (as determined by A_{235} assays) of enzyme from *E. coli* failed to macerate chrysanthemum leaf tissue or produce a band in activity-stained isoelectric focusing gels. The

result suggests that the enzyme is modified in *E. chrysanthemi*. Further
characterization of these new enzymes is necessary before their function
or role in the residual maceration activity of Δ*pelABCE* mutants can be
ascertained.

2.4. Cloned *out* genes from *E. chrysanthemi* enable *E. coli* to efficiently secrete multiple pectic enzymes

The *out* genes are responsible for the efficient extracellular secretion
of multiple plant cell wall-degrading enzymes by *E. chrysanthemi*. Out⁻
mutants are unable to secrete any of these proteins beyond the periplasm
and are reduced in virulence [1,21]. We have cloned a cluster of *out*
genes from *E. chrysanthemi* in the low copy-number cosmid pCPP19 (a mob^+
cos^+ derivative of pGB2 [4] constructed by D.W. Bauer) by
complementation of several *E. chrysanthemi* Out⁻ mutants.

One of the out^+ cosmids, pCPP2006, was chosen for further
analysis. The cosmid restored the ability of *E. chrysanthemi* Out⁻
mutants AC4198, AC4199, AC4200, AC4205, AC4206, and AC4207 to secrete
>90% of the total pectate lyase to the medium. When pCPP2006 was
transformed into *E. coli* strain MC4100 containing pPEL74, which carries
the *E. chrysanthemi* *pelA*, *pelE*, and *pem* genes [11], 91% of the pectate
lyase and 67% of the pectin methylesterase was secreted to the medium.
In cultures of MC4100 containing pPEL74 without pCPP2006, only 14% of
the pectate lyase and pectin methylesterase activity was found in the
medium. *E. coli* MC4100 containing pCPP2006 and pPEL3, which carries the
E. chrysanthemi *pelB* and *pelC* genes [11], secreted 66% of the total
pectate lyase to the medium. Without pCPP2006, only 7% of the pectate
lyase was released. Similar results were found with *E. coli* carrying
the *E. chrysanthemi* *pehX* gene encoding exo-poly-α-D-galacturonosidase.
The presence of pCPP2006 did not cause the release of the plasmid-
encoded, periplasmic marker protein, β-lactamase. The results indicate
that the cloned *E. chrysanthemi* *out* genes enable *E. coli* to selectively
secrete several *E. chrysanthemi* pectic enzymes to the medium.

The *out* loci in cosmid pCPP2006 were mapped by mutagenesis with
Tn*phoA* [14]. pCPP2006 mutations were initially screened in *E. coli*
cells carrying the *E. chrysanthemi* *pelE* gene. All *out* mutations were
clustered in a 12-kb region of the cloned DNA. When pCPP2006::Tn*phoA*
mutations were marker-exchanged into the *E. chrysanthemi* chromosome,
they produced the same phenotype as in *E. coli*. Four complementation
groups were defined by analysis of the Out phenotype of *E. chrysanthemi*
out::Tn*phoA* mutants harboring pCPP2006::Tn*phoA* plasmids.

The nucleotide sequence of the 2.4-kb *Eco*RI fragment which
contains most of complementation group C was determined. The sequenced
region contained two complete open reading frames (ORFs) and two
incomplete ORFs. These four ORFs shared extensive similarity at both
the nucleotide and amino acid levels and were arranged colinearly with
four *Klebsiella oxytoca* (=*pneumoniae*) pullulanase secretion genes *pulH*,
pulI, *pulJ*, and *pulK* [16]. The *pulI* and *pulJ* genes shared ca. 50%
identity in their amino acid sequences with the corresponding *out* genes.

Nonidentical residues were mostly conservative substitutes, and the similarity extended throughout the genes. The carboxyl-terminal region of *pulH* had 35% amino acid similarity with the limited region of the corresponding *out* gene that we sequenced. The similarity between *pulK* and the corresponding *out* gene was limited to the amino terminal region (110 residues with 46.4% similarity). The carboxyl termini were very different. A similar relationship between *E. carotovora out* gene sequences and the *Klebsiella* pullulanase secretion *pulC-O* operon has been observed by G.P.C. Salmond and coworkers (personal communication). The results suggest that these enterobacteria secrete polysaccharidases by a common mechanism. A better understanding of the secretion process may permit novel approaches to disease control through disruption of this essential aspect of soft-rot erwinia virulence.

3. ACKNOWLEDGEMENTS

We thank Noel T. Keen for providing pPEL74 and pPEL3. This work was supported by grants 87-CRCR-1-2352 and 85-CRCR-1-1770 from the Competitive Research Grants Office of the U.S. Department of Agriculture.

4. REFERENCES

1. Andro, T., Chambost, J.-P., Kotoujansky, A., Cattaneo, J., Bertheau, Y., Barras, F., Van Gijsegem, F., and Coleno, A. 1984. Mutants of *Erwinia chrysanthemi* defective in secretion of pectinase and cellulase. J. Bacteriol. 160:1199-1203.
2. Bishop, A.L., Katz, B.H., Burr, T.J., Kerr, A., and Ophels, K. 1988. Pectolytic activity in *Agrobacterium tumefaciens*. Phytopathology 78:1551 (abstract).
3. Burr, T.J., Bishop, A.L., Katz, B.H., Blanchard, L.M., and Bazzi, C. 1988. A root-specific decay of grapevine caused by *Agrobacterium tumefaciens* and *A. radiobacter* biovar 3. Phytopathology 77:1424-1427.
4. Churchward, G., Belin, D., and Nagamine, Y. 1984. A pSC101-derived plasmid which shows no sequence homology to other commonly used cloning vectors. Gene 31:165-171.
5. Collmer, A. and Keen, N.T. 1986. The role of pectic enzymes in plant pathogenesis. Annu. Rev. Phytopathol. 24:383-409.
6. Collmer, A., Ried, J.L., and Mount, M.S. 1988. Assay methods for pectic enzymes. Methods in Enzymol. 161:329-335.
7. Collmer, A., Schoedel, C., Roeder, D.L., Ried, J.L., and Rissler, J.F. 1985. Molecular cloning in *Escherichia coli* of *Erwinia chrysanthemi* genes encoding multiple forms of pectate lyase. J. Bacteriol. 161:913-920.
8. He, S.Y. and Collmer, A. 1990. Molecular cloning, nucleotide sequence and marker-exchange mutagenesis of the exo-poly-α-D-galacturonosidase-encoding *pehX* gene of *Erwinia chrysanthemi* EC16. J. Bacteriol. 172:4988-4995.
9. Hildebrand, D.C. 1971. Pectate and pectin gels for differentiation

of *Pseudomonas* spp. and other bacterial plant pathogens. Phytopathology 61:1430-1436.

10. Kado, C.I. and Heskett, M.G. 1970. Selective media for isolation of *Agrobacterium*, *Corynebacterium*, *Erwinia*, *Pseudomonas*, and *Xanthomonas*. Phytopathology 60:969-976.

11. Keen, N.T., Dahlbeck, D., Staskawicz, B., and Belser, W. 1984. Molecular cloning of pectate lyase genes from *Erwinia chrysanthemi* and their expression in *Escherichia coli*. J. Bacteriol. 159:825-831.

12. King, E.O., Ward, M.K., and Raney, D.E. 1954. Two simple media for the demonstration of pyocyanin and fluorescein. J. Lab. Med. 22:301-307.

13. Liao, C.H., Hung, H.Y., and Chatterjee, A.K. 1988. An extracellular pectate lyase is the pathogenicity factor of the soft- rotting bacterium *Pseudomonas viridiflava*. Mol. Plant-Microbe Int. 1:199-206.

14. Manoil, C. and Beckwith, J. 1985. Tn*phoA*: A transposon probe for protein export signals. Proc. Natl. Acad. Sci. U.S.A. 82:8129-8133.

15. Pohronezny, K., Larson, P.o., and Leben, C. 1978. Observations on cucumber fruit invasion by *Pseudomonas lachrymans*. Plant Dis. Reptr. 62:306-309.

16. Reyss, I. and Pugsley, A.P. 1990. Five additional genes in the *pulC-O* operon of the gram-negative bacterium *Klebsiella oxytoca* UNF5023 which are required for pullulanase secretion. Mol. Gen. Genet. 222:176-184.

17. Ried, J.L. and Collmer, A. 1988. Construction and characterization of an *Erwinia chrysanthemi* mutant with directed deletions in all of the pectate lyase structural genes. Mol. Plant-Microbe Int. 1:32-38.

18. Simon, R., Priefer, U., and Puhler, A. 1983. A broad host range mobilization system for in vivo genetic engineering: Transposon mutagenesis in gram-negative bacteria. Biotechnology 1:784-791.

19. Spaink, H.P., Okker, R.J.H., Wijffelman, C.A., Pees, E., and Lugtenberg, B.J.J. 1987. Promoters in the nodulation region of the *Rhizobium leguminosarum* Sym plasmid pRL1JI. Plant Mol. Biol. 9:27-39.

20. Stark, M.J.R. 1987. Multicopy expression vectors carrying the *lac* repressor gene for regulated high-level expression of genes in *Escherichia coli*. Gene 51:255-267.

21. Thurn, K.K. and Chatterjee, A.K. 1985. Single site chromosomal Tn*5* insertions affect the export of pectolytic and cellulolytic enzymes in *Erwinia chrysanthemi*. Appl. Environ. Microbiol. 50:894-898.

MOLECULAR ANALYSIS OF A GENE THAT AFFECTS EXTRACELLULAR POLYSACCHARIDE PRODUCTION AND VIRULENCE IN *PSEUDOMONAS SOLANACEARUM*

F. GOSTI, Y. HUANG[1], and L.SEQUEIRA
Department of Plant Pathology, University of Wisconsin, Madison WI 53706, USA.
[1]*Present address: Department of Plant Pathology, University of California, Davis CA 95616, USA.*

ABSTRACT. *Pseudomonas solanacearum*, agent of bacterial wilt, spontaneously gives rise to nonmucoid avirulent mutants under certain culture conditions. These mutants are affected in several traits including extracellular polysaccharide (EPS) production which results in altered colony morphology. An 8 kb DNA fragment isolated from a genomic library of such a mutant (strain B1) closely mimics the spontaneous shift in phenotype when it is introduced into a wild-type strain K60 in a low-copy number plasmid (pLAFR3). A functional DNA region called *eps*R of at least 1kb was established by saturation mutagenesis; maxicell experiments in *E. coli* indicated that a 25 kDa protein is the probable gene product (Huang and Sequeira, 1990). We report here on the isolation of the corresponding wild type *eps*R region and show by functional assay and sequencing that both units (isolated from either B1 or K60 strains) are capable of reproducing the phenotypic shift and that their nucleotide sequences are identical. Analysis of the coding capacity of the potential open reading frames suggests that translation of the *eps*R gene might not begin at a normal methionine initiation codon.

1. Introduction

Pseudomonas solanacearum is the causal agent of bacterial wilt of numerous important crops such as potato, tomato and tobacco in the warmer latitudes. A number of factors have been implicated in disease development by this bacterium. These include bacterial extracellular polysaccharides (EPS), plant cell wall degrading enzymes and plant growth substances. Although the mechanism of wilting is not presently understood, it is thought that production of extracellular polysaccharides by bacteria growing within the vascular system of the plant contributes to plugging of xylem vessels, thus interfering with water transport. This view is further supported by:
a) the correlation, first reported by Kelman in 1954, between colony mophology on a tetrazolium medium and virulence in tobacco, and
b) the ability of semi-purified EPS to wilt tomato cuttings (Husain and Kelman, 1958).
 The production of copious amounts of EPS is a characteristic common to all wild type, virulent strains of *P. solanacearum*. However, when grown in ordinary culture media without aeration, this bacterium undergoes a spontaneous shift from the large, slimy,

73

H. Hennecke and D. P. S. Verma (eds.),
Advances in Molecular Genetics of Plant-Microbe Interactions, Vol. 1, 73–77.
© 1991 *Kluwer Academic Publishers. Printed in the Netherlands.*

spreading colonies typical of the wild type to the small, slime-less, butyrous colonies of variants that are avirulent (Kelman, 1954). Although these variants are readily recognized in culture because of their lack of EPS, they encompass a vast array of biochemical phenotypes that are highly variable. The characteristics that may or may not be altered include: polygalacturonase and cellulase production, lipopolysaccharide structure, piliation, synthesis of indoleacetic acid and restoration of motility.

Athough this spontaneous shift in colony type occurs in all strains of *P. solanacearum*, the mechanism responsible for this change remains unknown. Recently, we have isolated a chromosomal DNA fragment from a variant strain (B1) which closely mimics (for the phenotypes studied: EPS production, virulence, polygalacturonase and tyrosinase activities), the spontaneous shift when it is introduced into wild type strains in pPLAFR3, a low copy number plasmid (Huang and Sequeira,1990). Saturation mutagenesis of this plasmid (pBE6), with the reporter gene construct Tn*3::gus* led to the identification of a 2 kb region encoding a 25 kd protein in a maxicell assay, suggesting that a single transcriptional unit, named *eps*R, is responsible for the phenotypic switch observed (Huang and Sequeira,1990). Given the potential importance of this sequence in the regulation of both EPS production and virulence, we have studied and now report on the structure and expression of the *eps*R gene in the wild type strain, K60.

2. Materials and methods

These experiments were performed on wildtype *Pseudomonas solanacearum* strain K60 and its spontaneous avirulent mutant, B1. Growth conditions and antibiotic concentrations for *P.solanacearum* and *Escherichia coli* strains were as described previously (Huang *et al.* 1990). Virulence assays were performed as described previously (Xu et al, 1988). Recombinant DNA techniques were as described by Maniatis *et al* (1982). Construction of genomic libraries and transformation of *P.solanacearum* by means of electroporation were as described previously (Allen *et al*, in press). Nucleotide sequencing by the dideoxy chain termination method was completed by special arrangement with the Cetus Corporation (Emeryville, CA) and the Novagen Corporation (Madison, WI). Computer analysis of the sequence was performed with the Genetic Computer Group package.

3. Results.

To isolate the wild type *eps*R sequence, we used colony hybridization to screen a K60 chromosomal library made in pLAFR5. As a probe, we used a restriction fragment from pBE6 containing the B1 *eps*R sequence. Two sets of non-overlapping clones were identified, but only one could reproduce the phenotypic switch when introduced into K60. This observation was confirmed in Southern blots using the same probe. In K60 as well as in B1 digests there was hybridization to two different DNA fragments (data not shown). We studied only one of the clones (pKL4) that was capable of reproducing the spontaneous switch. Subsequent subcloning led to the isolation of clone pKL44 in which a 1.3 kb fragment (Fig 1) was shown to retain the *eps*R function when introduced into K60.

The *eps*R-containing DNA fragments isolated from both strains, B1 and K60, were sequenced from pBluescript subclones (pBSK21, pKLS42 and pKLS43, respectively). Results showed that the two nucleotide sequences are identicals. Analysis of the coding capacity of the functional unit indicated the presence of a potential conventional open reading frame of 400 nucleotides.

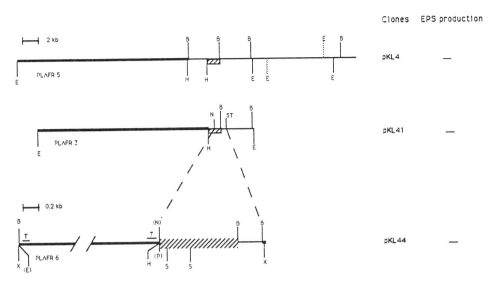

Figure 1. Illustration of the procedure used to subclone the *epsR* K60 gene. Abbreviations for restriction sites are as follows: B:BamHI, E:EcoRI, H: HindIII, N: NsiI, P: PstI, S: SphI, ST: StuI, X:XbaI. Transcription terminator sequences (T) are not shown to scale.

4. Discussion

We have confirmed that a DNA fragment from strain B1, when introduced into the wildtype strain K60, mimics the shift in phenotype that occurs spontaneously in K60. This sequence (*epsR*) appears to be a global regulatory element that affects both EPS production and virulence. Since the original sequence was from the variant B1 strain, it was essential to obtain information on the wild type sequence of *epsR*.

The wild type *epsR*-containing cosmid, pKL4, reproduced the phenotype shift (from EPS$^+$ to EPS$^-$ and from vir$^+$ to vir$^-$) when introduced into K60, indicating that the cause of the shift did not reside simply in the differential expression of altered sequences. The non-functional *epsR*-related sequence at a different location in the chromosome may be an example of a redundant or evolutionary related sequence but additional work is required to determine why it is not functional in our assay.

Subcloning of pKL4 led to the isolation of a 1.3 kb functional fragment when introduced in the vector pLAFR6. We deduce that expression of the gene encoded by this fragment is under the control of its own promoter, for the vector contains transcription termination signals at both sides of its polylinker site (Staskawicz *et al*, 1987). The nucleotide sequence determined for this region was shown to be identical in both K60 and B1, confirming the phenotype shift data. Analysis of the conventional ORF present in the strand previously shown to be transcribed (from the results of Tn*3-gus* mutagenesis) cannot account for the synthesis of the 25 kDa protein observed in the maxi-cell experiment (Huang and Sequeira, 1990) However, mesurement of codon usage frequency, together with the location of Tn*3-gus* insertions affecting the function (data not shown), suggest that translation of the the protein encoded by *eps*R probably begins with an unconventional initiation codon.

We have described the molecular characterization of a potential regulator of both EPS production and virulence.in *P solanacearum*. We could reveal its role only throught its presence in a low-copy number plasmid. This can be explained if, for example, *eps*R were normally subjected to *trans* negative regulation and the increased number of sequences in a plasmid titrates out the *trans*-acting factor. This working hypothesis is in good agreement with the fact that Tn*3-gus* marker exchange in the wild type K60 strain does not affect EPS production (data not shown). Alternatively, the region may have been physicaly removed from a *cis* acting regulator in the cloning process. Sequences with properties similar to *eps*R have been isolated from *Xanthomonas campestris* (Daniels *et al*, 1989) and the existence of a positive regulator of EPS production has been proposed by Denny *et al* (1989). In addition, expression of EPS synthesis in *Rhizobium sp.* was shown to depend on a correct balance between two different regulatory elements (Gray et al, 1990) Together with our results, these reports illustrate the complexity underlying this type of global regulation.

5. Acknowledgments

We sincerely thank Fred Reichert (Cetus Corporation) and Ken Barton (Agracetus Corporation) for providing the sequence of the B1 *eps*R gene We want to thank Merelee Atkinson for providing the K60 chromosomal library. This work was supported by grants from the National Science Fundation and the Department of Energy.

6. References

Allen, C., Huang, Y. and Sequeira, L. 1991 'Cloning of genes affecting polygalacturonase production in *Pseudomonas solanacearum'* Mol Plant. Microbe Interact. (in press).

Brumbley, S. M. and Denny, T. P. (1989). 'Molecular characterization of a locus regulating production of extracellular polysaccharide slime and virulence in *Pseudomonas solanacearum*. Phytopathol. 79, 1156 (abstract).

Daniels, M., J. Osbourn, A., E. and Tang, J.,L. (1989) 'Regulation in *Xanthomonas*-plant interactions' in B. J. J. Lugtemberg (ed), Signal Molecules in Plant and Plant-Microbe Interactions, Springer Verlag, Berlin. Heidelberg, pp 190-196.

Gray, J., X. Djordjevic, M., A.and Rolfe, B., G. (1990) "Two genes that regulate exopolysaccharide production in *Rhizobium sp.* strain NGR234: DNA sequences and resultant phenotypes', J. Bacteriol 172, 193-203.

Huang, Y. Xu, P. and Sequeira, L. (1990) 'A second cluster of genes that specify pathogenicity and host response in *Pseudomonas solanacearum* ', Mol.Plant-Microbe Interact 3, 48-53.

Huang, Y. and Sequeira, L. (1990) 'Identification of a locus that regulates multiple functions in *Pseudomonas solanacearum*. J. Bacteriol. 172, 4728-4731.

Husain, A. and Kelman, A. (1958) 'Relation of slime production to mechanism of wilting and pathogenicity of *Pseudomonas solanacearum*', Phytopathology 48, 155-164.

Kelman, A. (1954) ' The relationship of pathogenicity in *Pseudomonas solanacearum* to colony appearance on a tetrazolium medium.,' Phytopathology 44, 693-695.

Maniatis, T. Fritsch, E., F. and Sambrook, J. (1982) Molecular cloning: a laboratory manual, Cold Spring Harbor Press, Cold Spring Harbor. Harbor Laboratory, Cold Spring Harbor, N.Y.

Staskawicz, B., J, Dalhbeck, D., Keen, N and Napoli, C. (1987). 'Molecular characterization of cloned avirulence genes from race 0 and race 1 of *Pseudomonas syringae pv. glycinea'* .J. Bacteriol 169, 5789-5794.

Xu, P., Leong. S., A and Sequeira, L. (1988). 'Molecular cloning of genes specifying virulence in *Pseudomonas solanacearum* ' J. Bacteriol 170, 617-622.

INTERACTIONS BETWEEN *ARABIDOPSIS THALIANA* AND PHYTOPATHOGENIC *PSEUDOMONAS* PATHOVARS: A MODEL FOR THE GENETICS OF DISEASE RESISTANCE

Jeff Dangl, Hiltrud Lehnackers, Siegrid Kiedrowski, Thomas Debener, Christoph Rupprecht, Martin Arnold, and Imre Somssich*

Max-Delbrück-Laboratory in the MPG, and *Dept of Biochemistry, Max-Planck-Institute for Plant Breeding, Carl-von-Linné-Weg 10, D-5000 Köln 30, Germany

Abstract

The interaction between Arabidopsis thaliana and various phytopathogenic Pseudomonas pathovars presents an outstanding model to genetically define plant and bacterial loci necessary for generation of a hypersensitive resistance response (HR). Certain isolates of Pseudomonas syringae pv. maculicola are virulent on Arabidopsis, while others are not. We also tested 17 P. cichorii isolates, and found that all are avirulent. Variation exists among Arabidopsis ecotypes with respect to resistance/ susceptibility to P.s. maculicola isolates. We also identified candidate Arabidopsis mutants showing altered interaction phenotypes after inoculation with a P.cichorii isolate. Segregation analysis and RFLP mapping of the loci conditioning these altered responses is in progress. We also have cloned Arabidopsis homologs to genes induced by pathogen in another system. The function of many of these genes is unknown. Phenocopy mutants of one, which is highly induced by pathogen infiltration, will be constructed to test the gene product's involvement in the HR. Finally we isolated mutants in genes encoding bacterial membrane proteins unable to generate a normal response on Arabidodpsis. These mutants appear to evade the plant defense response, and the normal protein may be important in triggering the HR.

Introduction

Little is known about the mechanism by which plant resistance reactions are generated. In most cases, resistance is triggered by the presence of dominant genes encoding avirulence in the pathogen (avr) and a matching dominant function encoded by the plant (R genes) (1). Gene-for-gene relationships underlie all interactions in which a pathogen is fundamentally able to colonize a host, and may also be common in so called non-host resistance (2). It is also apparant that many induced events are

78

H. Hennecke and D. P. S. Verma (eds.),
Advances in Molecular Genetics of Plant-Microbe Interactions, Vol. 1, 78–83.

triggered after initial recognition in both compatible and incompatible plant-pathogen interactions (3). Several known biosynthetic pathways are activated, in many cases leading to the formation of protective proteins and secondary metabolites (4). However, many genes of unknown function are also activated, and their role, if any, in the eventual outcome of an interaction remains elusive.

While a great deal of knowledge regarding the structure of bacterial avr genes exists, their mode of action is not known. Even less information exists regarding the necessary signalling and activation steps taking place in the plant subsequent to recognition. The latter problem is largely due to the lack of a genetically simple model plant. Thus, while nearly 50 years have elapsed since the original description of an R-gene by Flor (5), and while the genetics of R-genes is characterized in a plethora of systems, no R-gene has been isoalted to date. In addition, although a great deal of induced gene activity accompanies plant-pathogen recognition, in no case can this induction be shown to be required for generation of a successful resistance response. We and others have turned to Arabidopsis thaliana as a model to define loci and isolate genes necessary for disease resistance. The genetic advantages of Arabidopsis have been oft extolled (6), and its usefulness as a model in plant-pathogen interaction is becoming apparant.

Here, we briefly introduce three types of experiments meant to genetically identify necessary components of plant defense responses. This, and other communications in this volume, show that paradigms of plant-pathogen interaction hold true for Arabidopsis with bacterial and fungal pathogens.

Materials and Methods

Plants: Arabidopsis seeds are sown in a 2:1 mixture of potting soil and sand and are germinated under high humidity. Between 20-22 days later, individual seedlings are picked to "jiffy pots" in the same soil mix. They are used for infection 14-17 days later, when rossette leaves are 1-3 cm long. Plants are maintained with short (10 hr) days under 22.000 lux from broad spectrum lamps, and watered sparingly. Temperature is 24° C day/ 20° C night, humidity 65% day/ 80% night.

Bacteria: Pseudomonas strains are grown at 28° - 30° C on King's Media B (KB) supplemented with rifampicin at 100 mg/l. Kanamycin is used to select for TnphoA at 30 mg/l. For infection experiments, overnight cultures are diluted 1:40 into fresh KB with antibiotics and grown to mid-log (OD_{600} =0.15-0.30). Bacteria are collected by centrifugation and resuspended in 10 mM $MgCl_2$ to OD_{600} =0.2 ($\sim 10^8$ cfu/ml). This concentration, or a 1:100 dilution from it ($\sim 10^6$ cfu/ml) are used as inoculation for either rapid assays or growth curves, respectively. Plants are infected by gently infiltrating the lower epidermis of rosette leaves using a 1-ml syringae, without needle. Typically 10-30 ul of suspension is inoculated on one side of the leaf vein. For in planta bacterial growth curves, a portion of infected leaf is harvested with a cork-borer (6 mm diameter). Either individual leaves or 4-5 pooled leaves are used per data point. Each leaf disc is ground in 200 ul 10 mM $MgCl_2$ and bacteria are titrated on KB/rif plates.

TnphoA mutagenesis was performed essentially as described (7). Molecular biological methods are all standard. Heterologous hybridizations between parsley probes and Arabidopsis DNA was performed at 60°C in 1 M NaCl/10% Detran Sulfate/1.0% SDS. Final washes were at 60°C in 2X SSC/0.5% SDS.

Results and Discussion

We are exploiting natural genetic variation between Arabidopsis ecotypes to define resistant and susceptible interaction phenotypes to a number of P.s. maculicola isolates. Figure 1A shows in planta growth kinetics from five ecotypes after infection with an isolate from Brassica olaraceae. The clearly resistant ecotype Oy-O rapidly develops a dry white tissue necrosis after infiltration at even at low density. We consider this phenotype a hypersensitive response. The ecotypes Mt-O and Nd-O are able to support high levels of bacterial growth. Interestingly, Nd-O leaves are phenotypically normal until around 8 days after infection, when they rapidly chlorose and collapse. Mt-O show obvious water soaking at around day 3. Col-O and C24 support intermediate levels of bacterial growth. Analysis of F1 progeny from crosses involving clearly susceptible phenotypes (Mt-O or Nd-O) and either intermediate (Col-O) or resistant (Oy-O) ecotypes exhibit both the phenotype and growth kinetics characteristic of the more resistant parent (Figure 1b). F2 progeny from the (Nd-O x Col-O) cross shows that the Col-O phenotype segregates as a single, dominant locus (111:43, $x^2 = 0.701$ for expected 3:1 segregation, p< 0.05).

We are also constructing degree of relatedness trees for 28 Arabidopsis ecotypes from broadly divergent geographical locations using RFLP analysis. Conventional analysis with restriction enzymes recognising 6 base sites is often sufficient to detect a large degree of polymorphism (figure 2a,b). When not, (e.g. figure 2c) we have employed the more refined RFLP detection offered by restriction enzymes recognizing 4 base pair sites (figure 2d). Phenograms from these analyses will be useful guides in choosing ecotypes for pathology experiments described above.

We are also analyzing the structure and pathogen-induced expression of Arabidopsis genes homologous to parsley genes known to be induced by fungal elicitor (8). The assumption for this work is that homologous genes from widely diverged species may be functionally important in generation of successful resistance reactions. These loci are also low or single copy in Arabidopsis. We are using them as probes of ecotype relatedness (see above) and will place them on the RFLP map. Most interestingly, we have chosen one clone, PIG3 (pathogen induced gene) as a

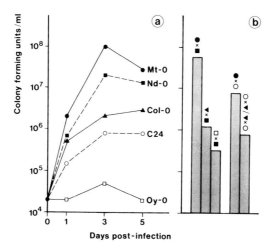

Figure 1

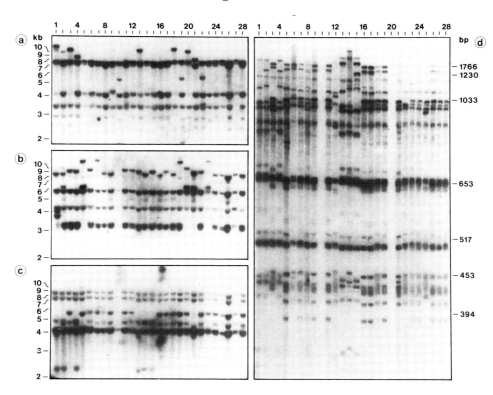

Figure 2

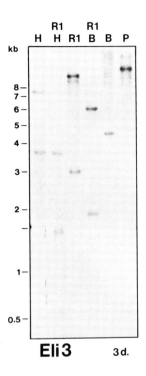

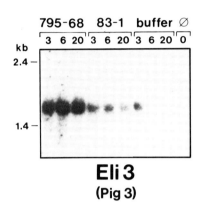

Eli 3
(Pig 3)

Figure 3A

Figure 3B

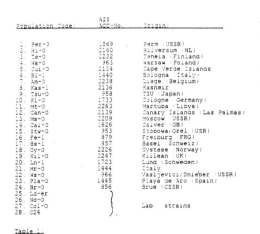

Table 1:

Origin of the ecotypes used in this study. All ecotypes were obtained from the Arabidopsis Information Service in Frankfurt.

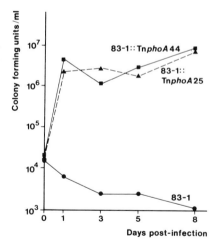

Figure 4

prototype for the construction of phenocopy mutants. Expression of this gene is highly induced by infiltration with avirulent and intermediate virulence Pseudomonas strains (figure 3a) and its organization is fairly simple (figure 3b). Anti-sense vectors are being constructed to create phenocopy mutants unable to accumulate large amounts of PIG3 mRNA after pathogen infiltration.

Finally, we are probing the bacterial membrane for proteins involved in triggering plant defense reactions. As mentioned, many bacterial avr genes have been cloned and analyzed. None encode obviously membrane bound proteins, thus prompting the question of how the avr gene product's signal reaches the plant. As well, it is not unreasonable to hypothesize that membrane partners of bacterial "two-component regulatory systems" (9) are involved in several aspects of the interaction between Pseudomonas and Arabidopsis.

We used the specialized transposon, TnphoA (10), to isolate mutants from a P.chiorii (83-1) which no longer are recognized effectively by Arabidopsis. From around 70,000 transposon insertion mutants, 550 were positive for alkaline phosphatase activity. This indicates TnphoA insertion into a gene normally encoding a membrane bound or periplasmic protein. From 135 screened to date, 2 mutants gave an altered phenotype after inoculation onto the Col-O ecotype. At high density (10^8 cfu/ml), a lesion appears between 24-36 h which is significantly faster than lesion appearance with the parent strain. Strikingly, each of these TnphoA insertion mutants is capable of rapid multiplication in planta after low density (10^6 cfu/ml) infiltration, in contrast to the parent strain (figure 4). The mutant bacteria grow to a high level, and slowly decline over a two week period (not shown). During this time course, few leaves develop disease symptoms, suggesting that Arabidopsis is now tolerant to these P.cichorii mutants. Cloning of the mutant locus and marker-exchange mutagenesis are in progress.

References
1. Keen N.T., and Staskawicz B.J. (1988) Ann. Rev. Microbiol. 42: 421-440.
2. Whalen M.C., Stall R.E., and Staskawicz B.J. (1988) Proc. Natl. Acad. Sci. USA 85: 6743-6747.
3. Dixon R.A., and Harrison M.J. (1990) Adv. Genet. 28: 165-234.
4. Scheel D., and Hahlbrock K. (1989) Ann. Rev. Plant Physiol. Plant Mol. Biol. 40: 347-369. 5.Flor H.H. (1947) J. Agricultural Research 74: 241-262.
6. Meyerowitz E.M. (1989) Cell 56: 263-269.
7. Long S., McCune S., and Walker G.C. (1988) J. Bacteriol. 170: 4257-4265.
8. Somssich I.E., et al (1989) Plant Mol Biol 12: 227-234
9. Stock J.B., Stock A.M., and Mottonen J.M. (1990) Nature 334: 395-400.
10. Manoil C.M., Mekalanos J.J., and Beckwith J. (1990) J. Bacteriol. 172: 515-518.

INTERACTION BETWEEN *ARABIDOPSIS THALIANA* AND *XANTHOMONAS CAMPESTRIS*

M.J. DANIELS, M.J. FAN, C.E. BARBER, B.R. CLARKE AND J.E. PARKER
The Sainsbury Laboratory
John Innes Centre for Plant Science Research
Norwich NR4 7UH
United Kingdom

ABSTRACT. After infiltration into leaves of *Arabidopsis thaliana*, *Xanthomonas campestris* pathovar *campestris* grew and caused symptoms similar to those incited in *Brassica* under similar conditions. *X. campestris* pathovars which do not infect crucifers and mutants of *X.c campestris* altered in pathogenicity to *Brassica* failed to grow and give symptoms in *A thaliana*. A survey of the interaction of many *A. thaliana* ecotypes with *X.c. campestris* wild type strains revealed some genetic variation for interaction phenotype (compatibility). In one case a gene-for-gene interaction involving a single dominant plant resistance locus and a bacterial avirulence gene was demonstrated. By screening M2 plants derived from mutagenised seed some mutants altered in response to *X.c. campestris* were identified.

1. INTRODUCTION

Understanding the molecular basis of microbial plant diseases requires parallel study of both the pathogens and host plants, but for technical reasons much more research has been carried out on pathogens, particularly bacteria, than on plants. The most widely used approach for studying the behaviour of plants in host-pathogen interactions is the analysis of patterns of gene expression in response to infection, for example by translation of mRNA *in vitro* and cDNA cloning (Collinge and Slusarenko 1987, Davis and Ausubel 1989). However this approach does not reveal the presence of disease-related genes which show constitutive expression. An alternative and potentially very powerful strategy of identifying plant genes by mutagenesis has been scarcely used in molecular plant pathology. In this paper we describe the interaction of the experimentally convenient crucifer *Arabidopsis thaliana* with the bacterial pathogen *Xanthomonas campestris* pathovar *campestris*, the agent of black rot, the most serious worldwide disease of brassicas (Williams 1980). The usefulness of this model pathosystem for molecular genetic studies of the resistance and response of plants to pathogen invasion is demonstrated by (a) the existence of a specificity-determining gene-for-gene system, involving a single dominant resistance locus in *A. thaliana* ecotype Columbia and an avirulence gene in *X.c. campestris* 1067, (b) other types of resistance determined by both dominant and recessive genes and (c) the feasibility of using a direct mutational approach to reveal plant genes involved in host-pathogen interaction.

84

H. Hennecke and D. P. S. Verma (eds.),
Advances in Molecular Genetics of Plant-Microbe Interactions, Vol. 1, 84–89.

2. METHODS

A. *thaliana* plants were grown in a growth chamber at 22°C with illumination for 8 hours per day. Under longer daylength conditions plants bolted early and the responses to bacterial infection were less consistent. Overnight cultures of bacteria were diluted with sterile water to 10^6 colony-forming units (cfu) per ml and suspensions were infiltrated into intercellular spaces of rosette leaves of plants through stomata, using a plastic syringe without a needle applied firmly but gently to the lower side of leaves (Collinge *et al.* 1987). Inoculated plants were maintained at 22°C under conditions of high humidity and observed daily. After symptoms had been recorded the infected leaves were removed and the plants were transferred to long-day conditions to induce flowering and seed production. The concentration of viable bacteria in leaves was determined by punching out at least three discs of diameter 0.2 cm from the infected area, homogenising the tissue and plating suitable dilutions on nutrient agar. Other methods have been described elsewhere (Daniels *et al.* 1984, Estelle and Somerville 1986).

3. INFECTION OF A. *THALIANA* BY X. *CAMPESTRIS*

A. *thaliana* plants were inoculated by forcing bacterial suspensions through stomata on the lower side of leaves so that about one third of the area of each half leaf was infiltrated. With *X.c. campestris* symptoms of spreading chlorosis followed by blackening and some rotting developed from 3 to 7 days at 22°C. However leaves inoculated in a similar manner with sterile water or broth, *X. campestris* pathovars which infect plants other than crucifers, or heat-killed *X.c. campestris* cells gave either no visible symptoms or slight localised chlorosis. Moreover *X.c. campestris* mutants with lesions in several classes of gene involved in pathogenicity to *Brassica* showed reduced symptom severity in A. *thaliana* leaves (Table 1). The concentration of *X.c. campestris* cells in leaves increased from an initial level of *ca.* 10^3 to 10^5 cfu.cm^{-2}. However some *X.c. campestris* mutants and and a selection of pathovars which infect non-crucifer crops showed reduced or no growth (Table 1). The similarities in the behaviour of the several bacteria in A. *thaliana* compared with *Brassica* in terms of symptoms, specificity and growth *in planta* indicate that A. *thaliana-X.c. campestris* can be regarded as a realistic "model" host-pathogen system. The physiological nature of the incompatible interactions (resistance) is not known. We have not observed rapid localised necrosis characteristic of the hypersensitive response induced in *Brassica* by some heterologous pathogens such as *X.c. vitians* (Collinge *et al.* 1987).

Although symptoms spread to cover the whole area of the leaf infected with wild type *X.c. campestris* bacteria did not move to other (uninoculated) leaves. It was possible to test several bacterial strains on the same plant, each on a separate leaf, and there was no indication of interference effects between the reactions of the separate leaves. After disease reactions had been scored inoculated leaves were removed and plants could be induced to flower. Seeds were not contaminated with bacteria. Thus the pathogenicity test was non-destructive and the plant was "preserved" for further study in the form of selfed progeny.

4. NATURAL VARIATION AMONG A. *THALIANA* AND X.C. *CAMPESTRIS* STRAINS

We examined the interaction of 25 ecotypes of A. *thaliana* from different geograhical locations,

Table 1

Behaviour of *X. campestris* strains in *A. thaliana* Col-0

Strain	Type	Symptom class	$\log_{10}$(population increase)
Pathovar *campestris*:			
8004	wild type	+++	2.05
1067	wild type	+/-	0.97
8004(*avr*Xca)	8004 carrying *avr* gene from 1067	+/-	0.88
8237	regulatory mutant	+	1.78
8288	enzyme export mutant	-	0
8409	endoglucanase⁻ mutant	++	1.97
ME-9	regulatory mutant	-	<0

Other pathovars (pvs. *graminis, malvacearum, phaseoli, translucens, vasculorum, vesicatoria, vitians*):

	wild type	+/-	<0 to 0.7

Symptoms were scored five days after inoculation of plants: -, no visible symptoms; +, chlorosis only at the site of inoculation; ++, limited spreading of chlorosis; +++, extensive spreading of chlorosis, inoculated area dark and beginning to rot. The population increase is the factor by which the number of viable bacteria per unit area of leaf increased over a period of four days. *X.c. campestris* strains are described in the following papers: 8004, 8237 and 8288: Daniels *et al.* 1984 and Dow *et al.* 1987; 8409: Gough *et al.* 1988; ME-9: Osbourn *et al.* 1990.

including ecotypes Columbia and Landsberg commonly used in genetic experiments, with 27 wild strains of *X.c. campestris* to find natural variation in interaction phenotype. Most ecotypes were susceptible to all the bacterial strains. However ecotype Columbia (Col-0) was resistant to *X.c campestris* strain 1067, but susceptible to other strains, and ecotype Oy-0 was resistant to many of the strains. For subsequent work we used ecotypes Col-0, Oy-0, the susceptible lines La-*er* (the *erecta* mutant of ecotype Landsberg) and JI-1 (a local isolate) , and *X.c. campestris* strains 1067 and 8004. 8004 is the parent isolate from which most of the strains used in this laboratory are derived (Daniels *et al.* 1984).

5. GENETIC BASIS OF INTERACTION PHENOTYPE

By analogy with many other host-pathogen systems (Keen and Staskawicz 1988) we postulated that the incompatibility of *X.c. campestris* 1067 with *A. thaliana* Col-0 results from a gene-for-gene interaction between an avirulence gene in 1067 and a resistance gene in Col-0. We have confirmed this by cloning the avirulence gene and identifying the resistance gene genetically. A genomic library of *X.c. campestris* 1067 DNA, partially digested with the restriction endonuclease *Sau*3A, was constructed in the broad host range cosmid vector pIJ3200 (Liu *et al.* 1990). Clones were transferred individually by conjugation into *X.c. campestris* 8004 (to which Col-0 is susceptible) and transconjugants were screened for virulence to Col-0. One clone rendered 8004 avirulent towards Col-0, but did not affect virulence to JI-1, La-*er* or *Brassica*. The specificity of the effect for Col-0 indicates that the clone contains an avirulence gene, tentatively designated *avr*Xca, and not for example a negative-acting regulatory gene which depresses pathogenicity when the copy number is increased by cloning (Tang *et al.* 1990). The avirulence determinant is located in a 4.5 kb *Hind*III fragment. The growth rate and final titre of 8004 in Col-0 were much higher than for 1067 and 8004 carrying the avirulence clone (Table 1).

A. *thaliana* Col-0 was crossed with La-*er* (susceptible to both *X.c campestris* strains 8004 and 1067). F1 plants were susceptible to 8004 and resistant to 1067. Sixty nine plants grown from seed taken at random from a pooled F2 seed lot were also tested. All were susceptible to 8004; 52 were resistant to 1067 and 17 were susceptible, the 3:1 ratio indicating that the resistant phenotype is determined by a single dominant locus. Thus the incompatibility of Col-0 with 1067 fulfils the criteria for gene-for-gene specificity determination in which the products of the host resistance gene and the bacterial avirulence gene probably interact to trigger events restricting pathogen growth and disease development. F3 plants grown from seed saved from the F2 plants have provided material for confirmation of the segregation pattern and for DNA preparation so that the resistance gene from Col-0 can be mapped with RFLP markers and subsequently cloned by genomic walking and transformation of a susceptible ecotype (Chang *et al.* 1988, Nam *et al.* 1989).

A. *thaliana* Oy-0 (which is resistant to both *X.c. campestris* strains 8004 and 1067) was crossed with JI-1 (susceptible to both). F1 plants were resistant to 1067 but susceptible to 8004. Segregation of phenotypes in the F2 progeny did not permit unequivocal deduction of the number of loci involved, but it is clear that whereas resistance of Oy-0 to 1067 is dominant, resistance to 8004 is recessive, or alternatively Oy-0 lacks a gene required to enable 8004 to establish and cause disease.

6. ISOLATION OF *A. THALIANA* MUTANTS

Approximately 1000 morphologically normal M2 Col-0 plants from ethylmethanesulphonate mutagenesis (Estelle and Somerville 1986) were inoculated on separate leaves with 8004 and 1067. One plant was found to have become sensitive to 1067, and was also susceptible to 8004(*avr*Xca). Crosses have been made to determine whether the mutated gene is allelic with the previously found resistance gene, and the mutant is being subjected to cycles of backcrossing to wild type Col-0 to give near-isogenic material for physiological study of the effect of the mutation. A mutant was also found which had acquired resistance to 8004.

Both *A. thaliana* and *X.c. campestris* have many desirable features for molecular genetic research and our findings, together with other reports indicating susceptibility of the plant to infection (Tsuji and Somerville 1988, Simpson and Johnson 1990), show that the combination provides an attractive system for fundamental studies of pathogenicity and specificity mechanisms. The steady coordinated development of tools and techniques for molecular analysis of *A. thaliana* means that it is relatively straightforward to clone a gene defined only by its phenotype, and the fact that genetic variation for attributes of interest to pathologists can be easily detected in both natural and mutagenised populations indicates that the plant contribution to the disease process can be dissected in detail.

ACKNOWLEDGMENTS. The Sainsbury Laboratory is supported by the Gatsby Charitable Foundation. This work was supported in part by the Agricultural and Food Research Council. MJF thanks the Royal Society for sponsorship.

REFERENCES

Chang, C., Bowman, J.L., DeJohn, A.W., Lander, E., and Meyerowitz, E.M. (1988) Restriction fragment length polymorphism linkage map for *Arabidopsis thaliana*. Proceedings of the National Academy of Sciences of the United States of America 85, 6856-6860.

Collinge, D.B., Milligan, D.E., Dow, J.M., Scofield, G., and Daniels, M.J. (1987) Gene expression in *Brassica campestris* showing a hypersensitive response to the incompatible pathogen *Xanthomonas campestris* pv. *campestris*. Plant Molecular Biology 8, 405-414.

Collinge, D.B., and Slusarenko, A.J. (1987) Plant gene expression in response to pathogens. Plant Molecular Biology 9, 389-410.

Daniels, M.J., Barber, C.E., Turner, P.C., Sawczyc, M.K., Byrde, R.J.W., and Fielding, A.H. (1984) Cloning of genes involved in pathogenicity of *Xanthomonas campestris* pv. *campestris* using the broad host range cosmid pLAFR1. EMBO Journal 3, 3323-3328.

Davis, K.R., and Ausubel, F.M. (1989) Characterisation of elicitor-induced defense responses in suspension-cultured cells of *Arabidopsis*. Molecular Plant-Microbe Interactions 2, 363-368.

Dow, J.M., Scofield, G., Trafford, K., Turner, P.C., and Daniels, M.J. (1987) A gene cluster in *Xanthomonas campestris* pv. *campestris* required for pathogenicity controls the excretion of polygalacturonate lyase and other enzymes. Physiological and Molecular Plant Pathology 31, 261-271.

Estelle, M.A., and Somerville, C.R. (1986) The mutants of *Arabidopsis*. Trends in Genetics 2, 89-93.

Gough, C.L., Dow, J.M., Barber, C.E., and Daniels, M.J. (1988) Cloning of two endoglucanase genes of *Xanthomonas campestris* pv. *campestris*: analysis of the role of the major endoglucanase in pathogenesis. Molecular Plant-Microbe Interactions 1, 275-281.

Keen, N.T., and Staskawicz, B.J. (1988) Host range determinants in plant pathogens and symbionts. Annual Review of Microbiology 42, 421-440.

Liu, Y.-N., Tang, J.-L., Clarke, B.R., Dow, J.M., and Daniels, M.J. (1990) A multipurpose broad host range cloning vector and its use to characterise an extracellular protease gene of *Xanthomonas campestris* pathovar *campestris*. Molecular and General Genetics 220, 433-440.

Nam, H-G., Giraudat, J., den Boer, B., Moonan, F., Loos, W.D.B., Hauge, B.M., and Goodman, H.M. (1989) Restriction fragment length polymorphism linkage map of *Arabidopsis thaliana*. Plant Cell 1, 699-705.

Osbourn, A.E., Clarke, B.R. Stevens, B.J.H., and Daniels, M.J. (1990) Use of oligonucleotide probes to identify members of two-component regulatory systems in *Xanthomonas campestris* pathovar *campestris*. Molecular and General Genetics 222, 145-151.

Simpson, R.B., and Johnson, L.J. (1990) *Arabidopsis thaliana* as a host for *Xanthomonas campestris* pv. *campestris*. Molecular Plant-Microbe Interactions 3, 233-237.

Tang, J.-L., Gough, C.L., and Daniels, M.J. (1990) Cloning of genes involved in negative regulation of production of extracellular enzymes and polysaccharide of *Xanthomonas campestris* pathovar *campestris*. Molecular and General Genetics 222, 157-160.

Tsuji, J., and Somerville, S.C. (1988) *Xanthomonas campestris* pv. *campestris*-induced chlorosis in *Arabidopsis thaliana*. Arabidopsis Information Service 26, 1-8.

Williams, P.H. (1980) Black rot: a continuing threat to world crucifers. Plant Disease 64, 736-742.

EXOPOLYSACCHARIDES IN THE INTERACTION OF THE FIRE-BLIGHT PATHOGEN *ERWINIA AMYLOVORA* WITH ITS HOST CELLS

Klaus Geider, Peter Bellemann, Frank Bernhard, Jeong-Rhan Chang, Gebhard Geier, Marianne Metzger, Armin Pahl, Thomas Schwartz, Richard Theiler

Max-Planck-Institut für medizinische Forschung, Jahnstr. 29, D6900-Heidelberg, Germany

Summary
The fireblight pathogen *Erwinia amylovora* depends on synthesis of exopoly-saccharides in order to prevent defense reactions of the host plants. Mutants in genes for capsular polysaccharides, for levan surcrase and in the galactose metabolism were created by mutagenesis with transposon Tn5. A deficiency for levan sucrase did not affect virulence on pears. Mutants in synthesis of acidic exopolysaccharides (EPS) were apathogenic. The *gal*-mutant induced phytoalexins and callose production in plant cells. Both defense reactions were suppressed by dihydrophenylalanine, a phenylalanine analog secreted by some *E. amylovora* strains. This compound is synthesized in the shikimic acid pathway of the bacteria and is inhibitory in plants for synthesis of aromatic amino acids. The other mutants without EPS-synthesis were genetically and physically characterized. The insertions were mapped in a gene cluster, which is controlled by regulatory genes like the cloned and sequenced *rcs*A-gene.

Introduction
Erwinia amylovora is a Gram-negative bacterium, which causes fireblight of many rosaceous plants, especially of the genus *Pomoidae* [1]. Infected tissue first appears water-soaked, then dry and becomes black with necrotic symptoms. At moderate temperature and high humidity the bacteria may produce ooze in the infected tissue which protects them against plant defense reactions and against damage in case of dry environmental conditions. The extracellular polysaccharide (EPS) consists of galactose and glucuronic acid. EPS deficient mutants are avirulent and may be agglutinated by the host. The hypersensitive reaction of the plant includes many responses against the invading pathogen like the production of phytoalexins and the deposition of extracellular barriers like callose and lignin [2]. The bacteria damage their host plants heavily by barely understood mechanisms.

90

H. Hennecke and D. P. S. Verma (eds.),
Advances in Molecular Genetics of Plant-Microbe Interactions, Vol. 1, 90–93.
© 1991 *Kluwer Academic Publishers. Printed in the Netherlands.*

Results and Discussion

Relatedness of E. amylovora *strains and bacterial dihydrophenylalanine synthesis.*
The fireblight pathogen *E. amylovora* is member of a homogeneous group within
the genus Erwinia. It does apparently not secrete plant cell wall degrading enzymes like cellulases or pectate lyases. Restriction fragment length polymorphisms
(RFLP) in *E. amylovora* strains are rare. Strain E9 (isolated in the USA) can be
distinguished from strains Ea7/74, Ea 1/79, Ea11/88 (Germany), Ea273 (USA),
and Ea1496-66 (New Zealand) by a missing *Pst*I site in the common plasmid
pEA29, which can be used to identify *E. amylovora* [3]. A cloned fragment from
the chromosomal *cps*-gene cluster did not produce an RFLP with genomic DNA
from these strains.

Some strains produce a zone of growth inhibition around colonies on a lawn of
agar embedded pear cells. We have identified the toxic substance secreted by
these bacteria as (L-)2,5- dihydrophenylalanine. It is not produced in the presence
of aromatic amino acids by the positive *E. amylovora*-strains E9 and Ea273. Its
toxic effect in the shikimic acid pathway of plant cells can be suppressed by phenylalanine. When an *E. amylovora* strain was labelled by a transposon carrying the
lux-operon of *Vibrio fischeri* bacterial spreading on a lawn of pear cells could be
easily monitored. Although DHP [4] inhibited growth of pear cells, which were
separated by a membran from secreting bacteria, pear cells were killed in contact
with *E. amylovora* cells regardless if they produced DHP or not. Mutants in DHP
synthesis had no effect on virulence of the bacterium.

Avirulent mutants with a deficiency in EPS-production. From more than 1600 Tn5-
mutants elven were found apathogenic on pear slices and pear seedlings. Nine
were deficient for EPS-synthesis in minimal medium. Partially purified EPS was
separated on HPLC sizing columns and had a small size in contrast to high molecular weight EPS from the parent strains. Staining of the capsule with polycationic
ferritin and visualization in the electronmicroscope (together with S. Berger, MPI
für Zellbiologie, Ladenburg) demonstrated the absence of the EPS-layer for these
avirulent mutant strains.

Classification of the EPS-mutants. The sites of the Tn5-insertions were localized on
the genome of *E. amylovora* by DNA hybridization and by recloning the transposon together with adjacent DNA regions of the bacterial chromosome. Insertions
were found in different *Eco*RI fragments. They could be complemented with a
cloned 15 kb insert in a cosmid, when they were deficient in EPS-production. This
fragment also complemented *cps*B, C and D mutants of *E. stewartii* [5]. Other avirulent transposon mutants of *E. amylovora* could not be complemented by this
cosmid clone. Such a mutant was able to partially restore virulence in a mixed
inoculation of pear slices with an EPS-deficient mutant.

Regulation of exopolysaccharide synthesis in E. amylovora. We have used a conjugative gene bank of *E. amylovora* to restore the mucoid colony type of EPS-defective mutants of *Erwinia stewartii* [6]. A DNA fragment with the *E. amylovora rcs*A gene complemented *rcs*A mutants of *E. stewartii* and of *E. coli*. By deletion analysis of the DNA-fragment the *rcs*A gene was localized and sequenced. An open reading frame corresponding to 211 amino acids was 55 % homologous to the *rcs*A gene of *Klebsiella aerogenes*. A Tn5-insertion in the *E. amylovora rcs*A gene produced a mutant with reduced EPS-synthesis in minimal medium and low virulence on pear slices. Synthesis of levan was also altered in this mutant. The mutation could be partially complemented by cloned *rcs*A-genes from *Escherichia coli* and *E. stewartii* for EPS and levan synthesis, but not for virulence on pear slices. The cloned *rcs*A-gene from *E. amylovora* restored both deficient properties [6].

A gal-*mutant affecting EPS-synthesis*. *E. amylovora* mutants were created with transposon Tn5, which was located on a pfd-plasmid and transformed into the bacteria by electroporation. One mutant failed to grow on galactose as the sole carbon source, but grew normally with other carbon sources like sorbitol and fructose. In media, allowing good EPS-synthesis for wildtype strains, this mutant strain was unable to produce a capsule. EPS-synthesis was restored by addition of galactose to the medium which had to contain also another carbon source than galactose. The genetic defect could be complemented by a plasmid carrying the *gal*E-gene of *E. stewartii* [7]. This indicated an insertion of Tn5 into the *gal*E-gene of the *E. amylovora* strain. Correspondingly, the mutant was deficient for UDP-galactose 4-epimerase activity, an enzyme of the Leloir pathway.

The mutant investigated caused no disease symptoms on plantlets and on immature pears. When the plant tissue was soaked with galactose, virulence could be restored. In contrast to wildtype strains the mutant heavily induced defense reactions on cultivated pear cells. PAL-activity could be measured first two hours after inoculation with the bacteria and reached its maximum at four hours. The enzyme is a key enzyme in the onset of plant defense reactions and converts phenylalanine to cinnamic acid. The response of plant cells to the bacteria was confirmed by production of callose in the pear cells after inoculation with the *E. amylovora* mutant. Synthesis of callose is a later step than PAL-production in the plant defense response. Wildtype *E. amylovora* strains do not induce callose synthesis on pear cells. Another reaction of cultivated pear cells to the mutant strain was rapid appearance of necrosis unlike mild necrosis observed with wildtype strains. This agrees with the inability of the mutant to produce acidic exopolysaccharide.

The role of levan synthesis in virulence of E. amylovora. Levan is polymerized from sucrose by the enzyme levan-sucrase which is constitutively secreted by *E. amylo-*

vora strains [8]. Levan is a ß-2,6-linked fructan. Tn5-mutants lacking this enzyme were assayed on immature pear slices and on pear seedlings. Whereas virulence on pears was normal. Some retardation of symptom development was observed on the seedlings. The mutations for levan-sucrase did not alter synthesis of acidic EPS.

A model for fireblight induction on Rosaceae. After infection of host plants *E. amylovora* tries to colonize the plant surface and to invade the tissue. In order to obtain sufficient nutrients from the host it may have to induce cell leakage in its environment. This may be achieved by acapsulated cells in the bacterial population which are in equilibrium with capsulated cells. If this balance is disturbed it seems possible to reduce fireblight on an infected plant 8 [9]. Nutrients from the plant cells support the pathogen in production of EPS. This shields the bacteria against plant defense reactions and allows *E. amylovora* to spread in the plant tissue. An inability to produce EPS would of course halt its propagation due to a massive HR of the infected plant. A lack in nutrients at a late stage of disease developement will also cause a decrease in EPS-production thus causing severe necotic plant reactions, the typical fireblight symptoms.

REFERENCES

[1] Van der Zwet, T., Keil, H.L.: United States Department of Agriculture, Agriculture Handbook Number 510, pp. 200 (1979)

[2] Hahlbrock, K., Scheel, D.: In: Innovative approaches to plant disease control (I. Chet, Editor), J. Wiley & Sons, Inc. (New York - London - Sydney - Toronto) pp. 229-254 (1987)

[3] Falkenstein, H., Bellemann, P., Walter, S., Zeller, W., Geider, K.: Applied and Environmental Microbiology **54**, 2798-2802 (1988)

[4] Feistner, G. J.: (L)-2,5-Dihydrophenylalanine from the fireblight pathogen *Erwinia amylovora*. Phytochemistry **27**, 3417-3422 (1988)

[5] Coplin, D.L., Majerczak, D.R.: Molecular Plant-Microbe Interactions, in press

[6] Bernhard, F., Poetter, K., Geider, K., Coplin, D.: Molecular Plant-Microbe Interactions, in press

[7] Dolph, P.J., Majerczak, D.R., Coplin, D.L.: J. Bacteriology **170**, 865-871 (1988)

[8] Gross, M., Geier, G., Geider, K., Klaus Rudolph, K.: Proceedings of the 7. International Conference on Plant Pathogenic Bacteria, Budapest (Hungary), in press

[9] Hignett, R.C.: Physiol. Mol. Plant Pathology **30**, 131-138 (1987)

Acknowledgement: This work was supported in part by grants from Stiftung Volkswagenwerk to P.B. and R.T.

IRON AS A MODULATOR OF PATHOGENICITY OF
ERWINIA CHRYSANTHEMI 3937 ON *SAINTPAULIA IONANTHA*

K. Sauvage, T. Franza and D. Expert
Laboratoire de Pathologie Végétale, INA-PG, 16 rue Claude Bernard
75231 Paris Cédex 05 France

To incite a systemic soft rot disease on saintpaulia plants, *E. chrysanthemi* requires a set of pectinolytic activities (PL) and a high affinity iron transport system to be functional. Chrysobactin, the siderophore of this system [10], is believed to provide iron to invading bacteria in plant vessels, an environment which is low in free iron. This uptake system is negatively controlled by the transcriptional regulator Cbr, effective under non restrictive iron conditions. Here, we show that iron stress also stimulates the production of all pectate lyase activities. In wild type cells, this fact is correlated with increased levels of PL specific mRNA. In the Cbr^- background, high levels of PL mRNA were also detected in replete iron conditions, but enzymatic activity was no longer found to be proportional.

Introduction

Pectinolytic erwinias, a group of the Enterobacteriacae, cause varied disorders on many plants, including soft rots, vascular wilts and parenchymental necrosis. *E. chrysanthemi* 3937 is a good candidate to study gene modulation during pathogenesis since a set of mutants unable to develop a systemic response on saintpaulia plants have been isolated from this strain. Mutants that have lost one or several of the 5 pectate lyase isozymes (PLa-e) or the pectin methyl esterase but are still able to macerate isolated organs [5, 4] might be useful especially since the relative importance of each isoenzyme seems to vary with the plant host (F. Van Gijsegem, pers. comm.). Mutants that are unable to acquire iron under restrictive conditions because of their failure to produce chrysobactin or to transport the ferric chelate of this siderophore across the enveloppe are also blocked in the early steps of the disease [6].

Iron is not available in most natural environments as for instance in extracellular locations of the mammalian body where this vital ion is bound to specific transport proteins (transferrin and lactoferrin). Consequently, production of siderophores by animal pathogenic bacteria appears to be important since iron availability may determine the final outcome of an infection [12]. Similarly, the low level of iron in the intercellular fluid of saintpaulia plant leaves requires a high affinity iron transport system from *E. chrysanthemi* as checked by *in planta* bacterial growth [7].

We are currently investigating the genetics of the chrysobactin system. We identified a genetic unit of about 8 kb in length that includes the genes for the 4 primary steps of chrysobactin biosynthesis (*cbsABCE*) and the gene for the outer membrane receptor. The *cbr* regulatory locus encodes a *trans* acting factor that in the presence of iron negatively controls the transcription of these loci. Here, we show that iron also regulates *pel* genes expression *via* the *cbr* network.

94

H. Hennecke and D. P. S. Verma (eds.),
Advances in Molecular Genetics of Plant-Microbe Interactions, Vol. 1, 94–98.
© 1991 *Kluwer Academic Publishers. Printed in the Netherlands.*

Material and Methods

The strains *E. chrysanthemi* 3937, and 3937 *cbr-2* a MudII1734 lysogenic derivative, as well as the media were described previously [6, 7]. Plasmids pUV70, a pelB$^+$ derivative of pUC9 [5], pB41, a pelADE$^+$ derivative of pUC8 [11] and pTF4, a *cbs*+ 8 kb *KpnI* fragment cloned in pTZ19R (Franza et al., in prep.)were used as [α32P] dCTP labeled DNA probes. PL isozymes were analysed by electrofocusing as described by Bertheau et al.[3]. PL activity was assayed at 235 nm using the method of Moran et al. [9] with polygalacturonate (PGA) as substrate. Specific activity is expressed as μmol of unsaturated oligogalacturonides produced per min per mg of cell dry weight. Total RNA was prepared by the method of Apel and Kloppstech [2] with minor modifications. For dot blots, 4 μl loopfuls of varying dilutions of an RNA preparation (2 μg/μl) denatured with formaldehyde were spotted onto nylon membranes. Blots were hybridized with DNA probes labeled by nicktranslation, washed under high stringency conditions. Autoradiograms were scanned using a Shimadzu DR2 recorder CS930.

Results

1) Iron limitation stimulates the production of total PL activity in 3937 but not in 3937 *cbr-2*.

We looked at the possible effect of iron on the production of global PL activity from wild type and mutant cells grown with 0.4 % PGA, the natural inducer of *pel* genes and in the presence of increasing amounts of FeCl3 up to 10 μM. Below 2 μM, iron is critical for the growth of erwinia cells. PL activity was assayed in culture supernatant fluids, over time during the late exponential and early stationary stages of bacterial growth as reported previously [1]. Figure 1 shows that for both strains and in any iron conditions, PL activity was externally released with the same kinetics. However, in the wild type strain the PL specific activity appeared to decrease with increasing levels of iron. The effect was particularly apparent in the 0-0.5 μM range that covers severe iron restrictive conditions. In the mutant, no variation of external iron significantly influenced the PL production which was always found to be 2-fold lower than in strain 3937. In all cases, the relative PL activity associated with the cell compartiments was the same, thus ruling out of a possible iron effect on the secretion of these enzymes. Analysis of the supernatants by gel electrofocusing failed to detect any qualitative change with respect to the 5 isoforms identified in strain 3937. Whereas with 2μM iron, each isozyme appeared to be produced in lower levels, with perhaps a major effect on PLb and PLe.

2) Effect of iron on *pel* genes transcription

The substantial decrease in PL production with increasing iron in the growth medium might result from a transcriptional control on *pel* expression. We checked this possibility by comparing PL mRNA from iron replete and deplete cells of both strains. PL mRNA was assessed by RNA/DNA dot blot hybridization analysis, using pUV70 and pBN41 DNA as probes. The basal level expression of *pel* genes was controlled using mRNA isolated from cells grown with no PGA. In parallel, mRNA homologous to *cbs* DNA was assessed with pTF4 as a DNA probe. The results of this analysis with the corresponding enzymatic activities are presented in figure 2. In iron deplete wild type cells grown with PGA, RNA homologous to *pelADE* and *pelB* DNA was 3-fold higher than in cells grown in replete iron, a result that corresponds to the increase in enzymatic activity. In mutant cells, levels of mRNA in both conditions were shown to be the same as in iron starved wild type cells though PL activity was 2-fold lower. Such a discrepancy between mRNA levels and protein activity was also found a lesser degree when cells were not induced by PGA.

Conclusion and Discussion

The presented data show the existence of a transcriptional control of *pel* genes by iron. This control appears to depend on the *cbr* locus since in the mutant strain PL mRNA was synthetized constitutively at high levels. Whether or not *cbr* directly acts on *pel* genes expression as a possible repressor in the presence of iron still remains to be demonstrated. In any case, the Cbr product would play the role of a global regulator, controlling in addition to iron transport functions *per se*, virulence functions that might be "advantageously" induced in low iron environments. Otherwise, the production of the pectin degrading enzymes is modulated by several regulatory elements to which no clear function has yet been assigned [8]. The presence of high quantities of mRNA homologous to *pel* DNA in strain 3937 *cbr-2* and to a lesser extent in the PGA non induced wild type strain that are not correlated with high level of protein activity may suggest the existence of some postranscriptional control.

Figure 1. PL activity in culture supernatant during bacterial growth with various amounts of iron.

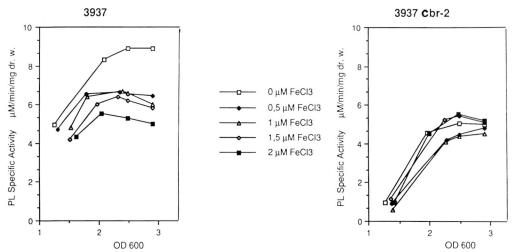

Figure 2. Effect of iron (+ Fe 5 μM, - Fe no Fe
added) on *pel* and *cbs* gene transcription.
Corresponding PL activities in culture medium (at
OD600 = 2) are shown.

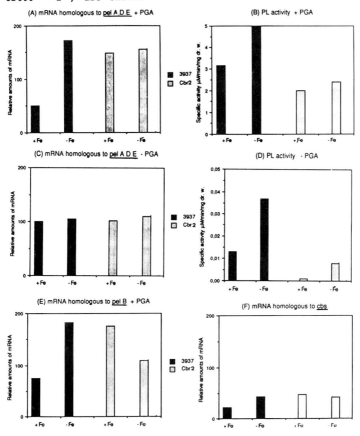

Acknowledgements:
We are pleased to thank here our close colleagues Y. Bertheau and M. Boccara as well as F. Van Gijsegem and the Robert-Baudouy's group for fruitful discussions. This work was supported by INRA funds.

References:
1 Andro T., Chambost J P., Kotoujansky A., Cattaneo J., Bertheau Y., Barras F., Van Gijsegem F. and Coleno A. (1984) Mutants of *E. chrysanthemi* defective in secretion of pectinases and cellulases, J. Bacteriol. 160, 1199-1203.
2 Apel K. and Kloppstech K. (1978) Eur. J. Biochem. 85, 581-588.
3 Bertheau Y., Madjidi-Hervan E., Kotoujansky A., Nguyen-The C., Andro T. and Coleno A. (1984) Detection of depolymerase isoenzymes after electrophoresis or electrofocusing or in titration curves, Anal. Biochem. 139, 383-389.
4 Boccara M. and Chatain V. (1989) Regulation and role of *E. chrysanthemi* 3937 pectin methylesterase, J. Bacteriol. 171, 4085-4087.
5 Boccara M., Diolez A., Rouve M. and Kotoujansky A. (1988) The role of individual pectate lyases of *E. chrysanthemi* in pathogenicity on saintpaulia plants, Physiol. Mol. Plant Pathol. 33, 95-104
6 Enard C., Diolez A. and Expert D. (1988) Systemic virulence of *E. chrysanthemi* 3937 requires a functional iron assimilation system J. Bacteriol. 170, 2419-2426.
7 Enard C., Franza T., Neema C., Gill P. R., Persmark M., Neilands J. B. and Expert D. (1990) The requirement of chrysobactin dependant iron transport for virulence incited by *E. chrysanthemi* on *Saintpaulia ionantha*. J. Plant Nutr. in press.
8 Hugouvieux-Cotte-Pattat N. and Robert- Baudouy J. (1989) Isolation of *E. chrysanthemi* mutants altered in pectinolytic enzyme production, Mol. Microbiol. 3, 1587-1597.
9 Moran F., Nasuno S. and Starr M. P. (1968) Extracellular and intracellular polygalacturonic acid trans-eliminases of *E. carotovora*. Arch. Biochem. Biophys. 123, 298-306.
10 Persmark M., Expert D.and Neilands J. B. (1989) Isolation, characterization and synthesis of chrysobactin, a compound with siderophore activity from *E. chrysanthemi*. J. Biol. Chem. 264, 3187-3196.
11 Reverchon S., Van Gijsegem F., Rouve M., Kotoujansky A. and Robert-Baudouy J. (1986) Organization of a pectate lyase gene family in *E. chrysanthemi,* Gene 29, 215-224.
12 Weinberg E. D. (1984) Iron withholding: a defense against infection disease, Physiol. Rev. 64, 65-102.

GENETIC AND PHYSIOLOGICAL ASPECTS OF THE PATHOGENIC INTERACTION OF *CLAVIBACTER MICHIGANENSE* SUBSP. *MICHIGANENSE* WITH THE HOST PLANT

R. EICHENLAUB, A. BERMPOHL AND D. MELETZUS.
Universität Bielefeld, Fakultät für Biologie,
Gentechnologie/ Mikrobiologie, Postfach 8640,
D-4800 Bielefeld 1, FRG.

ABSTRACT. The tomatopathogen Clavibacter michiganense subsp. michiganense NCPPB382 can be shown to produce a wilt inducing exopolysaccharide (EPS). The EPS has been purified. It consists of fucose (38.8%), galactose (30.1%), and glucose (31.1%). A plasmid curing was performed to determine whether the two plasmids (pCM1 and pCM2) carried by the strain were involved in pathogenicity. Derivatives of NCPPB382 lacking both plasmids were found to be apathogenic, exhibiting only a slight reduction of the biomass. However, they are able to effectively colonize the plant and can produce the toxic EPS in culture. Therefore, it appears possible that EPS production in planta is repressed under certain conditions. Our observations suggest that the plasmids may carry determinants responsible for the development of the disease. In order to be able to identify these gene loci a transformation procedure and a cloning vector for Clavibacter have been developed. This system has been successfully employed to clone the plasmid pCM1 encoded endocellulase.

1. Introduction

The coryneform, Gram-positive bacterium Clavibacter michiganense subsp. michiganense NCPPB382 is a pathogen specific for the tomato (Lycopersicon esculentum) causing bacterial wilt and canker. Infection occurs through wounds in the root or stem region followed by a colonization of the xylem vessels of the plant by the bacteria. First visuable disease symptoms are observed in susceptible tomato plants (cv. Moneymaker) about 12 days after infection. Wilt inducing toxins representing high molecular weight glycopeptides have been described as the causal agent of the disease (1, 2,). Although glycopeptides found in the culture fluid of C. m. subsp. michiganense have been shown to cause wilting in bioassays with tomato epicotyle cuttings, their production in planta has never been demonstrated.
The objectives of the present study were to reinvestigate the wilt inducing toxin, to test a possible involvement of endogenous plasmids in pathogenicity and to develop methods for gene cloning in Clavibacter michiganense.

2. Results and Discussion

2.1. CHARACTERIZATION OF THE EPS.

The cell free culture fluid of C. m. subsp. michiganense NCPPB382 contains a wilt inducing compound which can be detected in a bioassay. In this test 4 week old tomato cuttings (about 10 cm long with 4-6 leaves) are placed into 1 ml of test solution containing various

H. Hennecke and D. P. S. Verma (eds.),
Advances in Molecular Genetics of Plant-Microbe Interactions, Vol. 1, 99–102.

amounts of the culture fluid and are incubated for 8 h. The weight loss of the cuttings after the incubation, as compared to the control incubated in medium, gives an estimate for the toxic activity of a preparation. If the bacteria were grown on plates with potato extract medium they produced large amounts of exopolysaccharides allowing excellent yields of the wilt inducing activity. The EPS was extracted from the material collected from the plates with phosphate buffer and precipitated with ethanol (3). The wilt inducing EPS had a molecular weight of 100-300.000 as determined by ultrafiltration and gel permeation chromatography on a HPLC BioSil TSK-250 column and represented a polysaccharide with a protein contamination of less than 1%. For carbohydrate analysis the EPS was hydrolyzed with 66% trifluoracetic acid (4 h, 100°C). Sugar monomers were separated on a Carbopack-PA1 column by isocratic elution with 10mM NaOH and detected after postcolumn addition of 300 mM NaOH using a pulsed amperometric detector (Dionex PAD2). In contrast to earlier reports (4, 5) we detected only three sugar residues, fucose (38.8%), galactose (30.1%), and glucose (31.1%). The sugar composition was constant and the distribution did not vary by more than 5% in different preparations.

In tomato cuttings treated with the EPS water uptake was rapidly blocked which could be visualized by the addition of methyleneblue to the test solution. This suggests that wilting may result from plugging of the xylem vessels. Alternatively, the EPS could induce the formation of tyloses by the parenchyma cells of the xylem which block water transport thereby preventing the spreading of toxic compounds or pathogenic organisms in the plant (6).

2.2. INVOLVEMENT OF ENDOGENOUS PLASMIDS IN PATHOGENICITY.

Strain NCPPB382 carries two plasmids, termed pCM1 and pCM2, of 27.5 kb and 72 kb, respectively. In order to determine whether these plasmids are involved in pathogenicity a plasmid curing was performed. By growth at elevated temperature (33°C instead of 26°C) three classes of cured derivatives were obtained, one without any plasmid (reference strain CMM100) and two others carrying either one of the two plasmids (reference strains CMM101 pCM1 and CMM102 pCM2).

The cured derivatives were compared with the parent strain for their virulence after infection of tomato plants. The wilting index, the colonization, and the plant biomass were determined. Also the production of the wilt inducing EPS and an endocellulase were assayed. The results are summarized in Table 1.

The data show that the plasmid free derivative CMM100, although able to effectively colonize the tomato, is not causing any disease symptoms. The plants are healthy and the biomass is only slightly reduced. Surprisingly, the ability to produce the wilt inducing EPS in culture is not impaired in this strain. Therefore, it is possible that under certain conditions this component is not sufficiently expressed in planta.

The derivatives carrying either one of the two plasmids CMM101 and CMM102 exhibit a reduced virulence as compared to the wild type NCPPB382. Wilting occurs but with a retardation of 4-6 days, respectively. The data suggest that the two plasmids play an essential role in the development of the disease symptoms. Unfortunately, a restoration of the pathogenic phenotype in CMM100 by reintroduction of the plasmids is prevented by the size of the plasmids and the lack of any selective marker. At present the only potentially pathogenic determinant linked to the presence of a plasmid is the endocellulase encoded by plasmid pCM1. By cloning of genes located on the plasmids we hope to identify further plasmid factors involved in pathogenicity.

Table 1.   Phytopathogenic properties of strain NCPPB382 and
                three curing derivatives.

Strain	Toxin Activity	Wilting Index	Cellulase	Colonization	Biomass
NCPPB382	15.0/4.6	12	+	2.9×10^9	35.2 %
CMM100	19.2/6.0	-	-	1.2×10^9	75.0 %
CMM101 pCM1	17.1/7.8	18	+	3.6×10^9	62.2 %
CMM102 pCM2	18.7/8.1	16	-	4.9×10^9	49.8 %

Toxin activity: Weight loss of tomato cuttings in percent after incubation for 8 h. First value undiluted, second value 1:10 diluted culture filtrate. **Wilting index**: Days after infection at which 50 % of the plants show wilting symptoms (n = 30). **Cellulase**: Endocellulase activity determined on CMC-agar with indicator Congo Red. **Colonization**: CFU/g plant tissue five weeks after infection. **Biomass**: Dry weight of tomato plants five weeks after infection as compared to control plants (100%) (n = 10).

2.3. DEVELOPMENT OF A TRANSFORMATION PROCEDURE AND CONSTRUCTION OF A CLONING VECTOR.

In an attempt to develop a method for transformation of Clavibacter michiganense we first generated a series of hybrid plasmids consisting of pBR322 and pBR325 derivatives with the neomycin/kanamycin and gentamycin resistance genes carrying overlapping restriction fragments of plasmid pCM1. These were used in transformation experiments (protoplast/ PEG-method) with the plasmid free strain CMM100. With one hybdrid, pDM3212, we obtained transformants. Subsequently, the 13.5 kb insert of pDM3212 with the pCM1 replicon was reduced to a 3.2 kb DNA fragment and used to construct the E. coli - C. michiganense shuttle vector pDM100 (Figure 1).
Transformation rates with the protoplast-PEG method were always very poor (20 transformants/ug DNA). Therefore, we tested electroporation as an alternative method. A dramatic improvement was observed. A pulse of 12.5 kV/cm for 13.5 ms gave about 2×10^3 transformants/ug DNA. By this procedure it became possible to establish the vector pDM100 also in Clavibacter iranicus, Clavibacter michiganense subsp. nebraskense and C. m. subsp. insidiosum.

2.4. CLONING OF A PLASMID ENCODED ENDOCELLULASE GENE.

In a first test of the Clavibacter cloning system we succeeded to clone the endocellulase gene of plasmid pCM1. Clones expressing the enzyme were easily detected on carboxymethylcellulose agar plates by staining with Congo Red. A 3.1 kb DNA fragment has been identified to carry the endocellulase gene, which is expressed in Clavibacter michiganense CMM100 and also, however weaker, in E. coli. Sequencing revealed one open reading frame of 1.311 bp coding for a protein of 437 aminoacids and a molecular weight of 46.454. Since deletions extending into the carboxy-terminus of the hypothetical protein result in a loss of the endocellulase activity, we assume that the open reading frame indeed represents the endocellulase gene.

Comparison of the strains (Table 1) and the behaviour of CMM100 with the cloned endo-cellulase gene indicates that this gene has only a weak effect on pathogenicity.

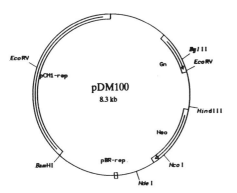

Figure 1. Physical map of the E. coli - Clavibacter shuttle vector pDM100. pCM-rep., internal 3.2 kb BamHI-BglII restriction fragment of pCM1 inserted into the single BamHI site of the pBR325 derivative pDM10. Gn - gentamycin resistance; Neo - neomycin resistance.

Since we have succeeded to establish a method for molecular cloning in Clavibacter it becomes feasible to identify essential factors for the pathogenic interaction between this microorganism and the plant in future studies.

ACKNOWLEDGEMENT. This work was supported by a grant from the Deutsche Forschungsgemeinschaft (SFB 223).

References

1. Rai, P.V. and Strobel, G.A. (1968) Phytotoxic glycopeptides produced by Corynebacterium michiganense. II. Biological properties. Phytopathol. 59, 53-57.

2. Van den Bulk, R.W., Löffler H.J.M. and Dons, J.J.M. (1989) Effect of phytotoxic compounds produced by Clavibacter michiganense subsp. michiganense on resistant and susceptible tomato plants. Neth. J. Pl. Path. 95, 107-117.

3. Coplin, D.L., Frederick, R.D., Majerczak, D.R. and Haas, E.S. (1986) Molecular cloning of virulence genes from Erwinia stewartii. J. Bacteriol. 168, 619-623.

4. Rai, P.V. and Strobel, G.A. (1968) Phytotoxic glycopeptides produced by Corynebacterium michiganense. I. Methods of preparation, physical and chemical characterization. Phytopathol. 59, 47-52.

5. Krämer, R. (1986) Darstellung und Teilcharakterisierung des Toxins von Corynebacterium michiganense pv. michiganense (Smith) Jensen. Zentralbl. Mikrobiol. 141, 327-335.

6. Wallis, F.M. (1977) Ultrastructural histopathology of tomato plants infected with Corynebacterium michiganense. Physiol. Plant Pathol. 11, 333-342.

DNA PROBES AS TOOLS FOR THE STUDY OF HOST-PATHOGEN EVOLUTION: THE EXAMPLE OF *PSEUDOMONAS SOLANAACEARUM*

DOUGLAS COOK, ELIZABETH BARLOW, and LUIS SEQUEIRA
Department of Plant Pathology, University of Wisconsin-Madison, WI, USA

1. Introduction

The use of DNA probes has made it possible to analyze the genomes of plant-pathogenic microorganisms isolated from virgin as well as cultivated areas, with the purpose of understanding how strains evolved in nature and how humans have perturbed the natural system. The bacterial plant pathogen, *Pseudomonas solanacearum*, represents a particularly useful system for these types of analyses because this bacterium has been found in pristine environments and the distribution of some strains on native plants and in virgin soils has been examined (Berg, 1971; Buddenhagen, 1960, 1985; Martin & French, 1977; Martin *et al* 1981; Sequeira & Averre, 1961). In addition, there is documented recent history of epidemics caused by: a) the introduction of new, susceptible hosts into virgin areas where the bacterium exists in association with native weeds and, b) the introduction of certain strains of the bacterium into geographic areas where these strains previously did not occur and where they developed the capacity to attack new host plants of great economic importance (Buddenhagen, 1985).

The purpose of this paper is to illustrate how the use of DNA probes that specify factors important in host-parasite interactions can help us understand the variability in *Pseudomonas solanacearum* and how this information can give us a glimpse of possible evolutionary trends in this species.

2. Variability of *Pseudomonas solanacearum*

In tropical, subtropical, and warm temperate regions of the world, bacterial wilt caused by *P. solanacearum* can be a major limiting factor in the cultivation of many agronomically-important plants. Over the past 50 yr there has been substantial effort to develop resistant varieties of tobacco, potato, tomato, eggplant, etc., but bacterial wilt is still a limiting factor in tropical agriculture because new and more virulent strains have appeared over the years. The difficulty in developing stable, improved resistant cultivars is due, in large part, to the extreme variability of the pathogen. This variability is only partially reflected in the present classification schemes for *P. solanacearum*, which divide the species into three major races (on the basis of host range) and four biovars (on the basis of physiological characteristics) (Buddenhagen *et al*, 1962; Hayward, 1964). Potato varieties developed with resistance to race 3, for example, often are not resistant to race 1 or even to other strains within race 3. In addition, diversity of the pathogen is often correlated with its geographic distribution. Thus, resistant potato varieties developed for use in Peru may not be suitable for cultivation in Kenya, Sri Lanka, or Australia.

The biovar system of classification (Hayward, 1964) divides the species into four major groups on the basis of biochemical differences, but each biovar contains strains with different host ranges, and host range transects the biovars. Only in the case of biovar 2 and race 3 is there an apparent correlation between the biovar and race systems of classification, but even in that case the correlation breaks down, as indicated later in this paper. The race system may indicate the existence of natural groupings because race 2 is restricted to members of the Musaceae and race 3 is mostly a pathogen of potato. Race 1, however, contains strains with an extremely wide host range and encompasses three different biovars (1, 3, and 4). It seems likely that there are several natural groupings within race 1 that cut across the biovar classification system. Thus, traditional methods of classification do not account for the natural variability in *P. solanacearum*. For this reason, we have utilized restriction fragment length

103

H. Hennecke and D. P. S. Verma (eds.),
Advances in Molecular Genetics of Plant-Microbe Interactions, Vol. 1, 103–108.
© 1991 *Kluwer Academic Publishers. Printed in the Netherlands.*

polymorphism (RFLP) analysis in an effort to resolve the apparent taxonomic complexity of the species.

3. RFLP Analysis of *P. solanacearum*

Recently, we began an analysis of restriction fragment polymorphism (RFLP) within *P. solanacearum* with the use of nine DNA probes, seven of which encode information essential for virulence and the hypersensitive response (*hrp* genes). In an initial study involving 62 strains of *P. solanacearum* we established 28 distinct RFLP patterns that could be grouped into two different divisions with similarity coefficients of only 13.5% between them (Cook, *et al*, 1989). Division I contains all members of race 1, biovars 3, 4, and 5; division II contains all members of race 1, biovar 1, and races 2 and 3. Within each division, similarity coefficients were high and many of the RFLP groups were highly correlated with geographical location and less frequently with host of origin. For example, groups 21-23 represented only Asiatic strains isolated from ginger; similarly, groups 19-20 included only strains isolated from mulberry in China (Table 1).

We have now extended the RFLP analysis to a total of 150 strains of *P. solanacearum* from a worldwide range of hosts and geographical locations, including 62 additional strains of Australasian and Latin American origin. In general, the RFLP patterns were as determined previously, except for a few additional RFLP groups that accounted for strains on new hosts or at different locations. For example, one additional group corresponded to strains originally isolated from bananas in Java and Sulawesi by Dr. Simon Eden-Green (Rothamsted Experimental Station, England). A significant finding, however, is the fact that most of the Australasian strains fell within the RFLP groups previously defined and that correspond to the three biovars (2, 3, and 4) that are represented in that region. In spite of obvious polymorphisms in strains from different hosts and geographic origin, the basic RFLP patterns of each of the biovars were maintained.

Similarity coefficients for all pairwise combinations of 30 of these RFLP groups were calculated by the procedure of Nei and Li (1979). Cluster analyses were completed by the SAS clustering program and are depicted as a phenogram in Fig. 1.

Table 1. Characteristics of RFLP groups.

RFLP group	division	race	host	geographical origin
1	I	1	t, p, z	NA, K
2	I	1	t	N A
3	I	1	p, tb, b	NA, CA
4	I	1	p	CA
5	I	1	mp	CA
6	I	1	t b	CA
7	I	1	t b	SA
8	II	1	pr	CA
9	II	1	t	SA
10	II	1	t, p	A
11	II	1	ol, pn, g, e, p	C, P
12	II	1	t b	A
13	II	1	eo	CA
14	II	1	p	CA
15	II	1	p	SL
16	II	1	m, pn, t	C
17	II	1	t b	N A
18	II	1	ol	C
19	II	1	m	C
20	II	1	m	C
21	II	1	g	A
22	II	1	g	A
23	II	1	g	C
24	I	2	plt, b, h	CA
25	I	2	plt	SA
26	I	3	p	SR, A, I
27	I	3	p	SA
28	I	2	plt, b	SA
29	I	*	p	SA
30	I	*	p	SA
31	I	*	p	SA
32	I	*	soil	SA
33	I	*	p	SA

Host: t, tomato; z, *Zebrina*; p, potato; t, tobacco; b, banana; mp, *Melampodium perfoliatum*; pr, pepper; ol, olive; p, peanut; g, ginger;, e, eggplant; eo, *Eupatorium odoratum*; m, mulberry; plt, plantain; h, *Heliconia*. Geographical origin: NA, North America; K, Kenya; CA, Central America; SA, South America; A, Australia; C, China; P, Philippines; SL, Sri Lanka; I, Israel.

Figure 1. Average linkage cluster analysis of 30 RFLP groups of *P. solanacearum*

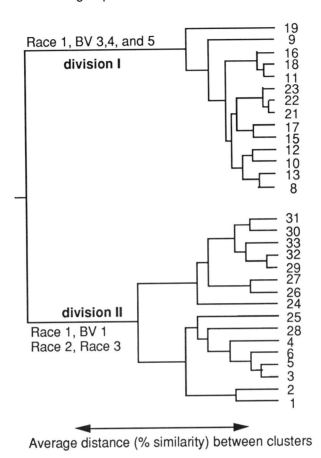

Average distance (% similarity) between clusters

Single RFLP groups include strains that have 100% similarity by RFLP analysis (far right of figure). Similarity decreases as the RFLP groups are clustered, moving leftward in the figure. Division I and II are joined at 13.5% similarity.

4. Phylogenetic and Evolutionary Relationships

The fact that *P. solanacearum* can be separated into two major divisions (Fig. 1) was not predicted by traditional methods of classification. It seemed likely that these two major divisions, as well as other subdivisions, reflected phylogenetic or evolutionary relationships. For example, 90% (45 out of 50) of the strains from division I were from Asia and Australia, while 98% (58/59) of strains from division II were from the Americas. This suggests that *P. solanacearum*, early in evolution, was divided into two geographically distinct populations. It is apparent that, within the New World and the Old World, multiple strains of the pathogen developed in isolated geographic areas, as represented by separate RFLP groups. Geographic isolation, therefore, was probably a significant factor in the divergent evolution of the species. This can be inferred from consideration of the RFLP groups that encompass race 3 and in race 2, as discussed below.

In the data regarding the geographical distribution of strains in the two divisions, we excluded from the calculations all race 3 strains originating from outside South America. The reason is that the RFLP data indicate clearly that these strains may have been introduced into this region from South America, probably via infected tubers. Race 3 is endemic in the Andean regions of South America and constitutes a homogenous group (RFLP groups 26 and 27, Fig. 1). The RFLP data support the notion that race 3 originated in geographic isolation in the Andean highlands, the center of origin of the potato (Buddenhagen, 1985). From a purely statistical point of view, it is highly unlikely that strains with the same RFLP patterns could have originated independently in South America and in Asia, as some researchers believe (Seneviratne, 1969).

The opposite argument can be made in the case of race 2. In this instance, the RFLP data provide evidence for parallel evolution of strains that are capable of attacking bananas in Central and South America, as well as in Asia. Most strains of race 2 apparently evolved as pathogens of *Heliconia* species in the Caribbean area, as evidenced by numerous instances of recovery of the pathogen from virgin areas (Buddenhagen, 1960; Sequeira & Averre, 1961). Several of these strains became pathogens of bananas and plantains when these crops were introduced into the Americas. The evolution of this race as geographically distinct clonal populations is exemplified by RFLP groups 24, 25, and 28, each associated with a recent, different epidemic on bananas and plantains in Central and South America (Fig. 2). Until very recently, it was thought that race 2 was not present outside of this area, except for a well-documented case of movement of infected banana rhizomes from Central America to the Philippines (Buddenhagen, 1985). However, RFLP analysis indicates that strains recently isolated from bananas in Southeast Asia (blood disease) constitute a novel RFLP group, clearly separable from the strains that attack bananas in Latin America and traditionally grouped in race 2. This suggests parallel evolution of strains capable of attacking members of the Musaceae at these two widely-separate geographic locations.

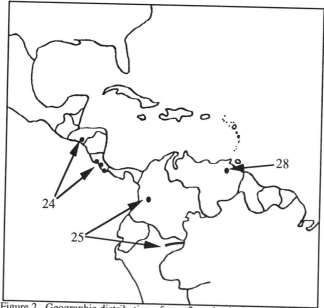

Figure 2. Geographic distribution of race 2 strains in South and Central America. Numbers represent RFLP groups in Table 1: 24, Honduras and Costa Rica; 25, Colombia and Peru; 28, Venezuela. Peruvian strains of RFLP group 25 were collected from the Amazon valley.

A possible complication in this picture of clonal evolution of the pathogen at isolated geographic locations is the existence of biovar 2 strains that are indigenous to the Amazon basin (Martin *et al.* 1981).

Since, according to the current dogma, biovar 2 is synonymous with race 3 and race 3 is indigenous to the Andean highlands, how do we explain the presence of biovar 2 in an extremely wide area of tropical lowlands? Is it merely a coincidence that the lowland strains belong to the same biochemical group as race 3? RFLP analysis has provided answers to these questions. All 25 lowland biovar 2 strains (received from Dr. E. R. French, International Potato Center, Lima, Peru) we examined had RFLP patterns (groups 29 to 33, Fig. 1) that were closely related but clearly distinguishable from the highland biovar 2 strains (groups 26 and 27, Fig. 1). Thus, the latter strains, which have a restricted host range and are mostly pathogens of potato, constitute a natural grouping, presently designated as race 3. It is tempting to speculate that these highland strains which are now adapted to potato may have originated from a common ancestor that is widely disseminated on native plants in the Amazon basin.

5. General Conclusions and Some Speculations

The initial results of RFLP analysis indicate that this method may be used as a basis for a future system of classification of *P. solanacearum* that will reflect the natural evolution within the species. Taken as a whole, the RFLP patterns confirm that the race designations in *P. solanacearum* constitute natural groupings of presumed phylogenetic significance. As Buddenhagen (1985) has pointed out, "research (on bacterial wilt) needs to be based on natural groupings (because) there are many bacterial wilts and there are many *Pseudomonas solanacearums* (and) the attempt to generalize bacterial wilt as a single entity tends to obscure the local reality". The plasticity of the species is evidenced by the recent appearance of new strains of race 1 that are very damaging to bean, eucalyptus, mulberry, and casuarina (Persley *et al*, 1985; Hayward, 1985; He, 1985; J. C. Dianese, personal communication). How do these new strains originate and what is the role of the host in the selection process?

These questions concerning the relationships between strains of *P. solanacearum* are, in broader terms, questions about the bacterium as it adapts to new hosts and differing environmental conditions, and as a population subjected to inevitable alterations of its genetic code. RFLP analysis provides a record of these evolutionary changes. In the simplest case, RFLPs arise from point mutations: the change of a single base pair resulting in the acquisition or loss of a restriction site. The evidence suggests that this has occurred within *P. solanacearum*. For example, within the biovar 2 strains of lowland origin, the RFLP patterns were variations of a general pattern that was typical of most division II strains.

RFLPs can also arise by insertion or deletion of DNA sequences. Evidence obtained by Southern blot analysis with one probe, pT13, indicates that several different deletions may have occurred within this region of the *P. solanacearum* genome. Hybridization patterns indicate that RFLP groups 1-5 and 30-31 lack approximately 4 kb of the probe DNA. Two additional possible deletions were found, one in strains of RFLP group 28 (race 2) and one in all strains of division I (race 1, biovars 3, 4, and 5). Since pT13 is part of a cluster of *hrp* genes that are important in host-parasite interactions (Xu *et al*, 1988; Boucher *et al*. 1987) it is likely that changes in this region affect host specificity.

The possibility that RFLP patterns may have originated by horizontal transfer of genes must also be considered, although admittedly this is a rare phenomenon in nature (Miller, 1988). For example, the *P. solanacearum* strain GMI1000 is naturally competent; it can acquire chromosomal markers simply by contact with lysed cells of a separate bacterial strain (C. Boucher, personal communication). GMI100 also exhibits an unusually high frequency of homologous recombination between chromosomal and extrachromosomal sequences, a trait that would facilitate the retention of transferred genes that cannot replicate autonomously.

The presence of common DNA sequences in distantly related strains that could not have arisen readily via mutation or common ancestry is an indication that horizontal gene transfer has occurred. For example, recently we have cloned a 2 kb DNA fragment that is specific to race 3 strains (Cook & Sequeira, unpublished data). The few (5/90) non-race 3 strains in which homology has been detected are distributed between division I and division II and, therefore, do not share a recent common ancestry. It seems possible that the race 3-specific DNA is a prophage associated primarily with race 3, but, upon induction, the phage may be able to infect other strains as well. Since race 3 has been moved from its apparent center of origin in the Andean highlands to other locations worldwide, this would explain the occasional occurrence of homology for this DNA in non-race 3 strains.

Ongoing research in our laboratory should provide answers to some of these questions concerning the evolution of *P. solanacearum*. The recent appearance of new strains on new host plants is an indication that this bacterium is constantly adapting to new environments. RFLP analysis provides an extremely powerful tool that gives us the means to understand how this evolutionary process takes place.

Acknowledgements

We thank A. C. Hayward, S. Eden-Green, A. Kelman, and E. R. French for providing many of the cultures of *P. solanacearum* used in this study. This work was supported by project 6070 from the College of Agricultural and Life Sciences, University of Wisconsin, Madison, and by a grant from the International Potato Center.

References

Berg, L.A. 1971. Weed hosts of SFR strains of *Pseudomonas solanacearum*, causal organism of bacterial wilt of bananas. Phytopathology 61:1314-1315.

Boucher, C. A., Van Gijsegem, F., Barberis, P. A., Arlat, M., and Zischek, C. 1987. *Pseudomonas solanacearum* genes controlling both pathogenicity on tomato and hypersensitivity on tobacco are clustered. J. Bacteriol. 169:5626-5632.

Buddenhagen, I. 1960. Strains of *Pseudomonas solanacearum* in indigenous hosts in banana plantations of Costa Rica, and their relationship to bacterial wilt of bananas. Phytopathology 50:660-664.

Buddenhagen, I., Sequeira, L., and Kelman, A. 1962. Designation of races in *Pseudomonas solanacearum*. Phytopathology 52:726.

Buddenhagen, I. 1985. Bacterial wilt revisited. Pages 126-143, *in*: Bacterial Wilt Disease in Asia and the South Pacific, ACIAR Proceedings No. 13. G. J. Persley, ed. ACIAR, Canberra.

Cook, D., Barlow, E., and Sequeira, L. 1989. Genetic diversity of *Pseudomonas solanacearum*: detection of restriction fragment length polymorphisms with DNA probes that specify virulence and the hypersensitive response. Mol. Plant-Microbe Interact. 2:113-121.

Hayward, A. 1964. Characteristics of *Pseudomonas solanacearum*. J. Appl. Bact. 27:265-277.

Hayward, A.C. 1985. Bacterial wilt caused by *Pseudomonas solanacearum* in Asia and Australia: an overview. Pages 15-24 in: Bacterial Wilt Disease in Asia and the South Pacific, ACIAR Proceedings No. 13. G.J. Persley, ed. ACIAR, Canberra.

He, L.Y. 1985. Bacterial wilt in the People's Republic of China. Pages 40-48 in: Bacterial Wilt Disease in Asia and the South Pacific, ACIAR Proceedings No. 13. G.J. Persley, ed. ACIAR, Canberra.

Martin, C., and French, E. R. 1977. *Pseudomonas solanacearum* affecting potatoes in the Amazon basin. Proc. Am. Phytopathol. Soc. 4:138 (abstract).

Martin, C., French, E. R., and Nydegger, U. 1981. Bacterial wilt of potatoes in the Amazon basin. Plant Disease 65:246-248.

Miller, R. V. 1988. Potential for transfer and establishment of engineered genetic sequences. Trends in Biotechnology. 3:s23-s27.

Nei, M. and Li, W. 1979. Mathematical model for studying genetic variation in terms of restriction endonucleases. Proc. Natl.Acad. Sci. USA 76:5269-5273.

Persley, G.J., Batugal, P., Gapasin, D., and Vander Zaag, P. 1985. Summary of Discussion and Recommendations. Pages 7-14, *in*: Bacterial Wilt Disease in Asia and the South Pacific, ACIAR Proceedings No. 13. G.J. Persley, ed. ACIAR, Canberra.

Seneviratne, A. 1969. On the occurrence of *Pseudomonas solanacearum* in the hill country of Ceylon. J. Hort. Sci. 44:393-402.

Sequeira, L. and Averre, C. 1961. Distribution and pathogenicity of strains of *Pseudomonas solanacearum* from virgin soils in Costa Rica. Plant Disease Reporter 45:435-440.

Xu, P., Leong, S., and Sequeira, L. 1988. Molecular cloning of genes that specify virulence in *Pseudomonas solanacearum*. J. Bacteriol. 170:617-622.

Section II

BACTERIA-PLANT INTERACTIONS (SYMBIOTIC)

OVERVIEW ON GENETICS OF NODULE INDUCTION: FACTORS CONTROLLING NODULE INDUCTION BY *RHIZOBIUM MELILOTI*

A. KONDOROSI
Institut des Sciences Végétales CNRS,
F-91198 Gif-sur-Yvette, France and
Institute of Genetics, Biological Research Center,
Hung. Acad. Sci.,H-6701 Szeged, Hungary

1. Introduction

The development of nitrogen-fixing nodules takes place in a series of stages, controlled by genes in the bacterium as well as in the plant. Molecular genetic and cell biological studies revealed three major, somewhat independent series of events : i.) nodule initiation and in many cases root hair curling ; ii.) infection and iii.) nodule development culminating in nitrogen-fixing symbiosis. In this overview the genetics of early *Rhizobium*-plant interactions leading to nodule initiation is discussed. Genes controlling these events have been identified almost exclusively from the bacterial partner, (except the work of Bisseling at al which is presented in this volume). Therefore, this short review deals primarily with bacterial genes controlling nodulation (*nod* genes). Due to the fairly high number of *nod* genes the new *nod* genes are designated as *nol* genes.

2. Location and organization of *nod* genes

In the three rhizobial genera, *Rhizobium*, *Bradyrhizobium* and *Azorhizobium* numerous mutations abolishing or influencing nodulation on any natural host plant have been recovered, leading to the identification of *nod* genes. By now over thirty *nod* genes have been sequenced. In all *Rhizobium* species the *nod* genes were localized on indigenous plasmids, often in clusters and in the vicinity of *nif* genes. In contrast, in *Bradyrhizobium* and *Azorhizobium* the *nod* genes appear to be located on the chromosome.

The organization of *nod* genes in several species is shown on Fig. 1. The *nod* genes form several transcriptional units and, as discussed below, the genes are coordinately regulated by the *nodD* product and specific plant signal molecules. Thus, the *nod* transcriptional units form one

111

H. Hennecke and D. P. S. Verma (eds.),
Advances in Molecular Genetics of Plant-Microbe Interactions, Vol. 1, 111–118.
© 1991 *Kluwer Academic Publishers. Printed in the Netherlands.*

112

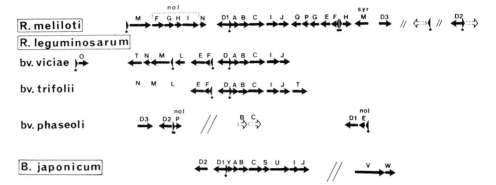

Figure 1. Organization of nodulation genes.
Full circles : *nod* box.

regulatory circuit, the *nod* regulon (for review Kondorosi,
1989 : Long, 1989).

3. Common and host specific nodulation genes

Different rhizobial species are able to form nodules on a
different range of legumes. One class of nodulation genes,
(the "common" *nodABC* genes) are conserved and have
universal function in all rhizobia. Mutations in these
genes abolish the early interactions with the plant, both
root hair curling (Hac⁻) and nodule induction (Nod⁻). They
can be complemented by the respective genes from other
species, indicating their functional conservation.

The specific recognition between the micro- and
macrosymbionts is controlled by specific genes (host
specific nodulation or *hsn* genes, although the gene
designation is generally *nod* or *nol*). The original
designation of *hsn* genes was based on the inability of
mutations to be complemented by *nod* genes of other
Rhizobium species (Kondorosi et al., 1984). The *hsn* genes
may be required for several or only one plant host species
or genotype. These genes in the majority of cases are not
essential for nodulation. This is reflected by their
mutational phenotypes which cause only some delay in
nodulation or affect slightly root hair curling. There are
however genes, such as *nodH* (*hsnD*) or *nodT*, which are
essential for extending the host range for *Medicago*
(Horvath et al., 1986) and for certain pea varieties (Surin
et al., 1989), respectively.

Several *nod* genes influence the efficiency of nodulation
only on certain plant hosts (p. ex. *nodM*, Baev et al.,
1990). Host specificity determinants may function also in
combination with each other and with common *nod* genes

(Downie and Surin, 1990). In some cases they prevent nodulation of nonhosts (Rolfe and Gresshoff, 1988).

The regulatory *nodD* gene is found in all rhizobial strains and when activated, it provides the same function as other common *nod* genes. However, the different NodD proteins have also a host specific character by recognizing only specific plant signal molecules.

4. Functions of *nod* genes : root hair curling and plant cell division.

The phenotypes of mutations in the common and many host specific *nod* genes clearly indicate that these genes control root hair curling and root cell division. The other *nod* genes are probably involved in optimizing either these interactions for particular strain-host combinations or for other yet unknown molecular interactions.

4.1. Biochemical functions of Nod proteins

There is little information about the biochemical functions of the *nod* gene products. Comparisons of amino acid sequences of the deduced *nod* gene products with existing protein sequence data banks have revealed only a few interesting hints for possible functions.(For references see Martinez et al., 1990 and this volume) . On this basis the *nodFE* (*hsn*AB) genes may be involved in fatty acid synthesis where NodF is an acyl carrier protein and NodE is fatty acid synthetase. The NodG (HsnC) resambles ribitol or glucose dehydrogenases and nodL may have acetyl transferase function . The NodM exhibits strong homology to glucosamine synthase (GlmS). Recently this function was demonstrated by showing that expressing *nodM* in an *E. coli* glucosamine synthase negative mutant, the mutant became able to grow without added glucosamine (Baev et al. 1990).

The NodI shows similarity to ATP-binding proteins and with nodJ it may play a role in transport through the membrane. The NodC is located in the outer membrane. NodO is exported outside and was shown to be Ca^{++} - binding protein (Downie et al., this volume).

Sequence analyses revealed that the NodU and NodW proteins of *B. japonicum* are members of the family of bacterial two-component regulatory systems where NodV responds to an environmental stimulus and may activate NodW. The nodV and nodW genes are host specificity determinants; perhaps nodW is required to activate yet unknown host specificity genes (Göttfert et al., 1990).

4.2. Production of extracellular Nod signals

In the past few years new bioassays have been developed to

look for *nod* gene functions. These bioassays are based on findings that *Rhizobium* cells induced to express *nod* genes excrete factors which induce root hair deformation and branching (Bhuvaneswari and Solheim (1986) or thick roots (Van Brussel et al. 1986) or stimulate mitotic cell division of plant protoplasts (Schmidt et al. 1988). It was shown that for the production of mitosis-stimulating factors the *nodAB* genes were sufficient (Schmidt et al., 1988), while for root hair deformation (Had) the *nodABC* genes were essential on all plants tested and for factors acting on numerous other plants *hsn* genes were also required (Faucher et al., 1988; Banfalvi and Kondorosi, 1989). The involvement of the *nodH* (*hsnD*) and *nodQ* in the production of *Medicago*-specific Had factor was found. The factor (Nod-Rm1) has been purified from *R. meliloti* exudates and its structure was determined as glucosamine tetrasaccharide, acylated with one unsaturated C_{16}-fatty acid at one end and containing a sulphate group at the other end (Lerouge et al. 1990). This sulphate group was not present in NodH$^-$ mutants. These observations and further details on the Nod factors are reported in this volume by Dénarié et al.. The Nod factors can induce the expression of early nodulins, such as ENOD12 (T. Bisseling et al., this volume). These factors act at very low concentrations like plant hormones. They may be recognized by specific plant receptors, and may affect the hormone balance of root cells leading to nodule morphogenesis.

5. *nod* gene regulation

5.1. The positive regulator NodD and plant signal molecules

Most *nod* genes do not express *ex planta*; only when bacteria are exposed to plant exudates, with the exception of NodD. In *R. meliloti nodD* expression is constitutive, while p. ex. in *R. leguminosarum* NodD autogenously regulates its own expression. In all rhizobia investigated so far the NodD was found to activate other *nod* genes in conjunction with the plant signal molecules found in plant exudates or extracts (for review, Long, 1989).

In different *Rhizobium*-plant systems the inducing (or inhibiting) molecules were identified as compounds of the phenylpropanoid pathway. In *R. meliloti* various flavonoids or chalcons, in *Bradyrhizobium* isoflavones were shown to be inducers. The NodD proteins of different species or strains confer distinct abilities to interact with compounds. In many strains or species there are multiple *nod* alleles with distinct flavonoid specificity, contributing to nodulation of various plant hosts to different extent.

The NodD sequences are highly conserved, particularly in the N-terminal region. Construction of hybrid *nodD* genes and their mutational analysis revealed that *nod* gene inducibility and its specificity are determined by the carboxy-part of the NodD (for review Kondorosi, 1989). The data suggest that the NodD directly interacts with the inducer. Chemical evidence, however, is still lacking. The N-terminal part of NodD carries a helix-turn-helix motif characteristic for DNA-binding protein domains. Gel retardation and foot printing experiments showed that the NodD binds to about a 50 bp long highly conserved sequence of the inducible *nod* promoters (*nod* box, Rostas et al., 1986). Based on sequence homology and gene activating ability, the NodD belongs to a family of prokaryotic activating proteins (LysR family, Henikoff et al., 1988).

The NodD was localized in the inner membrane and the flavonoids accumulate also at the inner membrane. Schlaman et al. (1989) suggested that in the inner membrane the NodD and the plant signal may interact directly, while the N-terminal part of NodD in the cytoplasm binds to the *nod* promoters.

The NodD protein is a receptor-like protein : it binds to DNA and probably binds the ligand, the plant signal molecule. Steroid receptors are known to accept various flavonoids as ligands. Interestingly, several steroid hormones were shown to activate *nod* genes in conjunction with NodD or had inhibitory or synergistic effects (Györgypal and Kondorosi, 1990). Comparison of the ligand-binding domain of steroid receptors with the putative flavonoid-binding NodD domain revealed significant sequence conservation. It was suggested that the ligand-binding domain of animal receptors and that of NodD may have common evolutionary origin.

5.2. Negative control of *nod* gene regulation

Recently in *R. meliloti* an additional transacting factor, a *nod* repressor protein was found which allows only a relatively low level of *nod* gene induction and the amount of NodD is also reduced (Kondorosi et al., 1989). This protein was detectable in 18 out of 22 *R. meliloti* strains of different geographical origin. Interestingly, it is not active in one of the widely used *R. meliloti* strains, SU47. Mutants of *R. meliloti* lacking the repressor were constructed which exhibited delay of nodulation, indicating that fine tuning of *nod* gene expression, mediated by positive and negative trans-acting factors, allows optimal nodulation of the plant host.

In several species autogenous regulation of nodD or other indications for negative control of *nod* gene expression were observable (Rossen et al., 1985). It seems that the signals produced by the *nod* gene products are required at optimal concentrations for nodulating the plant hosts.

116

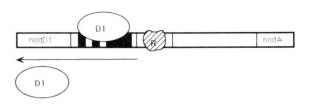

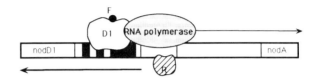

Figure 2. Regulation of *nod* genes in *R. meliloti*.
Black boxes : *nod* box sequence, F : flavonoid inducer,
R : repressor.

5.3. Model for *nod* gene regulation

A model depicting the possible interaction of the plant
signal molecule, the NodD and the repressor is shown on
Fig. 3. The NodD protein binds to the *nod* box, but without
inducer it cannot activate RNA polymerase to start
transcription. As proposed for *nod* gene regulation in *R.
meliloti* (Kondorosi et al., 1989), the repressor competes
with RNA polymerase for the binding site, therefore,
binding of RNA polymerase is occasional, resulting also in
lower amounts of NodD protein. Upon addition of the
inducer, the NodD interacts with the inducer, causing its
conformational change. Binding of RNA polymerase to this
modified protein-DNA complex competes with the repressor,
leading to the transcription of *nod* genes. In rhizobia
possessing autogenously regulated *nodD*, the autoregulation
may optimize *nod* gene expression.

5.4. Other factors controlling *nod* gene expression

These factors have been studied in *R. meliloti*.

5.4.1. SyrM

In *R. meliloti* a *nodD*-like locus *syrM* was detected
(Mulligan and Long, 1989). This gene together with *nodD3*,

when provided in several copies on a plasmid, allows inducer-independent expression of *nod* genes.

The SyrM protein is a positive regulator belonging to the LysR family. It controls also the production of extracellular polysaccharides and express at higher level in the nodule (S. Long, personal communication). When *syrM* and *nodD3* are present in several copies, the bacteria can nodulate those plants where the lack of nodulation is due to the inability of plant exudates activating the particular NodD (E. Kondorosi, unpublished).

5.4.2. Nitrogen control

High levels of ammonia decrease the level of *nod* gene expression in *R. meliloti*. It was found that this control is mediated primarily via NodD3 which is negatively regulated by a novel element of the global nitrogen regulation system. Mutation in the gene encoding the factor results in the escape of nitrogen control. Interestingly, the mutant is more competitive in nodulation than the wild type bacterium (Dusha et al., 1989).

5.4.3. Other factors controlling the expression of NodD, SyrM or the Nod repressor

Studies on the regulation of the *R. meliloti* genes encoding the transacting factors NodD1,D2,D3, SyrM and the Nod repressor, respectively, suggest that these genes are controlled by other factors (E. Kondorosi et al., unpublished) and mutation of the respective genes influence *nod* gene expression (S. Long, personal communication). These observations provide further support for the complexity and fine-turning of *nod* gene regulation.

6. Conclusions

In the last two years there has been a rapid progress in studies on the molecular genetics and biology of early *Rhizobium*-legume interactions. Not only numerous *nod* genes have been identified, but the function of *nod* genes is now better understood. The purification of Nod factors and determination of their chemical structures open up new possibilities to understand the molecular mechanism how *nod* gene products are involved in nodule induction.

7. References

Baev N., Endre G., Banfalvi Z. and Kondorosi (1990) (submitted).

Banfalvi Z., Kondorosi A. (1989) Plant Mol. Biol. 13, 1-12.

Bhuvaneswari TV, Solheim B (1985) Physiol. Plant. 63, 25-34.

Downie J.A. and Surin B.P.(1990) Mol. Gen. Genet. 222, 81-86.

Dusha I., Bakos A., Kondorosi A., de Bruijn F., Schell J. (1989) Mol. Gen. Genet. 219, 89-96.

Faucher C., Maillet F., Vasse J., Rosenberg C., van Brussel AAN, Truchet G., Dénarié J. (1988) J. Bacteriol. 170, 5489-5499.

Göttfert M, Grob P. and Hennecke H. (1990) Proc. Natl. Sci. U.S. 87, 2680-2684.

Györgypal Z., Kondorosi A. (1990) (submitted).

Henikoff S., Haughn G.W., Calvo J.M. and Wallace J.C. (1988) Proc. Natl. Acad. Sci. U.S. 85, 6602-6606.

Horvath B., Kondorosi E., John M., Schmidt J., Torok I., Gyorgypal Z. Barabas I., Wieneke U., Schell J., Kondorosi A. (1986) Cell 46, 335-343.

Kondorosi A.(1989)In : Plant-Microbe Interactions, Eds. Kosuge T., Nester E.W., Vol. III 383-420, McGraw-Hill P.Co. N.Y.

Kondorosi E., Banfalvi Z., Kondorosi A. (1984) Mol. Gen. Genet. 193, 445-452.

Kondorosi E., Gyuris J., Schmidt J., John M., Duda E., Hoffmann B., Schell J., Kondorosi A. (1989) EMBO J. 8, 1331-1340.

Lerouge P., Roche P. Faucher C., Maillet F.,Truchet, Promé J.C. & Dénarié J. (1990) Nature 344, 781-784.

Martinez E., Romero D. and Palacios R. (1990) Critical Reviews in Plant Sciences 9, 59-93.

Long S.R. (1989) Cell 56, 203-214.

Mulligan J.T. and Long S.R. (1989) Genetics 122,7-18.

Rolfe B.G., Gresshoff P.M. (1988) Ann. Rev. Plant Physiol. Plant Mol. Biol. 39, 297-319.

Rossen L., Shearman C.A., Johnston A.W.B. and Downie J.A. (1985) EMBO J. 4, 3369-3373.

Rostas K., Kondorosi E., Horvath B., Simoncsits A., Kondorosi A. (1986) Proc. Natl. Acad. Sci. U.S. 83,1757-1761.

Schlaman H., Spaink H.P., Okker R.J.H. and Lugtenberg B.J.J. (1989) J. Bacteriol. 171, 4686-4693.

Surin B.P., Watson J.M., Hamilton W.D.O., Economou A. and Downie J.A. (1990) Molecular Microbiology 4, 245-252.

Van Brussel A.A.N., Zaat S.A.J., Canter-Cremers H.C.J., Wijffelman C.A., Pees E., Tak T., Lugtenberg B.J.J. (1986) J. Bacteriol. 165, 517-522.

NODRM-1, A SULPHATED LIPO-OLIGOSACCHARIDE SIGNAL OF *RHIZOBIUM MELILOTI* ELICITS HAIR DEFORMATION, CORTICAL CELL DIVISION AND NODULE ORGANOGENESIS ON ALFALFA ROOTS

P. ROCHE, P. LEROUGE AND J.C. PROME
Centre de Recherche de Biochimie et de Génétique Cellulaire, CNRS-UPS, 118 route de Narbonne, 31062 Toulouse Cedex, France.
C. FAUCHER, J. VASSE, F. MAILLET, S. CAMUT, F. DE BILLY, J. DENARIE AND G. TRUCHET
Laboratoire de Biologie Moléculaire des Relations Plantes-Microorganismes, CNRS-INRA, BP27, Castanet-Tolosan Cedex, France.

ABSTRACT. We have addressed the questions of how nodulation (*nod*) genes of the symbiotic bacterium *R. meliloti* operate to determine host specificity, plant (alfalfa) infection and nodulation. Using a root hair deformation assay we have shown that the common *nodABC* genes and the host-range *nodH* and *nodQ* genes are all involved in the production of an extracellular signal. The structure of the major alfalfa-specific signal, NodRm-1, has been determined by mass spectrometry, NMR spectroscopy, radioactive labelling and chemical modifications. This signal is a sulphated and acylated glucosamine tetrasaccharide. NodRm-1 elicits root hair deformation, cortical cell divisions and nodule formation, on aseptically-grown alfalfa seedlings, at nanomolar concentrations. Host-range *nodH* or *nodQ* mutants are able to infect a non-homologous host, common vetch and were found to excrete another signal, NodRm-2, which differs from NodRm-1 by the absence of the sulphate group. Purified NodRm-2 elicits hair deformation on vetch but not on alfalfa. The function of nodH and nodQ is to make the signal alfalfa-specific by transfer of a sulphate group. We propose that the *nod* genes of *R.meliloti* determine host specificity, root hair curling and nodule formation via the production of NodRm-1.

1. Introduction

In addition to their agricultural importance, due to symbiotic nitrogen fixation, *Rhizobium*-legume associations provide good experimental systems to study the molecular basis of several problems of great interest in plant sciences, for example the specificity of recognition between plants and microbes and the elicitation of a major developmental switch, in this case the nodule organogenesis (for a review see Long 1989). The symbiosis can indeed be highly specific. For example fast growing rhizobia such as *R. meliloti* and *R. leguminosarum* have a narrow host range, with the former able to form nitrogen-fixing nodules only with *Medicago*, *Melilotus* and *Trigonella* species and the latter with *Pisum* and *Vicia*. In the case of the *R. meliloti*-alfalfa association, bacterial infection proceeds via the root hairs, and involves curling of the hair tip, followed by the formation of infection threads within the root hairs (Debellé et al. 1986). While these early stages of infection are proceeding, cell divisions are induced, at a distance in the inner cortical cell layers

119

H. Hennecke and D. P. S. Verma (eds.),
Advances in Molecular Genetics of Plant-Microbe Interactions, Vol. 1, 119–126.
© 1991 *Kluwer Academic Publishers. Printed in the Netherlands.*

of the root, giving rise to nodule primordia (Dudley et al. 1987). These primordia then develop into organs having characteristic traits, the nitrogen-fixing root nodules.

The genes of *R. meliloti* which are necessary for host recognition and nodule induction (*nod* genes, see Figure 1) can be classified into three categories: (i) the regulatory *nodD* genes which activate the transcription of the other *nod* operons, (ii) the "common" *nodABC* genes which are structurally and functionally conserved among *Rhizobium* species, and (iii) the "host-specific" *nodFEG*, *nodH* and *nodPQ* genes which determine the host-range. For example, a mutation in the *R. meliloti* *nodH* gene causes a shift in host-range from alfalfa to vetch and a mutation in the *nodQ* gene results in an extension of the host-range to vetch (Horvath et al. 1986; Faucher et al. 1989; Cervantès et al. 1989).

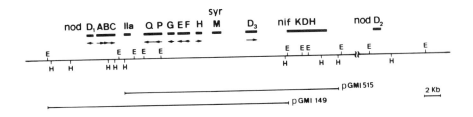

Figure 1. Genetic and physical map of the nodulation region of *R. meliloti* 2011. The horizontal line represents the restriction map (E, *Eco*R1; H, *Hind*III). The plasmids pGMI149 and pGMI515 are shown below the map, and the arrows indicate the direction of transcription of *nod* genes (Long, 1989).

The early interaction between alfalfa and *R. meliloti* involves the transcriptional activation of the bacterial *nod* genes by the combined action of specific flavonoid compounds present in plant exudates and the regulatory *nodD* gene products (Long, 1989). But once activated, what could be the mechanisms by which the *Rhizobium nod* genes operate? An important advance to answering this question came with the discovery that sterile supernatants of *R. leguminosarum* cultures are able to elicit a thick and short root (Tsr) reaction and root hair deformations (Had) on seedlings of the *R. leguminosarum* host, common vetch (*Vicia sativa* subsp. *nigra*) (Van Brussel et al. 1986; Zaat et al. 1987). The authors showed that the *nodD* and *nodABC* genes are required for the Tsr and Had responses, and furthermore, that a flavonoid *nod* gene inducer must also be present in the bacterial growth medium. These results indicated that, in *R.leguminosarum*, at least some *nod* genes are involved in the production of extracellular factors.

In this article we would like to present the interdisciplinary approach, including molecular genetics, cytology and chemistry, that we have used to identify and characterize *R. meliloti* extracellular symbiotic signals whose production requires *nod* genes. The structure of the major alfalfa-specific signal, NodRm-1, has been determined. This molecule elicits root hair deformations, cortical cell divisions and nodule formation on alfalfa. The role of major *R. meliloti* host-range genes *nodH* and *nodQ* in the synthesis of NodRm-1 will also be described.

2. Results

2.1. ROOT HAIR ASSAYS FOR EXTRACELLULAR SIGNALS

The Tsr reaction, which can be observed on *Vicia sativa* subsp. *nigra*, could unfortunately not be detected on any of a range of *R. meliloti* hosts tested, but it was observed that treating alfalfa with the sterile supernatants of flavonoid-activated cultures can induce a generalised deformation of root hairs (Had) (Faucher et al. 1988 and 1989). This bioassay which can be used also for common vetch (Zaat et al. 1987)) and white clover (Bhuwaneswari and Solheim, 1985; Hollingsworth et al., 1989) is more rapid than the Tsr assay and significantly more sensitive (Zaat et al., 1987; Faucher et al. 1989). As with Tsr, the Had reaction is dependent on functional *nodDABC* genes. The supernatants of *R. meliloti* 2011 were Had$^+$ on alfalfa but not on common vetch, and *vice versa* for the supernatants of *R. leguminosarum* cultures (Faucher et al. 1988, 1989). These results indicated that there is a correlation between host specificity and extracellular factor activity and suggested that host-range *nod* genes might be involved in the production of the symbiotic extracellular factors.

2.2. PURIFICATION AND STRUCTURE OF NodRm-1 ALFALFA-SPECIFIC SIGNAL

The first attempts to analyse the hair deformation factors present in filtrates of luteolin-induced *R. meliloti* cultures had shown that most of the Had activity could be extracted in a butanol-soluble fraction. Unfortunately, the concentration of the active molecule(s) proved to be insufficient for further structural analysis. In order to amplify factor production by *R. meliloti*, we introduced into the strain Rm 2011 an Inc-P1 plasmid derivative, pGMI149 (Fig.1), carrying the common and host specific *nod* genes, as well as the three regulatory genes *nodD1*, *nodD3* and *syrM*. The presence of this plasmid led to an increase of more than a hundred-fold in the Had activity on alfalfa. As a result of this amplification it was now possible to observe two far-UV (220 nm) absorbing peaks when the butanol-soluble fraction was further fractionated by HPLC on a C_{18} reverse-phase column (Lerouge et al. 1990a). Tn5 insertions into either the *nodA* or *nodC* genes resulted in the simultaneous disappearance of the Had activity and of these two UV-absorbing peaks.

Subsequent large-scale culture supernatant processing was carried out with the exopolysaccharide-deficient *R. meliloti* strain EJ355 (containing plasmid pGMI149) by virtue of its non-mucoid characteristics (Finan et al. 1985). The major signals produced by the Exo$^+$ strain Rm 2011 (pGMI149) and the Exo$^-$ strain Rm EJ355 (pGMI149) were found to be identical as assessed by both chemical criteria (HPLC, NMR and mass spectrometry) and the biological Had assays. Filtrates were extracted and extracellular factors purified by a combination of reverse-phase C_{18} HPLC, gel permeation on a Sephadex LH20 column and by ion-exchange chromatography on a DEAE column (Lerouge et al. 1990a). Ten litres of a luteolin-induced *R. meliloti* culture yielded around 4 mg of purified factors, which showed Had activity on alfalfa, but not on common vetch.

Mass spectrometry, NMR spectroscopy, chemical modification and radioactive labelling were used to establish the structure of the compounds present in the two peaks as N-acyl tri N-acetyl ß-1,4 D-glucosamine tetrasaccharides, bearing a sulphate group on C-6 of the reducing sugar. The aliphatic chain carried by the non-reducing terminal sugar residue is a 2,9-hexadecadienoic N-acyl group (Fig. 2). The two peaks correspond to α and ß anomers of the same molecule at the C-1 position of the reducing end sugar: we propose to call this molecule NodRm-1. That NodRm-1 is indeed the plant specific symbiotic signal previously characterised in the hair deformation assays is supported by the following evidence: (i) NodRm-1 elicited root hair deformations on alfalfa, and not on vetch, at extremely low concentrations, between 10^{-8} and 10^{-11} M; (ii) we have always observed a strict correlation both qualitatively and quantitatively between the presence of this molecule and the specific Had activity (Had^{+} on alfalfa and Had^{-} on vetch); (iii) in the active fractions possible aromatic contaminants, such as the phytohormones auxins and cytokinins, could not be detected. It is worth noting that most phytohormones are active in the 10^{-7} M range when added exogenously (Zeroni and Hall, 1980), and only certain fungal oligosaccharide elicitors which trigger the plant defense response are known to be active in the nanomolar range (Darvill and Albersheim 1984). Other experiments described in the following paragraphs will further support that NodRm-1 is the major active Nod factor.

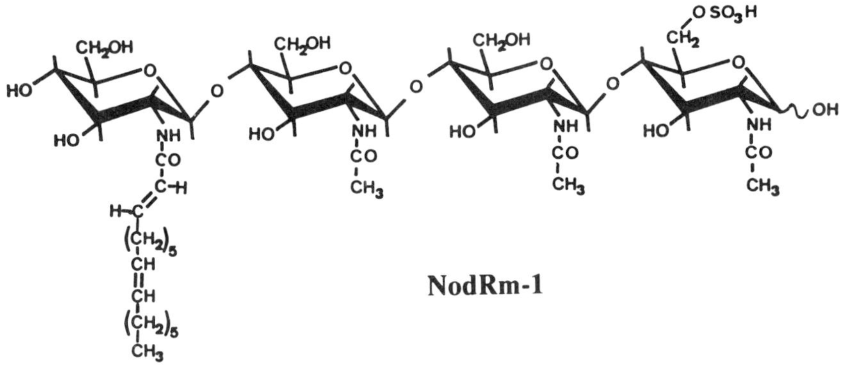

NodRm-1

Figure 2. Structure of the sulphated and acylated glucosamine oligosaccharide symbiotic signal (NodRm-1) which has been purified from supernatants of flavonoid-induced cultures of *R. meliloti*. This structure has been established from a combination of NMR spectroscopic, mass spectrometric, ^{35}S-labelling and methylation analyses (Lerouge et al. 1990a).

2.3. NodRm-1 ELICITS CELL DIVISIONS IN THE ALFALFA ROOT CORTEX

In addition to root hair deformation, purified NodRm-1 was found to elicit cortical cell divisions in roots of aseptically grown alfalfa seedlings. At concentrations of 10^{-9}-10^{-10} M, discrete foci of meristematic cells were observed in the inner cortex of secondary roots. At higher concentrations of NodRm-1 the mitotic activity extended over the entire cortex and even reached the epidermis over large areas of the secondary root. Thus NodRm-1 is a potent mitogenic factor. That during early steps of symbiosis *R. meliloti* might trigger cortical cell divisions via NodRm-1 is supported by the observations that (i) in the course of infection rhizobia elicit mitosis in the host cortex at a distance (Dudley et al. 1987) and (ii) mutations in the *nodABC* operon result in the loss of the ability to both induce the first round of cell division (Dudley et al. 1987) and to produce NodRm-1 (Lerouge et al. 1990). An extracellular factor whose synthesis requires the *nodAB* genes of *R. meliloti* was shown to trigger divisions of plant protoplasts (Schmidt et al. 1988) and factors produced by *R. trifolii* were reported to elicit foci of cortical cell divisions which resemble nodule primordia (Hollingsworth et al. 1989). The chemical nature of these factors has not yet been determined.

2.4. NodRm-1 ELICITS ALFALFA NODULE ORGANOGENESIS

The NodRm-1-elicited mitotic activity in the cortex gave rise to various root deformations, varying from root bumps to elongate or multilobate structures having the appearance of nodules. Cytological observations revealed that most of these structures could not be classified as secondary roots but rather exhibited the following ontological, anatomical and physiological features of genuine alfalfa nodules: they originated from the inner cortex, endodermis and vascular bundles were peripheral, and their formation was totally repressed by the addition of combined nitrogen (15 mM potassium nitrate).

Several lines of evidence support the hypothesis that in the course of infection *R. meliloti* induces nodule organogenesis by a diffusible compound, the "nodule organogenesis inducing principle" (NOIP): (i) bacterial mutants altered in the infection process are able to elicit nodule formation at a distance (Truchet et al. 1980; Finan et al. 1985); (ii) bacterial colonies separated from the host roots by a filter membrane can elicit bacteria-free nodules (Kapp et al. 1990); (iii) alfalfa clones may develop nodules in the absence of rhizobia indicating that the host plant possesses the entire genetic programme for nodule organogenesis and that, under normal circumstances, the role of the microsymbiont is to switch on this programme (Truchet et al. 1989). That NodRm-1 is the NOIP diffusible signal is supported by the results described in this section and by the fact that only two classes of *R.meliloti* mutants, those altered in the *nodABC* and *nodH* genes, have so far been found to be unable to elicit nodule formation on alfalfa (Debellé et al. 1986; Long et al. 1989); these two operons are absolutely required for NodRm-1 production (Faucher et al. 1988; Lerouge et al. 1990a and 1990b).

2.5. *R. meliloti nodH* AND *nodQ* GENES DETERMINE THE HOST-SPECIFIC MODIFICATION OF EXTRACELLULAR SIGNALS.

The observation that supernatants of *R. meliloti* 2011 were able to induce root hair deformations on alfalfa but not on common vetch, and vice versa for the supernatants of *R. leguminosarum* cultures (Faucher et al. 1988 and 1989) showed that there is a correlation between host specificity and extracellular factor Had activity. To investigate the possible role of host specificity genes of *R. meliloti* in the production of extracellular signal(s), we examined the Had activities of the sterile supernatants of various *R. meliloti* mutants on both homologous and heterologous hosts. Mutations in *nodH* led to a shift in signal activity, Had⁻ on alfalfa and Had⁺ on vetch (Faucher et al. 1988), and mutations in *nodQ* led to an extended activity, Had⁺ on both alfalfa and vetch (Faucher et al. 1989). These modifications precisely mirror the changes in host specific nodulation of the corresponding mutated strains (Debellé et al. 1986 ; Horvath et. al. 1986 ; Cervantès et al. 1989). It should be pointed out that in all the above experiments the same flavonoid inducer, luteolin, was used to induce *nod* activity.

The introduction of the *R. meliloti* host range genes into *R. leguminosarum* (plasmid pGMI515 containing *nodPQ, nodFEG, nodH, nodD3* and *syrM*, Fig. 1) results in the production of an extracellular factor that is now able to deform alfalfa hairs, and at the same time has decreased Tsr activity on vetch (Faucher et al. 1989). Most significantly, mutations in either *nodH* or *nodQ* restore the original phenotype of *R. leguminosarum* supernatants (Had⁻ on alfalfa and Tsr⁺ on vetch). This clearly shows that the introduction of the *R. meliloti nodH* and *nodQ* genes into *R. leguminosarum* results in a modification of the specificity of the *R. leguminosarum* extracellular Had factors. The additional observation that a sterile filtrate of a *R. meliloti* strain carrying only *nodD1* and the common *nodABC* genes is able to deform root hairs on vetch, but not on alfalfa (Faucher et al. 1988) has led us to propose the following model. Both *R. leguminosarum* and *R. meliloti* common *nod* genes lead to the synthesis of a factor which is Had⁺ on vetch, but Had⁻ on alfalfa. The *nodH* and *nodQ* genes of *R. meliloti* then convert this common factor to an alfalfa-specific signal, which is no longer recognised by vetch root hairs. The reason why the *nodQ* gene is apparently required for the production of the alfalfa-specific signal in *R. leguminosarum*, but not in *R. meliloti* itself could be due to the presence of more than one *nodQ* gene in *R. meliloti* (Schwedock and Long 1989), and that this second gene has weak activity.

2.6. *nodH* AND *nodQ* ARE INVOLVED IN THE SULPHATION OF A PRECURSOR NodRm-2

To study the role of the *nodH* host-range gene in the synthesis of NodRm-1, we constructed a derivative of the Rm EJ355(pGMI149) strain carrying *nodH*::Tn5 insertions in both the pSym and the pGMI149 plasmids. No NodRm-1 peaks could be detected in the butanol extract of the *nodH* mutant supernatant, after reverse phase HPLC analysis. Instead, peaks corresponding to more hydrophobic compounds were observed and these fractions were Had⁻ on alfalfa and Had⁺ on vetch. Mass spectrometry, NMR spectroscopy and methylation studies revealed that the new peaks correspond to the α and ß anomers of a lipo-oligosaccharide which we propose to call NodRm-2 and which differs from NodRm-1 only by the absence of the sulphate group.

A Rm EJ355(pGMI149) *nodQ⁻* derivative was constructed in the same way and the HPLC profile of the butanol extract of its supernatant showed two clusters of peaks corresponding to the NodRm-1 and NodRm-2 compounds. Further purification and chemical characterization confirmed that the *nodQ⁻* mutant produced both NodRm-1 and NodRm-2 signals.

Thus the *R. meliloti* wild-type strain, which infects and nodulates alfalfa, produces NodRm-1; a *nodH⁻* mutant which infects and nodulates common vetch produces NodRm-2; and a *nodQ⁻* mutant which has an extended host-range (alfalfa and vetch) produces both signals. A *nodA⁻* mutant, which does not infect either host, does not produce either of these signals. There is therefore a striking correlation in

R. meliloti between the production of NodRm factors and the ability to specifically infect and nodulate a particular legume. These results indicate that the *nodH* and *nodQ* genes are responsible for the transfer of a sulphate group onto the lipo-oligosaccharide NodRm-2 factor, and by these means determine host-specificity.

3. Conclusions

Current research into the earliest stages of the interaction between *Rhizobium* and its legume host is now giving us a clearer view of how these two organisms exchange information at the molecular level. Firstly, plant signals in the form of flavonoids present in root exudates, activate bacterial *nod* genes in conjunction with an appropriate regulator *nodD* gene. Some specificity is expressed at this stage by a requirement for a correct matching between the *nodD* gene product and the flavonoid content of the plant exudate (Long 1989).

The second, and highly specific stage of the bacteria-plant interaction occurs with the production of rhizobial extracellular signals, of a lipo-oligosaccharidic nature, by a combination of common and host specific *Rhizobium nod* gene activities. The simplest hypothesis for the mechanism by which the *nodABC*, *nodH* and *nodQ* genes operate in determining host-range, infection and nodulation is by coding for proteins which contribute to the synthesis of the NodRm-1 signal.

New approaches towards studying the molecular basis of specific plant-microbe interactions are opening up, and the availability of molecules such as NodRm-1 and NodRm-2 will surely prove invaluable in our search to understand some of the mechanisms by which signalling can take place between *Rhizobium* and its legume hosts.

Future objectives of research on Nod factors will concern (1) their production by the microsymbiont and the role of each of the *nod* genes in the synthesis and transport of the signal; (2) the specific perception of the signal(s) by the host, and the isolation and characterization of plant receptor(s); (3) the plant reactions to the signal(s): elucidation of the signal transduction pathway(s) leading to the regulation of the expression of the symbiotic plant genes which control the dramatic developmental switch to nodule initiation.

4. Acknowledgements

We thank David Barker and Julie Cullimore for reviewing the manuscript.

5. References

Bhuvaneswari, T.V. and Solheim, B. (1985) "Root hair deformation in the white clover-*Rhizobium trifolii* symbiosis", Physiol. Plant. 63, 25-34.

Cervantes, E., Sharma, S.B., Maillet, F., Vasse, J., Truchet, G. and Rosenberg, C. (1989) "The product of the host specific *nodQ* gene of *Rhizobium meliloti* shares homology with translation elongation and initiation factors", Molec. Microbiol. 3, 745-755.

Darvill, A.G. and Albersheim, P. (1984) "Phytoalexins and their elicitors - A defense against microbial infection in plants", Ann. Rev. Plant. Physiol. 35, 243-298.

Debellé, F., Rosenberg, C., Vasse, J., Maillet, F., Martinez, E., Dénarié, J. and Truchet, G. (1986) "Assignment of symbiotic developmental phenotype to common and specific nodulation (*nod*) genetic loci of *Rhizobium meliloti*", J. Bacteriol. 168, 1075-1086.

Dudley, M.E., Jacobs, T.H. and Long, S.R. (1987) "Microscopic studies of cell divisions induced in alfalfa roots by *Rhizobium meliloti*", Planta, 171, 289-301.

Faucher, C., Maillet, F. Vasse, J. Rosenberg, C., van Brussel, A.A.N., Truchet, G. and Dénarié, J. (1988) "*Rhizobium meliloti* host range *nodH* gene determines production of an alfalfa-specific extracellular signal", J. Bacteriol. 170, 5489-5499.

Faucher, C., Camut, S., Dénarié, J., and Truchet, G. (1989) "The *nodH* and *nodQ* host range genes of *Rhizobium meliloti* behave as avirulence genes in R. *leguminosarum* bv. *viciae* and determine changes in the production of plant-specific extracellular signals", Mol. Plant-Microbe Interact. 2, 291-300.

Finan, T.M., Hirsch, A.M. Leigh, J.A., Johansen, E., Kuldau, G.A., Deegan, S., Walker, G.C. and Signer, E.R. (1985) "Symbiotic mutants of *Rhizobium meliloti* that uncouple plant and bacterial differentiation", Cell 40, 869-877.

Hollingsworth, R., Squartini, A., Philip-Hollingsworth, S. and Dazzo, F. (1989) "Root hair deforming and nodule initiating factors from *Rhizobium trifolii*", in Signal molecules in Plant and Plant-Microbe Interactions, B.J.J. Lugtenberg, ed., (Springer-Verlag, Berlin/Heidelberg), pp. 387-393.

Horvath, B., Kondorosi, E., John, M., Schmidt, J., Torok, I., Gyorgypal, Z., Barabas, I., Wieneke, U., Schell, J. and Kondorosi, A. (1986) "Organization, structure and symbiotic function of *Rhizobium meliloti* nodulation genes determining host specificity for alfalfa", Cell 46, 335-343.

Kapp, D., Niehaus,K., Quandt, J., Müller, P. and Pühler, A. (1990) "Cooperative action of *Rhizobium meliloti* nodulation and infection mutants during the process of forming mixed infected alfalfa nodules", The Plant Cell 2, 139-151.

Lerouge, P., Roche, P., Faucher, C., Maillet, F., Truchet, G., Promé, J.C. and Dénarié, J. (1990a) "Symbiotic host-specificity of *Rhizobium meliloti* is determined by a sulphated and acylated glucosamine oligosaccharide signal", Nature 344, 781-784.

Lerouge, P., Roche, P., Promé, J.C., Faucher, C., Vasse, J.,Maillet, F., Camut, S., de Billy, F., Barker, D., Dénarié, J. and G. Truchet (1990b) "*Rhizobium meliloti* nodulation genes specify the production of an alfalfa-specific sulphated lipo-oligosaccharide signal", in P. M. Gresshoff, W. E. Newton, E. L. Roth and G. Stacey (eds.) Nitrogen Fixation: Achievements and Objectives, Chapman-Hall Publishers, New York, in press.

Long, S.R. (1989) "*Rhizobium*-legume nodulation: life together in the underground" Cell 56, 203-214.

Schmidt, J., Wingender, R., John, M., Wieneke, U. and Schell, J. (1988) "*Rhizobium meliloti nodA* and *nodB* genes are involved in generating compounds that stimulate mitosis of plant cells", Proc. Nat. Acad. Sci. USA 85, 8578-8582.

Schwedock, J. and Long, S.R. (1989) "Nucleotide sequence and protein products of two new nodulation genes of *Rhizobium meliloti, nodP* and *nodQ*", Molec. Plant-Microbe Interact. 2, 181-194.

Truchet, G., Michel, M. and Dénarié, J. (1980) "Sequential analysis of the organogenesis of lucerne (*Medicago sativa*) root nodules using symbiotically-defective mutants of *Rhizobium meliloti*", Differentiation 16, 163-173.

Truchet, G., Barker, D.G., Camut, S., de Billy, F., Vasse, J. and Huguet, T. (1989) "Alfalfa nodulation in the absence of *Rhizobium*", Mol. Gen. Genet. 219, 65-68.

van Brussel, A.A.N., Zaat, S.A.J., Canter Cremers, A.C.J., Wijfellman, C.A., Pees, E., Tak, T., and Lugtenberg, B.J.J. (1986) "Role of plant root exudate and Sym plasmid-localized nodulation genes in the synthesis by *Rhizobium leguminosarum* of Tsr factor, which causes thick and short roots on common vetch", J. Bacteriol. 165, 517-522.

Zaat, S.A.J., van Brussel, A.A.N., Tak, T., Pees, E.E. and Lugtenberg, B.J.J. (1987) "Flavonoids induce *Rhizobium leguminosarum* to produce *nodDABC* gene-related factors that cause thick short roots and root hair responses on common vetch", J. Bacteriol. 169, 3388-3391.

Zeroni, M. and Hall, M.A. (1980) in Hormonal regulation of development 1. Molecular aspects, cMacMillan, J., ed., (Springer-Verlag, Berlin/Heidelberg), pp. 511-586.

RHIZOBIUM MELILOTI NODULATION GENE REGULATION AND MOLECULAR SIGNALS

S.R. Long, R.F. Fisher, J. Ogawa, J. Swanson, D.W. Ehrhardt,
E.M. Atkinson, J.S. Schwedock.
Department of Biological Sciences, Stanford University,
Stanford CA 94305-5020 U.S.A.

Abstract

We have been studying the genes used by *Rhizobium meliloti* to form
nodules on host alfalfa plants. In recent work, we have characterized
the interaction of the NodD protein with upstream *nod* box promoters,
and have determined the circuit of analysis for the *nodD3* and *svrM*
genes. We have found that a locus on the chromosome, mutated in
strain B4, is required for full NodD activity and have discovered the
B4 gene to be at least partly homologous to a family of chaperonin
proteins. We have also used the homology of the *nodP* gene to *E. coli*
as a means to trace the function of this gene, and have examined the
response of single plant root hairs to bacterial signals by means of
electrophysiological monitoring.

Background

The ability of *Rhizobium meliloti* to form nodules on plants such as
alfalfa and its relatives (*Medicago* species) is dependent on a series
of nodulation (*nod*) genes. Our group and others have described a
number of *nod* genes that are present on the *nod-nif* megaplasmid of *R.
meliloti*. Of these, the *nodABC* genes are functionally conserved in
other *Rhizobium* species, and the *nodH*, *nodG* and *nodPQ* genes are so far
uniquely found in *R. meliloti*. The *nodFE* genes are an intermediate
type, in that they are found in other *Rhizobium* species but appear to
affect host range at a number of levels (Long, 1989 a, b; Young and
Johnston, 1989).
 Upstream of several *nod* genes is found a highly conserved DNA
segment termed the *nod* box (Rostas *et al.* , 1986). We mapped the
transcripts of nodABC, nodFE and nodH to be 26 to 28 bases downstream
of the 3' end of each corresponding *nod* box (Mulligan and Long, 1989;
Fisher *et al.*, 1987). The expression of these operons is dependent on
activation by a member of the NodD regulatory protein family (Mulligan
and Long, 1985, 1989; Honma and Ausubel 1987, 1990; Güttfert *et al.*,
1986; Kondorosi *et al.*, 1989). NodD1, located next to and divergent
from the *nodABC* genes, activates *nod* gene expression when cells are

127

H. Hennecke and D. P. S. Verma (eds.),
Advances in Molecular Genetics of Plant-Microbe Interactions, Vol. 1, 127–133.
© 1991 *Kluwer Academic Publishers. Printed in the Netherlands.*

presented with inducers such as luteolin (Peters et al., 1986) or
related flavonoids (Peters and Long, 1988; Hartwig et al., 1989, 1990;
Maxwell et al., 1989). The NodD1 and NodD3 proteins both bind to the
nod box, displaying an extensive (about 60 bp) footprint (Fisher and
Long, 1989). In strain Rm41, but not in strain SU47, a repressor is
found that binds upstream of nodD1 adjacent to the NodD footprint on
the nodABC nod box (Kondorosi et al., 1989).

Interaction of NodD with nod box

The footprint of NodD on the nod promoters is distinctive and somewhat
large. We asked what might be the basis for a relatively small
protein (about 35 kDa) to establish such a large binding site. It
remains unknown whether NodD acts as a monomer or as a multimer, and
because there is a lack of outstanding symmetry in the nod box site,
it is difficult to infer a model for the binding site based on the
mechanisms known for other proteins and promoters.

The sequence of the nodH nod box includes two fortunately placed
sites for restriction enzyme recognition, a ClaI site near the 3' end
and a BamHI site in the middle. We found that cutting at the ClaI
site and removing two bases created a nod box that was completely
unable to interact with NodD (as measured by a mobility-shift assay).
However, the result of altering the sequence in the middle of the nod
box was interestingly different. Here, we found that if 4 bases were
either removed or added at the BamHI site, the nod box was again
rendered unable to interact with NodD. However, if 10 bases --
corresponding to a full turn of the B-form of the double helix -- were
added at the BamHI site, then the nod box was actually improved in its
ability to bind NodD. This suggests that the NodD binding is
sensitive to the face of the helix on which certain bases are located,
and implies that there may be cooperativity of binding by NodD to the
two halves of the nod box.

One model that could account for this would invoke the bending of
the nod box DNA to contact two sides of a NodD protein or multimer.
We constructed nod boxes placed at varying positions on fragments of
constant length. These fragments were bound to NodD and tested for
mobility in an electrophoretic separation. The results indicated that
NodD does cause substantial bending of the nod-box DNA, above the
slight bending that the nod box shows in the absence of any
interacting protein.

The pattern of NodD-nod box contacts can be estimated by a
methylation interference assay. In this experiment, we made randomly
altered guanine residues (aiming for one change per molecule) in a
population of nod box molecules and carried out a mobility-shift
procedure. We then asked, using sequencing cleavage and analysis,
what was the difference between the molecules in the shifted
population -- in which base alterations must not interfere with NodD
binding -- and the unshifted population, in which the alteration in a
residue must have prevented the nod-box from binding NodD.
Methylations seen preferentially in the unshifted population therefore
indicate bases which recognize NodD. We found, using this procedure

on each of the three *nod* boxes, a pattern that suggests binding of NodD to two places, each on the same side of the helix, in a direct-repeat arrangement.

Role of the B4 locus in NodD action

In a screen for mutants altered in their response to luteolin induction, we recovered a Tn5-induced mutant, B4, that gave very low expression of a *nodC-lacZ* fusion on plates. Further analysis showed that this mutant appeared to have about 10-fold lower expression of this *nod* fusion, compared to wild type, in response to luteolin induction. The Tn5 co-transduced with the mutant phenotype, but did not show transductional linkage to any known *nod* genes on pSym-a. Both physical and genetic analysis now place this not on either Sym plasmid, but on the chromosome of *R. meliloti*, in the segment between *trp-33* and *pvr-49*.

We asked what was the role of the gene mutated in strain B4. This locus might be important for the response to luteolin, or it might be involved in the expression activity of NodD. Because of the presence in *R. meliloti* of the NodD3 protein, which acts independently of luteolin or other inducers, we had available a control for effects specific to inducer utilization. We found that the activity of NodD3, as well, appeared to be lessened in a B4 mutant background.

The picture became more complicated with tests for expression of the two *nodD* genes. We found the B4 mutation did not lower the expression of *nodD1*, but that the abundance of NodD3 protein was substantially lessened. When we expressed *nodD3* under the control of the *trp* promoter in expression vector pTE3 (Egelhoff and Long, 1985), we observed that the level of NodD3 protein was rather less in the B4 background than in wild type, and that the level of NodD3-caused *nod* gene activation was also lower. It is possible that effects on *nodD3* expression and/or stability, as well as on its activity, may be responsible for the mutant effects. In the case of the *nodD1* gene product, the B4 mutation may affect stability and/or activity.

To begin understanding the mechanism of the B4 mutation, we cloned the Tn5-interrupted DNA fragment from strain B4 and have established part of the nucleotide sequence. We found a potential open reading frame at the position of the Tn5 insertion, which could be part of a longer gene. By computer analysis we found a striking similarity of this portion of an ORF with the known sequence for the *GroEL* gene product of *E. coli*, and other members of this family of chaperonin proteins (Hemmingsen, *et al.* 1988). This homology suggests that a *GroEL*-like protein could be required as a participant in the folding or assembly of NodD into an active form. The full determination of the B4 gene sequence, and tests of whether this gene may play any other role in symbiosis, are currently in progress.

The biochemical function of *nodPQ*

In a large-scale mutagenesis of the region including *nodFEG* and *nodH*, we observed a single mutation about 1 kb downstream of *nodG*, that displayed a substantial delay in nodulation (Swanson *et al.*, 1987). Sequencing this region and that further downstream showed the existence of two open reading frames, *nodP* and *nodQ* (Schwedock and Long, 1989; Cervantes *et al.*, 1989). We found that the *nodP* sequence, while not homologous to anything in the computer data banks of gene sequence, did show enough homology to *E. coli* to hybridize on a Southern blot to total *E. coli* DNA (Schwedock and Long, 1989).

We set out to identify the *E. coli* homolog to *nodP* as a possible route to finding a function. The genome of *E. coli* has been cloned in an ordered Lambda library by Kohara and colleagues (1987). We screened this bank by preparing DNA from each phage, and observed hybridization to phage number 451 and 452. This corresponds to 58 minutes of the *E. coli* chromosome. No genes were shown in the Kohara map for this region, and many *E. coli* genes have not been precisely mapped. We began examining the restriction maps for genes in reported to be at 58 minutes, then progressively further from that genetic location. We found that the restriction map for *cysDNC*, previously reported to map at 59 minutes, corresponded to the restriction map for the fragment we found to be homologous to *nodP*.

The *E. coli cysDN* genes encode the proteins that constitute ATP sulfurylase (Leyh *et al.*, 1988). This enzyme carries out the reaction of ATP and inorganic sulfate to produce APS. In most organisms, APS is subsequently converted to PAPS by APS kinase, encoded in *E. coli* by the *cysC* gene (Leyh *et al.*, 1988). PAPS is then subjected to steps that reduce the sulfur to be used in cystein biosynthesis; PAPS is also the donor for activated sulfate in certain sulfurylation reactions. We tested the relationship of *nodPQ* to these reactions by a series of experiments. First, we found that cloned *nodPQ* genes controlled by an exogenous promoter could complement an *E. coli cysDN* strain to prototrophy on sulfate. Secondly, we prepared extracts of *E. coli* grown in rich medium (so that chromosomal *cys* genes would not be expressed), using strains that carried either no plasmid, the cloned *E. coli cysDN* genes, or *R. meliloti nodPQ* genes. These extracts were assayed for *in vitro* ATP sulfurylation, using TLC separation of products followed by scanning densitometry of autoradiograms. We observed that extracts made from *nodPQ*-containing strains showed the same *in vitro* ATP sulfurylase activity as those containing the *cysDN* genes. We conclude from this that *nodPQ* encode ATP sulfurylase in *Rhizobium meliloti*. Because the *nod* gene factor NodRm-1 structure reported by Lerouge *et al.* (1990) is a sulfated molecule, it seems likely that the function of *nodPQ* is to synthesize an activated sulfate intermediate that subsequently serves as sulfate donor for NodRm-1 synthesis.

Several questions remain to be answered. First, we have previously reported that there are two highly conserved copies of *nodPQ* in *Rhizobium meliloti*. The role, and regulation, of the second copy remains to be determined. Secondly, we do not know what is the relationship of either *nodP1Q1* or *nodP2Q2* to routine cysteine biosynthesis in *R. meliloti*. Third, and very importantly, we do not yet know whether there is an analog of *cysC* in *R. meliloti*, or whether the sulfur donor for NodRm-1 synthesis is APS or PAPS. There is enough DNA downstream of *nodQl* for a *cysC* analog to fit, but this has not yet been sequenced.

Reaction of root hairs to bacterial supernatants.

Faucher *et al.* (1988, 1989) have described the behavior of alfalfa root hairs when exposed to supernatants of *R. meliloti*. Several characteristic reactions, including root hair deformation, are dependent on *nod* genes being expressed in the *Rhizobium* cells, and this has now been correlated with the production of the factor NodRm-1. Bacterial factors are also invoked as a cause of cell division in the nodulation response. A major question about the response of plants to *Rhizobium* is whether it involves secondary messengers, or whether cell division and root hair curling are each caused directly by the bacterial factors.

We have examined the reaction of single plant cells to *Rhizobium* factors by monitoring the transmembrane potential of alfalfa root hair cells. Using microelectrodes pulled to 0.2 u outside diameter, we impale single root hairs on intact roots of alfalfa plants placed in a simple perfusion chamber. We observed a steady resting potential of -135 mV in conditions of 5 mM external potassium (K^+) ions. When these plants are presented with the supernatants of wild type *Rhizobium meliloti*, grown with luteolin to induce *nod* genes, the membrane potential in the root hairs shows a rapid (within 1-2 minutes) depolarization. The membrane potential recovers after about 10-15 minutes and hyperpolarizes slightly. The hair cell is then refractory to further stimulation by supernatants.

No depolarization of plant cell membranes occurs when cells are treated with supernatants of *nodA*::Tn5 cells, indicating that the factor causing depolarization is not present without action of *nod* genes. We surmised that the agent causing depolarization might be the modified tetra-NAc-glucosamine molecule, NodRm-1 (LeRouge *et al.*, 1990). Using purified NodRm-1 provided by the CNRS groups at Toulouse, we found that this molecule does cause the same depolarization and hyperpolarization when supplied to roots at 10^{-7} M. We have not yet determined how low the concentration can be dropped while maintaining the effect. Also, we do not know whether NodRm-1 is the only *Rhizobium*-produced molecule that can cause this reaction in the plant host. For now, though, it is clear that NodRm-1 is at least sufficient to cause plant membrane depolarization. The relationship of this rapid cell-autonomous reaction to the overall nodulation process, if any, now needs to be established.

Acknowledgements

This work is supported by NIH grant 5-R01-GM30962 and by Department of Energy contract DE-AS03-82ER12084. We thank K. Faull, B. Rushing, L. Zumstein, R. Scheller and M. Barnett for their contributions to these studies, and A. Bloom for preparing the manuscript.

References

Cervantes, E., Sharma, S. B., Maillet, F., Vasse, J., Truchet, G., and Rosenberg, C. (1989) "The Rhizobium meliloti host-range nodQ gene encodes a protein which shares homology with translation elongation and initiation factors", Molec. Microbiol. 3, 745-755.

Faucher, C., Camut, S., Denarie, J., and Truchet, G. (1989) "The nodH and nodQ host range genes of Rhizobium meliloti behave as avirulence genes in R. leguminosarum bv. viciae and determine changes in the production of plant-specific extracellular signals", Mol. Plant-Microbe Interactions 2, 291-300.

Faucher, C., Maillet, F., Vasse, J., Rosenberg, C., Van Brussel, A. A. N., Truchet, G., and Denarie, J. (1988) "Rhizobium meliloti host range nodH gene determines production of an alfalfa-specific extracellular signal", J. Bacteriol. 170, 5489-5499.

Fisher, R., and Long, S. R. (1989) "DNA footprint analysis of the transcriptional activator proteins NodD1 and NodD3 on inducible nod gene promoters", J. Bacteriol. 171, 5492-5502.

Fisher, R. F., Swanson, J., Mulligan, J. T., and Long, S. R. (1987) "Extended region of nodulation genes in Rhizobium meliloti 1021. II. Nucleotide sequence, transcription start sites, and protein products", Genetics 117, 191-201.

Göttfert, M., Horvath, B., Kondorosi, E., Putnoky, P., Rodriguez-Quinones, F., and Kondorosi, A. (1986) "At least two nodD genes are necessary for efficient nodulation of alfalfa by Rhizobium meliloti", J. Mol. Biol. 191, 411-420.

Hartwig, U. A., Maxwell, C. A., Joseph, C. M., and Phillips, D. A. (1989) "Interactions among flavonoid nod gene inducers released from alfalfa seeds and roots", Plant Physiol. 91, 1138-1142.

Hartwig, U. A., Maxwell, C. A., Joseph, C. M., and Phillips, D. A. (1990) "Chrysoeriol and luteolin released from alfalfa seeds induce nod genes in Rhizobium meliloti", Plant Physiol. 92, 116-122.

Honma, M., and Ausubel, F. M. (1987) "Rhizobium meliloti has three functional copies of the nodD symbiotic regulatory gene", Proc. Natl. Acad. Sci. USA 84, 8558-8562.

Honma, M. A., Asomaning, M., and Ausubel, F. M. (1990) "Rhizobium meliloti nodD genes mediate host-specific activation of nodABC", J. Bacteriol. 172, 901-911.

Kohara, Y., Akiyama, K., and Isono, K. (1987) "The physical map of the whole E. coli chromosome: Application of a new strategy for rapid analysis and sorting of a large genomic library", Cell 50, 495-508.

Kondorosi, E., Gyuris, J., Schmidt, J., John, M., Duda, E., Hoffmann, B., Schell, J., and Kondorosi, A. (1989) "Positive and negative control of nod gene expression in Rhizobium meliloti is required for optimal nodulation", EMBO J. 8, 1331-1340.

Lerouge, P., Roche, P., Faucher, C., Maillet, F., Truchet, G., Promée, J. C., and Dénarié, J. (1990) "Symbiotic host-specificity of Rhizobium meliloti is determined by a sulphated and acylated glucosamine oligosaccharide signal", Nature 344, 781-784.

Leyh, T. S., Taylor, J. C., and Markham, G. D. (1988) "The sulfate activation locus of Escherichia coli K12: Cloning, genetic, and enzymatic characterization", J. Biological Chem. 263, 2409-2416.

Long, S. R. (1989 a) "Rhizobium genetics", Annual Review of Genetics 23, 483-506.

Long, S. R. (1989 b) "Rhizobium-legume nodulation: Life together in the underground", Cell 56, 203-214.

Maxwell, C. A., Hartwig, U. A., Joseph, C. M., and Phillips, D. A. (1989) "A chalcone and two related flavonoids released from alfalfa roots induce nod genes of Rhizobium meliloti", Plant Physiol. 91, 842-847.

Mulligan, J. T., and Long, S. R. (1985) "Induction of Rhizobium meliloti nodC expression by plant exudate requires nodD", Proc. Natl. Acad. Sci. USA 82, 6609-6613.

Mulligan, J. T., and Long, S. R. (1989) "A family of activator genes regulates expression of Rhizobium meliloti nodulation genes", Genetics 122, 7-18.

Peters, N. K., Frost, J. W., and Long, S. R. (1986) "A plant flavone, luteolin, induces expression of Rhizobium meliloti nodulation genes", Science 233, 917-1008.

Peters, N. K., and Long, S. R. (1988) "Alfalfa root exudates and compounds which promote or inhibit induction of Rhizobium meliloti nodulation gene", Plant Physiology 88, 396-400.

Rostas, K., Kondorosi, E., Horvath, B., Simoncsits, A., and Kondorosi, A. (1986) "Conservation of extended promoter regions of nodulation genes in Rhizobium", Proc. Natl. Acad. Sci. USA 83, 1757-1761.

Schwedock, J., and Long, S. R. (1989) "Nucleotide sequence and protein products of two new nodulation genes of Rhizobium meliloti, nodP and nodQ", Mol. Plant-Microbe Interactions 2, 181-194.

Swanson, J., Tu, J. K., Ogawa, J. M., Sanga, R., Fisher, R., and Long, S. R. (1987) "Extended region of nodulation genes in Rhizobium meliloti 1021. I. Phenotypes of Tn5 insertion mutants", Genetics 117, 181-189.

Young, J. P. W., and Johnston, A. W. B. (1989) "The evolution of specificity in the legume-rhizobium symbiosis", TREE 4, 331-349.

GENETIC AND BIOCHEMICAL STUDIES ON THE NODULATION GENES OF *RHIZOBIUM LEGUMINOSARUM* BV. *VICIAE*

J. A. DOWNIE, C. MARIE, A-K. SCHEU, J. L. FIRMIN, K. E. WILSON,
A. E. DAVIES, T. M. CUBO, A. MAVRIDOU, A. W. B. JOHNSTON[*] AND
A. ECONOMOU
John Innes Institute
John Innes Centre for Plant Science Research
Norwich NR4 7UH, U.K. and [*]School of Biological Sciences, University of East
Anglia, Norwich NR4 7IJ, U.K.

ABSTRACT. A strain of *Rhizobium leguminosarum* bv *viciae* deleted for the *nodFELMNTO* genes (but retaining *nodDABCIJ*) is unable to nodulate vetch. The deletion mutant can be partially corrected for nodulation by plasmids carrying *nodO* or *nodFE*. However the structures and predicted functions of the *nodO* and *nodFE* gene products are very different. *nodO* is a host-specific nodulation gene and encodes a protein that is secreted by a specialised mechanism analogous to that required for haemolysin section. In contrast, the *nodFE* genes encode enzymes that appear to be involved in the formation of the host-specific root-hair curling factor. It appears that the *nodM* gene product is involved in the formation of glucosamine phosphate which is probably a precursor of the root hair curling molecule.

1. Introduction

In *Rhizobium leguminosarum* biovar *viciae*, thirteen *nod* genes have been identified as playing a role in the nodulation of legumes such as peas or vetch. These genes (Figure 1) are arranged in five operons, four of which are under the control of the constitutively-expressed *nodD* gene product which functions as a positive transcriptional regulatory protein. The NodD protein binds (Hong et al. 1987) to the highly conserved promoter elements (about 50 nucleotides in length) that precede the *nodA*, *nodF*, *nodM* and *nodO* genes and NodD probably interacts directly with flavonoid molecules (Burn et al. 1987) secreted from legume roots thereby stimulating transcription of the other *nod* operons. The expression of the *nodABCIJ*, *nodFEL*, *nodMNT* and *nodO* operons leads to the formation of signal molecules that allow nodulation to proceed. Not all of the induced *nod* genes shown in Figure 1 are necessary for nodulation to occur (Downie and Johnston, 1988). Whereas mutation of the *nodABC* genes totally blocks nodulation, mutation of the *nodF* or *nodE* genes reduces and delays nodulation (Downie et al. 1985; Spaink et al. 1989), mutation of *nodL* inhibits nodulation of peas, but not of some vetch plants (Surin and Downie, 1988) and mutations in the other *nod* genes only slightly affect nodulation (Downie et al. 1985; Economou et al. 1989; Surin et al. 1990). The lack of phenotypic effect or only partial inhibition of nodulation observed in these mutant strains can be explained in a number of possible ways; (a) specific

134

H. Hennecke and D. P. S. Verma (eds.),
Advances in Molecular Genetics of Plant-Microbe Interactions, Vol. 1, 134–141.
© 1991 *Kluwer Academic Publishers. Printed in the Netherlands.*

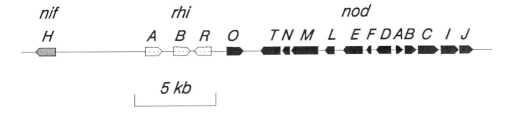

Figure 1. Map of the nodulation gene region of R. leguminosarum biovar *viciae* on the symbiotic plasmid pRL1JI. The nodulation (*nod*) and rhizosphere expressed (*rhi*) genes are shown as arrows which illustrate their orientations. The *nod* genes are in five operons (*nodABCIJ, nodD, nodFEL, nodMNT* and *nodO*) while two *rhi* operons (*rhiAB* and *rhiR*) have been identified, with the *rhiAB* genes under the control of the *rhiR* gene.

nod genes may be required for the nodulation of specific legume cultivars; (b) there could be multiple pathways of recognition encoded by *nod* genes and in the absence of one of these nodulation can still proceed; (c) some *nod* gene products could be involved in the formation of metabolites which are also produced by "housekeeping" enzymes during normal growth of the rhizobia.

In an attempt to address these hypotheses, we have carried out a coordinated genetic and biochemical study of the nodulation genes of *R. leguminosarum* bv *viciae*.

2. Results

2.1 *nodFE* AND *nodO* ENCODE COMPLEMENTARY BUT STRUCTURALLY DISTINCT PRODUCTS

Individual mutations in the *nodF,E,L,M,N,T* or *O* genes do not block nodulation. In order to determine if this is due to complementary effects of different gene products a mutant strain (A69) was made which lacks the *nodFELMNTO, rhiABR* and *nifH* gene regions due to a ≈ 20 kb deletion of the symbiotic plasmid pRL1JI. When this mutant strain was inoculated onto vetch plants no nodulation was observed (Downie and Surin, 1990) even after prolonged periods of growth. This result indicates that although mutations in the individual *nod* genes in this region cause leaky phenotypes, these genes are collectively essential for nodulation. The mutant strain could be partially restored for nodulation by recombinant plasmids carrying the *nodFE* or *nodFEL* genes. Moreover the mutant could also be partially complemented for nodulation by a cosmid clone carrying the *nodO* gene (but lacking the *nodFELMNT*) genes and this complementation was shown to be due to the presence of the NodO protein, initially by mutation of the *nodO* gene. Subsequently the subcloned *nodO* gene was shown partially to complement strain A69 for nodulation, although the level of nodulation (10-30%) was somewhat less than that observed with the cosmid (pIJ1088) carrying *nodO* and other genes to the left of *nodO* (e.g. *rhi* genes) as drawn on Figure 1.

To confirm these observations a double mutant strain was made carrying both the *nodO93*::Tn*3HoHo1* allele and the *nodE68*::Tn*5* allele on pRL1JI. When this mutant was inoculated onto vetch plants, the level of nodulation was significantly less than that observed with strains carrying single mutations in either *nodO* or *nodE*. Therefore these genes must have a synergistic effect during the nodulation process. Nevertheless, a low level of nodulation was observed with the double mutant strain, indicating that even in the absence of *nodO* and *nodE* other genes in the region deleted in strain A69 can allow a low level of nodulation of vetch. The genes that could play a role in this residual nodulation include *nodFLMT* or the *rhi* genes. The *rhi* genes appear to be specific to bv. *viciae* and three *rhi* genes have been identified (Figure 1); the *rhiAB* genes are under the control of the *rhiR* gene product which is a positively acting regulator whose transcription appears to be repressed by flavonoids.

Although the *nodE* and *nodO* genes appear to encode complementary nodulation functions, it is clear that they are biochemically and structurally distinct. The predicted protein sequence of the *nodE* gene product suggests that it is a membrane associated, (Spaink et al. 1989) protein that is homologous to the condensing enzyme subunit of the fatty-acid synthetase (Bibb et al. 1989); as such it probably functions in conjunction with the *nodF* gene product, an acyl carrier-like protein (Shearman et al. 1986). In contrast, *nodO* encodes a secreted (de Maagd et al. 1989; Economou et al. 1990), Ca^{2+} binding protein (Economou et al. 1990) which has been suggested to interact directly with plant cells. Thus it appears that there are at least two parallel nodulation processes, one encoded by *nodO* and the other by *nodFE* and these are additive during nodulation.

2.2. THE *nodFEL* GENE PRODUCTS ARE INVOLVED IN THE SYNTHESIS OF HOST-SPECIFIC ROOT HAIR CURLING FACTOR

As mentioned above, the *nodFE* gene products are likely to be involved in reactions analogous to fatty-acid biosynthesis and additionally protein sequence homologies suggest that the *nodL* gene product is probably an acetyltransferase (Downie, 1989). On the basis of these observations it appeared likely that the formation of their metabolic products could be followed using acetate labelling experiments. After ^{14}C-acetate or ^{14}C-glucose labelling of the wild-type strain used as a control, the growth medium supernatant was passed through a preparative C-18 column which was eluted with methanol and then fractionated on an analytical C-18 column using a methanol gradient. At about 40% methanol a radioactive peak was found to co-elute with those fractions that had vetch root hair curling activity. Both the root hair curling activity and peak of radioactivity were absent from strains carrying the *nodC128*::Tn*5* mutation confirming that the radioactively-labelled peak corresponds with the root hair curling factors made by the *nodABC* gene products. When strains carrying mutations in *nodE* or *nodL* were used in similar experiments, the elution profile was altered indicating that their gene products modify the root hair curling molecule.

Lerouge et al. (1990) have described the structure of the root hair curling factor made by *R. meliloti* and it is an acylated, sulphated N-acetyl-glucosamine tetramer. It is proposed that *R. leguminosarum* bv *viciae* makes a related molecule and that the *nodL* gene product is involved in the acetylation of the glucosamine while the *nodFE* genes are involved in determining the type of acyl group that is present.

It appears that the *nodABC* gene products in the deletion mutant A69 (lacking the *nodFELMNTO* genes) make a partially functional root hair curling molecule since the mutant can induce root hair deformation on vetch. However the properties of this molecule are altered since the root hair deformation activity is not retained by a hydrophobic (C18) reverse phase column. Presumably, this molecule is not sufficient to enable the deletion mutant strain of bv *viciae* to nodulate vetch. Therefore we can conclude that nodulation requires additional signals delivered by either of two routes; either by direct modification of the factor by the *nodFEL* gene products or surprisingly by the secretion of a protein (NodO) by the bacteria.

2.3 THE *nodO* GENE IS SPECIFIC FOR BIOVAR *viciae* AND THE SECRETION OF NodO REQUIRES A SPECIALISED MECHANISM

When a *nodO*-specific probe was hybridised (at low stringency) to DNA from *R. meliloti*, *B. japonicum* and to several strains of the *R. leguminosarum* biovars *viciae*, *trifolii* and *phaseoli*, no significant hybridisation was found except with DNA from strains of bv *viciae*. In all, 10 strains of bv. *viciae* were tested and all showed homology to the *nodO* gene. Therefore it is concluded that *nodO* is a host-specific nodulation gene.

It was shown previously (Economou et al. 1990) that the *nodO* gene product, which is homologous to haemolysin and related secreted proteins, is secreted even in strains of *R. leguminosarum* lacking a *sym* plasmid. Furthermore, the genes involved in NodO secretion must be regulated differently from the *nodO* gene itself. When a strain of *R. leguminosarum* carrying a constitutively expressed *nodO* gene (expressed from a vector promoter) was grown in the presence or absence of flavonoids, similar amounts of NodO protein was secreted. Therefore it appears that the genes required for NodO export are expressed even in the absence of flavonoids. When the constitutively expressed *nodO* gene was transferred to *E. coli*, no secretion of NodO was found. However, a strain of *E. coli* carrying the haemolysin secretion genes (*hylBD*) together with *nodO* was able to secrete the NodO protein into the growth-medium supernatant. Therefore we conclude that in bv. *viciae*, other genes (analogous to *hlyBD*) are required for NodO export, and these genes must be genetically unlinked to *nodO* and expressed constitutively.

2.4 *nodM* ENCODES AN ENZYME THAT IS INVOLVED IN GLUCOSAMINE SYNTHESIS

The DNA sequence of the *nodM* gene was established previously and when a translation of this sequence was used to screen a translation of the DNA sequence databases, strong homology was found to the *E. coli* gene *glmS*. The reaction catalysed by the *glmS* gene product is the formation of glucosamine-6-phosphate and glutamate from glutamine plus fructose-6-phosphate. It is possible to measure this reaction by measuring the glucosamine-6-phosphate product as described by Badet et al. (1987). A wild-type and NodM⁻ mutant strain of bv, *viciae* were each grown in the presence or absence of the flavonoid eriodictyol to induce *nod* gene expression. In the presence of the flavonoid, it was possible to measure a higher level of glucosamine-6-phosphate formation than that seen for the mutant strains grown under similar conditions. However in the mutant strain there was a residual level of glucosamine-6-phosphate formation. Significantly, similar low rates of glucosamine-6-phosphate formation could be measured in both the wild-type and NodM mutant strain grown in the absence of eriodictyol. Therefore, it is concluded that this strain of bv. *viciae* has two glucosamine-phosphate synthase genes, one of which (*nodM*) is induced in the presence of flavonoids.

Significantly, when DNA from the wild-type strain was probed at low stringency with the *nodM* gene, two hybridising bands were found and it is thought that the weaker hybridising band corresponds to a "housekeeping" *glmS* homologue. In strains of *R. leguminosarum* cured of their *sym* plasmid the weaker hybridising band was retained suggesting that this gene may be chromosomally located.

3. Discussion

The aim of this work was to determine why mutations in some *nod* genes have partial or minimal effects on nodulation and three proposals were tested to explain these observations.

Firstly it has been shown that mutations in the *nodL* gene strongly inhibit the nodulation of peas and lentils but had little effect on the nodulation of vetch (Surin and Downie, 1988). Furthermore it has also been established that some strains of *R. l.* bv. *viciae* carry an additional nodulation gene (*nodX*) that is not necessary for nodulation of some cultivars of peas but is essential for the nodulation of others (Götz et al. 1985; Davis et al. 1988). Therefore it is clear that some *nod* genes are important for the nodulation of specific legumes or legume cultivars.

Secondly it is probable that nodulation may occur via multiple recognition pathways since either of two *nod* gene regions (*nodO* or *nodFE*) can partially restore a *nod* deletion mutant for nodulation. Since the *nodO* and *nodFE* gene products are quite different it is likely that their modes of actions are very different. It is proposed that the *nodFE* gene products modify the root hair curling molecule, probably by determining the type of acyl group whereas the NodO protein is secreted and probably interacts directly with legume root cells. Interestingly, in a strain lacking *nodO* and *nodE* a low level of nodulation was still seen. This observation underlines the importance of other genes such as the *nodLMNT*, and *rhi* genes and current work is aimed at identifying their functions in nodulation.

Thirdly it was proposed that some *nod* gene products may form metabolites that are also made by rhizobial "housekeeping" enzymes. This appears to be the case for the *nodM* gene product which may be involved in the formation of glucosamine phosphate. It is most likely that this product is used in the formation of the root hair curling molecule whose synthesis also requires the products of the *nodABC* genes.

Given that Lerouge et al. (1990) have identified the structure of the root-hair-curling molecule made by *R. meliloti*, it is possible to draw parallels with *R. leguminosarum* and propose working hypotheses for the possible biochemical functions of some of the *nod* gene products. As shown in Figure 2, the molecule described by Lerouge et al. (1990) is a tetraglucosamine polymer that is substituted with one N-acyl group, three N-acetyl groups and a sulphate group.

It is possible that the *nodM* gene product is involved in increasing the pool of glucosamine (or derivatives of glucosamine) that could act as precursors for the synthesis of such a root hair curling molecule (Figure 2). The observation that mutation of *nodM* does not block root-hair curling can be explained because there appears to be a second "housekeeping" gene homologous to *nodM*; it is apparent then that even in the absence of *nodM* there would be sufficient precursors that would have the potential to form the root-hair-curling molecule. Since a *Rhizobium* strain carrying the *nodABC* genes (but lacking other *nod* genes) can induce root-hair deformation on vetch (Knight et al., 1985) it is reasonable to propose that the

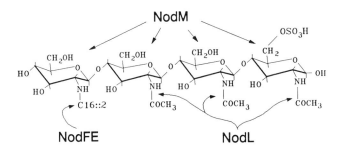

Figure 2. Structure of the root-hair-curling molecule described by Lerouge et al. (1990) indicating possible roles for *nod* gene products.

nodABC gene products are involved in the polymerisation of glucosamine derivatives to form a backbone precursor of the mature root-hair-curling molecule described by Lerouge et al. (1990).

Other *nod* gene products could be involved in other substitutions on this structure. Thus for example, since NodL is homologous to acetyl transferases (Downie, 1989), a possible role for NodL could be to acetylate the glucosamine residues (Figure 2). Further, since the *nodFE* gene products are both homologous to enzymes involved in fatty-acid biosynthesis, it is possible that they play a role in determining the type of fatty acid present (Figure 2). Significantly these *nodFE* genes determine the nodulation specificity of the *R. leguminosarum* biovars *viciae* and *trifolii* (Surin and Downie, 1989; Spaink et al., 1989). A prediction of this hypothesis would be that these two biovars make root-hair-curling molecules that differ in their acyl substituents.

Since the *nodIJ* gene products are homologous to other active transport proteins (Evans and Downie, 1986) it is possible that they may be involved in the efficient export of the root-hair-curling factor.

Thus it is possible to propose potential biochemical functions for most of the *nod* genes identified in *R. leguminosarum* bv. *viciae* and these proposals can be tested experimentally. However, if this model of *nod* gene function does hold, we are faced with the intriguing question of how a strain deleted for the *nodFELMNTO* gene region can be partially restored for nodulation by either *nodFE* or *nodO*, genes which encode proteins of apparently very different functions.

ACKNOWLEDGEMENTS

This work was supported by the AFRC. Additional financial support was provided by the European Community with Sectoral Training Grants (M.T.C., A.M. and A.E.), the Spanish Ministry of Education (M.T.C.) and the John Innes Foundation (C.M.). We are grateful to M-A. Barny for comments on the manuscript.

REFERENCES

Badet, B., Vermoote, P., Haumont, P-Y., Lederer, F., and Le Goffic, F. (1987) 'Glucosamine synthetase from *Escherichia coli*: purification, properties and glutamine-utilizing site location', Biochemistry **26**, 1940-1948.

Bibb, M.J., Biro, S., Motamedi, H., Collins, J.F., and Hutchinson, C.R. (1989) 'Analysis of the nucleotide sequence of the *Streptomyces glaucescens tcm1* genes provides information about the enzymology of polyketide antibiotic biosynthesis', EMBO J. **8**, 2727-2736.

Burn, J.E., Rossen, L., and Johnston, A.W.B. (1987) 'Four classes of mutations in the NodD gene of *Rhizobium leguminosarum* and *R. phaseoli*', Genes Dev. **1**, 456-464.

Davis, E.O., Evans, I.J., and Johnston, A.W.B. (1988) 'Identification of *nodX*, a gene that allows *Rhizobium leguminosarum* biovar *viciae* strain TOM to nodulate Afghanistan peas', Mol. Gen. Genet. **212**, 531-535.

De Maagd, R.A., Wijfjes, A.H.M., Spaink, H.P., Ruiz-Sainz, J.E., Wijffelman, C.A., Okker, R.J.H., and Lugtenberg, B.J.J. (1989b) '*nodO*, a new *nod* gene of the *Rhizobium leguminosarum* biovar *viciae* Sym plasmid encodes a secreted protein', J. Bacteriol. **171**, 6764-6770.

Downie, J.A. (1989) 'The *nodL* genes from *Rhizobium leguminosarum* is homologous to the acetyl transferases encoded by *lacA* and *cysE*', Mol. Micro. **3**, 1649-1651.

Downie, J.A., and Johnston, A.W.B. (1988) 'Nodulation of legumes by *Rhizobium*', Plant, Cell and Environment, **11**, 403-412.

Downie, J.A., Knight, C.D., Johnston, A.W.B., and Rossen, L. (1985) 'Identification of genes and gene products involved in nodulation of peas by *Rhizobium leguminosarum*', Mol. Gen. Genet. **198**, 255-262.

Downie, J.A., and Surin, B.P. (1990) 'Either of two *nod* gene loci can complement the nodulation defect of a *nod* deletion mutant of *Rhizobium leguminosarum* bv. *viciae*', Mol. Gen. Genet. **222**, 81-86,

Economou, A., Hamilton, W.D.O., Johnston, A.W.B., and Downie, J.A. (1990) 'The *Rhizobium* nodulation gene *nodO* encodes a Ca^{2+} binding protein that is exported without N-terminal cleavage and is homologous to haemolysin and related proteins', EMBO J. **9**, 349-354.

Economou, A., Hawkins, F.K.L., Downie, A.J., and Johnston, A.W.B. (1989) 'Transcription of *rhiA*, a gene on a *Rhizobium legumonosarum* bv. *viciae* Sym plasmid, requires *rhiR* and is repressed by flavonoids that induce *nod* genes', Mol. Microbiol. **3**, 87-93.

Evans, I., and Downie, A.J. (1986) 'The *nodI* product of *Rhizobium leguminosarum* is closely related to ATP-binding bacterial transport proteins: nucleotide sequence of the *nodI* and *nodJ* genes', Gene **43**, 95-101.

Götz, R., Evans, I.J., Downie, J.A., and Johnston, A.W.B. (1985) 'Identification of the host-range DNA which allows *Rhizobium leguminosarum* strain TOM to nodulate cv. Afghanistan peas', Mol. Gen. Genet. **201**, 296-300.

Hong, G-F., Burn, J.E., and Johnston, A.W.B. (1987) 'Evidence that DNA involved in the expression of nodulation *nod* genes in *Rhizobium*, binds to the product of the regulatory gene *nodD*', Nucl. Acids Res. **15**, 9677-9690.

Knight, C.D., Rossen, L., Robertson, J.G., Wells, B., and Downie, J.A. (1986) 'Nodulation inhibition by *Rhizobium leguminosarum* multicopy *nodABC* genes and analysis of early stages of plant infection', J. Bacteriol. **166**, 552-558.

Lerouge, P., Roche, P., Faucher, C., Maillet, F., Truchet, G., Prome, J.C., and Denarie, J. (1990) 'Symbiotic host-specificity of *Rhizobium meliloti* is determined by a sulphated and acylated glucosamine oligosaccharide signal', Nature **344**, 781-784.

Shearman, C.A., Rossen, L., Johnston, A.W.B., and Downie, J.A. (1986) 'The *Rhizobium* gene *nodF* encodes a protein similar to acyl carrier protein and is regulated by *nodD* plus a factor in pea root exudate', EMBO J. **5**, 647-652.

Spaink, H.P., Okker, R.J.H., Wijffelman, C.A., Pees, E., and Lugtenberg, B.J.J. (1987) 'Promoters in the nodulation region of the *Rhizobium leguminosarum* Sym plasmid pRL1JI', Plant Mol. Biol. **9**, 27-39.

Surin, B.P., and Downie, J.A. (1988) 'Chracterization of the *Rhizobium legumonosarum* genes *nodLMN* involved in efficient host specific nodulation. Mol. Microbiol. **2**, 173-183.

Surin, B.P., and Downie, A.J. (1989) '*Rhizobium leguminosarum* genes required for expression and transfer of host specific nodulation', Plant Mol. Biol. **12**, 19-29.

Surin, B.P., Watson, J.M., Hamilton, W.D.O., Economou, A., and Downie, J.A. (1990) 'Molecular characterization of the nodulation gene *nodT* from two biovars of *Rhizobium leguminosarum*', Mol. Microbiol. **4**, 245-252.

THE BIOCHEMICAL FUNCTION OF THE *RHIZOBIUM LEGUMINOSARUM* PROTEINS INVOLVED IN THE PRODUCTION OF HOST SPECIFIC SIGNAL MOLECULES

H.P. Spaink[1,2], O. Geiger[1], D.M. Sheeley[3], A.A.N. van Brussel[2], W.S. York[4], V.N. Reinhold[3], B.J.J. Lugtenberg[2], and E.P. Kennedy[1]
1: Harvard University Medical School, Boston, MA, U.S.A.; 2: Leiden University, Leiden, The Netherlands; 3: Harvard University School of Public Health, Boston, MA, U.S.A.; 4: Complex Carbohydrate Research Center, Athens, GA, U.S.A.

ABSTRACT

In *R.leguminosarum* the NodF and NodE proteins play an important role in the determination of host specificity of nodulation [1]. We have shown that the NodF protein contains a 4'-phosphopantetheine prosthetic group. This result suggests that NodF protein is involved in the synthesis of polyketide derivatives. After labeling with ^{14}C-acetate we have isolated the compounds produced by this nodulation protein. We have also investigated the role of the other nodulation proteins. In addition to the regulatory NodD protein, the NodABC and NodFEL proteins appear to be sufficient to produce the five detected wild-type Nod metabolites. Physical and chemical studies indicate that these *R.leguminosarum* compounds differ significantly from the reported *R.meliloti* signal compound [2]. The biological activity of several Nod metabolites was studied in three different bioassays on *Vicia sativa* plants showing a biological functionality of the O-acetyl modification produced by NodL protein.

Introduction

Bacteria of the genus *Rhizobium* are able to interact symbiotically with specific legume host plants leading to the formation of nitrogen-fixing root nodules. During the onset of the symbiosis flavonoids, secreted by the plant host, cause the induction of rhizobial nodulation (*nod*) genes. The *nod* genes are essential for the symbiotic interaction and consist of the common *nodABCIJ* genes which are functionally interchangeable between different *Rhizobium* species and of host-specific *nod* genes determining the host range of nodulation [3]. To establish the difference in host range between *R.leguminosarum* and *R.trifolii* the *nodE* gene, which is part of the *nodFEL* operon, is of greatest importance (Figure 1) [1]. A region of the NodE protein has been localized which determines its host-specific properties [1]. The NodF and NodE proteins are involved, together with the common nodulation genes, in the production of a host-specific rhizobial signal which activates the transcription of host plant genes [4].

An indication of the function of the NodFEL proteins comes from homology studies. NodL protein appears to be homologous to the acetyl

142

H.Hennecke and D. P. S. Verma (eds.),
Advances in Molecular Genetics of Plant-Microbe Interactions, Vol. 1, 142–149.
© 1991 *Kluwer Academic Publishers. Printed in the Netherlands.*

transferases LacA and CysE [5] [for *R.trifolii* NodL: J.Weinman, personal communication]. NodE protein shares homology with a group of β-ketoacyl synthases like the *E.coli* condensing enzyme of fatty acid biosynthesis, FabB [6], and those presumed to be involved in the synthesis of β-ketide antibiotics in *Streptomyces* species [6]. NodF was found to be homologous to acyl carrier proteins (ACP) [7]. The ACP of *E.coli* is a small, anionic protein that functions as an essential component of enzyme systems for the biosynthesis of fatty acids, membrane phospholipids and of lipopolysaccharides [8]. Each of these functions involves an acyl residue linked to the sulfhydryl group of the 4'-phosphopantetheine prosthetic group of ACP. A very different function for the ACP of *E.coli* in the enzymic synthesis of MDO (membrane-derived oligosaccharides) from UDP-glucose was reported by Therisod *et al.*[9]. This function in the synthesis of a cell-surface carbohydrate does not require the 4'-phosphopantetheine prosthetic group [10].

Homologies in protein primary structures ,however, are only weak indications since little is known about structure-function relations of proteins. Furthermore, multiple functions of proteins, like in the case of ACP, make it difficult to speculate which functional characteristics could be expected. In this paper biochemical studies are presented which indicate a function for the NodFEL proteins. This was done by studying the proteins themselves or the metabolites produced by means of these Nod proteins.

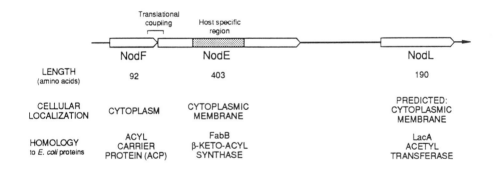

	NodF	NodE	NodL
LENGTH (amino acids)	92	403	190
CELLULAR LOCALIZATION	CYTOPLASM	CYTOPLASMIC MEMBRANE	PREDICTED: CYTOPLASMIC MEMBRANE
HOMOLOGY to *E. coli* proteins	ACYL CARRIER PROTEIN (ACP)	FabB β-KETO-ACYL SYNTHASE	LacA ACETYL TRANSFERASE

Figure 1. Characteristics of the NodFEL operon. The position and size of the NodFEL products are indicated. The NodF and NodE protein are translationally coupled (H.P. Spaink *et al.* in prep.). The cellular localization of the proteins and homologies to known *E.coli* proteins are indicated below their position in the genetic map. The cellular localization of the NodE protein has been reported [1]. The cellular localization of the NodF protein has been determined using antibodies and radiolabeled NodF protein (data not shown). The localization of the NodL protein is predicted by its primary structure [11]

Isolation and characterization of the NodF protein; NodF carries a 4'-phosphopantetheine

To learn more about the biochemical function of NodF in creating a host-specific signal we have purified the NodF protein of *R.leguminosarum*. The purified NodF protein migrates as a protein of apparent molecular weight of 5000 Dalton, considerably faster than its predicted molecular weight of 9946 Dalton. Furthermore, antibodies have been produced against the native NodF protein which were used to localize the NodF protein in the cytoplasm. To test whether NodF could indeed be involved in the production of polyketide derived compounds we investigated whether NodF carries a 4'-phosphopantetheine prosthetic group which is essential for an acyl carrier function. If NodF carries a 4'-phosphopantetheine, it should become labeled with β-alanine which is a precursor of this prosthetic group. After labelling with radioactive ß-alanine a protein has been purified which possesses a N-terminal amino acid sequence which is expected for the NodF protein as determined by Edman degradation analysis. Treatment of NodF protein with mild alkali releases labeled 4'-phosphopantetheine by β-elimination. These experiments show that NodF protein carries 4'-phosphopantetheine.

We have constructed several mutants of NodF using site directed mutagenesis. Future study of these mutants will give more information on the structure-function relation of the NodF protein and the biological relevance of the 4'-phosphopantetheine.

Isolation and characterization of the rhizobium signal compounds produced by means of the nodulation proteins

RADIOACTIVE LABELING OF *Rhizobium* COMPOUNDS PRODUCED BY MEANS OF THE nod GENES

The presence of a 4'-phosphopantetheine on NodF suggests that it is indeed involved in the synthesis of β-ketides. It is therefore expected that the catabolites produced by means of NodF as a carrier protein can be efficiently labeled by radio-labeled precursors of the fatty acid synthetic pathway. Efficient labeling of *R.leguminosarum* cells was achieved using ^{14}C-acetate in minimal growth medium. As shown in Figure 2 (left panel) five compounds were isolated from the culture medium of flavonoid induced *R.leguminosarum* cells. These five compounds could be detected in culture medium within 2 hours after induction. Five flavonoid inducible compounds with similar mobility on reverse phase TLC plates were also present in the growth medium of *R.trifolii* strain ANU843 (data not shown).

ANALYSIS OF THE GENETICAL REQUIREMENTS FOR Nod METABOLITE PRODUCTION

Using the rapid test system described in figure 3 a large number of *Rhizobium* strains containing Tn5 insertions or cloned nodulation genes were tested. Some of the results are also shown in figure 3. The results of all tested strains can be summarized as follows: (1) In cells containing a polar Tn*5* insertion in the *nodD* or *nodABC* genes no flavonoid-inducible compounds could be detected in any of the cellular

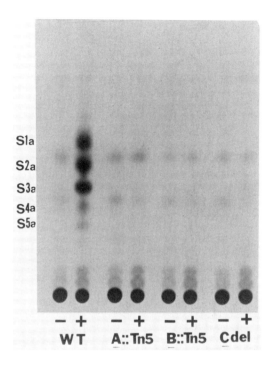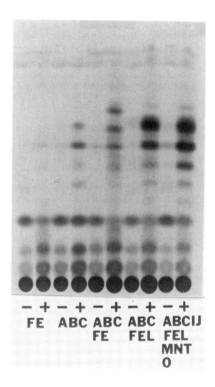

Sla
S2a
S3a
S4a
S5a

— + — + — + — +
WT   A::Tn5 B::Tn5 Cdel

— + — + — + — + — +
FE ABC ABC ABC ABCIJ
    FE FEL FEL
        MNT
        O

Figure 2. Analysis of flavonoid-induced production of compounds on reverse phase (C18) TLC plates. *Rhizobium* strains were grown in the absence (-) or presence (+) of the flavonoid naringenin. Bacteria were removed by centrifugation and the growth medium was subsequently extracted with n-butanol. In the panel on the left a wild type (WT) *R.leguminosarum* (RBL5560) or *nod* mutants were studied. The strain indicated as Cdel contains a non-polar deletion in *nodC*. In the panel on the right sym plasmid-cured *Rhizobium* strains (LPR5045) containing cloned *nod* genes were used. In the latter strains the regulatory *nodD* gene was always present.

compartments nor in the growth medium. Therefore we can conclude that at the least the NodD and NodC proteins, are essential for the production of all five detected compounds. The detected flavonoid-inducible compounds have therefore be designated as Nod metabolites. (2) Tn5 insertion in the *nodFEL* genes results in production of Nod metabolites with a different migration on TLC plates. (3) A qualitative or quantitative influence of Tn5 insertions in the other *nod* genes (*nodMNT*, *nodO* and *nodIJ* in *R.leguminosarum*) could not be detected. (4) A *Rhizobium* strain containing the cloned *R.leguminosarum* genes *nodABC* and *nodFEL* produced five flavonoid-inducible compounds with identical mobility as the compounds produced by the wild type strain on reverse phase TLC plates (Figure 2, right panel) and in two dimensional TLC using silica gel. (5) Four flavonoid-inducible compounds could be detected in a *Rhizobium* strain containing the cloned *nodABC* and *nodD* genes and no other *nod* genes present (Figure 2, right panel). The presence of the *nodFE* genes in the latter strain cause the production of at least one new Nod metabolite (Figure 2). (6) In *Rhizobium* strain containing the *nodFE* and *nodD* genes but no other *nod* genes, no flavonoid inducible compounds could be detected neither in the growth medium nor in any of the cellular compartments. These results suggests that for the production of at least one wild type Nod metabolite the presence of the NodFE together with NodABC is essential. (7) The presence of the *nodL* gene modifies the produced compounds in strains containing NodD, NodABC and NodFE proteins (Figure 2) and in strains containing only NodD and NodABC (results not shown). (8) The addition of other *nod* genes does not alter the detected Nod metabolites in their mobility in TLC (Figure 2).

CHEMICAL CHARACTERIZATION OF Nod METABOLITES

Several of the Nod metabolites or intermediates detected in Figure 3 have been purified with column liquid chromatography using (1) silica gel, (2) reverse phase (C18) silica gel, (3) ion exchanger, (4) sephadex LH20, and (5) reverse phase HPLC. The isolated compounds were compared in reverse phase TLC and two dimensional silica TLC (UV detection) with the radioactively labeled compounds of Figure 2 and were designated accordingly (Figure 2). They were also designated as Nod metabolites since they could not be isolated from a *Rhizobium* strain which did not contain *nod* genes. Of six isolated compounds the molecular mass could be assigned with mass spectroscopy using fast atom bombardment (FAB) in the positive mode (Figure 3). From these results it can be concluded that the compounds designated S1a and S2a of the wild type are identical to products produced by a strain containing NodD, NodABC and NodFEL and no other Nod proteins. Furthermore, the S1a and S2a products differ from S1b and S2b by 42 mass units which suggests a difference of one acetyl group. Chemical analysis and NMR studies indicate that the extra acetyl group is an O-acetyl substitution of a N-acetylglucosamine residue. Since the NodL protein is the only difference in the strains producing these two sets of compounds it can be concluded that the NodL protein is responsible for the addition of an O-acetyl group to S1b and S2b. A molecular mass difference of 203 mass units is observed between S1a and S2a and also between S1b and S2b indicative for a difference of one N-acetyl hexosamine.

Further analysis of compound S1a indicates that it is similar to the reported signal compound of *R.meliloti* in that it contains a β-linked poly-N-acetylglucosamine moiety [2] and a fatty acyl chain. However, besides the extra O-acetyl substitution the following other differences have been observed as well. (1) The compound is not ionic and therefore does not contain a sulfate substitution. (2) The fatty acyl chain is mono-unsaturated and contains a double bond at a different, namely the fifth, position. (3) The oligosaccharide backbone contains more sugar units all of which are probably β-linked N-acetylglucosamines. Further analysis is in progress to discover more details on the molecular structure of all characterized Nod metabolites. As in the case with NodL this might lead to a further understanding of the biochemical function of the NodABC and NodFE proteins.

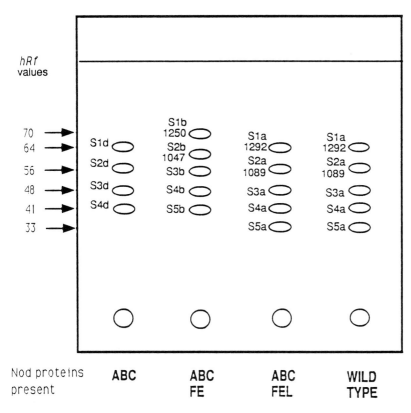

Figure 3. Designation and molecular mass assignment of Nod metabolites. The molecular mass was determined by mass spectroscopy using FAB and detection in the positive mode.

BIOLOGICAL ACTIVITY OF THE Nod METABOLITES

Five purified Nod metabolites have been tested in several bio-assays on *Vicia sativa* plants [12]. The results, presented in table 1, indicate a biological functionality of the O-acetyl substitution produced by NodL. The lack of TSR activity of S1d at the tested concentration could be a result of the absence of a modification by the NodFE proteins. Other results indicate that the absence of activity of S1d could be a result of different solubility in water compared to S1b and S2b. A further characterization of the other Nod metabolites and a comparison with the activity of compounds which have been isolated from *R.trifolii* is in progress.

TABLE 1. Results of bioassays on *Vicia* plants using the Nod metabolites indicated in Figure 4. Concentrations tested were 2 x 10^{-8} M as determined by UV absorption of the purified compounds. Abbreviations used for the bioassays: HAD, Root hair deformation; TSR, Thick and short roots; INI, increased *nod* gene inducing activity [12]. -: no significant activity; +: significant activity. As a control was used a neighboring fraction in the final HPLC purification step of S1a.

tested compound	HAD	TSR	INI
S1a	+	+	+
S2a	+	+	+
S1b	+	+	−
S2b	+	+	−
S1d	+	−	−
control	−	−	−

Discussion

We have shown that the host specificity-determining NodFEL proteins together with the common NodABC proteins are involved in the production of five compounds, designated Nod metabolites, at least two of which have biological activity in low concentrations. These can therefore be designated as signal compounds. Molecular characterization of these compounds indicate several structural similarities and also several differences with the reported *R.meliloti* signal compound. Since the NodF and NodE proteins are likely to be involved in polyketide synthesis it is tempting to speculate that they are involved in the production of the fatty acyl chain of the signal compounds, especially since these are different in S1a and the reported *R.meliloti* signal compound. However care must be taken in this consideration since also the compounds S1d to S4d (Figure 3) which are produced in the absence of the NodFE proteins are similarly hydrophobic and are likely to contain a fatty acyl chain as well. A difference in the oligosaccharide chain length of S1a and S2a, the former of which is relatively much better produced in NodFE overproducing strains (Figure 2) could even be an indication for a transglucosylation activity of the NodF protein. A conclusion on the function of the NodFE proteins should therefore await further biochemical studies and the characterization of the other detected Nod metabolites (Figure 3). The role of the NodL protein as a transacetylase is confirmed by our observation of the O-acetyl substitution on S1a and S2a. The results of bio-assays with isolated compounds (table 1) imply that this O-acetyl substitution results also in a different biological activity between the *R.leguminosarum* and *R.meliloti* compounds. However, since the relevance of the INI phenotype for the symbiosis is still not understood, other bio-assays should be developed for *Vicia* plants to test the symbiotic importance of the O-acetyl moiety.

Presently we are testing the isolated Nod metabolites for binding with the host plant lectin. Preliminary results indicate that Sla indeed binds to the pea lectin Psl. This binding could be partially reversed by the addition of high concentrations of mono-saccharide lectin haptens. Further study and a comparison with the lectin binding capacity of *R.trifolii* and *R.meliloti* signal compounds are in progress to test the specificity of the binding reaction.

Acknowledgements

We thank H.R.M. Schlaman (Leiden University, The Netherlands) for critically reading the manuscript and T. Tak (Leiden University, The Netherlands) for technical assistance. This research was supported by grants GM19822 and GM22057. H.P. Spaink was supported by a NATO fellowship from the Netherlands Organization for Scientific Research. Otto Geiger is a Feodor Lynen Fellow of the Alexander von Humboldt Foundation (FRG).

References

1. Spaink, H.P., Weinman, J., Djordjevic, M.A., Wijffelman C.A., Okker R.J.H., and Lugtenberg, B.J.J. (1989) *EMBO J.* **8**, 2811-2818.
2. Lerouge, P., Roche, P., Faucher, C., Maillet, F., Truchet, G., Promé, J.C., and Dénarié, J. (1990) *Nature* **344**, 781 -784.
3. Long, S.R. (1989) Cell **56**, 203-214.
4. Scheres, B.C., van de Wiel, A., Zalensky, A., Horvath, B., Spaink, H.P., van Eck, H., Zwartkruis, F., Wolters, A.-M., Gloudemans, T., van Kammen A., and Bisseling, T. (1990) Cell **60**, 281-294.
5. Downie, J.A. (1989) Mol. Microbiol. **3**, 1649-1651.
6. Bibb, M.J., Biro, S., Motamedi, H., Collins, J.F., and Hutchinson, C.R. (1989) EMBO J. **8**, 2727-2736.
7. Shearman, C.A., Rossen, L., Johnston, A.W.B., and Downie, J.A. (1986) EMBO J. **5** , 647-652.
8. Cronan, J.E., and Rock, C.O. (1987) In: *Escherichia coli* and *Salmonella typhymurium*; cellular and molecular biology, pp. 474-497 (Eds., Ingraman J.L. *et al.*) American Society for Microbiology, Washington.
9. Therisod, H., Weissborn, A.C., and Kennedy, E.P. (1986) Proc.Natl. Acad. Sci. USA **83** , 7236-7240.
10. Therisod, H., and Kennedy, E.P. (1987) Proc. Natl. Acad. Sci. USA **81**, 8235-8238.
11. Canter Cremers, H.C.J., Spaink, H.P., Wijfjes, A.H.M., Pees, E., Wijffelman, C.A., Okker, R.J.H., and Lugtenberg, B.J.J. (1989) Plant Mol. Biol. **13** , 163-174.
12. van Brussel, A.A.N., Recourt, K., Pees, E., Spaink, H.P., Tak, T., Wijffelman, C.A., Kijne, J.W., and Lugtenberg, B.J.J. (1990) J.Bacteriol. **172** , 5394-5401.

Send all correspondence to: H.P. Spaink, Department of Plant Molecular Biology, Leiden University, Nonnensteeg 3, 2311 VJ Leiden, The Netherlands.

STUDIES ON THE FUNCTION OF *RHIZOBIUM MELILOTI* NODULATION GENES

J. SCHMIDT[1], M. JOHN[1], U. WIENEKE[1], G. STACEY[1,2],
H. RÖHRIG[1], and J. SCHELL[1]

[1]*MPI für Züchtungsforschung* [2]*Dept. of Microbiology*
Carl-von-Linné-Weg 10 *The University of Tennessee*
D - 5000 Köln 30, FRG *Knoxville, TN 27996-0845, USA*

ABSTRACT. The common nodulation genes *nodABC* and the host range genes of *Rhizobium meliloti* are involved in generating extracellular factors that control plant morphogenesis. The NodA and NodB proteins produce compounds that stimulate the mitosis of various plant protoplasts. To study the biological role of the bacterial signals in the plant we transformed tobacco with the *nodAB* genes expressed from plant promoters in different combinations. The biological effects caused by these gene products in the transgenic plants are discussed. Since the NodA and NodB proteins are not sufficient to induce root-hair curling the presence of the NodC cell-surface protein is additionally required. To evaluate the particular role of the NodC transmembrane protein in the plant-bacteria interaction we used immunochemical techniques for the analysis of the functional domains of this protein.

1. Introduction

The soil bacterium *Rhizobium* releases different signal molecules which control in the leguminous plant such functions as growth and differentiation so that an organ, the root nodule, is formed in which the bacteria are able to fix nitrogen. For the synthesis of these signal molecules in *Rhizobium* the common *nodABC* genes [1,2,3,4] and host-range genes [5,6,7,8] are essential.

The expression of these genes is activated by the product of the *nodD* gene [9,10] in the presence of flavonoid compounds which are exuded from the roots of the host plants [11]. Recently it was shown that in *R. meliloti* 41 the expression of the common *nod* genes is not only under positive but also under negative control [12]. In the proposed model a negative transacting factor regulates the concentration of the NodD activator protein in the cells. This dual control of *nod* gene expression provides a mechanism which allows a more successful interaction of *Rhizobium* with the host plant [12]. It is known that overexpression of the common *nod* genes inhibits nodulation [13]. Furthermore, the signal mole-

150

H. Hennecke and D. P. S. Verma (eds.),
Advances in Molecular Genetics of Plant-Microbe Interactions, Vol. 1, 150–155.

cules produced by the Nod proteins are active on the host plant only in very low amounts and within a rather narrow range of concentration [14,15] so that their production must be highly regulated. Recently an alfalfa-specific Nod signal has been purified from the supernatant of a luteolin-induced culture of *R. meliloti* [15]. Both the common *nod* genes and host-range genes are required for the production of this extracellular signal molecule. Structural analysis revealed that this Nod factor (Nod Rm-1) is a sulphated ß-1,4-linked tetrasaccharide containing a glucosamine and three N-acetylglucosamine residues. The oligosaccharide is modified at the non-reducing end by a long-chain bis-unsaturated fatty acid [15].

Recently we have shown that the NodA and NodB proteins are sufficient to produce compounds that stimulate the mitosis of various plant protoplasts [14]. In this paper we have characterized the biological effects caused by the *nodA* and *nodB* genes in transgenic tobacco. Furthermore, domain specific antibodies were used to define some functional domains of the NodC transmembrane protein.

2. Results and Discussion

2.1 ROLE OF NODA AND NODB PROTEINS

We generated antibodies against these *R. meliloti nod* gene products which are essentially required for the nodulation of the host plant alfalfa. These antibodies were used to localize the appropriate proteins in cell fractions of *R. meliloti*. We found that the NodA and NodB proteins are present in the cytosol [14,16] whereas NodC is associated with the bacterial membrane [17]. The localization of the NodA and NodC proteins was confirmed by immunogold labelling and subsequent EM studies [18].

Recently we found that mitosis of cultured soybean protoplasts is stimulated by the presence of *Rhizobium* cells. Using various *R. meliloti* strains with mutations in the *nod* region, we found that *nod* gene expression affects cell division of plant protoplasts, and we used this as a bioassay to elucidate the function of the *nod* gene products [14]. Our data indicate that the *nodA* gene product alone can produce a factor which shows some activity in the cell division assay. This compound may be modified by NodB to a compound with enhanced biological activity. The factor produced by the cytosolic proteins NodA and NodB stimulates not only mitosis of soybean protoplasts but also cell division of protoplasts from alfalfa, barley, carrots and tobacco. The NodAB factor is heat stable, partially hydrophobic and has a molecular weight of less than 1000 [14].

2.2 BIOLOGICAL EFFECTS OF *NODA* AND *NODB* GENES IN TRANSGENIC TOBACCO PLANTS

To study the biological effects of the mitosis stimulating Nod factors in plants, we transformed tobacco (SR1) with the *nodA* and *nodB* genes from *R. meliloti* either singly or in combination. For the expression of the single *nodA* or *nodB* gene we used the plant cloning vector pPCV702 [19] in which the genes were under control of the cauliflower mosaic virus 35S promoter. For combined expression of *nodA* and *nodB* the vector pPCV701 [19] was used in which both genes were under control of the dual 1',2'TR promoter. Transgenic plants raised via leaf disc transformation were regenerated. The regenerants were self-pollinated and the seeds were used for cultivation of the F1 generation. The integration of the *nod* genes into the plant genome was checked by Southern blot analysis. Northern blot analysis of total RNA extracted from leaves revealed that the *nod* genes were expressed in the transgenic tobacco plants.

A comparison of SR1 control plants with transgenic tobacco plants carrying only the *nodA* gene shows slightly reduced growth, a reduction in the internode distance and an altered leaf morphology. Expression of the single *nodB* gene driven by the CaMV35 promoter is responsible for strongly reduced growth and a compact inflorescence. Many flowers have only four petals and four anthers. In all cases we find heterostyly with increased stigma size, so that the plants are unable to self-pollinate.

Transgenic tobacco plants expressing *nodA* and *nodB* show other phenotypic alterations. The mitosis stimulating NodAB factor, which is apparently produced in these transgenic plants, seems to have an effect on cell differentiation which leads to the formation of bifurcated leaves. Probably due to the low activity of the phytohormone regulated TR promoter, we found these bifurcated leaves only in the lower part of the plants. This effect on organogenesis can also lead to the formation of two or more stems emerging independently of the leaf axle. Our data indicate that the factors produced by the NodA or NodB protein alone or in combination are active in non-legumes, and the expression of these genes affects the phytohormone balance in transgenic tobacco.

2.3 ANALYSIS OF DIFFERENT NOD FACTORS

The mitosis stimulating factor produced by NodA and B is not sufficient to induce root hair curling, which is one of the early steps in the nodulation process. For the synthesis of a factor which causes root hair deformation the NodC cell surface protein is additionally required. Using a reversed-phase column and a methanol or acetonitrile gradient we analyzed different Nod factors produced in *E. coli* containing the *nodABC* genes and the host-range gene *hsnD* (*nodH*). The product of the host-range gene *hsnD* has been recently localized in the bacterial membrane (John et al., unpublished). Depending on the Nod proteins involved in factor synthesis we obtained three different active fractions which were tested with the cell division assay [14] and the root hair deformation (Had) assay on clover and alfalfa [20]. According to the chromatographic data the NodAB factor is less hydrophobic than the common Had factor or the alfalfa-specific Had factor which both require a higher solvent concentration for elution. Apparently the mitosis stimulating factor produced by the cytoplasmic NodA an B proteins is modified by the two membrane associated proteins NodC and HsnD to more hydrophobic compounds causing deformation on clover or alfalfa root hairs, respectively.

2.4 ROLE OF THE NODC CELL SURFACE PROTEIN

The 46.8 kDa NodC protein of *R. meliloti* is a cell-surface protein [17] which is essential for nodulation. Based on several experiments we proposed a model of NodC in which a large transmembrane anchor domain separates a long extracellular domain containing an unusual cysteine-rich cluster from a short putative C-terminal intracellular domain [21]. The proposed structure shows striking similarities with various eukaryotic cell-surface receptors with diverse functions.

During the initial stages of nodulation the NodC protein is processed to a smaller molecule of ca. 34 kDa. The truncated NodC is also present in the nodules of various other legumes and we have shown that the amount of this truncated protein increases during nodule development [21]. Using domain specific antibodies directed against amino-terminal epitopes of NodC we could show by immunoblotting that the N-terminal portion of the NodC membrane protein is truncated in the nodule. In a previous experiment we added anti-NodC antibodies directed against a highly conserved region of this protein to *Rhizobium* cells and the appropriate host plant. Due to the addition of these antibodies nodule formation was strongly inhibited [17]. Using antibodies directed against N-terminal sequences of NodC we did not observe an inhibition of nodulation but an increase in the number of nodules.

We speculate that the binding of the appropriate antibodies to the N-terminal part of the putative NodC receptor mimics the binding of an extracellular ligand molecule to that region, and thereby the catalytic properties may be allosterically modulated. We assume that the functional domain of the NodC protein is located further downstream of the N-terminus, probably within the highly conserved region. It seems likely that the truncation of the N-terminal region of NodC serves to modulate the activity of the NodC membrane protein and thereby the signal flow in the nodule.

3. Acknowledgements

This project was supported by a grant from Bundesministerium für Forschung und Technologie (BCT 03652/Projekt 8). G.S. was a recipient of a Humboldt Fellowship.

4. References

1. Kondorosi, E., Banfalvi, Z., and Kondorosi, A. (1984), Physical and genetic analysis of a symbiotic region of *Rhizobium meliloti*: identification of nodulation genes', Mol. Gen. Genet. 193, 445-452.
2. Djordjevic, M.A., Schofield, P.R., Ridge, R.W., Morrison, N.A., Bassam, B.J., Plazinski, J., Watson, J.M., and Rolfe, B.G. (1985) '*Rhizobium* nodulation genes involved in root hair curling (Hac) are functionally conserved', Plant Mol. Biol. 4, 147-160.
3. Downie, J.A., Knight, C.D., Johnston, A.W.B., and Rossen, L. (1985). 'Identification of genes and gene products involved in the nodulation of peas by *Rhizobium leguminosarum*', Mol. Gen. Genet. 198, 255-262.
4. van Brussell, A.A.N., Zaat, S.A.J., Canter Cremers, H.C.J., Wijffelman, C.A., Pees, E., Tak, T., and Lugtenberg, B.J.J. (1986), 'Role of plant root exudate and sym plasmid-localized nodulation genes in the synthesis of *Rhizobium leguminosarum* Tsr factor, which causes thick and short roots on common vetch, J. Bacteriol. 165, 517-522.
5. Horvath, B., Kondorosi, E., John, M., Schmidt, J., Török, I., Györgypal, Z., Barabas, I., Wieneke, U., Schell, J., and Kondorosi, A. (1986), 'Organization, structure and symbiotic function of *Rhizobium meliloti* nodulation genes determining host specificity for alfalfa', Cell 46, 335-343.
6. Debellé, F., and Sharma, S.B. (1986), 'Nucleotide sequence of *Rhizobium meliloti* RCR2011 genes involved in host specificity of nodulation', Nucl. Acids Res. 14, 7453-7471.
7. Fisher, R.F., Swanson, J.A., Mulligan, J.T., and Long, S.R. (1987), 'Extended region of nodulation genes in *Rhizobium meliloti* 1021 II. Nucleotide sequence, transcription start sites, and protein products', Genetics 117, 191-201.
8. Faucher, C., Maillet, F., Vasse, J., Rosenberg, C., van Brussel, A.A.N., Truchet, G., and Denarie, J. (1988), '*Rhizobium meliloti* Host Range *nodH* Genes Determines Production of an Alfalfa-Specific Extracellular Signal', J. Bacteriol. 170, 5489-5499.
9. Mulligan, J.T., and Long, S.R. (1985), 'Induction of *Rhizobium meliloti nodC* expression by plant exudate requires *nodD*', Proc. Natl. Acad. Sci. USA 82, 6609-6613.
10. Rossen, L., Shearman, C.A., Johnston, A.W.B., and Downie, J.A. (1985),'The *nodD* gene of *Rhizobium leguminosarum* is autoregulatory and in the presence of plant exudate induces the *nodABC* genes', EMBO J. 4, 3369-3373.
11. Peters, N.K., Frost, J.W., Long, S.R. (1986), 'A plant flavone, luteolin, induces expression of *Rhizobium meliloti* nodulation genes', Science 233, 977-980.
12. Kondorosi, E., Gyuris, J., Schmidt, J., John, M., Duda, E., Hoffmann, B., Schell, J. and Kondorosi,A. (1989), 'Positive and negative control of *nod* gene expression in *Rhizobium meliloti* is required for optimal nodulation', EMBO J. 8, 1331-1340.

13. Knight, C.D., Rossen, L., Robertson, J.G., Wells,B. and Downie, J.A. (1986), 'Nodulation inhibition of *Rhizobium leguminosarum* multicopy *nodABC* genes and analysis of early stages of plant infection', J. Bacteriol. 166, 552-558.

14. Schmidt, J., Wingender, R., John, M., Wieneke, U., and Schell, J. (1988), '*Rhizobium meliloti nodA* and *nodB* genes are involved in generating compounds which stimulate mitosis of plant cells', Proc. Natl. Acad. Sci. USA 85, 8578-8582.

15. Lerouge, P., Roche, P., Faucher, C., Maillet, F., Truchet,G., Promé, J.C., and Denarié, J. (1990), 'Symbiotic host-specificity of *Rhizobium meliloti* is determined by a sulphated and acylated glucosamine oligosaccharide signal', Nature 344, 781-784.

16. Schmidt, J., John, M., Wieneke, U., Krüßmann, H.-D., and Schell, J. (1986), 'Expression of the nodulation gene *nodA* in *Rhizobium meliloti* and localization of the gene product in the cytosol', Proc. Natl. Acad. Sci. USA 83, 9581-9585.

17. John, M., Schmidt, J., Wieneke, U., Kondorosi, E., Kondorosi, A., and Schell, J. (1985), 'Expression of nodulation gene *nodC* of *Rhizobium meliloti* in *Escherichia coli*: role of the *nodC* gene product in nodulation', EMBO J. 4, 2425-2430.

18. Johnson, D., Roth, E.L., and Stacey, G. (1989) 'Immunogold localization of the NodC and NodA proteins of *Rhizobium meliloti*', J. Bacteriol. **171**, 4583-4588.

19. Koncz, C., Mayerhofer, R., Koncz-Kalman, Z., Nawrath, C., Reiss, B., Redei, G.P., and Schell, J. (1990), 'Isolation of a gene encoding a novel chloroplast protein by T-DNA tagging in *Arabidopsis thaliana*', EMBO J. 9, 1337-1346.

20. Banfalvi, Z. and Kondorosi, A. (1989), 'Production of root hair deformation factors by *Rhizobium meliloti* nodulation genes in *Escherichia coli*: HsnD (NodH) is involved in the plant host-specific modification of the NodABC-factor', Plant Mol. Biol. 13, 1-12.

21. John, M., Schmidt, J., Wieneke, U., Krüßmann, H.-D., and Schell, J. (1988), 'Transmembrane orientation and receptor-like structure of the *Rhizobium meliloti* common nodulation protein NodC', EMBO J. 7, 583-588.

GENETICS OF HOST SPECIFIC NODULATION BY *BRADYRHIZOBIUM JAPONICUM*

G. STACEY, M.G. SCHELL, A. SHARMA, S. LUKA, G. SMIT, AND S.-P.
WANG
Center for Legume Research
Department of Microbiology and Graduate Program of cology
The University of Tennessee
Knoxville, TN 37996-0845
USA

ABSTRACT. Rapid progress has been made in identifying and characterizing nodulation genes in *Bradyrhizobium japonicum*. The results of this
research indicate that many similarities exist between *Bradyrhizobium*
and the taxonomically distinct *Rhizobium* spp. However, clear differences are also emerging. Unique *nod* genes have recently been identified
in *B. japonicum* and further DNA sequence data suggest that this list
will grow. Further examination of *Bradyrhizobium* genetics focusing on
these differences will greatly add to and broaden our understanding of
the molecular mechanism of *(Brady)Rhizobium*-plant interaction.

1. Introduction

The genus *Bradyrhizobium* contains only one named species, *B. japonicum*,
with additional strains lumped together in a miscellaneous group
(Jordan, 1982). Bradyrhizobia are differentiated by slow growth (>8 h
generation time), an alkaline reaction on yeast extract-mannitol
medium, the presence of one to two polar or subpolar flagella, a high
G+C content (63-66%), the ability of some strains to grow autotrophi-
cally on H_2 as an energy source, and the ability of some strains to
induce nitrogenase *ex planta*.

Each species of *Rhizobium* and *Bradyrhizobium* has a defined symbiotic
host range determined, in part, by host specific nodulation (HSN)
genes. Generally, *Bradyrhizobium* spp. have a broader host range than
Rhizobium. *Bradyrhizobium* spp. nodulate some of the most important
leguminous crops grown worldwide (e.g., soybean, peanut, mungbean,
cowpea, etc.). Therefore, the study of *Bradyrhizobium* can easily be
justified due to their agricultural importance. Furthermore, due to
their broad host range, genetic study of *Bradyrhizobium* spp. could
provide important insights into the mechanisms of host selection and
infection. Recent results from investigations of *B. japonicum* have

156

H. Hennecke and D. P. S. Verma (eds.),
Advances in Molecular Genetics of Plant-Microbe Interactions, Vol. 1, 156–161.

identified important HSN genes.

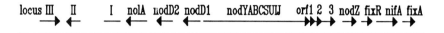

Figure 1. Genetic map of the nodulation (*nod*) genes of *B. japonicum* USDA110. The locations of the genes mentioned within the text are shown.

2. Locus III

A number of HSN genes have been identified in *Rhizobium* spp. Current models concerning the functions of these genes suggest that they are involved in modifying a plant signal factor produced as a result of the *nodABC* gene products (e.g., Faucher et al, 1988; Banfalvi and Kondorosi, 1989; Cervantes et al, 1989; Lerouge et al, 1990; Long et al, this volume). Locus III is a candidate to encode analogous genes in *B. japonicum*. This assertion is based solely on the fact that mutations in Locus III result in a Nod⁻, Coi⁻ (i.e., no induction of plant cortical cell division) phenotype on soybean, with other plant species nodulated normally (Deshmane and Stacey, 1989). Studies with *Rhizobium* spp. have clearly shown that the NodABC factor is required for cortical cell division and the specificity of this process is controlled by the HSN gene products (e.g., Debelle et al., 1986; Faucher et al., 1988; Banfalvi and Kondorosi, 1989).

We have recently determined the DNA sequence of the Locus III region (Sharma and Stacey, unpublished). Two putative open-reading-frames (ORFs) were found; one ORF is preceded by a consensus *nod* box sequence, the presumptive NodD binding site. The presence of a *nod* box is consistent with previous results showing that genes in this region are inducible by plant produced isoflavones, similar to the *nodYABC* operon (Deshmane and Stacey, 1989). As previously suggested by hybridization data, comparison of the DNA sequence to that of known *Rhizobium* HSN genes indicated no significant similarity. Therefore, the function of these genes remains uncertain, but they appear to represent novel *nod* genes.

3. NolA

Recently, Sadowsky et al. (1990) reported the identification of the *nolA* gene (Figure 1). Conjugation of the *nolA* gene from *B. japonicum* USDA110 to *B. japonicum* strains within the 123 serogroup allows these strains to nodulate soybean genotypes that normally restrict nodulation (c.f., Cregan and Keyser, 1986; Cregan et al., 1989). Therefore, the *nolA* gene is an example of a genotype specific nodulation (GSN) gene that appears to distinguish between different soybean genotypes. Another example of a GSN gene is *nodX* found in certain strains of *R. leguminosarum* bv. *viciae*, which is essential for nodulation of Afghanistan pea (c.f., Davis et al., 1988).

Expression of *nolA* is induced by soybean excreted isoflavones (Sadowsky et al., 1990). The deduced N-terminal amino acid sequence of NolA shows a reasonable match to the helix-turn-helix, DNA binding motif. Therefore, NolA could act as a transcriptional regulatory protein.

4. NodD

The *nodD* gene encodes a positive transcriptional regulatory protein required for expression of other *nod* genes (reviewed in Long, 1989). Activation of *nod* gene expression is postulated to depend on the specific recognition by NodD of flavonoid inducers excreted by the host plant. Therefore, inducer specificity is apparently determined by NodD and, therefore, is an important determinant for host selection.
Nod gene induction in *B. japonicum* requires the presence of isoflavones; genistein and daidzein give the best response (Kosslak et al., 1987; Banfalvi et al., 1988; Gottfert and Hennecke, 1988). The *nodD₁* gene of *B. japonicum* is required for induction of the *nodYABC* operon (Banfalvi et al., 1988). Transfer of a *nodY-lacZ* fusion (lacking a *nodD* gene) to several strains of *B. japonicum* indicated that all respond qualitatively the same to a variety of inducing compounds.

The *nodD* genes of *R. leguminosarum* bvs. *viciae* and *trifolii*, as well as *R. meliloti* (excepting *nodD₃*, Mulligan and Long, 1989), are constitutively expressed (reviewed by Long, 1989). An interesting feature of *B. japonicum* is that *nodD₁* expression is autoregulated and inducible by isoflavones (Banfalvi et al., 1988). Recently, the *nodD* genes in *Rhizobium leguminosarum* bv. *phaseoli* have been shown to be similarly regulated (Davis and Johnston, 1990). The fact that the *nodD₁* gene of *B. japonicum* was inducible led to an examination of its promoter and the identification of a divergent *nod* box-like sequence. Deletions removing this sequence were used to show that it is essential for *nodD1* expression (Wang and Stacey, 1990). It is unclear what advantage the organism may gain by this further level of regulatory complexity.

However, recently we have shown that soybean roots excrete specific substituted isoflavones capable of specifically inducing *B. japonicum* *nodD₁* expression; the *nodYABC* operon is not expressed (Smit et al., 1990). Therefore, inducible expression of *nodD₁* may be required under, as yet unidentified, conditions in which induction of other *nod* genes is not advantageous.

5. NodZ

The *nodZ* gene of *B. japonicum* was first identified by hybridization to cloned HSN genes isolated from *Rhizobium* sp. strain MPIK3030 (Nieuwkoop et al., 1987). Subsequent mutation of this region indicated that it was essential for nodulation of siratro, a normal host of *R.* sp. MPIK3030, but not for nodulation of soybean. NodZ⁻ mutants are apparently not affected in nitrogen fixation activity. Yet, recent data indicates that *nodZ* expression *in planta* requires NifA, a key transcriptional regulatory protein for *nif/fix* gene expression (Schell and Stacey, unpublished). Therefore, *nodZ* appears to provide a, as yet unidentified, link between regulation of nitrogen fixation and nodulation.

The DNA sequence of *nodZ* predicts a 35 kd protein with 4 putative metal coordination sites. As yet, there are no clues as to the possible biochemical function of NodZ. However, recent results suggest that NodZ may be phosphorylated *in vivo*. Antibody made against a peptide (CS-15) made from the deduced amino acid sequence of NodZ reacts on Western blots with a protein which comigrates with a major phosphoprotein in bacteroids. Furthermore, the CS-15 peptide is the substrate for a *B. japonicum* protein kinase (Schell and Stacey, unpublished). Further analysis of the covalent modification of NodZ may reveal the role of this protein in the nodulation process.

6. Conclusions

The understanding of the genetics of *B. japonicum* is advancing at a good rate. The results thus far point to similarities between *Bradyrhizobium* and *Rhizobium* spp. (e.g., *nodD*, *nodABC*). However, there are notable differences and a few novel genes have been identified (e.g., *nodZ*, *nolA*). Further examination of *B. japonicum* genes promises to broaden our understanding of the molecular mechanisms of rhizobia/bradyrhizobia-plant interaction.

6.1. ACKNOWLEDGEMENTS

Recent results from the authors' laboratory were supported by Public

Health Service grants GM33494 and GM40183 from the National Institutes of Health and grant 62-600-1636 from the U.S. Department of Agriculture.

7. References

Banfalvi, Z. and Kondorosi, A. (1989) 'Production of root hair deformation factors by *Rhizobium meliloti* nodulation genes in *Escherichia coli*: HsnD (NodH) is involved in the plant host-specific modification of the NodABC factor', Plant Mol. Biol. 13, 1-12.

Banfalvi, Z., Nieuwkoop, A.J., Schell, M.G., Besl, L. and Stacey, G. (1988) 'Regulation of *nod* gene expression in *Bradyrhizobium japonicum*', Mol. Gen. Genet. 214, 420-424.

Cervantes, E., Sharma, S.B., Maillet, F., Vasse, J., Truchet, G., and Rosenberg, C. (1989) 'The *Rhizobium meliloti* host range *nodQ* gene encodes a protein which shares homology with translation elongation and initiation factors', Mol. Microbiol. 3, 745-755.

Cregan, P.B. and Keyser, H.H. (1986) 'Host restriction of nodulation by *Bradyrhizobium japonicum* strain USDA123 in soybean', Crop Sci. 26, 911-916.

Cregan, P.B., Keyser, H.H., and Sadowsky, M.J. (1989) 'Soybean genotype restricting nodulation of a previously unrestricted serocluster 123 *Bradyrhizobium*', Crop Sci. 29, 307-312.

Davis, E.O., Evans, I.J. and Johnston, A.W.B. (1988) 'Identification of *nodX*, a gene that allows *Rhizobium leguminosarum* bv. *viciae* strain TOM to nodulate Afghanistan peas', Mol. Gen. Genet. 212, 511-535.

Davis, E.O. and Johnston, A.W.B. (1990) 'Regulatory functions of the three *nodD* genes of *R. leguminosarum* bv. *phaseoli*'. Mol. Microbiol. 4, 933-941.

Debelle, F., Rosenberg, C., Vasse, J., Maillet, F., Martinez, E., Denairie, J. and Truchet, G. (1986) 'Assignment of symbiotic developmental phenotypes to common and specific nodulation (*nod*) genetic loci of *Rhizobium meliloti*', J. Bacteriol. 168, 1075-1086.

Deshmane, N. and Stacey, G. (1989) 'Identification of *Bradyrhizobium nod* genes involved in host-specific nodulation', J. Bacteriol. 171, 3324-3330.

Faucher, C., Maillet, F., Vasse, J., Rosenberg, C., van Brussel, A.A.N., Truchet, G., and Denarie, J. (1988) '*Rhizobium meliloti* host range *nodH* gene determines production of an alfalfa-specific extracellular signal', J. Bacteriol. 170, 5489-5499.

Gottfert, M. Weber, J. and Hennecke, H. (1988) 'Induction of a *nodA-lacZ* fusion in *Bradyrhizobium japonicum* by an isoflavone', J. Plant Physiol. 132, 394-397.

Jordan, D.C. (1982) 'Transfer of *Rhizobium japonicum* to *Bradyrhizobium*, slow-growing root nodule bacterium from leguminous plants', Int. J. Syst. Bacteriol. 32, 136-139.

Kosslak, R.M., Bookland, R., Barkei, J., Paaren, H.E., and Appelbaum, E.R. (1987) 'Induction of *Bradyrhizobium japonicum* common *nod* genes by isoflavones isolated from *Glycine max*', Proc. Natl. Acad. Sci. (USA) 84, 7428-7432.

Lerouge, P., Prome, J.C. and Denarie, J. (1990) 'Symbiotic host-speci
ficity of *Rhizobium meliloti* is determined by a sulphated and
acylated glucosamine oligosaccharide signal', Nature 344, 781-784.

Long, S.R. (1989) '*Rhizobium*-legume nodulation: Life together in the
underground', Cell 56, 203-214.

Mulligan, J.T. and Long, S.R. (1989) 'A family of activator genes
regulates expression of *Rhizobium meliloti* nodulation genes',
Genetics 1221, 7-18.

Nieuwkoop, A.J., Banfalvi, Z., Deshmane, N., Gerhold, D., Schell, M.G.,
Sirotkin, K.M., and Stacey, G. (1987) 'A locus encoding host range
is linked to the common nodulation genes of *Bradyrhizobium japonicum*', J. Bacteriol. 169, 2631-2638.

Sadowsky, M.J., Cregan, P.B., Gottfert, M., Sharma, A., Gerhold, D.,
Rodriquez-Quinones, F., Keyser, H.H., Hennecke, H. and Stacey, G.
(1990) 'The *Bradyrhizobium japonicum nolA* gene and its involvement
in the genotype-specific nodulation of soybeans. Proc. Natl. Acad.
Sci. (USA), in review.

Smit, G., Puvanesarajah, V., Carlson, R.W., and Stacey, G. (1990)
'*Bradyrhizobium japonicum nodD* can specifically be induced by
soybean seed compounds which do not induce the *nodYABC* genes', in P.
Gresshoff, E. Roth, G. Stacey, and W.E. Newton (eds.), Nitrogen
Fixation: Achievements and Objectives, Chapman and Hall, New York.

Wang, S.-P. and Stacey, G. (1990) 'A divergent *nod* box sequence is
essential for *nodD*$_1$ induction in *B. japonicum*', in P.M. Gresshoff,
E. Roth, G. Stacey, and W.E. Newton (eds.), Nitrogen Fixation:
Achievements and Objectives, Chapman and Hall, New York.

SIGNAL EXCHANGE MEDIATES HOST-SPECIFIC NODULATION OF TROPICAL LEGUMES BY THE BROAD HOST-RANGE *RHIZOBIUM* SPECIES NGR234

W.J. BROUGHTON*, A. KRAUSE*, A. LEWIN*, X. PERRET*, N.P.J. PRICE*, B. RELIC*,[1], P. ROCHEPEAU*, C.-H. WONG*,[2], S.G. PUEPPKE*[3], AND S. BRENNER[4]. L.B.M.P.S., University of Geneva, 1 ch. de l'Impératrice, 1292 Chambésy, Genève. Permament addresses: [1]Institute of Biology, University of Novi Sad, Ilije Djuricia 6, 21000 Novi Sad, Yugoslavia; [2]School of Biological Sciences, Universiti Sains Malayisa, Pulau Pinang, Malaysia; [3]Department of Plant Pathology, University of Missouri, Columbia, MO 65211, U.S.A., and; [4]Molecular Genetics Unit, Medical Research Council, Hills Road, CB2-2QH, Cambridge, U.K.

ABSTRACT Rhizobium species NGR234 nodulates at least 35 diverse genera of legumes as well as the non-legume Parasponia andersonii. Initially, three host-range determinants [HsnI, nodSU (= HsnII), and HsnIII] were identified by their ability to extend the nodulation capacity of heterologous rhizobia to include Vigna unguiculata. These loci, along with nodABC, nodD1, nodD2 (which is probably not involved in nodulation), RegionII, nolB, etc., are located on the 500 kb symbiotic plasmid, pNGR234a. In addition to nodulation of Vigna, HsnI confers upon heterologous transconjugants the ability to nodulate Glycine max, Macroptilium atropurpureum, as well as Psophocarpus tetragonolobus, nodSU controls nodulation of Leucaena leucocephala, while HsnIII complements (at low efficiency) mutations in nodH of R. meliloti. We propose that broad host-range in Rhizobium sp. NGR234 is controlled by : (a) a nodD1-gene that responds to a range of plant-excreted flavonoids [Rolfe, B.G., Biofactors 1(1988)3-10]; (b) that the nodD1-gene activates transcription of the nodABC-operon producing a "common" nodulation factor capable of deforming legume root-hairs; (c) that host-specificity loci/genes tailor this "common" Nod-factor to specific plants, and; (d) this now completely specific Nod-factor is released from the Rhizobium where it causes root-hair deformation and the onset of nodulation. Chemical analyses have shown that the Nod-factor(s) of NGR234 belong to the same family of sulphated ß-1,4-polysaccharides of D-glucosamine [in which one of the terminal amino-groups is acetylated with a C_{16} bis-unsaturated fatty acid while the rest are acetylated] that mediate signal transduction in R. meliloti [P. Lerouge et al, Nature 344(1990)781-784]. Exogenously applied Nod-factor (NodNGR2) provokes root-hair deformation on numerous plants. Five "early nodulins" are induced in the root-hairs of V. unguiculata 24 h after inoculation with NGR234.

H. Hennecke and D. P. S. Verma (eds.),
Advances in Molecular Genetics of Plant-Microbe Interactions, Vol. 1, 162–167.

Introduction

Legume-rhizobia associations are characterised by varying degrees of symbiotic specificity. We have shown that <u>Rhizobium</u> sp. strain NGR234 has a broader host-range than any other <u>Rhizobium</u> species reported, and is currently known to nodulate 35 different legume genera (Lewin et al, 1987; S.G. Pueppke and W.J. Broughton, unpublished). Similarly, <u>Vigna unguiculata</u> (L.) Walp. forms nodules with a broader spectrum of rhizobia than any other legume (Lewin et al, 1987). Genetic analysis of the NGR234 host-range determinants showed that at least three different loci on the Sym-plasmid (HsnI, HsnII, and HsnIII) are involved (Broughton et al, 1986; Lewin et al, 1987; Nayudu and Rolfe, 1987). Currently, we are investigating the molecular and biochemical function of these Hsn loci, both in relation to other <u>nod</u>-genes and to the early response of <u>V. unguiculata</u> to inoculation with NGR234. This report summarises our findings.

Materials and Methods.

All the methods used in this report have been described previously (Broughton et al, 1984; 1986; Lewin et al, 1987; 1990; Lerouge et al, 1990; Perret et al, 1990).

Results and Discussion

SYMBIOTIC ORGANISATION OF THE BACTERIAL GENOME

Despite the fact that mobilisation of pNGR234<u>a</u> into <u>Agrobacterium</u> as well as heterologous rhizobia (Broughton et al, 1984) and curing NGR234 of pNGR234<u>a</u> (Morrison et al, 1983) suggest that most symbiotic genes are located on the plasmid, sequences homologous to <u>nodE</u>, <u>nodG</u>, <u>nodP</u>, and <u>nodQ</u> hybridise to another replicon (Perret et al, 1990). All other symbiotic genes (<u>nodABC</u>, <u>nodD1</u>, <u>nodD2</u>, <u>nodSU</u>, <u>nolB</u>, HsnI, HsnIII, RegionII, <u>nifKDH</u> etc.) are widely dispersed over pNGR234<u>a</u> (Fig. 1). Although the roles of RegionII, <u>nolB</u>, etc., have still to be elucidated, it seems fairly clear that HsnI helps control nodulation of <u>G. max</u>, <u>M. atropurpureum</u>, and <u>P. tetragonolobus</u> (Lewin et al, 1987). Nodulation of <u>L. leucocephala</u> is determined by the <u>nodSU</u> operon [= HsnII (Lewin et al, 1990)], while HsnIII influences the interaction of NGR234 with <u>M. sativa</u> (Lewin et al, 1990). Apparently only one copy of the <u>nodABC</u> genes exists, since deletion of this locus produces NGR234 derivatives that are Nod[-] on <u>V. unguiculata</u> (B. Relic and S.G. Pueppke unpublished). The situation with regard to the <u>nodD</u>-genes is more complicated. Hybridisation experiments reveal two <u>nodD</u>-loci (Perret et al, 1990), yet insertion of an Omega fragment in the <u>nodD1</u>-gene results in a NGR234 derivative that is Nod[-] on <u>L. leucocephala</u>, <u>M. atropurpureum</u>, <u>V. unguiculata</u>

etc (A. Lewin, unpublished). Thus we conclude that nodD2 does not play an active role in the control of nodulation by NGR234.

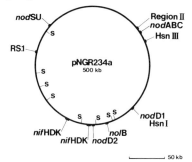

Fig. 1. Schematic organisation of pNGR234a (after Perret et al, 1990). S within the circle represents the recognition site for the restriction enzyme SpeI. Only the approximate locations of the various genes/loci are shown. Note: nodD2 and nolB have been identified by homology only. No function has yet been assigned to them.

BIOCHEMICAL AND GENETIC CONTROL OF HOST-SPECIFICITY

Since Peters et al (1986) first showed that flavonoids excreted by plant roots induce expression of the nodABC-operon, it has become widely accepted that flavonoids interact with the product of the regulatory nodD-gene. In turn, the NodD-protein binds to "nod-boxes" in the 5'-region of nodD-regulated genes (e.g. nodABC) so regulating their expression. A major difference between "narrow host-range" Rhizobium species and NGR234, is that the nodD1-gene of NGR234 responds to a large number of flavonoids (Bassam et al, 1988), and thus the NGR234 nod-genes are activated in the rhizoshpere of many plants. Activation of the nodABC-genes is not sufficient to explain the broad host-range of NGR234 however, since deletions/mutations in the nodS-gene are Nod⁻ on L. leucocephala (Lewin et al, 1990).

For this reason we examined the suggestion of Faucher et al (1988) that the nodABC-operon determines the production of a "common" hair deformation factor (Had- or Nod-factor) which is modified by the product of the R. meliloti host-specificity gene nodH into a M. sativa specific Had-factor. Although the structural analysis of Had-factors excreted into the growth medium after apigenin induction of NGR234 cultures is not complete, fast-atom bombardment mass-spectrometry, proton nuclear-mass resonance spectrometry, high-pressure liquid chromatography, ^{35}S-sulphate-labelling experiments etc., show that the Had-factors of NGR234 belong to the same class of compounds as those produced by R. meliloti (Fig. 2). Proof that Had-factor production is under control of the nod-genes came from the demonstration that deletion of the nodABC-genes abolished Had-factor activity (N.P.J. Price, and B. Relic, unpublished).

EARLY NODULINS IN THE ROOT-HAIRS OF V. UNGUICULATA

Microscopically, the first visible response of legumes to inoculation with rhizobia is deformation and curling of the root-hairs. These

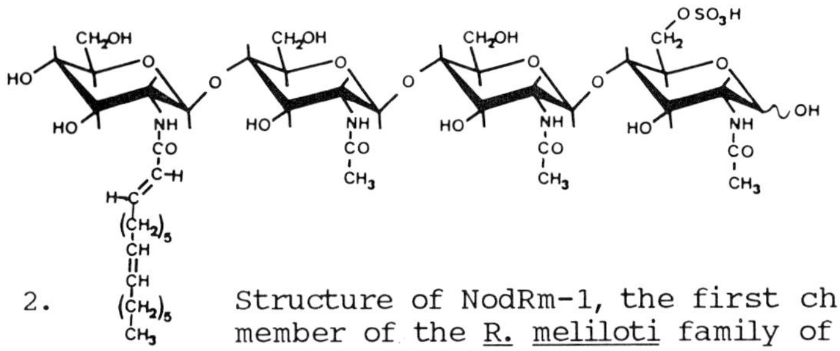

Fig. 2. Structure of NodRm-1, the first characterised member of the R. meliloti family of Had-factors (from Lerouge et al, 1990). Wild-type Nod-factor(s) of NGR234 also possess N-acetyl D-glucosamine residues, an unsaturated fatty-acid side-chain, and a sulphate group (N.P.J. Price, unpublished).

observations suggest that plant genes are induced by rhizobia only a few hours after inoculation. Indeed, 2-D gel electrophoresis of proteins isolated from infected V. unguiculata root hairs shows that five NGR234-induced proteins are visible 24 h after inoculation (Table 1). Of these, spots #3 and #5, which are only present in the root-hairs, seem to play a crucial role in the onset of nodulation. Four additional proteins ranging in size from 36 to 45 kDa were visible two to four days after infection (three at day two, an additional one at day four). Furthermore, three other proteins (including one that is root-hair specific) are repressed during nodule development (A. Krause unpublished).

Table 1. Putative "early nodulins" of V. unguiculata root-hairs induced by Rhizobium sp. NGR234 and its derivatives.

Spot # (Size – kDa)	Days after inoculation with NGR234			in nodules	in NGR234 mutants	
	Day 1	Day 2	Day 4		NodABC⁻	NodD⁻
3 (15)	+	+	±	−	−	−
5 (31)	+	+	+	−	−	−
6 (37)	+	+	+	+	−	?
7 (38)	+	+	+	+	−	−
27 (37)	+	+	+	+	−	−

Note: ± this spot was visible in some experiments, not in others; ? the exact location of the missing spot is uncertain.

Conclusions

We suggest that control of nodulation in broad host-range associations is mediated by signal exchange between legumes and Rhizobium. Legumes (of which there are currently 35 known genera of hosts for NGR234) secrete a variety of flavonoids, some of which interact with the DNA-binding nodD1-gene product. Then, the flavonoid NodD1-product complex binds to cis-acting nod-boxes in the 5'-region of nodD-regulated genes. De-repression of the nodABC-operon results in the production of a "common" Had-factor. In turn, this "common" Had-factor is tailored to meet the needs of specific plants by the host-specificity genes, the expression of which is often also regulated by flavonoids and the NodD1-product. As a result, a Had-factor is excreted from the Rhizobium. The host-specific Had-factor (also called Nod-factor) induces hair-deformation and the first stages of nodulation in homologous legumes (Fig. 3). Host-specificity is thus regulated at three different levels: first, by plant-excreted flavonoids interacting with the NodD1-product; second, by a series of hsn-genes which modify the "common" Had-factor, and; third, by interaction between the Nod-factors and receptors on the root hairs of the host legumes.

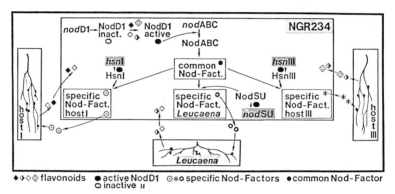

Fig. 3. Proposed model for the control of host-specificity in broad host-range legume-Rhizobium associations (see text).

Acknowledgements

We wish to thank J. Dénarié for many helpful discussions and D. Gerber for her constant support. Also, we are deeply indebted to the Erna och Victor Hasselblads Stiftelse, the European Molecular Biology Organisation, the Fonds national suisse de la recherche scientifique (#3.180.088), and the Université de Genève for supporting this research. Funds to attend this congress were generously donated by the H. Dudley Wright Foundation.

References

Bassam, B.J., Djordjevic, M.A., Redmond, J.W., Batley, M., and Rolfe, B.G. (1988) 'Identification of a nodD-dependent locus in the Rhizobium strain NGR234 activated by phenolic factors secreted by soybeans and other legumes.' Mol. Plant-Microbe Inter. 1, 161-168.

Broughton, W.J., Heycke, N., Meyer z.A., H., and Pankhurst, C.E. (1984) 'Plasmid-linked nif and "nod" genes in fast-growing rhizobia that nodulate Glycine max, Psophocarpus tetragonolobus, and Vigna unguiculata.' Proc. Natl. Acad. Sci. 81, 3093-3097.

Broughton, W.J., Wong, C.H., Lewin, A., Samrey, U., Myint, H., Meyer z.A., H, Dowling, D.N., and Simon, R. (1986) 'Identification of Rhizobium plasmid sequences involved in recognition of Psophocarpus, Vigna, and other legumes.' J. cell Biol. 102, 1173-1182.

Faucher, C., Maillet, F., Vasse, J., Rosenberg, C., van Brussel, A.A.M., Truchet, G., and Dénarié, J. 'Rhizobium meliloti host range nodH gene determines production of an alfalfa specific extracellular signal'. J. Bacteriol. 170, 5489-5499.

Lerouge, P., Roche, P., Faucher, C., Maillet, F., Truchet, G., Promé, J.C., and Dénarié, J. (1990) 'Symbiotic host-specificity of Rhizobium meliloti is determined by a sulphated and acylated glucosamine oligosaccharide signal.' Nature 344, 781-784.

Lewin, A., Rosenberg, C., Meyer z.A., H., Wong, C.H., Nelson, L., Manen, J.-F., Stanley, J., Dowling, D.N., Dénarié, J., and Broughton, W.J. (1987) 'Multiple host-specificity loci of the broad host-range Rhizobium sp. NGR234 selected using the widely compatible legume Vigna unguiculata', Plant mol. Biol. 8, 447-459.

Lewin, A., Cervantes, E., Wong, C.H., and Broughton, W.J. (1990) 'nodSU, two new nod genes of the broad host range Rhizobium strain NGR234 encode host-specific nodulation of the tropical tree Leucaena leucocephala', Mol. Plant-Microbe Inter. 3, in press.

Morrison, N.A., Hau, C.Y., Trinick, M.J., Shine, J., and Rolfe, B.G. (1983) 'Heat curing of a Sym plasmid in a fast-growing Rhizobium sp. that is able to nodulate legumes and the nonlegume Parasponia sp'. J. Bacteriol. 153, 527-531.

Nayudu, M., and Rolfe, B.G. (1987) 'Analysis of R-primes demonstrates that genes for broad host-range nodulation of Rhizobium strain NGR234 are dispersed on the Sym plasmid'. Mol. gen. Genet. 206, 326-337.

Perret, X., Broughton, W.J., and Brenner, S. (1990) 'Canonical ordered cosmid library of the symbiotic plasmid of Rhizobium species NGR234'. Proc. Natl. Acad. Sci., submitted.

Peters, N.K, Frost, J.W., and Long, S.R. (1986) 'A plant flavone, luteolin, induces expression of Rhizobium meliloti nodulation genes'. Science 233, 977-980.

THE USE OF THE GENUS *TRIFOLIUM* FOR THE STUDY OF PLANT-MICROBE INTERACTIONS

J.J. WEINMAN, M.A. DJORDJEVIC, P.A. HOWLES, T. ARIOLI, W. LEWIS-HENDERSON, J. McIVER, M. OAKES, E.H. CREASER AND B.G. ROLFE
Plant-Microbe Interaction Group
Research School of Biological Sciences, Australian National University
PO Box 475, Canberra 2601, ACT, AUSTRALIA
Fax: +61 62 249 0754

ABSTRACT. We have been developing *Trifolium repens* and *Trifolium subterraneum* for the study of both symbiotic and pathogenic interactions between plants and bacteria. Extensive research into the symbiosis formed between both of these species and *Rhizobium leguminosarum* bv. *trifolii* has now been broadened to examine the molecular events which take place during the induction of a disease response in these important Australian pasture species. *T. subterraneum* is proving particularly good as a model system for such study due to a number of valuable genetic and biological features.

1. Introduction

The study of the symbiotic relationship between *Rhizobium leguminosarum* bv. *trifolii* and white and subterranean clovers (*Trifolium repens* and *subterraneum*, respectively) has allowed us to develop a range of ways to perturb the normal developmental pattern of the clover roots, and to elicit defense responses. This paper describes ways in which we are now investigating the nature of these responses.

2. Results

2.1. ADVANTAGES OF *TRIFOLIUM* SPECIES AS MODEL SYSTEMS

The use of *Trifolium* species, in particular *Trifolium repens* and *Trifolium subterraneum*, has enabled much genetic information about the symbiotic interaction with *Rhizobium* to be discerned. These clover species also are now proving useful in the examination of plant genes induced both during symbiosis and during the induction of a disease response. Our intention is to examine the similarities and differences existing between a successful symbiotic interaction and a pathogenic infection.

We have chosen to concentrate on a particular cultivar of subterranean clover, Karridale, for this purpose. As is detailed in Table 1, subterranean clover has a number of features which make it ideal for this purpose. Both sub- and white clovers are small seeded legumes, rapidly and easily germinated, and relatively fast to nodulate. Both species possess small genomes (approximately 1×10^9 nucleotides for *T. subterraneum*) which have been cloned into genomic libraries. Transformation and regeneration is possible for white clover (White and Greenwood, 1987) and a considerable research effort is being undertaken to perfect the transformation and regeneration of subterranean clover. A significant reason for this is the considerable heterogeneity of the white clover genome, due to its outcrossing nature. Subterranean clovers

168

H. Hennecke and D. P. S. Verma (eds.),
Advances in Molecular Genetics of Plant-Microbe Interactions, Vol. 1, 168–173.

are self-fertilizing and a wide variety of certified cultivars are available through the Australian National Sub Clover Improvement Programme coordinated by the West Australian Department of Agriculture. Seed is available for varieties with less than a 1% contamination frequency.

TABLE 1.   COMPARISON OF WHITE AND SUBTERRANEAN CLOVERS

ATTRIBUTE	WHITE CLOVER	SUBTERRANEAN CLOVER
seed size	very small	small
ploidy	diploid	diploid
genome size	small	small
breeding characteristics	outcrossing	self-fertilizing
cultivar availability	many different lines	many certified cultivars
life cycle	perennial	annual
Agrobacterium transformation	yes	probable
tissue culture propagation	yes	yes
regeneration	yes	yes
pathogens available:		
bacterial	many	many
fungal	limited	many
viral	numerous characterized isolates	numerous characterized isolates
rapidity of nodulation	3-5 days	9-12 days
minimum genetic requirement for nodulation	*nodDABCIJFEL*	*nodDABCIJL*
genomic library available	yes	yes
spectrum of symbiotic flavonoid signals	simple	complex
bioactivity of *Rhizobium* EPS	demonstrated	demonstrated

Rhizobium strain ANU843 shows both *Trifolium* species and cultivar specificity with different *nod* and host-specificity *(hsn)* gene combinations. Through our research on the genetics of the symbiotic interaction we have been able to develop a model to explain the different legume host ranges which result from combinations of different *nod* and *hsn* genes. This is shown in Figure 1.

3.2. *RHIZOBIUM-TRIFOLIUM* INTERACTIONS INVOLVE A GENE-FOR-GENE INTERACTION

We have investigated a cultivar specific interaction between *R.l.* bv. *trifolii* and subterranean clover. The *R.l.* bv. *trifolii* strain TA1 is capable of infecting a broad range of white and red clovers and *T. subterraneum* cultivars and induces nitrogen-fixing nodules. It is, however, incapable of inducing nodules on the *T. subterraneum* cultivar Woogenellup. In contrast to strain TA1, the well characterised *R.l.* bv. *trifolii* strain, ANU843, is capable of inducing nitrogen-fixing nodules on several *T. subterraneum* cultivars including Woogenellup. The ability of *R.l.* bv. *trifolii* strain TA1 to induce root hair curling and infection thread initiation is not impaired. Microscopic observations show that nodule foci are initiated by strain TA1 on cv. Woogenellup but these fail to develop. The inability of TA1 to nodulate cv. Woogenellup is temperature sensitive: non-nodulation occurs at a root temperature of 22oC but normal nodulation occurs between 26oC and 29oC (Gibson, 1968).

3.2.1. Negatively-Acting Genes.
Random mutagenesis of strain TA1 with Tn5:*mob* (15,000 mutants screened) resulted in the generation of two mutant types which showed enhanced nodulation capacity on cultivar Woogenellup. Two genes were identified as negatively-acting determinants, *nodM* and a novel gene designated *csn1* (cultivar specific nodulation). Separate mutations in the TA1 *nodM* gene both resulted in strains which gave well developed nitrogen-

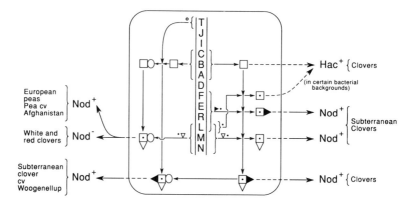

Figure 1. The effect of the interaction of *R.l.* bv. *trifolii* nodulation genes on the infection of homologous and heterologous hosts. When induced by plant signals the *Rhizobium* strain produces extracellular compounds which either cause the root hair curling (Hac⁺) phenotype or aid in the infection and nodulation (Nod⁺) of different host plants. We propose that a variety of extracellular products are formed as a result of different parallel pathways, each under the control of the various host-specific nodulation genes, acting on the factors produced after the expression of the *nodABC* genes. Extracellular factors produced as a result of *nodABC* gene expression (□) usually cause the Hac⁺ phenotype on all tested clovers (white, red and subterranean clovers). However, in certain *R.l.* bv. *trifolii* strains a combination of *nodABC* plus *nodL* (⊡) is required for the Hac⁺ phenotype. The Sym-plasmid cured *R. l.* bv *trifolii* strain ANU845, containing combinations of *nod* genes, *nodDABC* plus *nodFERL* or *nodDABC* plus *nodLMN*, can induce the Hac⁺ Inf⁺ (infection thread formation) phenotype; however, significant, but relatively poor nodulation results on subterranean clovers. The complete *nod* gene combination, *nodDABCFERLMN*, produces a more rapid and efficient nodulation of all tested clovers. The host specific nodulation of *R. l.* bv. *trifolii* also show the phenomenon of dominance or epistasis. The *nodE⁻* mutants are Nod⁻ on white and red clovers, Nod⁺ on subterranean clovers and have an extended host range, as they are able to nodulate (Nod⁺) peas but not vetches. Finally, certain strains of *R. l.* bv. *trifolii* exhibit cultivar specificity; the positively acting *nodT* gene counteracting the negatively acting *nodM* gene to enable nodulation of subterranean clover cv. Woogenellup.

fixing nodules. These nodules were induced as rapidly as nodules induced by the *R.l.* bv. *trifolii* strain ANU843 and the nodulation phenotype was clearly distinguishable from that induced by strain TA1. Mutations in *nodM* were confirmed by DNA sequencing analysis and a reversed mutagenesis technique using chromosomal mobilisation of the mutated allele to another TA1 background strain.

Hybridization analysis showed that the *csn1* locus is not located on the pSym and has no homology to known *nod* genes. Strain TA1 carrying a mutated *csn1* gene gave a markedly enhanced nodulation response when compared to nodulation induced by the parent strain. Partial DNA sequence analysis verifies that *csn1* is novel.

3.2.2 Positively-Acting Genes. The *nodT* gene from *R.l.* bv. *trifolii* strain ANU843 acts as a dominant suppressor of the TA1 *nodM*. Cloned DNA fragments carrying *nodT* confer rapid, well-developed tap root nodulation to strain TA1 on cv. Woogenellup. However, cosmid clones from ANU843 which encode both ANU843 *nodT* and *nodM* were incapable of conferring cv. Woogenellup nodulation to TA1. Further analysis indicated that the ANU843 *nodM* gene could also confer non-nodulation on cv. Woogenellup in the background of strain TA1. Different effects of introduced cosmid clones carrying *nod* genes could be observed in different genetic backgrounds.

In addition, the introduction into strain TA1 of *nod* genes coded on cosmid clones from foreign rhizobia (e.g. pRmSL26, pIJ1089, pIJ1095) suppresses the non-nodulation phenotype on cv. Woogenellup

Strain TA1 was shown to be naturally deleted for *nodT* but complementation and hybridization experiments indicated that TA1 possessed all other recognised nodulation genes present in strain ANU843. A cosmid clone spanning the TA1 *nod* region confers to strain ANU845 nodulation of cvs. Geraldton and Mt. Barker but not cv.Woogenellup.

3.2.3. Involvement Of Host Genes. A single recessive host gene trait, designated rWT1, was shown to be responsible for cultivar specificity. When the F_2 progeny of crosses between cv. Woogenellup and cv. Geraldton (which is nodulated normally by strain TA1) were examined it was found that 75% of the germinated seedlings were rapidly nodulated by strain TA1 while 25% were recalcitrant to nodulation.

3.3. INDUCTION OF A DEFENSE RESPONSE

As a consequence of the involvement of *Trifolium* flavonoid biosynthesis in both symbiotic signaling with *Rhizobium* and in the induction of a defense response we have been able to develop a bioassay in which *Rhizobium nod:lac* fusions are induced as a result of pathogenic attack.

3.3.1. The Use of Pseudomonas Strains To Induce A White Clover Defense Response.
Pseudomonas andropogonis strains 2990 and 0350 (Moffett *et al*, 1986), normally pathogenic on white clover, cause the accumulation in the roots of flavonoid compounds which can activate *nod* gene transcription in *R.l.* bv. *trifolii*. (see Figure 2). This supports previous results (Weinman *et al*, 1988) which showed increased production of *nod* gene inducing compounds when white clovers were presented with incompatible *Rhizobium* strains or with *hsn* gene mutants of strain ANU843.

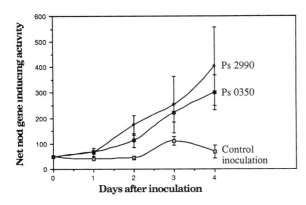

Figure 2. Freshly germinated seedlings were placed onto nutrient agar plates onto which had been placed a lawn of ~10^7 cells of the *Pseudomonas andropogonis* strains indicated. At appropriate intervals, 40 seedlings from each treatment were removed, their root flavonoids extracted, and the induction of strain ANU843 *nod:lac* fusion activity assayed as described in Weinman *et al*, 1988. 4 replicates per treatment were used for each time point.

3.4. CLONING OF GENES INDUCED BY MICROBIAL INTERACTION

We have been keen to obtain homologous probes to monitor two responses induced in legumes by microbes. Symbiotic nodulation begins with the induction of a number of plant genes (early nodulins; ENOD's). From a genomic library of cultivar Karridale we have isolated a number of lambda clones which hybridize with the pea ENOD2 probe (Van De Wiel *et al*, 1990). These are being further characterized. In order to better understand the regulation of the synthesis of flavonoids/isoflavonoids seen when *Trifolium* is subjected to pathogen and *Rhizobium* attack we have used our genomic library of variety Karridale to obtain a number of clones to the enzyme Chalcone Synthase (CHS). The enzyme coded for by these genes catalyses the first (committed) reaction in the branch pathway of plant phenylpropanoid metabolism leading to the biosynthesis of the flavonoids/isoflavonoids. .More interestingly, as different members of the bean CHS gene family are induced by different stimuli (Ryder *et al*, 1987) we aim to determine if CHS induction observed with *Rhizobium* infection differs from that due to other pathogens.

Figure 3 shows a cloned region of *Trifolium subterraneum* cv. Karridale genome encoding a cluster of four CHS genes. (At least 5 and perhaps as many as seven copies of CHS exist in this cultivar). Sequencing has revealed strong conservation between the coding segments and variability in the single intron and the promoter region. This work is continuing.

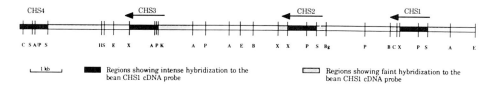

Figure 3. Detailed restriction map for the 14.5kb of *Trifolium subterraneum* DNA containing four CHS genes. The regions of hybridization to the bean CHS1 probe (Ryder *et al*, 1987) are indicated. The position and direction of transcription of three of these genes, as determined by DNA sequencing, is shown by arrows. A- *Acc*I, B-*Bam*HI, Bg-*Bgl*II, C-*Cla*I, E-*Eco*RI, H-*Hin*dIII, P-*Pst*I, S-*Sac*I, X-*Xba*III.

3.5. THE ISOLATION OF CINNAMYL ALCOHOL DEHYDROGENASE

We are also interested in the biosynthesis of lignin in clover, especially in relation to its control by environmental parameters and as a defence mechanism. To this end we are studying an essential enzyme in the pathway-cinnamyl alcohol dehydrogenase [CAD]. It was found that roots were the richest sources of CAD enzymes and Figure 4 shows the pattern of dehydrogenases revealed in them. There are several cinnamyl dehydrogenase bands evident, of which three are major and about eight minor. Several of the enzymes appear to have differential specificities for the three alcohols tested. Of the major bands reacting with NAD the more intense one also reacts well with ethanol and could be the clover alcohol dehydrogenase. The major CAD reacting with NADP has an Mr of 8000 and could thus have a subunit of 40000 which would be similar to the CAD enzymes found in many other plants (Mansell *et al*, 1974). Preliminary experiments showing increases in CAD activity due to wounding and elicitation indicate that this is the species of CAD which could be an inducible defence protein.

Figure 4. Root extracts subjected to gel electrophoresis and stained for dehydrogenase enzyme activity. Six sections of the gel were made and differentially stained. Sections a,b and c had NAD; d.e and f had NADP. Substrates used were; - ethanol with a and f, cinnamyl alcohol with c and d, and coniferyl alcohol with sections b and e.

4. Discussion

The choice of *T. repens* and *T. subterraneum* for the examination of plant-microbe interactions is proving to be a prudent decision. The small size of these clover species makes them most amenable for use in laboratory assays of induced responses; large numbers can be screened rapidly under defined growth conditions. These systems benefit from extensive research into their symbiotic partner, *R.l.* bv. *trifolii,* which has provided a genetically defined range of strains able to elicit specific plant developmental responses by the production of extracellular factors analogous to the *R.meliloti* NodRm-1 (Lerouge *et al,* 1990). A characterized *Rhizobium* expolysaccharide with known biological activity in host-specific nodulation (Djordjevic *et al,* 1987) is also available. In addition, a range of microbes which interact with these *Trifolium* species can be used as bioprobes to induce a plant disease response.

The discovery of loci in *R.l.* bv. *trifolii* which confer cultivar specific nodulation demonstrates a gene-for-gene interaction between plant and symbiont. The defined genetic nature of the *Rhizobium* genes responsible is now being used to isolate the bioactive factor(s). We hope to use this as a probe to identify the mechanism or receptor whereby information from the microbe is perceived.

Isolation of genes responding to microbial factors (CHS, ENOD2, CAD) will permit, through the use of *in situ* hybridization, or later via the use of transgenic clovers carrying fused reporter genes, studies on the regulation of these genes. Central to our inquiry is the question of what similarities and differences exist when a symbiotic interaction and a pathogenic infection are compared. Molecular manipulations and analyses of these plant genes will be expedited through the use of the genetic homogeneity and tractability of *T. subterraneum.*

5. Acknowledgements

The authors thank Chris Lamb for his gift of the bean CHS1 cDNA clone, Ton Bisseling for the gift of the pea ENOD2 probe, Bill Collins and Phil Nichols of the National Sub-Clover Improvement Program for providing advice and seeds from cultivar crosses, Sharon Long and Roger Innes for *Rhizobium* clones, and Chris Hayward for *Pseudomonas* strains. TA and WL-H are recipients of Australian Postgraduate Research Awards, and TA also acknowledges support from the Wenkart Foundation. JJW, MAD and BGR are recipients of research grants from the Wool Research and Development Fund.

6. References

Djordjevic, S., Chen, H., Batley, M., Redmond, J. and Rolfe, B. (1987) 'Nitrogen fixation ability of exopolysaccharide synthesis mutants of *Rhizobium* sp. strain NGR234 and *Rhizobium trifolii* is restored by the addition of homologous exopolysaccharides'. J. Bacteriol. 169, 53-60.

Gibson, A.H. (1968) 'Nodulation failure in *Trifolium subterraneum* L. cv. Woogenellup (syn. Marrar)'. Aust. J. Agric. Res. 19, 907-918.

Lerouge, P., Roche, P., Faucher, C., Mailet, F., Truchet, G., Prome, J.C. and Denarie, J. (1990) 'Symbiotic host-specificity of *Rhizobium meliloti* is determined by a sulphated and acylated glucosamine oligosaccharide signal'. Nature 344, 781-784.

Mansell, R.L., Gross, G.G., Stöckigt, J. Franke, H. and Zenk, M.H. (1974) 'Purification and properties of cinnamyl alcohol dehydrogenase from higher plants involved in lignin biosynthesis'. Phytochemistry 13, 2427-2435.

Moffett, M.A, Hayward, A.C. , and Fahy, P.C. (1986) 'Five new hosts of *Pseudomonas andropogonis* occurring in eastern Australia: host range and characterization of isolates'. Plant Pathol. 35, 34-43.

Ryder, T.B., Hedrick, S.A., Bell, J.N., Liang, X., Clouse, S.D., and Lamb, C.J. (1987) 'Organization and differential activation of a gene family encoding the plant defense enzyme chalcone synthase in *Phaseolus vulgaris*'. Mol. Gen. Genet. 210, 219-233.

Van De Wiel, C., Scheres, B., Franssen, H., Van Lierop M.-J., Van Lammeren, A., Van Kammen, A., and Bisseling, T. (1990) 'The early nodulin transcript ENOD2 is located in the nodule parenchyma (inner cortex) of pea and soybean root nodules'. EMBO J. 9, 1-7.

Weinman, J.J., Djordjevic, M.A., Sargent, C.L., Dazzo, F.B., and Rolfe, B.G. (1988) 'A molecular analysis of host range genes of *Rhizobium trifolii*', in R. Palacios and D.P.S. Verma (eds.), Molecular Genetics of Plant-Microbe Interactions, American Phytopathological Society, St. Paul, MN, pp.33-34.

White, D.W.R. and Greenwood, D. (1987) 'Transformation of the forage legume *Trifolium repens* L. using binary *Agrobacterium* vectors'. Plant Molec. Biol. 8, 461-469.

ROLES OF LECTIN IN THE *RHIZOBIUM*-LEGUME SYMBIOSES

Ben J.J. Lugtenberg, Clara Díaz, Gerrit Smit, Sylvia de
Pater, and Jan W. Kijne
*Leiden University, Dept. of Plant Molecular Biology and
RUL/TNO Center for Phytotechnology, Botanical Laboratory,
Nonnensteeg 3, 2311 VJ Leiden,
The Netherlands*

ABSTRACT. A recent publication of Díaz et al. has convincingly shown
the role of lectin as a major host-specificity determinant of the
Rhizobium-legume symbiosis. In this paper we present the presently
known *in vivo* and *in vivo* properties of pea lectin. We discuss
possible bacterial lectin receptors and propose important future
experiments.

1. Introduction

A lectin (from the Latin *legere:* to select) is currently defined as a
sugar-binding protein or glycoprotein of non-immune origin which
agglutinates cells and/or precipitates glycoconjugates. Thus, lectins
have at least two sugar binding sites. Sugar binding is specific, and
it is defined in terms of monosaccharides or simple oligosaccharides
which are able to inhibit lectin-induced agglutination, precipitation
or aggregation.

Lectins have been found in plants as well as in many other organisms
varying from bacteria to vertebrates. Critics point out that the
current definition of lectins is based on *in vitro* biological
processes which are complex and poorly understood, namely cell
agglutination and receptor precipitation which might not be related to
the yet unknown physiological function(s) of lectin *in vivo* [1,2].

The seeds of leguminous plants of agricultural importance are a rich
source of lectin. Lectin is also present, at low levels, on the root
surface and in the rhizosphere. The latter property, in combination
with that of carbohydrate-binding specificity, made that lectins were
the first molecules proposed to play a role in the determination of
host-plant-specificity in the *Rhizobium-legume* symbiosis [3,4].

Lectins have been attributed a variety of functions which are
compatible with their location. Lectins in seeds might function as
storage proteins, in the packaging and mobilization of storage
materials and as mitogenic stimulators of plant embryonic cells.

174

H. Hennecke and D. P. S. Verma (eds.),
Advances in Molecular Genetics of Plant-Microbe Interactions, Vol. 1, 174–181.
© 1991 *Kluwer Academic Publishers. Printed in the Netherlands.*

Lectins may also be involved in carbohydrate transport, cell wall extension, growth regulation, tissue differentiation and cell-to-cell recognition. Lectins, especially those that are secreted or surface-bound, have also been implied in defence mechanisms of plants against bacteria, fungi, nematodes, insects and other animal predators. In addition, lectins may act as receptors for elicitors which trigger the accumulation of phytoalexins.

Until recently, the specific carbohydrate-binding properties of lectins were generally believed to determine their function(s) inside and outside plant cells. However, many lectins, including those of leguminous plants, have been shown to possess hydrophobic sites which are able to bind molecules with phytohormone activity. The non-carbohydrate binding site of some lectins has also been shown to recognize specific peptide sequences. Whereas lectins from legumes representing members of different cross-inoculation groups differ in their sugar-binding specificity, differences in high-affinity binding of hydrophobic molecules by different lectins have not been found (yet), and the hydrophobic pocket in different lectin molecules is fairly well conserved.

In the following paragraphs we will discuss the role of pea lectin in nitrogen-fixing symbiosis with the microsymbiont *Rl viciae* (*Rhizobium leguminosarum* bv. *viciae*). Pea lectin is supposed to play at least two roles: (i) a non-biovar specific role in accumulation of rhizobia on root hairs (cap formation) and, (ii) a host-plant-specific role in a later infection step, presumably at the time of initiation of infection thread formation.

2. Structure of pea lectin

Pea lectin (*Pisum sativum* lectin, Psl) consists of two pairs of different polypeptides, designated as a 2α2ß structure. The molecular weights of the subunits are 5.8 and 17 kD, respectively, for α and ß. In contrast to some other lectins, pea lectin is not a glycoprotein. It contains two high-affinity binding sites for oligosaccharides that contain a trimannoside core with a neighbouring fucosyl-α-1,6-N-acetylglucosamine group. Binding of pea lectin is inhibited by several monosaccharides, including 3-0-methyl-D-glucose, mannose and glucose. Amino acid residues of both an α and a ß subunit contribute to the structure of one sugar-binding site.

In mature pea seeds, Psl constitutes a mixture of two isolectins, Psl1 and Psl2. In addition, a small amount of the unprocessed Psl precursor is present. Psl1 and Psl2 have indistinguistable molecular weights and subunit structures, and display the same specificity of sugar-binding. However, the isolectins differ in isoelectric point, pI's being 7.2 for Psl1 and 6.1 for Psl2. This charge difference corresponds with electrophoretic differences of the subunits of the isolectins, and probably results from the additional removal of a small peptide at the C-terminus of the Psl1-α subunit, yielding the α subunit of Psl2 [9].

3. Biosynthesis of pea lectin

The pea (*Pisum sativum*) genome contains one functional lectin gene and, dependent on the pea variety, one or more pseudo-genes [5,6,16]. The functional gene is expressed in seeds as well as in roots, leaves and stems. The steady state mRNA level in seeds harvested 20 days after flowering is approximately three orders of magnitude higher than in other parts of the plant [16]. Pea lectin is synthesized as a single translation product and cleaved into an α - and a ß-subunit [7]. Two α- and two ß- subunits together form the 49 kD mature protein [8,15]. Seed Psl is a mixture of two major lectins, due to further processing of part of the α- subunit peptides. In non-seed tissue the processing seems to be complete, resulting in the absence of one of the isolectins [9].

4. Localization of pea lectin

By far most of the plant's lectin is found in seeds. However, lectin is also present in other parts of the plant and part of it is secreted. Psl1 is only found in seeds whereas a lectin similar, but not identical, to Psl2 is present in pea roots and in other parts of the pea plant and is also secreted into the pea rhizosphere [9].

Using indirect immunofluorescence, presence of Psl could be demonstrated on the surface of pea roots. Remarkably, the pattern of lectin location corresponds entirely with the pattern of susceptibility of pea root epidermal cells to infection by *Rl. viciae* [9,17]. Lectin was not found at the surface of pea root tips, in the mature root hair zone or in the zone between root hairs and the cotyledons, i.e. the areas which are resistant to rhizobial infection. Lectin appears on the surface of epidermal cells at the onset of their differentiation into trichoblasts. Trichoblasts with emerging root hair tips constitute the majority of of Psl-positive root epidermal cells [9]. Despite large differences in the surface localization of lectin between various parts of the root, crude protein abstracts of these parts all contain Psl [9]. These observations point at developmental control of Psl secretion. The similarity of the patterns of susceptibility to rhizobial infection and of Psl location on the root surface suggests that either both processes are regulated by the same developmental mechanisms or that a causal relationship exists.

Using a novel binding assay with rat IgE antibodies as high affinity ligands, Psl was localized on the pea root to the same extent and at the same site as described previously with monospecific polyclonal antibodies. These results indicate that part, if not all, of the Psl molecules at the outer surface of the target cells for rhizobial infection are functional. Using slightly plasmolyzed trichoblasts, it was further shown that functional Psl is predominantly present at the external surface of the plasma membrane and not at the cell wall surface of the root hair [9].

5. Role of lectin in attachment of *Rhizobium leguminosarum* bv. *viciae* to pea root hair tips

Attachment of *Rl viciae* cells to pea root hair tips is supposed to be a crucial step in the nodulation process. So far, two steps have been distinguished in attachment. The first step, attachment of single *Rl viciae* cells to the surface of the root hair tip, is mediated by bacterial rhicadhesin, a small Ca^{2+}-binding protein, and an unknown but common plant receptor [11]. The second step, aggregate formation, apparently depends on bacterial growth conditions. When bacteria are grown under carbon-limitation, aggregate formation is mediated by bacterial cellulose fibrils [12]. In contrast, limitation for manganese ions results in cellulose fibril- as well as lectin-mediated attachment, the latter indicated by partial inhibition of this process by the presence of the pea lectin hapten 3-0-methyl-D-glucose [10]. Interestingly, only when lectin is involved in attachment, this process is followed by infection thread formation, suggesting that the involvement of lectin is essential for infectivity.

The observation that direct binding of single *Rhizobium* cells to the pea root hair tip is not inhibited by the Psl-haptenic monosaccharide 3-0-methyl-O-glucose [10,11] is consistent with the absence of Psl at the root hair cell wall surface.

Several lines of evidence suggest that bacterial receptors for pea lectins are extracellular polysaccharides (EPS) or derivatives thereof. Psl binds to and can precipitate EPS's of various fast-growing *Rhizobium* species [18]. This observation is consistent with the observation that Sym plasmid-encoded *nod* genes are not involved in lectin-enhanced aggregation [10]. Furthermore, EPS is only poorly produced by rhizobia grown under carbon-limitation, which attach to pea root hair tips without involvement of lectin. However, it is hard to imagine which part of the lectin molecule would be responsible for binding to EPS, since neither the hydrophobic binding site nor the sugar-specific binding site are likely candidates. In other words, it is hard to reconcile binding of lectin to EPS with the reported [13] role of lectin in host-specific nodulation.

6. Role of root lectin as a determinant of host plant-specificity

In the seventies, it has been proposed that lectin plays a role in determination of host specificity of nodulation. However, several laboratories subsequently published results which were apparently inconsistent with this hypothesis. Recently we decided to test the involvement of lectin in host plant-specifity by using transgenic white clover hairy roots, which efficiently can be nodulated [13].

White clover (hairy) roots can be nodulated by *Rle trifolii*. When normal or hairy clover roots are inoculated with *Rle viciae*, root hair curling is observed but infection threads are not formed. We asked the question whether the introduction of the pea lectin gene, *psl*, into

white clover could cause a progress of infection by *Rle viciae* beyond
the stage of root hair curling.

For technical reasons, white clover hairy roots were used instead of
transgenic white clover plants. In independent experiments, both the
complete genomic *psl* gene as well as a c-DNA construct were used. Both
experiments gave essentially the same results, showing that, whereas
the homologous *Rl trifolii* nodulated 100% of the plants, also *Rl
viciae* was able to induce nodules albeit on only 30 to 40 percent of
the plants. In the latter case, nodulation was delayed and often
abnormal with respect to nodule morphology. However, red, functional
nodules appeared on thirty percent of the nodulated transgenic roots,
indicating that a significant number of infections by *Rl viciae* had
been successful [13].

This result is the first example showing that the host specificity
barrier in a plant-bacterium interaction can be broken by genetic
engineering of the host plant. Moreover, the results show that the
host range is at least partially determined by interactions between
the bacteria and root lectin of the plant.

7. Is the role op psl in attachment essential for nodulation?

Secreted pea lectin plays a role in the accumulation of *Rl viciae*
cells, grown under Mn^{2+}-limitation, to the tips of root hairs [9,10].
However, conclusive evidence that this accumulation is a prerequisite
for infection, is lacking. It can even be argued that secreted lectin,
by causing agglutination of bacteria at some distance from the root
hair tips, inhibits nodulation.

It should be noted however, that evidence has been presented that
secreted soybean lectin increases the infectivity of *B. japonicum*
[14]. These studies should be extended to include other symbiont
combinations. In pea root hair tips, lectin has only occasionally been
shown to be present on the outer cell wall surface. Since membrane-
bound pea lectin is shielded from rhizosphere bacteria by a cell wall,
lectin in pea root hair tips is unlikely to play a role in attachment.
In conclusion, evidence for an essential role for secreted lectin in
nodulation of pea is lacking.

The situation in clover is different since the clover lectin
trifoliin A is present at the outer cell wall surface of tips of
growing white clover root hairs where it mediates attachment.

8. Possible roles of membrane-bound pea lectin in nodulation

Data from Díaz et al. [13] have clearly shown that the *psl* gene plays
a role in host plant-specificity of nodulation. Bacterial counterparts
from which a host-specific role in nodulation are known are *nodD* [19]
and *nodE* [20]. Since this role of *nodD* is at the level of *nod* gene
activation in combination with plant flavonoids, it is unlikely that

nodD is the bacterial counterpart of pea lectin.

Rl viciae nodE is known to play a role in the production of a secreted hostplant-specific low Mw signal that elicits the activation of nodulin PsENOD12 [21]. This signal molecule is likely to be related to that described by the Toulouse group [22]. The latter *R. meliloti* signal molecule (Mw 1102) is a sulphated ß-1,4-tetrasaccharide of D-glucosamine in which three amino groups are acetylated and one is acylated with a C16 bis-unsaturated fatty acid. It is not known whether this molecule interacts with *Medicago* lectin. Characterization of the structures of similar *Rl viciae* signals should be awaited before an interaction between these molecules and pea lectin can be discussed.

It should be kept in mind that other signal molecules which are candidates for an interaction with pea lectin are derivatives of LPS (lipopolysaccharide) [18]. The observation that SDS-PAGE patterns of LPS of induced and uninduced *Rl viciae* cells are indistinguishable does not exclude that *nodE* modifies a small fraction or a released fraction of the LPS molecule in a host-specific way [23].

Summarizing, we speculate that a host-specific oligosaccharide released by *Rl viciae* cells, attached to the pea root hair tip, interacts with pea lectin. This interaction somehow results in internalization of the bacterial cells and in subsequent infection thread formation. Whether, and if so how, this interaction is linked to the activation of PsENOD12 is unknown.

9. References

1. Kocourek, J. and Horejsi, V. (1981) 'Defining a lectin', Nature 290, 188.
2. Kocourek, J. and Horejsi, V. (1983) 'A note on the recent discussion on the definition of the term lectin ', in B09-Hansen, T.C. and Spengler, G.A. (eds.), Lectins, biology, biochemistry, clinical biochemistry, vol 3, pp. 3-6.
3. Bohlool, B.B. and Smidt, E.L. (1974) 'Lectins: a possible basis for specificity in the *Rhizobium* - legume root nodule symbiosis', Science 185, 269-271.
4. Dazzo, F.B. and Hubbell, D.H. (1975) 'Cross-reactive antigens and lectin as determinants of symbiotic specificity in the *Rhizobium-clover* association', Appl. Microbiol. 30, 1017-1033.
5. Gatehouse, J.A., Bown, D.J, Evans, I.M., Gatehouse, L.N., Jabes, D.,Preston, P., Croy R.R.D., (1987) 'Sequences of the seed lectin gene from pea (*Pisum sativum* L.)', Nucl. Acid Res. 15, 7642.
6. Kamiski, P.A., Buffard, D., Strosberg, A.D. (1987) 'The pea lectin gene family contains only on functional lectin gene', Plant Mol. Biol. 9, 497-507.
7. Higgins, T.J.V., Chandler, P.M., Zurawski, G., Button S.C. and Spencer D. (1983) 'The biosynthesis and primary structure of pea

seed lectin', J. Biol. Chem. 258, 9544-9549.

8. Van Driessche, E., Beckmans, S., Dejacqere, R. and Kanarek, L. (1988) 'Isolation of pea lectin precursor and characterization of its processing products' in Leed, D.L.J. and B09-Hansen, T.C. (eds.), Lectins, Biology, Biochemistry, Clinical Biochemistry, vol. 6, pp. 355-362.

9. Díaz, C.L., (1989) ' Root lectin as a determinant of host-plant specificity in the *Rhizobium*-legume symbiosis' Ph D Thesis, Leiden University.

10. Kijne,J.W., Smit, G., Díaz, C.L., and Lugtenberg, B.J.J. (1988) 'Lectin-enhanced accumulation of manganese-limited Rhizobium Leguminosarum cells on pea roothair tips', J. Bacteriol. 170, 2994-3000.

11. Smit, G. 91988) 'Adhesins from *Rhizobiaceae* and their role in plant-bacterium interactions', Ph D Thesis, Leiden University.

12. Smit, G., Kijne, J.W. and Lugtenberg, B.J.J. (1986) 'Involvement of both cellulose fibrils and a Ca^{2+}-dependent adhesin in the attachment of *Rhizobium leguminosarum* to pea root hair tips', J.Bacteriol. 168, 821-827.

13. Díaz, C.L., Melchers, L.S., Hooykaas, P.J.J., Lugtenberg, B.J.J., and Kijne, J.W. (1989) 'Root lectin as a determinant of host-plant specificity in the *Rhizobium-legume* symbiosis', Nature 338, 579-581.

14. Halverson, L.J. and Stacey, G. (1985) 'Host recognition in the *Rhizobium*-soybean symbiosis. Evidence for the involvement of lectin in nodulation', Plant Physiol. 77, 621-625.

15. Trowbridge, I.S., (1974) 'Isolation and chemical characterization of a mitogenic lectin from *Pisum Sativum*', J.Biol.Chem. 249, 6004-6012.

16. De Pater, S. et al., In preparation.

17. Díaz, C.L., Van Spronsen, P.C., Bakhuizen, R., Logman, G.J.J., Lugtenberg, E.J.J., and Kijne, J.W. (1986) 'Correlation between infection bij *Rhizobium leguminosarum* and ledtin on the surface of *Pisum sativum* L. roots', Planta 168, 530-539.

18. Kamberger, W. (1979) 'Role of cell surface polysaccharides in the *Rhizobium*-pea symbiosis', FEMS Microbiol. Letters 6, 361-365.

19. Spaink, H.P., Wijffelman, C.A., Pees, E., Okker, R.J.H. and Lugtenberg, B.J.J. (1987) '*Rhizobium* nodulation gene *nodD* as a determinant of host specificity', Nature 328, 337-340.

20. Spaink, H.P., Weinman, J., Djordjevic, M.A., Wijffelman, C.A., Okker, R.J.H. and Lugtenberg, B. (1989) 'Genetic analysis and cellular localization of the *Rhizobium* host specificity-determining NodE protein', EMBO J. 8, 2811-2818.

21. Scheres, B., Van de Wiel, C., Zalensky, A., Horvath, B., Spaink, H., Van Eck, H., Zwartkruis, F., Wolters, A.-M., Gloudemans, T., Van Kammen, A. and Bisseling, T. (1990) 'The ENOD 12 gene product is involved in the infection process during the pea-*Rhizobium* interaction', Cell 60, 281-294.

22. Lerouge, P., Roche, P., Faucher, C., Maillet, F., Truchet, G., Promé, J.C. and Dénarie, J. (1990) 'Symbiotic host-specificity

of *Rhizobium meliloti* is determined by a sulphated and acylated glucosamine oligosaccharide signal', Nature 334, 781-784.

23. De Maagd, R.A. and Lugtenberg, B.J.J. (1989) 'Lipopolysaccharide: a signal in the establishment of the *Rhizobium*-legume symbiosis ?' in Lugtenberg, B. (ed.) Signal molecules in plants and plant-microbe interactions (Springer Verlag, Heidelberg, NATO ASI Series H36) pp. 337-344.

ANALYSES OF THE ROLES OF *R.MELILOTI* EXOPOLYSACCHARIDES IN NODULATION

T. L. REUBER, J. W. REED, J. GLAZEBROOK, A. URZAINQUI,
and G. C. WALKER
Department of Biology
Massachusetts Institute of Technology
Cambridge, MA 02139
USA

ABSTRACT. Genetic experiments have indicated that succinoglycan (EPS I), the acidic Calcofluor-binding exopolysaccharide, of the nitrogen-fixing bacterium *Rhizobium meliloti* strain Rm1021 is required for nodule invasion and possibly for later events in nodule development on alfalfa and other hosts. Thirteen *exo* loci on the second megaplasmid have been identified that are required for, or affect, the synthesis of EPS I. Mutations in certain of these loci completely abolish the production of EPS I and result in mutants that form empty Fix⁻ nodules. We have identified two loci, *exoR* and *exoS*, that are involved in the regulation of EPS I synthesis in the free-living state. Certain *exo* mutations which completely abolish EPS I production are lethal in an *exoR95* or *exoS96* background. Histochemical analyses of the expression of *exo* genes during nodulation using *exo*::Tn*phoA* fusions have indicated that the *exo* genes are expressed most strongly in the invasion zone. In addition, we have discovered that *R. meliloti* has a latent capacity to synthesize a second exopolysaccharide (EPS II) that can substitute for the role(s) of EPS I in nodulation of alfalfa but not of other hosts. Possible roles for exopolysaccharides in symbiosis are discussed. Evidence indicating that the *R. meliloti exoD* gene encodes a novel function needed for nodule invasion is summarized.

1. The succinoglycan exopolysaccharide of R. meliloti is required for nodule invasion

Our laboratory has obtained strong genetic evidence that succinoglycan (EPS I), the acidic Calcofluor-binding exopolysaccharide of *Rhizobium meliloti* strain Rm1021 (1) is required for nodule invasion and quite possibly for later events in nodule development (5,12,13). EPS I is a high molecular weight polymer composed of polymerized octasaccharide subunits. Each octasaccharide consists of a backbone of three glucoses and one galactose, a side chain of four glucoses, and 1-carboxyethylidene (pyruvate), acetyl, and succinyl modifications in a ratio of approximately 1:1:1 (1). We isolated a set of mutants (*exo*) of *R. meliloti* Rm1021 on the basis of their failure to fluoresce under UV light on medium containing Calcofluor and showed that these mutants did not synthesize EPS I (13). Alfalfa seedlings inoculated with these *exo* mutants formed ineffective (non-nitrogen fixing) nodules that contained few if any bacteria and no bacteroids (13). More detailed characterizations of nodules elicited by an *exoB* mutant revealed that, in one genetic background, no infection threads are formed following inoculation with this mutant and that the plant cells in the interior of nodules elicited by this strain do not contain bacteria or bacteroids (5). Subsequent studies (A. Hirsch, personal communication) of a series of *exo* mutants generated by insertion of the transposon Tn5 in *R. meliloti* strain Rm1021 have shown that root hair curling is

182

H. Hennecke and D. P. S. Verma (eds.),
Advances in Molecular Genetics of Plant-Microbe Interactions, Vol. 1, 182–188.
© 1991 *Kluwer Academic Publishers. Printed in the Netherlands.*

significantly delayed and that infection threads can form but abort at a very early stage so that the bacteria are never able to reach the interior of the nodule. Furthermore, these *exo* mutants are only able to elicit the synthesis of two of the nodule-specific plant proteins termed nodulins as opposed to the seventeen that are elicited by infection with a wild type *R. meliloti* (10,12,18; A. Hirsch, personal communication). Thus these *R. meliloti exo* mutants uncouple the ability of the bacteria to signal the plant to initiate nodule formation from the ability of the bacteria to invade the nodule and establish an effective symbiosis.

2. Genetic analyses of the synthesis of EPS I by R. meliloti

We were able to subdivide our initial set of *R. meliloti exo* mutants into five different genetic classes (13). Three of these classes were then shown to be located on the second symbiotic megaplasmid, pRmeSU47B, of *R. meliloti* and the other two classes (*exoC* and *exoD*) were shown to be located on the chromosome (6). A detailed genetic analysis of the cluster of *exo* genes located on the second symbiotic megaplasmid has led to the identification of a number of loci that are required for the synthesis of the *R. meliloti* EPS I (15). Mutations in *exoA*, *exoB*, *exoF*, *exoL*, *exoM*, *exoP*, *exoQ* and *exoT* completely abolish production of the EPS I and result in mutants that form Fix⁻ nodules (13,15,20). Analyses of the properties of fusions of various *exo* genes to alkaline phosphatase that were generated using Tn*phoA* (16) suggest that the *exoF*, *exoP*, *exoQ* and *exoA* gene products are membrane proteins (14,20).

In addition, we have found that mutations in *exoG*, *exoJ* and *exoN* (15), diminish the production of Calcofluor-binding material. *exoG* and *exoJ* mutants form effective nodules with decreased efficiency, whereas plants inoculated with *exoN* mutants fix nitrogen normally. The *exoG* mutants are of particular interest since they produce no detectable high molecular weight exopolysaccharide yet the low molecular weight Calcofluor-binding material they do produce seems to be sufficient to allow some amount of nodule invasion and development to proceed. We are in the process of characterizing this low molecular weight material produced by *exoG* mutants but do not yet know its structure. However, it appears that this mutant may represent an example of an derivative of *R. meliloti* that produces a biologically active oligosaccharide or low molecular weight polysaccharide but no high molecular weight polysaccharide. Another class of *exo* mutation, *exoK*, that causes a decrease in exopolysaccharide production also causes a delay in the appearance of a fluorescent halo on Calcofluor plates but no nodulation defect (12,15).

A very interesting class of *exo* mutants that synthesize a structural variant of EPS I was originally identified on the basis of the failure of colonies of such mutants to form a fluorescent halo under UV light when grown on medium containing Calcofluor (12). These mutations defined a locus termed *exoH* which mapped in the middle of a cluster of *exo* genes on the second symbiotic megaplasmid. Alfalfa seedlings inoculated with *exoH* mutants form ineffective nodules that do not contain intracellular bacteria or bacteroids. Root hair curling is significantly delayed and infection threads abort in the nodule cortex. In other words, despite the fact that these mutants made Calcofluor-binding material, their behavior on plants was the same as those *exo* mutants that produced no EPS I. Analyses of exopolysaccharide secreted by *exoH* mutants have shown that it is identical to the Calcofluor-binding exopolysaccharide secreted by the parental *exoH* strain except that it completely lacks the succinyl modification. *In vitro* translation of total RNA isolated from nodules induced by an *exoH* mutant has shown that, as in the case of the exopolysaccharide-deficient exo mutants, only two of the plant-encoded nodulins are induced, as compared to the 17 nodulins induced by the wild type strain. These observations raise the possibility that succinylation of the bacterial exopolysaccharide is important for its role(s) in nodule invasion and possibly nodule development (12).

3. Regulation of EPS I biosynthesis in the free-living state

We initially observed that the synthesis of EPS I by *R. meliloti* is greatly increased if the cells are limited for nitrogen, phosphorus, or sulfur in the presence of a good carbon source. We have identified two unlinked loci, *exoR* and *exoS*, whose products play a role in regulating the synthesis of EPS I by *R. meliloti* strain Rm1021 (4). Tn5-generated mutations in these loci are recessive and lead to substantial increases in the amount of exopolysaccharide synthesized indicating that the *exoR* and *exoS* gene products play negative roles in regulating exopolysaccharide synthesis. Introduction of an *exoR95*::Tn5 or *exoS96*::Tn5 mutation into strains containing Tn*phoA*-generated fusions to *exoA*, *exoF*, *exoP*, *exoQ*, or *exoT* results in a 2-5 fold increase in the level of expression of alkaline phophatase activity suggesting that they negatively regulate *exo* expression. A fundamental difference between the *exoR95*::Tn5 and *exoS96*::Tn5 mutants is that the *exoR95*::Tn5 mutant synthesizes its EPS I at a high constitutive level regardless of the presence or absence of nitrogen in the medium whereas the *exoS96*::Tn5 mutant undergoes a further increase in the rate of synthesis upon nitrogen starvation. It seems that the *exoR* gene product is either involved directly in sensing the level of nitrogen in the medium or else that it acts later in a putative regulatory cascade than the element(s) that actually does the sensing. The relationship of *exoS* action to *exoR* action is not yet clear.

exoS96::Tn5 mutants formed Fix$^+$ nodules on alfalfa. In contrast, we found that, on alfalfa, *exoR95*::Tn5 mutants formed both empty Fix$^-$ nodules and also Fix$^+$ nodules that contained widely varying numbers of bacteria and bacteroids. All the bacteria we isolated from the Fix$^+$ nodules induced by the *exoR95*::Tn5 strain had acquired unlinked suppressors that reduced the amount of exopolysaccharide produced (4) suggesting that the bacteria need to control either how much EPS I they synthesize or when they synthesize it it order to invade nodules.

4. Certain exo mutations which completely abolish EPS I production are lethal in an exoR or exoS background.

While investigating the regulation of *exo*::Tn*phoA* fusions by *exoR* and *exoS*, we discovered that the *exoR95*::Tn5-233 or *exoS96*::Tn5-233 mutations could not be introduced into certain *exo*::Tn*phoA* or *exo*::Tn5 mutant strains unless a cosmid complementing the exo mutation was present in the strain. The *exoR95* and *exoS96* mutations are lethal in combination with *exoL*, *exoM*, *exoQ*, and *exoT* mutations, all of which completely eliminate EPS I production. However, the *exoA*, *exoB*, and *exoF* mutations, which also completely eliminate EPS I production, are not lethal in combination with these regulatory mutations. Mutations in the *exoP* locus fell into two classes: Tn5 and Tn*phoA* insertions in the upstream region of the locus were lethal in *exoR95* or *exoS96* backgrounds, but insertions farther downstream were not. None of the mutations which reduce the quantity of EPS I produced (*exoG*, *exoJ*, *exoN*, *exoK*) or alter its structure (*exoH*), are lethal in *exoR95* or *exoS96* backgrounds. These results suggest that certain mutations blocking exopolysaccharide production may cause a toxic accumulation of intermediates in these EPS I-overproducing strains. It is known that EPS I biosynthesis takes place on polyprenyl lipid carriers in the cytoplasmic membrane (22,23). These same lipid carriers are also used for peptidoglycan and lipopolysaccharide biosynthesis (21). An attractive hypothesis is that the exo mutations which are lethal in combination with *exoR95* or *exoS96* cause accumulation of incomplete, unpolymerizable EPS I subunits on the lipid carriers, and therefore diminish the pool of lipid carriers to the point where cell wall and lipopolysaccharide biosynthesis can no longer take place. The *exoA*, *exoB*, and *exoF* mutations, which are not lethal in combination with *exoR95* or *exoS96*, may block EPS I biosynthesis at an earlier stage.

5. Histochemical staining shows exo::Tn_phoA_ fusions are active primarily in the early symbiotic zone of the nodule.

We found that it is possible to stain nodules specifically for the alkaline phosphatase activity present in the inducing bacteria by staining at pH9. At this pH, nodules induced by a Pho⁻ strain show no staining. Nodules induced by a strain carrying the _exoF369_::Tn_phoA_ fusions and a plasmid which complements the _exoF_ mutation to allow normal nodulation show staining primarily in the early symbiotic or invasion zone of the nodule where the bacteria are invading the plant cells. In the late symbiotic zone, which contains mature bacteroids, little staining was seen. Nodules induced by strains carrying _exoP_::Tn_phoA_ and _exoA_::Tn_phoA_ fusions showed a much lower degree of staining than that in nodules induced by the _exoF_::Tn_phoA_ strain, but faint staining of the invasion zone was seen after long incubations. These results suggest that little or no new EPS I synthesis is needed after nodule invasion.

6. R. meliloti can produce a second exopolysaccharide that can substitute for the role of EPS I in nodule invasion

The symbiotic defects of _exo_ mutants can be suppressed by the presence of a mutation, _expR101_, which causes overproduction of a second exopolysaccharide, EPS II (7). Genetic analyses have shown that the products of a cluster of at least six _exp_ genes located on the second symbiotic megaplasmid, as well as the product of the _exoB_ gene, are required for EPS II synthesis. The presence of the _expR101_ mutation causes increased transcription of the _exp_ genes, resulting in overproduction of EPS II. As a consequence of this genetic analysis, we were able to construct strains which produced EPS I or EPS II exclusively, or neither EPS. _Medicago sativa_ plants inoculated with an EPS II producing strain formed nitrogen-fixing nodules. However, the EPS II producing strain formed empty, ineffective nodules on four other plant species which were effectively nodulated by the EPS I producing strain. Thus, it appears that exopolysaccharides are involved in determining host range at the level of nodule invasion, rather than at the level of primary recognition of the host.

The structure of EPS II has been determined (7,8). The independent discovery of this exopolysaccharide has been reported by Zhan et al. (25) and the partial structure they report is in agreement with this structure. Both exopolysaccharides are acidic, contain glucose and galactose, and have acetyl and pyruvate (1-carboxyethylidene) modifications. However, the structures differ in many respects: i) EPS II does not contain any succinate groups, ii) EPS II contains more galactose than EPS I, iii) EPS II is unbranched, iv) EPS II has both α and β glycosidic linkages, while EPS I has only β glycosidic linkages, and v) The pyruvate group in EPS I is linked to glucose, while in EPS II it is linked to galactose. However, it is interesting to note that each exopolysaccharide has a single Glc- β(1,3)-Gal linkage in its backbone and that these two rather diverse exopolysaccharides may share the common structural motif of O-6-acetylglucose- β(1,3)-galactose.

7. Possible roles for exopolysaccharides in nodulation

Exopolysaccharides may have more than one function in nodulation. Since _exoG_ mutants, which make low molecular weight EPS I, but not high molecular weight EPS I, form effective nodules at reduced efficiency relative to wild-type, but much better than mutants which make no exopolysaccharide, it is possible that both the high and low molecular weight forms of EPS have symbiotic functions. One of the most intriguing possibilities for the role of low molecular weight forms of the exopolysaccharides in nodulation is that they act as signals to the plant during the process of nodule invasion and development. Carbohydrates have previously been shown to function as signal molecules in plants (2). Several lines of evidence suggest that oligosaccharide fragments of EPS may have a symbiotic function. First, *exoG* mutants, which produce mainly low molecular weight EPS I can invade nodules, albeit at reduced efficiency (15). Second, both we and John Leigh and his colleagues (personal communication) have obtained preliminary evidence that a low molecular weight fraction of EPS I can partially suppress the symbiotic deficiencies *R. meliloti exo* mutants. A similar finding has been reported previously by Djordjevic *et al.* (3) who found that *exo* mutants of *Rhizobium* sp. NGR234 and *Rhizobium trifolii* can form effective nodules if either high molecular weight EPS or oligosaccharide subunits of EPS are supplied exogenously. However there appears to be some difference between the *R. meliloti*-alfalfa system and those studied by Djordjevic *et al.* (3) since neither we nor others (17) have been able to suppress the symbiotic deficiencies of *R. meliloti exo* mutants by the addition of purified high molecular weight exopolysaccharide isolated from their *exo⁺* parent Rm1021.

In the course of these experiments, we observed that we could also partially suppress the symbiotic deficiencies of *exoH* mutants, which fail to succinylate their EPS I, by the addition of a low molecular weight fraction of EPS I. This observation is consistent with the report of Leigh and Lee (11) that *exoH* mutants produce less low molecular weight EPS I than the wild-type; perhaps the symbiotic defects of these mutants result from a failure to process the non-succinylated EPS I into an oligosaccharide signal molecule. Furthermore, this suppression of *exoH* mutants appeared to be somewhat more efficient than that of *exoA* mutants. This observation is consistent with the possibility that low and high molecular weight forms of the exopolysaccharide may have different symbtiotic functions and suggests the additional possibility that there may be different structural requirements for these two types of roles.

Other possible roles for exopolysaccharides include serving as a carrier for extracellular enzymes or signal molecules, forming part of the infection thread matrix, constraining descendents of attached bacteria to the immediate vicinity of the plant, or helping to evade or suppress plant defense responses (7). With respect to this later possibility, it is interesting that $\beta(1,3)$-glucanases are among the major hydrolytic enzymes induced when plants are exposed to pathogens (9) and that each of the two *R. meliloti* exopolysaccharides has a single $\beta(1,3)$ linkage in its backbone. Furthermore, this linkage is known to be modified in the case of EPS II and likely to be modified in the case of EPS I.

8. The R. meliloti exoD gene encodes a novel function needed for nodule invasion.

We originally identified *exoD* mutants because they produced less exopolysaccharide than wild-type (i.e. they had a "dim" fluorescence phenotype when grown on medium containing Calcofluor and viewed under UV light) (13). They were initally presumed to belong to the *exo* mutant class because they induced the same small white empty nodules on alfalfa plants as other strains mutant in *exo* genes. However, we have found that the exopolysaccharide produced by *exoD* strains appears at the level of ¹H-NMR analysis to be identical to that produced by wild-type (19). In addition, we have observed that whereas most *exo* strains occasionally induce Fix ⁺ nodules, *exoD* strains almost never do. Both observations suggested that *exoD* strains might have a defect different from that of other *exo* mutants, prompting us to investigate the properties of *exoD* mutations in greater detail.

We obtained genetic evidence that *exoD* is in a class distinct from other *exo* mutations from three sources (19). First, we and others have found that, whereas the majority of *exo* mutations map to a cluster of *exo* genes on the second symbiotic megaplasmid, *exoD* mutations map on the chromosome. Second, we found that two different suppressors of the nodule invasion defect of other *exo* mutations, *expR101* (7) and *lpsZ⁺* (24), failed to suppress the invasion defect of *exoD* mutations, although we observed partial suppression by *lpsZ⁺*. Third, we found that in coinoculation experiments *exoD* strains could not be helped to invade nodules by an *exo⁺* helper strain, whereas other *exo* strains were. All of these findings lend credence to the suggestion that *exoD* strains have a defect different from that of other *exo* mutants.

To determine what that defect might be, we looked more carefully at the physiology of *exoD* strains (19). Interestingly, we found that *exoD* strains formed substantially smaller colonies than the wild-type parent on yeast-extract agar plates. More careful investigation of this phenomenon revealed that the growth defect was particularly severe at alkaline pH, and that when any of several different buffers was included in the yeast extract plates, the *exoD* strains formed colonies as large as those formed by the wild-type strain. These observations suggested that the defect of *exoD* strains might be in some function required for survival or growth at alkaline pH. Furthermore, it was possible that the nodule invasion defect of *exoD* strains arose from this same susceptibility to nonoptimal growth conditions.

It therefore seemed possible that the nodule invasion deficiency of *exoD* mutants might be alleviated by modifications of the nodulation medium similar to those which improved the growth of *exoD* strains on yeast extract plates. That is, inclusion of buffer in the Jensen's medium used in plant inoculation experiments might allow *exoD* strains to invade. We found that this was indeed the case, as buffering the medium between pH 6.0 and 6.5 allowed *exoD* strains to invade nodules and fix nitrogen (19). This pH dependence of nodulation paralleled the pH dependence of growth, as in both cases *exoD* strains performed well at acidic pH and poorly at neutral or alkaline pH. It thus seems likely that *exoD* strains fail to invade nodules effectively because they are sensitive to whatever conditions prevail in the infection thread. It is tempting to conclude that the infection thread may be alkaline.

Acknowledgments

We thank the other members of the laboratory for their support and encouragement. This work was supported by Public Health Service Grant GM31030 to G. C. W. J. W. R., J. G., and T. L. R. were supported by National Science Foundation Predoctoral Fellowships.

References

1. Aman, P., M. McNeil, L.-E. Franzen, A. G. Darvill, and P. Albersheim. 1981. Structural elucidation, using HPLC-MS and GLC-MS, of the acidic polysaccharide secreted by *Rhizobium meliloti* strain 1021. Carbohydr. Res. 95:263-282

2. Darvill, A.G., P. A. Albersheim, P. Bucheli., S. Doares, N. Doubrava, S. Eberhard, D.J. Gollin, M. G. Hahn, V. Marfa-Riera, W. S. York, and D. Mohnen. 1989. Oligosaccharins - plant regulatory molecules. In "NATO ASI Series, Vol. H36, Signal Molecules in Plants and Plant-Microbe Interactions", B. J. J. Lutgenberg (Ed.), Springer-Verlag, Berlin Heidelberg, pp 41-48.

3. Djordjevic, S., H. Chen, M. Batley, J. W. Redmond, and B. G. Rolfe, 1987. Nitrogen fixation ability of exopolysaccharide synthesis mutants of *Rhizobium* sp. strain NGR234 and *Rhizobium trifolii* is restored by the addition of homologous exopolysaccharides. J. Bacteriol. 169:53-60.

4. Doherty, D., J. A. Leigh, J. Glazebrook, and G. C. Walker. 1988. Mutants of *Rhizobium meliloti* that overproduce its acidic Calcofluor-binding exoplysaccharide. J. Bacteriol. 170:4249-4256.

5. Finan, T. M., A. M. Hirsch, J. A. Leigh, E. Johanson, G. A. Kulda, S. Deegan, G. C. Walker, and E. R. Signer. 1985. Symbiotic mutants of *Rhizobium meliloti* that uncouple plant from bacterial differentiation. Cell 40:869-877.

6. Finan, T. M., B. Kunkel, G. F. De Vos, and E. R. Signer. 1986. Second symbiotic megaplasmid in *Rhizobium meliloti* carrying exopolysaccharide and thiamine synthesis genes. J. Bacteriol. 167:66-72.
7. Glazebrook J., and G. C. Walker. 1989. A novel exopolysaccharide can function in place of the Calcofluor-binding exopolysaccharide in nodulation of alfalfa by *Rhizobium meliloti*. Cell 56:661-672.
8. Her, G. R., J. Glazebrook, G. C. Walker, and V. N. Reinhold. 1990. Structural studies of a novel exopolysaccharide produced by a mutant of *Rhizobium meliloti* strain Rm1021. Carbohydr. Res. 198:305-312.
9. Kombrink E., M. Schroeder, and K. Hahlbrock. 1988. Several "pathogenesis-related" proteins in potato are 1,3- β-glucanases and chitinases. Proc. Natl. Acad. Sci. USA 85:782-786.
10. Lang-Unnasch, N., and F. M. Ausubel. 1985. Nodule-specific polypeptides from effective alfalfa nodules and from ineffective root nodules lacking nitrogenase. Plant Physiol. 77:833-839.
11. Leigh, J.A., and C. C. Lee. 1988. Characterization of polysaccharides of *Rhizobium meliloti exo* mutants that form ineffective nodules. J. Bacteriol. 170:3327-3332.
12. Leigh, J. A., J. W. Reed, J. F. Hanks, A. M. Hirsch, and G. C. Walker. 1987. *Rhizobium meliloti* mutants that fail to succinylate their Calcofluor-binding exopolysaccharide are defective in nodule invasion. Cell 51:579-587.
13. Leigh, J. A., E. R. Signer, and G. C. Walker. 1985. Exopolysaccharide-deficient mutants of *Rhizobium meliloti* that form ineffective nodules. Proc. Natl. Acad. Sci. USA 82:6231-6235.
14. Long, S., S. McCune, and G. C. Walker. 1988. Symbiotic loci of *Rhizobium meliloti* identified by random Tn*phoA* mutagenesis. J. Bacteriol. 170: 4257-4265.
15. Long, S., J. W. Reed, J. W. Himawan, and G. C. Walker. 1988. Genetic analysis of a cluster of genes required for the synthesis of the Calcofluor-binding exopolysaccharide of *Rhizobium meliloti*. J. Bacteriol. 170:4239-4248.
16. Manoil, C., and J. Beckwith. 1985. Tn*phoA*: a transposon probe for protein export signals. Proc. Natl. Acad. Sci. USA 82:8129-8133.
17. Muller, P., M. Hynes, D. Kapp, K., Niehaus, and A. Puhler. 1988. Two classes of *Rhizobium meliloti* infection mutants differ in exopolysaccharide production and in coinoculation properties with nodulation mutants. Mol. Gen. Genet. 211:17-26.
18. Norris, J. H., L. A. Marcol, and A. M. Hirsch. 1988. Nodulin gene expression in effective alfalfa nodules and in nodules arrested at three different stages of development. Plant Physiol. 88:321-328.
19. Reed, J. W., and G. C. Walker. The *exoD* gene of *Rhizobium meliloti* encodes a novel function needed for alfalfa nodule invasion. Submitted.
20. Reuber, T.L., and G. C. Walker. Regulation of *Rhizobium meliloti exo* genes in free-living cells and *in planta* analyzed using Tn*phoA* fusions. Submitted
21. Sutherland, I. W. 1982. Biosynthesis of microbial exopolysaccharides. Adv. Microbiol. Physiol. 23:79-150.
22. Tolmasky, M. E., R. J. Staneloni, R. A. Ugalde, and L. F. Leloir. 1980. Lipid-bound sugars in *Rhizobium meliloti*. Arch. Biochem. & Biophys. 203:358-364.
23. Tolmasky, M. E., R. J. Staneloni, and L. F. Leloir. 1982. Lipid-bound saccharides in *Rhizobium meliloti*. J. Biol. Chem. 257:6751-6757.
24. Williams, M. N. V., R. I. Hollingsworth, S. Klein, and E. R. Signer. 1990. The symbiotic defect of *Rhizobium meliloti* is suppressed by *lpsZ*+, a gene involved in lipopolysaccharide biosynthesis. J. Bacteriol., 172:2622-2632.
25. Zhan, H., S. B. Levery, C. C. Lee, and J. A. Leigh. 1989. A second exopolysaccharide of *Rhizobium meliloti* strain SU47 that can function in root nodule invasion. Proc. Natl. Acad. Sci. USA 86:3055-3059.

THE ROLE OF THE *RHIZOBIUM MELILOTI* EXOPOLYSACCHARIDES EPS I AND EPS II IN THE INFECTION PROCESS OF ALFALFA NODULES

A. Pühler, W. Arnold, A. Buendia-Claveria, D. Kapp,
M. Keller, K. Niehaus, J. Quandt, A. Roxlau, and W.M. Weng
University of Bielefeld, Faculty of Biology
Department of Genetics
POB 8640, D-4800 Bielefeld, F.R.G.

ABSTRACT

Rhizobium meliloti is able to produce two different exopoly-saccharides, the succinoglucan EPS I and the alternative EPS II. R. meliloti mutants deficient in EPS I production are unable to form infection threads. They induce alfalfa nodules which lack bacteroids. Six exo genes which were found to be involved in EPS I production as well as in nodule infection were localized on megaplasmid 2. The regulatory gene rexA which effects the conversion from EPS I to EPS II production, was identified. EPS II proved unable to substitute EPS I in the alfalfa infection process. R. meliloti mutants deficient in EPS I biosynthesis were found to elicit the plant defence mechanisms at the contact zone between the microsymbiont and the surface of the alfalfa nodule. We suggest that a signal molecule from the EPS I biosynthetic pathway is necessary for the invading R. meliloti strain to be recognized as a symbiont.

RESULTS AND DISCUSSION

Structural and functional analysis of an R. meliloti DNA fragment of megaplasmid 2 carrying genes for EPS I production

Previously we described the isolation of a 7.8 kb EcoRI-fragment of megaplasmid 2 of R. meliloti 2011 complementing mutants defective in EPS production and nodule infection (Müller et al. 1988b). Mutagenesis of this fragment with the Tn5-lacZ transposon revealed several regions involved in EPS synthesis and nodule infection (Keller et al., 1988). A further EPS⁻/Inf⁻ mutant (RmH36) carrying a deletion in a DNA fragment adjacent to this EcoRI-fragment (Fig. 1) indicated that this region is also involved in EPS synthesis and symbiosis (Müller et al., 1988a). To determine the genetic organization of these fragments in more detail, a 7 kb ClaI-BamRI-fragment (Fig. 1) was sequenced.

189

H. Hennecke and D. P. S. Verma (eds.),
Advances in Molecular Genetics of Plant-Microbe Interactions, Vol. 1, 189–194.
© 1991 *Kluwer Academic Publishers. Printed in the Netherlands.*

The sequence revealed six open reading frames (OFR's) which were designated exoX, exoY, exoV, exoW, exoZ, and exoB.

The protein encoded by the exoX gene which is located at the left end of the sequenced region (Fig. 1) showed 73% homology to the ExoX protein of Rhizobium sp. strain NGR234 (Gray et al, 1990). The previously described mutant Rm124 (Keller et al., 1988) is mutated in exoX. It is characterized by EPS overproduction. Therefore, we considered the R. meliloti exoX gene to be responsible for a downregulation of EPS I production.

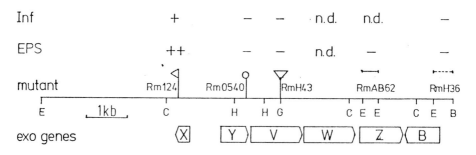

Figure 1. Genetic organization of a DNA fragment of the R. meliloti megaplasmid 2. The location of six exo genes are indicated by open bars below the restriction map of the analyzed DNA fragment. Mutations in the exo genes are indicated by a flag (Tn5-B20), a pin (Tn5), a triangle (interposon), by a dashed bar (deletion and replacement by an interposon), and by a thin bar (fragment used for plasmid integration mutagenesis). The phenotype of the corresponding mutants is presented for EPS synthesis (EPS) and nodule infection (Inf).
Abbreviations: E, EcoRI; C, ClaI; H, HindIII; B, BamHI; G, BglII; n. d. = not determined.

Four further genes reading divergently from exoX could be identified upstream of exoX. The first gene, exoY, is mutated in the previously described EPS⁻/Inf⁻ mutant Rm0540 (Müller et al, 1988b). Mutants in exoV (e.g. RmH43) also caused an EPS⁻/Inf⁻ phenotype (Müller et al., 1988a). The R. meliloti ExoY protein showed 84% homology to the ExoY protein of Rhizobium sp. strain NGR234 (Gray et al., 1990). The proteins encoded by R. meliloti exoV and by ORF1 of Rhizobium sp. strain NGR234 (Gray et al., 1990) were also highly homologous. For the protein encoded by R. meliloti exoW we found no homology to any known protein held in the NBRF databank. The hydrophobicity plot of ExoW suggested that this protein is membrane bound.

The gene exoZ downstream of exoW is involved in EPS synthesis, since a mutant obtained by plasmid integration mutagenesis was Cellufluor dark when grown on agar containing Cellufluor. Further right of exoZ the gene exoB is located which runs in opposite direction when compared to exoZ. The exoB gene got its name from the exoB locus, previously identified by Long et al. (1988). The phenotype of the deletion mutant RmH36 fully supported this idea. The deduced amino acid sequence of the ExoB protein is homologous (39%) to the E. coli galE gene product encoding an UDP-galactose-4-epimerase. Since the activity of this enzyme is severely reduced in mutant RmH36 exoB seems to encode this enzyme. An operon analysis of the identified genes showed that exoY, exoZ and exoB form single transcriptional units.

The R. meliloti rexA gene regulates the biosynthesis of EPS I and EPS II

After Tn5 mutagenesis of R. meliloti 2011 we isolated a mutant, called Rm3131 with a mucoid colony morphology but only weak UV fluorescence when grown on agar containing Cellufluor. We found that Rm3131 produced exclusively an alternative EPS, called EPS II (Keller et al., 1990). This mutant still induced and infected alfalfa nodules, although it was defective in the production of the Cellufluor-stainable EPS I. We complemented this mutant with an R. meliloti chromosomal DNA fragment. Sequencing of the complementing 3.6 kb EcoRI-fragment revealed several open reading frames. The Tn5 insertion of mutant Rm3131 could be related to one open reading frame (ORF) forming a single transcriptional unit. We designated this ORF rexA for regulation of exopolysaccharide production since the mutation caused a switch in EPS production from the succinoglucan EPS I to the alternative exopolysaccharide EPS II (Fig. 2).

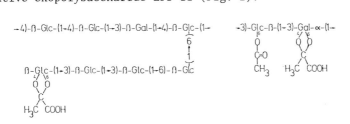

Figure 2. Structure of the repeating units of the exopolysaccharides EPS I (left) and EPS II (right) of R. meliloti. Abbreviations: Glc, glucose; Gal, galactose.

We determined the role of the rexA gene concerning the expression of the described exo genes by construction of lacZ fusions concerning the genes exoV and exoW. These fusions were transferred into the R. meliloti rexA mutant and into the R. meliloti wild type strain Rm2011. In the rexA background the expression of the monitor gene was reduced to ca. 30% of the level expressed in the wild type background. Therefore, it is concluded that rexA is necessary for the effective transcription of the exo genes involved in EPS I biosynthesis.

The alternative exopolysaccharide EPS II cannot replace EPS I in the alfalfa infection process

In order to test whether the alternative exopolysaccharide EPS II can replace EPS I in the alfalfa infection process, we constructed R. meliloti double mutants carrying mutations in the rexA and the exoY gene. Two resulting double mutants RmJQ46 (exoY::Tn5-Gm, rexA::Tn5) and RmJQ461 (exoY::Tn5, rexA::Tn5-Gm) showed the expected phenotye. They were very mucoid, and did not fluoresce on Cellufluor containing medium. We measured the amount of exopolysaccharides produced by these double mutants and found that they were able to synthesize EPS II in a range of 65% when compared to the R. meliloti wild type which of course produced EPS I.

R. meliloti mutants that fail to synthesize the succinoglucan EPS I, are no longer able to infect the plant. Nevertheless, these mutants induce meristematic cell divisions in the host plant root, leading to the formation of a noninfected nodule (Müller et al. 1988b). We tested the abilities of Rm2011 (WT), Rm0540 (exoY), Rm3131 (rexA), RmJQ46 (rexA exoY), and RmJQ461 (rexA exoY) to induce effective nodules on Medicago sativa, M. truncatulata and Trigonella sp.. While plants inoculated with Rm2011 or Rm3131 formed fully effective nodules, the strains Rm0540, RmJQ46 and RmJQ461 induced noninfected Fix⁻ nodules. The last three strains have the mutations in the exoY gene in common. They are unable to synthesize EPS I. In contrast, RmJQ46 and RmJQ461 synthesize the exopolysaccharide EPS II in a high amount. Therefore it is concluded that EPS II did not replace EPS I in the infection process of alfalfa nodules. The signal for nodule infection is obviously synthesized by the EPS I biosynthetic pathway. From the fact that Rm3131, but not RmJQ46 and RmJQ461, showed a dim fluorescence on Cellufluor agar it is speculated, that oligosaccharides from the EPS I pathway may represent the signal for infection thread formation.

R. meliloti mutants deficient in EPS I production induce the plant defence system

R. meliloti mutants that fail to synthesize EPS I, e.g. Rm0540, induce on M. sativa roots pseudonodules, which do not contain infection threads and bacteroids. A detailed light and electron microscopic analysis of these pseudonodules showed, that their cortical cell walls were abnormally thick and incrusted. By the use of fluorescence microscopy, a strong autofluorescence of incrusted cortical cell walls which were in contact with the R. meliloti mutant Rm0540, became visible (Fig. 3 B). This autofluorescence indicated an incrustation with polyphenolic material. In some cells, changes in the cell wall structure led to the formation of papillae. In older pseudonodules, infection thread like structures were observed in cortical cells. These structures contained numerous bacteria. In comparison to an infection thread induced by the wildtype R. meliloti 2011, these threads are thicker and surrounded by a rigid wall (Fig. 3 A). Electron microscopy showed that these papillae and part of the threads are composed of dense fibrillar material. Staining of nodule thin sections with the fluorochrome Sirofluor and subsequent fluorescence microscopy proved the presence of ß-1,3 glucan, presumably callose, in the cell walls, in papillae and in parts of the infection thread like structures. These morphological data provide a strong evidence for plant defence against the R. meliloti EPS⁻ mutant Rm0540.

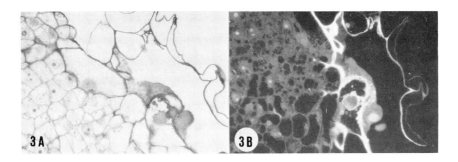

Figure 3 **A,B.** Microscopical evidence for plant defence of M. sativa against the R. meliloti EPS⁻ mutant Rm0540.
(A) Semithinsection through the cortex region of a pseudonodule, three weeks after inoculaton with Rm0540. Thickened cell walls and papillae show defence reactions against the adhered bacteria.
(B) The same region seen with blue-violet fluorescence microscopy. Strong autofluorescence in cortical cell walls indicates incrustation with polyphenolic material.

To prove whether EPS⁻ mutants of R. meliloti induce plant defence reactions, we carried out a biochemical analysis of nodules induced by the R. meliloti strains Rm2011 (wild type) and Rm0540 (exoY). HPLC analyses of extracts from three weeks old nodules showed no difference in the major M. sativa phytoalexins daidzein, formononetin and genistein. In contrast, the extraction and a subsequent HPLC separation and determination of the phenolic compounts of the plant cell walls from Rm2011 and Rm0540 induced nodules showed characteristic differences. Nodules of the EPS⁻ mutant Rm0540 showed a strongly increased pattern of phenolic compounds. An increase in the amount of p-coumaric acid, which plays an important role in lignification of plant cell walls was noticed.
As proved by microscopical and biochemical data, EPS⁻ mutants of

R. meliloti induce a strong local plant defence against the former symbiont. This necrosis like defence reaction could be the primary cause for the inability of EPS⁻ mutants to invade the developing nodules. From this data, we speculate that carbohydrates derived from the succinoglucan pathway act as a suppressor of plant defence.

References:

- Gray, J.X., Djordjevic, M.A., and Rolfe, B.G. (1990) Two genes that regulate exopolysaccharide production in Rhizobium sp. strain NGR234: DNA sequences and resultant phenotypes. J. Bacteriol. **172**: 193-203.
- Keller, M., Arnold, W., Kapp, D., Müller, P., Niehaus, K., Schmidt, M., Quandt, J., Weng, W.M., and Pühler, A. (1990) Rhizobium meliloti genes involved in exopolysaccharide production and infection of alfalfa nodules, in Silver, S., Chakrabarty, A.M., Iglewski, B., and Kaplan, S. (eds.), Pseudomonas: Biotransformations, Pathogenesis, and Evolving Biotechnology, ASM, Washington, pp. 91-97.
- Keller, M., Müller, P., Simon, R., and Pühler, A. (1988) Rhizobium meliloti genes for exopolysaccharide synthesis and nodule infection located on megaplasmid 2 are actively transcribed during symbiosis. Molec Plant-Microbe Interact **1**: 267-274.
- Long, S., Reed, J.W., Himawan, J., and Walker, G.C. (1988) Genetic analysis of a cluster of genes required for synthesis of the calcofluor-binding exopolysaccharide of Rhizobium meliloti. J. Bacteriol. **170**: 4239-4248.
- Müller, P., Enenkel, B., Hillemann, A., Kapp, D., Keller, M., Quandt, J. and Pühler, A. (1988a) Genetic analysis of two DNA regions of the Rhizobium meliloti genome involved in the infection process of alfalfa nodules, in Palacios, R. and Verma, D.P.S. (eds.), Molecular Genetics of Plant-Microbe Interactions 1988, APS Press, St. Paul, MN, USA, pp. 26-32.
- Müller, P., Hynes, M., Kapp, D., Niehaus, K., and Pühler, A. (1988b) Two classes of Rhizobium meliloti infection mutants differ in exopolysaccharide production and in coinoculation properties with nodulation mutants. Mol. Gen. Genet. **211**: 17-26.

REGULATION OF NITROGEN FIXATION GENES IN *RHIZOBIUM MELILOTI*

P. BOISTARD, J. BATUT, M. DAVID, J. FOURMENT, A.M. GARNERONE, D. KAHN[1], P. DE PHILIP, J.M. REYRAT AND F. WAELKENS.
Laboratoire de Biologie Moléculaire des Relations Plantes-Microorganismes, CNRS-INRA, BP 27
31326 Castanet-Tolosan Cedex, France.
[1] Present address : Department of Molecular Biology - Netherlands Cancer Institute - H5 Plesmanlaan- 121 1066CX Amsterdam - The Netherlands.

ABSTRACT. *Rhizobium meliloti nif* and *fix* genes are expressed in symbiotic conditions or in microaerobic cultures under the control of a cascade regulatory pathway. One regulon is activated by the oxygen sensitive activator NifA. The NifA independent regulon is activated by the Fnr homologue FixK. Both *nifA* and *fixK* are activated by the pair of regulatory proteins FixL and FixJ.

We report that FixL is responsible for oxygen sensitive activation by FixJ. The upstream sequence of *fixK* carries control elements needed for its activation by FixJ as well as a region which could be responsible for the negative autoregulation of *fixK* expression. In spite of its homology to Fnr, FixK lacks a 21 N-terminal amino acid stretch which is thought to be responsible for oxygen sensitivity of Fnr transcriptional activity, which raises the question of the existence and nature of the physiological effector of FixK. We provide data indicating that activation of *nifA* and *fixK* expression requires phosphorylation of FixJ aspartate residues.

Introduction : The cascade regulatory pathway of *nif* and *fix* genes of *Rhizobium meliloti*.

Symbiotic nitrogen fixation provides an example of how the ecological niche provided by the plant regulates the expression of bacterial genes whose products in turn modify the host plant metabolism.

In the recent years, the regulation of the expression of *R. meliloti* genes needed for symbiotic nitrogen fixation has been the subject of intense studies and as a consequence an overall picture of what constitutes a cascade regulatory pathway is now available (fig. 1).

There are two categories of nitrogen fixation genes in *R. meliloti* : *nif* genes are homologous to *Klebsiella pneumoniae* genes ; in this enterobacteria, nitrogen fixation occurs in pure bacterial cultures under anaerobic conditions in response to fixed nitrogen limitation. On the other hand *R. meliloti fix* genes are needed together with *nif* genes for symbiotic nitrogen fixation but have no homologues in *K. pneumoniae*.

195

H. Hennecke and D. P. S. Ver. .u (eds.),
Advances in Molecular Genetics of Plant-Microbe Interactions, Vol. 1, 195–202.
© 1991 *Kluwer Academic Publishers. Printed in the Netherlands.*

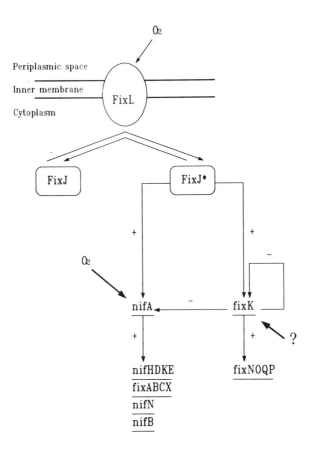

Figure 1. Cascade regulatory pathway of *nif* and *fix* genes of *R. meliloti*.
The position of the arrow indicating the interaction of O_2 with FixL does not imply
that this interaction necessarily takes place in the periplasmic space.

R. meliloti nif and *fix* genes belong to two regulons : *nifHDK E*, and *N*, *fixABCX*,
nifA,B, *fdxN* which constitute one regulon are under the positive control of the
transcriptional activator NifA (Masepohl *et al.* (1990)). A high degree of homology
is found between *R. meliloti* NifA and *K. pneumoniae* NifA proteins. However,
whereas the *K. pneumoniae* NifA is inactivated by NifL in the presence of oxygen,
the *R. meliloti* NifA as well as *Bradyrhizobium japonicum* NifA protein possess an
additional domain in which cystein residues are thought to be involved in oxygen
sensitivity of the transcriptional activity (Fischer *et al.* (1988)).

A second difference between *K. pneumoniae* and *R. meliloti nifA* is at the level of
gene expression. Whereas *K. pneumoniae nifA* is expressed under the control of

ntrB/C in response to nitrogen limitation, a *R. meliloti ntrC* mutant is not affected in *nifA* expression (Szeto *et al.* (1987)). Furthermore *R. meliloti nifA* is expressed in microaerobic bacterial cultures (Ditta *et al.* (1987)).

The second *fix* regulon is represented by the *fixN* regulon (David *et al.* (1988)), a reiterated transcription unit in which *fixP* codes for a predictably cytochrome c-like protein. Like the *nif-fix* regulon the *fixN* regulon is actively expressed in symbiotic conditions but instead of being activated by *nifA* it requires the product of the reiterated *fixK* gene for its expression either in symbiosis or in microaerobic culture (Batut *et al.* (1989)). *fixK* as well as *nifA* expression can be induced in microaerobic cultures. Microaerobic and symbiotic expression of both *nifA* and *fixK* depend on a pair of regulatory genes *fixL* and *fixJ* (David *et al.* (1988)) whose products show homology respectively to the sensor and regulator components of the so called two-component regulatory systems which, in prokaryotes, allow adaptation to various environmental conditions (see for example the review of Stock *et al.* (1989)) and references therein).

In this paper we report data providing evidence that FixL is responsible for oxygen sensitivity of the expression of *nifA* and *fixK* using *Escherichia coli* as a heterologous system for *in vivo* regulation studies. We describe oligonucleotide-directed mutagenesis of *fixJ* to study the role of phosphorylation in the transduction of the regulatory signal. Interaction of FixJ with its target genes has been approached by performing deletions in the *fixK* upstream sequences. Finally, we discuss the biological significance of the regulatory gene *fixK* as well as of its functional reiteration.

FixL mediates oxygen sensitive activation of *nifA* **and** *fixK* **in** *Escherichia coli.*

We have shown before that FixJ is able to activate *fixK* and *nifA* expression in *E. coli*. Activation was observed only when *fixJ* was overexpressed under the control of the IPTG inducible *lac* promoter. In these conditions expression of *nifA* and *fixK* was independent of the aeration status of the culture and of FixL (Hertig *et al.* (1989)).

In order to test whether the sensor protein FixL was responsible for oxygen sensitivity we cloned *fixL* downstream of the Nm constitutive promoter of pML123 and introduced the resulting plasmid into the *E. coli* strains containing the *Plac-fixJ* construct and the *fixK* or *nifA-lacZ* fusions.

In the absence of IPTG, 8 fold microaerobic induction of *fixK* and *nifA* expression was observed only in the strains expressing FixL (table 1).

The fact that FixL mediates microaerobic activation of FixJ dependent promoters in the heterologous host *E. coli* implies that either FixL itself is sensitive to oxygen concentration or that it interacts with redox sensitive cellular components which are present in *E. coli* as well as in *R. meliloti*.

TABLE 1. Oxygen-sensitive expression of *fixK-lacZ* and *nifA-lacZ* fusion in the presence of *fixL* and *fixJ*.

E. coli TG1 strains	ß-GALACTOSIDASE UNITS	
	Aerated culture	Microaerobic culture
fixK-lacZ	0	0.07
fixJ, fixK-lacZ	3	4
fixL, fixJ, fixK-lacZ	4	33
nifA-lacZ	0.2	0.6
fixJ, nifA-lacZ	1.5	1.5
fixL, fixJ, nifA-lacZ	1	8

The role of conserved aspartate residues of FixJ.

Signal transduction by two component regulatory systems is thought to be mediated by transfer of a phosphoryl group from a histidine residue of the sensor molecule to an aspartate of the regulator (Stock *et al.*(1989)). The phosphorylated form of the regulator is the active molecule able to interact with its target, a DNA regulatory sequence and/or a protein.

In the CheY molecule, crystallographic studies have shown that 3 aspartate residues Asp 12.13 and 57 are located in close vicinity in an acidic pocket (Stock *et al.* (1989)). The homologous FixJ residues are Asp 10.11 and Asp 54. In order to study the role of FixJ phosphorylation we substituted asparagine for aspartate at positions 10.11 on one hand and at position 54 on the other hand yielding $FixJ_{10/11}$ and $FixJ_{54}$ respectively. We then studied the activation properties of these mutated FixJ on *fixK-lacZ* and *nifA-lacZ* expression in *E. coli* (table 2).

It appears that both mutated FixJ proteins have a decreased transcriptional activity on *nifA* as well as *fixK* promoters.

However, *fixK* residual activation by $FixJ_{54}$ is relatively higher than *nifA* activation. The reverse is true with $FixJ_{10/11}$ which suggests that FixJ interaction with both promoters might be different.

In CheY, the Asp 57 residue homologous to the Asp 54 mutated in $FixJ_{54}$ has been shown to be phosphorylated by CheA. Therefore we addressed the question whether $FixJ_{54}$ residual activity was due to phosphorylation of another aspartate residue or to a conformational change of the protein which resulted in the ability to activate its target promoters at a reduced rate independently of phosphorylation. Circumstantial evidence that the latter hypothesis was more likely to be correct came from experiments in which we introduced *fixL* cloned downstream of the Nm promoter into an *E. coli* strain expressing the mutated $FixJ_{54}$ protein and carrying the *nifA* or *fixK-lacZ* fusions. In these strains *nifA* and *fixK* expression was not affected by the aeration contrary to the strain expressing the wild type FixJ.

TABLE 2. *nifA-lacZ* and *fixK-lacZ* expression in the presence of wild-type and mutant *fixJ*.

E.coli TG1 strains	IPTG (mM)	ß-galactosidase units	Relative activity
fixJ, fixK-lacZ	0	10	
	2	192	100
fixJ$_{10/11}$, fixK-lacZ	0	1	
	2	22	12
fixJ$_{54}$, fixK-lacZ	0	3	
	2	64	33
fixJ, nifA-lacZ	0	0.8	
	2	27	100
fixJ$_{10/11}$, nifA-lacZ	0	0.7	
	2	8.7	33
fixJ$_{54}$, nifA-lacZ	0	0.4	
	2	3.7	13

Deletion studies of the *fixK* promoter

As an approach to the study of interactions between FixJ and its target genes, we performed sequential deletions in the *fixK* promoter.

The deleted promoters were tested either in *E. coli* or in *R. meliloti*. It can be seen (fig. 2) that in *E. coli* the -41 -79 region from the *fixK* transcription start contains regulatory elements essential for FixJ activated transcription.

It is interesting to note that similarly in the *nifA* upstream sequence, the -45 -62 region has been shown to be involved in FixJ mediated activation of the promoter (Virts *et al.*(1988)).

The same minimal promoter was sufficient for *fixK* expression in *R. meliloti*. However deletion of the region upstream of residue -450 led to an enhanced expression of the *fixK-lacZ* fusion. Because *fixK* autoregulates its own expression negatively (Batut *et al.* (1989)), we compared the expression of various deleted *fixK-lacZ* fusions in the wild type and in a *fixK* mutant. In the *fixK* mutant, *fixK* expression was activated at a higher level than in the wild type strain and this level was not influenced by the presence of the upstream silencing region. This would argue in favour of a repressing effect of FixK due to its binding in this region. However there is no clear Fnr box upstream of the *fixK* promoter (see following paragraph).

200

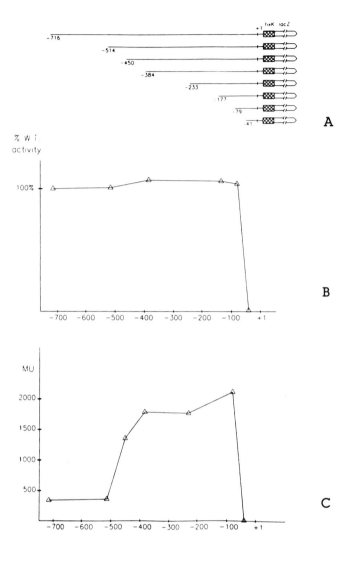

Figure 2. Deletion mapping of upstream regulatory sequences of the *fixK* promoter.
A - Structure of the various deletions constructed.
B - Expression of the deleted fusions in *E. coli* carrying *fixLJ*.
C - Expression of the deleted fusions in *R. meliloti* 2011 *fixK*⁺.

Role of FixK

Sequence data have allowed J. Batut *et al.* (1989) to predict that the activator FixK is homologous to the *E. coli* Fnr protein which activates genes coding for anaerobic respiratory pathways.

A sequence homologous to the Fnr box has been found upstream of the *fixN* gene which requires FixK in order to be expressed (Colonna-Romano *et al.* (in press)). An Fnr box has also been found upstream of the *fixLJ* operon which could explain the negative effect of FixK on the expression of both *nifA* and *fixK* (see preceding paragraph).

The similarity between FixK and Fnr was further confirmed by the fact that a *fixN-lacZ* fusion was expressed in an *fnr*$^+$ *E. coli* strain 40 fold more actively than in a *fnr* mutant (Cherfils *et al.* (1989)).

In a *fixK* mutant *nifA* and *fixK* are still induced in microaerobic conditions which implies that FixK does not play a major role in oxygen sensitivity of nitrogen fixation gene expression. In this respect it is interesting to note that a *fnr* like gene, *fnrN*, has been identified in *R. leguminosarum*. In a *R. meliloti fixJ* mutant, *fnrN* is able to activate *fixN* expression in microaerobic conditions (Colonna-Romano *et al.* (in press)).

This means either that this gene, whose product is highly homologous to FixK, is expressed in microaerobic conditions in *R. meliloti* independently of *fixJ* or that the FnrN protein itself is sensitive to oxygen concentration. In favour of this latter hypothesis, the FnrN protein, like Fnr, has been shown to contain a N-terminal sequence of 21 residues which is not present in FixK. Both FnrN and Fnr contain a cystein motif at position 20 which is thought to play a role in the oxygen response of Fnr, possibly as part of a metal binding site .

If FnrN responds to oxygen, the question arises whether FixK responds to a regulatory effector and if so what is the nature of the signal.

Another question concerns the possible biological significance of the functional reiteration of the *fixK* gene (Batut *et al.* (1989)). As an approach to this question we have started the sequencing of the *fixK* repeat. Preliminary sequence data indicate a high degree of conservation of the coding sequence (more than 95%) at the 5 prime end as well as of the upstream regulatory region (about 85%). It will be interesting to see if some of the non conserved nucleotide residues are located in sequences playing a role in the interaction with the FixJ activator thereby allowing differential expression of the two *fixK* copies.

Conclusions

Although the main components of the cascade regulatory pathway of *nif* and *fix* gene expression have been identified in *R. meliloti*, two types of questions remain to be elucidated.

What are the molecular mechanisms of oxygen perception by FixL and of gene activation by FixJ ?

Does *R. meliloti fixK* respond to a regulatory effector and what is the nature of this effector ?

The answer to this latter question may not be easy to obtain. One approach could be to look for *R. meliloti* mutants unable to activate *fixN* expression.

Acknowledgements

This work was supported by a contract with the Commission of the European Communities in the frame of the Biotechnology Action Programme. Franciska

202

Waelkens was supported by a postdoctoral fellowship from the D. Collen Foundation (Louvain, Belgium).

Pascale de Philip was supported by a predoctoral fellowship from Ministère de la Recherche et de la Technologie.

We thank A. Pühler and U. Priefer for providing us with unpublished strains and results and for stimulating discussions.

References

Batut, J., Daveran-Mingot, M.L., David, M., Jacobs, J., Garnerone, A.M. and Kahn, D. (1989) "*fixK*, a gene homologous with *fnr* and *crp* from *Escherichia coli*, regulates nitrogen fixation genes both positively and negatively in *Rhizobium meliloti*", EMBO J., 8, 1279-1286.

Cherfils, J., Gibrat, J.F., Levin, J., Batut, J. and Kahn, D. (1989) "Model building of Fnr and FixK DNA-binding domains suggests a basis for specific DNA recognition", J. Molec. Recognition, 2, 114-121.

Colonna-Romano, S., Arnold, W., Schlüter, A., Boistard, P., Pühler, A. and Priefer, U.B. (in press) "An Fnr-like protein encoded in *Rhizobium leguminosarum* biovar *viciae* shows structural and functional homology to *Rhizobium meliloti* FixK", Mol. Gen. Genet.

David, M., Daveran, M.L., Batut, J., Dedieu, A., Domergue, O., Ghai, J., Hertig, C., Boistard, P. and Kahn, D. (1988) "Cascade regulation of *nif* gene expression in *Rhizobium meliloti*", Cell, 54, 671-683.

Ditta, G., Virts, E., Palomares, A. and Kim, C.H. (1987) "The *nifA* gene of *Rhizobium meliloti* is oxygen regulated", J. Bacteriol., 169, 3217-3223.

Fischer, H.M., Bruderer, T. and Hennecke, H. (1988), Essential and non-essential domains in the *Bradyrhizobium japonicum* NifA protein identification of indispensable cysteine residues potentially involved in redox reactivity and/or metal binding", Nucleic Acids Res., 16, 2207-2224.

Masepohl, B., Krey, R., Riedel, K.U., Reiländer, H., Jording, D., Prechel, H., Klipp, W. and Pühler, A. (1990) "Analysis of the *Rhizobium meliloti* main *nif* and *fix* gene cluster" in P. Nardon, V. Gianinazzi-Pearson, A.M. Grenier, L. Margulis and D.C. Smith (eds.), Endocytobiology IV, INRA Paris, pp. 63-68.

Stock, J.B., Ninfa, A.J. and Stock, A.M. (1989) "Protein phosphorylation and regulation of adaptive responses in bacteria", Microbiol. Reviews, 53, 450-490.

Szeto, W.W., Nixon, B.T., Ronson, C.W. and Ausubel, F.M. (1987) "Identification and characterization of the *Rhizobium meliloti ntr* gene : *R. meliloti* has separate regulatory pathways for activation of nitrogen fixation genes in free-living and symbiotic cells", J. Bacteriol., 169, 1423-1432.

Virts, E.L., Stanfield, S.W., Helinski, D.R. and Ditta, G.S. (1988) "Common regulatory elements control symbiotic and microaerobic induction of *nifA* in *Rhizobium meliloti*", Proc. Natl. Acad. Sci. USA, 85, 3062-3065.

COMPLEX REGULATORY NETWORK FOR *NIF* AND *FIX* GENE EXPRESSION IN *BRADYRHIZOBIUM JAPONICUM*

H.M. Fischer, D. Anthamatten, I. Kullik, E. Morett, G. Acuña and H. Hennecke
Mikrobiologisches Institut
Eidgenössische Technische Hochschule
Schmelzbergstrasse 7
CH-8092 Zürich
Switzerland

ABSTRACT. In the nitrogen-fixing soybean symbiont, *Bradyrhizobium japonicum*, the expression of *nif* and *fix* genes is regulated predominantly, if not exclusively, in response to the cellular oxygen status. Activation of these genes occurs during symbiosis in root nodules or in free-living cells grown at low oxygen concentrations. We have identified three regulatory elements that are involved in oxygen control at different levels. First, with regard to *nif* and *fix* gene expression, the positive regulatory protein NifA plays a crucial role. It was demonstrated that the binding of NifA to its target site in the upstream region of the *B.japonicum nifD* promoter was regulated by oxygen and required metal ions. Second, *fixLJ*-like genes that had been shown to activate the *Rhizobium meliloti nifA* gene under low oxygen conditions (1, 2), were identified also in *B.japonicum*. However, they were probably not involved in the expression of *nifA*, but in the regulation of other oxygen-controlled genes. Third, we found two functional genes (*rpoN1*, *rpoN2*) encoding specific σ factors (σ^{54}) and demonstrated that *rpoN1* expression was oxygen regulated via a mechanism involving *fixLJ* whereas *rpoN2* was subject to negative autoregulation.

INTRODUCTION

Bradyrhizobium japonicum is a nitrogen-fixing endosymbiont of soybean (*Glycine max* L. Merr.). Upon infection of its host plant, root nodules are formed and within the infected plant cells the bacteria differentiate into bacteroids which are then able to reduce atmospheric dinitrogen. Synthesis and activity of nitrogenase, the key enzyme complex in nitrogen fixation, are tightly regulated by oxygen. In soybean root nodules, low oxygen conditions optimal for nitrogenase function are maintained by a specialized oxygen-binding protein named leghemoglobin. Expression of genes involved in nitrogen fixation (*nif* and *fix* genes) not only occurs in symbiosis but is also induced in free-living *B.japonicum* cells grown microaerobically or anaerobically with nitrate as terminal electron acceptor.

Our laboratory is interested in understanding the molecular mechanisms underlying oxygen control of *nif* and *fix* gene expression in *B.japonicum*. Here we summarize recent

H. Hennecke and D. P. S. Verma (eds.),
Advances in Molecular Genetics of Plant-Microbe Interactions, Vol. 1, 203–210.

results obtained during the analysis of three regulatory elements involved in this type of control.

BINDING OF NIFA TO THE *nifD* UAS IS OXYGEN REGULATED AND SENSITIVE TO CHELATORS

Background

Six *nif/fix* operons of *B.japonicum* have been mapped by transcript analysis and shown to be preceded by conserved promoter sequences in the -24/-12 region (3). Recognition of these promoters by RNA polymerase requires a special σ factor (σ^{54}, RpoN) and their transcriptional activation depends on the regulatory protein NifA. The crucial regulatory role of NifA is reflected by the pleiotropic phenotype of *nifA⁻* mutants. Interestingly, recent results indicated that, in *B.japonicum*, NifA is not only involved in the control of *nif* and *fix* gene expression but also in the regulation of the synthesis of two proteins whose NH_2-terminal ends show significant homology to the *Escherichia coli* chaperonins GroES and GroEL.

About 100 bp upstream of the transcriptional start site, some -24/-12 promoters carry the conserved sequence element 5'-TGT-10 bp-ACA-3'. For the *Klebsiella pneumoniae nifH* promoter this 'upstream activator sequence' (UAS) was shown to be the target site for NifA (4) which is able to bind to this motif by means of its COOH-terminal DNA binding domain. It is believed that transcriptional activation is brought about by interaction of NifA bound to the UAS with the RNA polymerase holoenzyme interacting with the downstream -24/-12 core promoter region. In contrast to *K.pneumoniae* NifA, the activity of the *B.japonicum* NifA protein was found to be oxygen-labile and sensitive to the presence of chelating agents such as *o*-phenanthroline (5, 6). This functional difference may be reflected also at the structural level by the presence of an extra protein domain (interdomain linker) located between the highly conserved central domain and the COOH-terminal DNA binding domain of the *B.japonicum* NifA protein. Within the interdomain linker two cysteine residues were shown to be essential for NifA activity and they were proposed to be involved in metal ion binding (6, 7). In order to understand in more detail the mechanism of oxygen regulation we addressed the question which is the step where oxygen and chelators interfered with NifA activity (NifA binding or *nif* promoter activation). For this purpose the interaction of NifA with the UAS of the *B.japonicum nifD* promoter was analyzed under different conditions by *in vivo* dimethylsulphate footprinting experiments.

Experimental

The *B.japonicum nifD* promoter cloned on a plasmid in *E.coli* strain MC1061 was analyzed essentially as described (4) by *in vivo* dimethylsulphate (DMS) footprinting in the presence or absence of a second plasmid that expressed *B.japonicum nifA* constitutively from the pBR329 *cat* promoter. The cultures were grown aerobically or anaerobically and, in some experiments, *o*-phenanthroline was added at 30 µM final concentration. After exposing the cells for 1 min to DMS, plasmid DNA was isolated, cleaved at the methylated positions and used as template for *in vitro* primer extension analysis. The elongation products were analyzed by polyacrylamide gel electrophoresis.

Results

As an example the results of the DMS footprinting analysis of one of the two *B.japonicum nifD* UAS elements is shown in Fig. 1. Under anaerobic growth conditions the guanine residue at position -95 showed a clear NifA-dependent protection. Aerobic growth, or metal depletion by the addition of *o*-phenanthroline resulted in loss of protection, indicating that the DNA binding function of NifA required metal ions and was controlled by the oxygen status of the cell. The same result was obtained when the second UAS of the *nifD* promoter was analyzed. Similar footprinting experiments in the presence of rifampicin showed that UAS-bound NifA led to the formation of an open *nifD* promoter-RNA polymerase complex. For this process metal ions and a low O_2 tension were also essential. In conclusion, we propose that under microaerobic conditions the coordination of a reduced metal ion induces a favourable NifA conformation for the DNA-binding and activating functions of this protein.

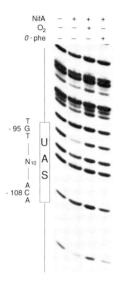

Fig. 1. *In vivo* dimethylsulphate footprinting analysis of a *B.japonicum nifD* UAS. The open circle with the arrow indicates the guanine residue at position -95 which is protected from methylation under microaerobic conditions in the presence of NifA. *o*-Phenanthroline (*o*-phe) was added at 30 μM concentration to the cultures.

EXPRESSION OF THE *B.JAPONICUM fixRnifA* OPERON SEEMS TO BE INDEPENDENT OF *fixLJ*

Background
NifA in *B.japonicum* was found to be the promoter distal gene in a operon consisting of *fixR* and *nifA* (8). Although the FixR protein shows significant sequence similarity to NAD-dependent oxidoreductases (9) no function could be attributed so far to this protein. FixR is not required for symbiotic N_2 fixation. Unlike *nifA* of *R.meliloti*, the *fixRnifA* operon of *B.japonicum* is expressed aerobically. This expression depends on the presence of an upstream activator sequence located around position -66 with respect to the transcriptional start site (10). A protein present in crude extracts of *B.japonicum* was shown by gel retardation experiments to bind to this DNA region, and it is speculated that this protein is responsible for aerobic activation of the *fixRnifA* operon (10). Under anaerobic or symbiotic conditions expression of *fixRnifA* is increased at least five-fold by a mechanism that involves *nifA* and does not depend on the upstream DNA. In *R.meliloti*, *nifA* expression is activated under conditions of low oxygen and in symbiosis

by FixJ which is the response regulator in the two-component regulatory system represented by *fixLJ* (1, 2). Despite these differences we were interested to know if *fixLJ*-like genes were present in *B.japonicum*, and if they were, whether they were involved in the control of *fixRnifA* expression.

Experimental

Using an *R.meliloti fixLJ*-specific probe, a homologous region was cloned from the *B.japonicum* genome and sequenced. To study the functions of the *B.japonicum fixLJ*-like genes they were disrupted by marker exchange mutagenesis, and the symbiotic properties as well as the growth characteristics of the resulting mutant strains were analyzed.

Results

B.japonicum contained *fixLJ*-like genes which probably formed an operon. The predicted FixL and FixJ proteins were homologous over almost their entire lengths to the corresponding *R.meliloti* proteins (approx. 50 % identity). Strains mutated either in *fixL* or *fixJ* showed a drastically reduced symbiotic Fix activity but clearly differed from *nifA*⁻ strains both in Fix activity and nodulation phenotype (Table 1). In addition they were unable to grow anaerobically with nitrate as terminal electron acceptor. This suggested that the *fixLJ*-like genes of *B.japonicum* were involved in the control of genes important for symbiosis and anaerobic respiration. Recent preliminary data from our laboratory indicates that there exists a *fixK*-like gene in *B.japonicum* whose expression may depend on *fixLJ* similarly as it was shown in *R.meliloti* (11).

Table 1. Symbiotic nitrogen fixation activity (C_2H_2 reduction) of *B.japonicum fixL* and *fixJ* mutants, and aerobic expression of a *nifA'*-*'lacZ* fusion in these backgrounds.

Strain	Fix activity	β-Galactosidase activity derived from *nifA'*-*'lacZ*
Wild type	100 %	100 %
fixL⁻	5 %	86 %
fixJ⁻	4 %	81 %
nifA⁻	0 %	N.D.[a]

[a] Not determined

To analyze the potential role of *fixLJ* in the regulation of *fixRnifA* expression the activity of a *nifA'*-*'lacZ* fusion was assayed in *fixL*⁻ and *fixJ*⁻ backgrounds (Table 1). Under aerobic conditions expression *nifA* was not affected significantly by either mutation. Since the *fixLJ*⁻ mutant strains did not grow anaerobically it was not possible to test a potential influence of *fixLJ* on *nifA* expression under these conditions. In conclusion, it seems that *B.japonicum* and *R.meliloti* differ with respect to the involvement of *fixLJ* in control of *nifA* expression but there may exist other *fixLJ* dependent genes common to both organisms (possibly *fixK*).

B.JAPONICUM HAS TWO FUNCTIONAL *rpoN* GENES ONE OF WHICH IS OXYGEN REGULATED

Background

As mentioned above, the initiation of transcription at -24/-12 promoters of *nif and fix* genes requires RNA polymerase and the specific σ factor σ^{54} (RpoN). For the *B.japonicum nifH* and *nifD* promoters this has been shown indirectly by expression studies with corresponding *lacZ* fusion in *rpoN*$^+$ and *rpoN*$^-$ *E.coli* backgrounds (12). Unfortunately, analogous experiments could not be performed with the *B.japonicum fixRnifA* promoter since it was not at all activated in *E.coli*. Although sequence analysis and transcript mapping of this promoter indicated the presence of a classical -24/-12 promoter only a mutation in the -12 but not in the -24 region resulted in a reduced activity of this promoter in *B.japonicum* (8). We were interested in identifying the *B.japonicum* gene encoding σ^{54}, particularly in order to test the presumed *rpoN* dependence of the *fixR* promoter in the homologous background.

Experimental

Using the *R.meliloti rpoN* gene as a probe, two *rpoN*-like genes were cloned from the *B.japonicum* genome and characterized further by sequencing and mutational analysis. The symbiotic phenotype of *rpoN* mutant strains was determined in a plant infection test, and the activities of a *nifH'*- and a *fixR'-'lacZ* fusion were assayed in *rpoN*$^-$ backgrounds. *RpoN'-'lacZ* fusions were used to study expression and regulation of both *rpoN* genes.

Results

Unexpectedly, two highly homologous *rpoN* genes (*rpoN1*, *rpoN2*) were identified in *B.japonicum*. The deduced amino acid sequences of the RpoN proteins were 87 % identical. Individual mutations in *rpoN1* and *rpoN2* did not affect the symbiotic properties but a strain mutated in both *rpoN* genes had a Nod$^+$Fix$^-$ phenotype. Under aerobic conditions the growth rate of the *rpoN2* single mutant and the *rpoN1/2* double mutant but not of the *rpoN1* mutant strain was reduced when KNO$_3$ was provided as the sole N-source. Under microaerobic conditions only the *rpoN1/2* double mutant showed this phenotype. Similarly, under aerobic conditions only *rpoN2* of *B.japonicum* was able to complement a *R.meliloti rpoN* mutant whereas both *rpoN* genes could complement under microaerobic conditions. These observations indicated that *B.japonicum rpoN1* was expressed preferentially under microaerobic conditions. To test this hypothesis, the expression in *B.japonicum* of chromosomally integrated *rpoN1'*- and *rpoN2'-'lacZ* fusions was assayed under different oxygen conditions (Table 2). We found that *rpoN1* was hardly expressed aerobically and was induced about 25-fold under microaerobic conditions whereas expression of *rpoN2* was constant under both oxygen conditions tested.

Since in *R.meliloti* the *fixLJ* genes are mediators for oxygen control of gene expression (see above) and since we had identified homologous genes in *B.japonicum* we next asked the question whether in *B.japonicum* the *fixLJ*-like genes were involved in oxygen control of *rpoN1*. To this end the activity of a *rpoN1'-'lacZ* fusion was assayed in *B.japonicum fixL*$^-$ and *fixJ*$^-$ backgrounds (Table 3). In both mutant backgrounds the microaerobic activation of *rpoN1* observed in the wild type was abolished completely indicating a direct or indirect oxygen control of *B.japonicum rpoN1* by *fixLJ*. Expression of *rpoN2* was not affected by mutations in *fixLJ* but was increased approx. fivefold in a *rpoN2* background (data not shown). Thus, in contrast to *rpoN1*, the expression of

Table 2. β-Galactosidase activities (Miller units) derived from chromosomally integrated rpoN'-'lacZ fusions in B.japonicum grown aerobically or microaerobically.

Growth conditions	rpoN1'-'lacZ	rpoN2'-'lacZ
Aerobic	8	152
Microaerobic	220	150

Table 3. β-Galactosidase activities (Miller units) derived from a plasmid borne rpoN1'-'lacZ fusion in B.japonicum wild-type, fixL⁻ and fixJ⁻ backgrounds grown under the conditions indicated.

Background	Expression of rpoN1'-'lacZ	
	Aerobic	Microaerobic
Wild type	41	281
fixL⁻	13	13
fixJ⁻	15	13

rpoN2 was negatively autoregulated. For the future work it will be of interest to identify in more detail the cis- and trans-acting factors involved in the regulation of rpoN1 and rpoN2 expression in B.japonicum.

Finally we measured the activities of the B.japonicum nifH and fixR promoter in the rpoN1/2⁻ background. As expected, the nifH promoter was not activated in this background under anaerobic or symbiotic conditions. However, the aerobic expression from the fixR promoter was not affected by the rpoN mutations, and under anaerobic conditions a two- to threefold increase was observed similarly as in the wild type. Thus, although it was possible that the apparent -24/-12 promoter in front of fixR was used in the wild type our data implied the presence of an alternative, yet to be identified fixR promoter which is independent of rpoN.

Our current understanding of the oxygen regulation of nif and fix gene expression in B.japonicum is summarized in Fig. 2. The existence of two functional, distinctly regulated rpoN genes and the aerobic expression of fixRnifA represent clear differences to R.meliloti. Our future efforts will be focused on a detailed structural and functional analysis of the complex fixR promoter including the search for the postulated activator and the identification of *fixLJ*-controlled genes other than *rpoN1*.

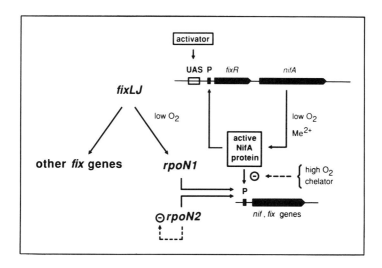

Fig. 2. Current model on *nif* and *fix* gene regulation in *B.japonicum*. The scheme is a modified and updated version of that presented in Ref. (9). Details are discussed in the text. Abbreviations: P, promoter; UAS, upstream activator sequence of the *fixR* promoter; Me²⁺, divalent metal ion.

ACKNOWLEDGEMENTS

This work was supported by grants from the Swiss National Foundation for Scientific Research and the Federal Institute of Technology Zürich.

REFERENCES

1. David, M., Domergue, O., Pogonec, P. and Kahn, D. (1987) 'Transcription patterns of *Rhizobium meliloti* symbiotic plasmid pSym: identification of *nifA*-independent *fix* genes', J. Bacteriol. 169, 2239-2244.
2. Virts, E.L., Stanfield, S.W., Helinski, D.R. and Ditta, G.S. (1988) 'Common regulatory elements control symbiotic and microaerobic induction of *nifA* in *Rhizobium meliloti*', Proc. Natl. Acad. Sci. USA 85, 3062-3065.
3. Hennecke H. (1990) 'Nitrogen fixation genes involved in the *Bradyrhizobium japonicum*-soybean symbiosis', FEBS Lett. 268, 422-426.
4. Morett, E. and Buck, M. (1989) 'NifA-dependent *in vivo* protection demonstrates that the upstream activator sequence of *nif* promoters is a protein binding site', Proc. Natl. Acad. Sci. USA 85, 9401-9405.
5. Fischer, H.M. and Hennecke, H. (1987) 'Direct response of *Bradyrhizobium japonicum nifA*-mediated *nif* gene regulation to cellular oxygen status', Mol. Gen. Genet. 209, 621-626.

6. Fischer, H.M., Bruderer, T. and Hennecke, H. (1988) 'Essential and non-essential domains in the *Bradyrhizobium japonicum* NifA protein: identification of indispensable cysteine residues potentially involved in redox reactivity and/or metal binding', Nucl. Acids Res. 16, 2207-2224.
7. Fischer, H.M., Fritsche S., Herzog, B. and Hennecke, H. (1989) 'Critical spacing between two essential cysteine residues in the interdomain linker of the *Bradyrhizobium japonicum* NifA protein', FEBS Lett. 255, 167-171.
8. Thöny, B., Fischer, H.M., Anthamatten, D., Bruderer, T. and Hennecke, H. (1987) 'The symbiotic nitrogen fixation regulatory operon (*fixRnifA*) of *Bradyrhizobium japonicum* is expressed aerobically and is subject to a novel, *nifA*-independent type of activation', Nucl. Acids Res. 15, 8479-8499.
9. Hennecke, H, Bott, B., Ramseier, T., Thöny-Meyer, L., Fischer, H.M., Anthamatten, D., Kullik, I. and Thöny, B. (1990) 'A genetic approach to analyze the critical role of oxygen in bacteroid metabolism', in P.G.M. Gresshoff, L.E. Roth and W.E. Newton (eds.), Nitrogen fixation: achievements and objectives, Chapman and Hall, New York, London 1990 in press
10. Thöny, B., Anthamatten, D. and Hennecke, H. (1989) 'Dual control of the *Bradyrhizobium japonicum* symbiotic nitrogen fixation regulatory operon *fixRnifA*: analysis of *cis*- and *trans*-acting elements', J. Bacteriol. 171, 4162-4169.
11. Batut, J., Daveran-Mingot, M.L., David, M., Jacobs, J., Garnerone, A.M. and Kahn, D. (1989) '*FixK*, a gene homologous with *fnr* and *crp* of *Escherichia coli*, regulates nitrogen fixation genes both positively and negatively in *Rhizobium meliloti*', EMBO J. 8, 1279-1286.
12. Alvarez-Morales, A. and Hennecke, H. (1985) 'Expression of *Rhizobium japonicum* *nifH* and *nifDK* operons can be activated by the *Klebsiella pneumoniae* NifA protein but not by the product of *ntrC'*, Mol. Gen. Genet. 199, 306-314.

GEONOMIC INSTABILITY IN *RHIZOBIUM:* FRIEND OR FOE?

D. ROMERO, S. BROM, J. MARTINEZ-SALAZAR, M.L. GIRARD,
M. FLORES, L. DURAN, A. GARCIA DE LOS SANTOS, R. PALACIOS
AND G. DAVILA.
Department of Molecular Genetics, Nitrogen Fixation Research
Center, UNAM. Ap. Postal 565-A, Cuernavaca, Mor., México.

INTRODUCTION.

For over a century, Rhizobium species have been studied due to their
ability to establish nitrogen-fixing symbioses with leguminous plants.
During the past decade, intensive biochemical and molecular genetics
studies have paved the way for an intimate understanding of the nodulation and nitrogen fixation processes (1,2).

However, we think that these studies should be accompanied by a
thorough investigation of the genomic characteristics of Rhizobium.
Internal, as well as external genomic dynamics may have far-reaching
consequences not only for the stability and present functioning of the
elaborate genetic systems devoted to nodulation and nitrogen fixation,
but also to understand the transmission of relevant characters. As a
whole, these studies may illuminate, in the long term, the mechanisms
responsible for the evolutionary plasticity of this genus.

In this context, there are three genomic characteristics that makes
Rhizobium an interesting organism to study. First, a substantial portion of the genome is located on large plasmids, ranging in size from
100 kb to 1200 kb. In fact, for the fast-growing Rhizobium species, determinants important for the symbiotic process are located on such large plasmids (2,3). Second, Rhizobium species possess a large amount of
reiterated DNA sequences (2,4); some of them corresponds to genes relevant for the symbiotic process (2,5-8). Third, the Rhizobium genome is
subject to frequent genomic rearrangements (9). This characteristic is
specially relevant, because genomic rearrangements may impair the symbiotic abilities of Rhizobium (9-12).

We have undertaken a research program devoted to the study of genomic rearrangements in Rhizobium phaseoli CFN42, the symbiont of the
common bean. High-frequency genomic rearrangements, with a cumulative
frequency of 10^{-2} have been observed. Rearrangements belonging at least
to four different classes were detected. A mechanism responsible for
the generation of two of these classes is presented.

H. Hennecke and D. P. S. Verma (eds.),
Advances in Molecular Genetics of Plant-Microbe Interactions, Vol. 1, 211–214.
© 1991 *Kluwer Academic Publishers. Printed in the Netherlands.*

RESULTS AND DISCUSSION.

We have employed three approaches for the isolation and characterization of genomic rearrangements. The first approach, termed the "random clone" approach, consists in the isolation of a large number of
colonies under non-stressful conditions and screening these isolates
for any alteration in plasmid profiles or hibridization patterns
against specific plasmidic or chromosomal probes (9). This approach
allows the isolation of high-frequency genomic rearrangements, with a
cumulative frequency of 10^{-2}. Among the rearrangements observed, we
have found curing of specific plasmids, cointegration events between
two different plasmids as well as amplification and deletion events
altering the pSym (S. Brom et al., manuscript in preparation).

The second approach, termed the "cotransfer" approach, consists in
labeling two different elements of the Rhizobium genome (for instance,
two different plasmids), with different antibiotic resistance markers
and then search for conjugal cotransfer of both resistance markers.
This system has allowed the isolation of cointegrates between two plasmids of R. phaseoli, and may be potentially useful for the isolation of
high-frequency cointegrates between any pair of replicons, as well as
transposition of a specific region to a new replicon.

The third approach makes use of a specific construction. This construction contains a dose-dependent kanamycin resistant determinant
(which allows the easy selection for amplifications) as well as the
sacB gene of Bacillus subtilis, which confers a sucrose-dependent letha
lity to gram-negative bacteria (13). Therefore, this cassette is potentially useful to screen any genomic region for the occurrence of amplifications, deletions, insertions and curing of plasmids. Using this
system, we have detected high-frequency amplification and deletion
events (in the 10^{-3} to 10^{-4} range) that affect the structure of the
pSym of R. phaseoli CFN42. These rearrangements alter a large 120 kb re
gions that contains essential nodulation and nitrogen fixation genes.
The mechanism responsible for the generation of both rearrangements is
through homologous recombination between two naturally-ocurring nifHDK
reiterations flanking the 120 kb region (D. Romero et al., submitted
for publication). Other pairs of reiterated elements are also used for
the generation of amplifications or deletions, albeit at a lower frequency. Preliminary evidence obtained by using this approach, suggest
that amplifications may occur in other regions of the R. phaseoli genome.

Insofar as symbiotic characteristics are concerned, strains carrying
deletions of the 120 kb region are unable to nodulate. Experiments are
under way to explore the saprophytic and symbiotic abilities of strains
carrying different rearrangements.

One interesting characteristic of the amplifications described,
which maybe also holds for cointegrations, is the reversibility of
these rearrangements. Under non-selective conditions, amplified derivatives are unstable structures that return to the original state at high
frequency. If these rearrangements confer any positive selective value
to the cell, this mechanism could be favorable to adapt to harsh environmental conditions without irreversibly compromising the bacterial

genome. In this regard, it is important to recall that in other bacteria tandem amplifications are useful adaptations for the explotation
of scarce or even novel nutrients (14,15).

During its life cycle, Rhizobium passes through different microen-
vironments. We believe that specific genomic rearrangements must be
evaluated carefully for any positive selective value, either as a sa-
prophyte or as a symbiont. During the past years, the study of genomic
rearrangements has been tainted as the study of laboratory oddities or
as a fastidious character that must be controlled in order to obtain
better inoculants. However, if the occurrence of genomic rearrangements
confer advantages to the bacteria, it could turns out to be more a
Friend than a Foe.

ACKNOWLEDGMENTS.

The authors are grateful to O. López, R.M. Ocampo and V. Quinto for
technical assistance, and to A. Córdova for typing the manuscript. Partial finantial support for this research was provided by Grant N° 936-
5542.01-523-8.600 from the US Agency for International Development.

REFERENCES.

1.- Long, S.R. (1989) 'Rhizobium genetics', Ann. Rev. Genet.
    23, 483-506.
2.- Martínez, E., Romero, D., and Palacios, R. (1990) 'The Rhizobium
    genome'. Crit. Revs. in Plant Sci. 9, 59-93.
3.- Prakash, R.K., and Atherly, A.G. (1986) 'Plasmids of Rhizobium and
    their role in symbiotic nitrogen fixation', Int. Rev. Cytol.
    104, 1-24.
4.- Flores, M., González, V., Brom, S., Martínez, E., Piñero, D.,
    Romero, D., Dávila, G., and Palacios, R. (1987) 'Reiterated DNA
    sequences in Rhizobium and Agrobacterium spp.', J. Bacteriol.
    169, 5782-5788.
5.- Quinto, C., de la Vega, H., Flores, M., Leemans, J., Cevallos, M.
    A., Pardo, M.A., Azpiroz, R., Girard, M.L., Calva, E., and
    Palacios, R. (1985) 'Nitrogenase reductase: a functional multigene
    family in Rhizobium phaseoli', Proc. Natl. Acad. Sci. U.S.A.
    82, 1170-1174.
6.- Honma, M.A., and Ausubel, F.M. (1987) 'Rhizobium meliloti has
    three functional copies of the nodD symbiotic regulatory gene',
    Proc. Natl. Acad. Sci. U.S.A. 84, 8558-8562.
7.- Gyorgypal, Z., Iyer, N., and Kondorosi, A. (1988) 'Three regula-
    tory nodD alleles of diverged flavonoid specificity are involved
    in host-dependent nodulation by Rhizobium meliloti', Mol. Gen.
    Genet. 212, 85-92.
8.- Renalier, M.H., Batut, J., Ghai, J., Terzaghi, B., Gherardi, M.,
    David, M., Garnerone, A.M., Vasse, J., Truchet, G., Hughet, T.,
    and Boistard, P. (1987) 'A new symbiotic cluster on the pSym
    megaplasmid of Rhizobium meliloti 2011 carries a functional gene
    repeat and a nod locus', J. Bacteriol. 169, 2231-2238.
9.- Flores, M., González, V., Pardo, M.A., Leija, A., Martínez, E.,

214

Romero, D., Piñero, D., Dávila, G., and Palacios, R. (1988) 'Genomic instability in Rhizobium phaseoli', J. Bacteriol. 170, 1191-1196.

10.- Djordjevic, M.A., Zurkowski, W., and Rolfe, B.G. (1982) 'Plasmids and stability of symbiotic properties in Rhizobium trifolii', J. Bacteriol. 151, 560-568.

11.- Soberón-Chávez, G., Nájera, R., Olivera, H., and Segovia, L. (1986) 'Genetic rearrangements of a Rhizobium phaseoli symbiotic plasmid', J. Bacteriol. 167, 487-491.

12.- Hahn, M., and Hennecke, H. (1987) 'Mapping of a Bradyrhizobium japonicum DNA region carrying genes for symbiosis and an asymmetric accumulation of reiterated sequences', Appl. Environ. Microbiol. 53, 2247-2252.

13.- Gay, P., Le Coq, D., Steinmetz, M., Berkelman, T., and Kado, C.I. (1985) 'Positive selection procedure for entrapment of insertion sequence elements in gram-negative bacteria', J. Bacteriol. 164, 918-921.

14.- Mortlock, R.P. (1982) 'Metabolic acquisitions through laboratory selection', Ann. Rev. Microbiol. 36, 259-284.

15.- Sonti, R.V., and Roth, J.R. (1989) 'Role of gene duplications in the adaptation of Salmonella typhimurium to growth on limiting carbon sources', Genetics 123, 19-28.

CYTOKININ PRODUCTION BY RHIZOBIA

B. J. TALLER AND D. B. STURTEVANT
Department of Biology
Memphis State University
Memphis, Tennessee 38152, USA

ABSTRACT. Culture media from a number of rhizobial strains, including the type-strains of 8 major cross-inoculation groups, were analyzed for cytokinin content. Cytokinins were partially purified by chromatography on Amberlite XAD-2, then on Sephadex LH-20. The tobacco callus assay, HPLC and/or immunoassay were used for cytokinin analysis. All strains of rhizobia examined produced at least 2 cytokinin-active compounds, with total cytokinin activity ranging from 1 to several µg kinetin equivalents per liter of culture filtrate. There were both qualitative and quantitative differences between rhizobial species. The cytokinin profiles of most strains included zeatin or its derivatives. Addition of adenine, seed extract or flavonoid inducers to the culture medium altered cytokinin synthesis. Preliminary genetic data do not support *nod* gene involvement in cytokinin synthesis under noninducing conditions. Growth pouch experiments showed increased nodulation of soybean plants treated with *trans*-zeatin.

Introduction

Over thirty years ago, Arora, Skoog and Allen suggested that the action of rhizobia in nodule formation likely involved the supply or activation of several different growth factors [2]. Circumstantial evidence indicates that cytokinins may be involved in nodule development. Exogenous application of cytokinin induces cortical cell division in roots of soybean, cowpea and alfalfa [3] and produces pseudonodules on tobacco roots [2]. *E. coli* or *nod* gene mutants of *R. meliloti* containing the *trans-zeatin* secretion gene (*tzs*) from *Agrobacterium* induces cell division in alfalfa roots [9]. Also an early nodule gene (ENOD2) is induced in cytokinin-treated soybean roots [10]. Despite considerable evidence for the involvement of cytokinins in legume nodulation, information concerning cytokinin production by rhizobia is fragmentary and contradictory [11]. Thus, it was of interest to characterize cytokinin production by rhizobia.

Methods

Methods used for cytokinin isolation and identification have been described [11] and are briefly summarized below. For routine analysis, one-liter cultures were grown to early stationary phase. Cells were removed by centrifugation and the culture supernatant applied to an Amberlite XAD-2 column. The XAD eluate was then fractionated on a Sephadex LH-20 column in 35% ethanol, which separates many commonly occurring cytokinins. Individual fractions were pooled according to the elution positions of cytokinin standards, and the pooled fractions were analyzed in the tobacco bioassay. Bioassay activity is expressed as µg kinetin equivalents (KE), defined as the µg of kinetin required to give the same activity as the test sample.
   In some cases the cytokinin-active fractions were further analyzed by LH-20 chromatography in

H. Hennecke and D. P. S. Verma (eds.),
Advances in Molecular Genetics of Plant-Microbe Interactions, Vol. 1, 215–221.
© 1991 *Kluwer Academic Publishers. Printed in the Netherlands.*

water, HPLC and/or enzyme-linked immunospecific assay. Affinity chromatography on phenylboronate agarose (PBA) was used for purification of cytokinin ribosides [5]. The column eluate was analyzed for cytokinins by reversed-phase HPLC and immunoassay.

Growth pouch assays were performed as described by Bhuvaneswari *et al* [4].

The following type strains were examined for cytokinin synthesis: *R. fredii* USDA 205; *R. leguminosarum* bv. *phaseoli* ATCC 14482, bv *trifolii* ATCC 14480, bv. *viceae* ATCC 10004; *R. loti* ATCC 33669; *R. meliloti* ATCC 9930; *B. japonicum* ATCC 10324; *B.* sp.(*lupini*) ATCC 10319.

Results

CYTOKININ ANALYSIS

Each of the rhizobial strains examined, including the type-strains from each of the major cross-inoculation groups, produced at least two cytokinin-active compounds; some produced as many as four. The total cytokinin activity ranged from 0.5 to about 4 µg kinetin equivalents per liter of culture medium (Fig. 1). Labels on the figures denote cochromatography with the indicated cytokinin standards, not necessarily identity. Yields represent minimum values, as they are not corrected for loss. Any cytokinin nucleotides are retained on the XAD column under the conditions used, and thus would not be detected. The cytokinin profiles of most strains included zeatin (Z) or its derivatives. There were both qualitative and quantitative differences among species of rhizobia. For example, none of the *Bradyrhizobium* strains examined produced isopentenyladenine (IP) or its riboside (IPR), while most *Rhizobium* strains did. Among *Rhizobium* strains there was considerable variation in the amount of IP derivatives. Variation between species appeared to be greater than within a species, but more strains need to be examined.

Work has focused on cytokinin synthesis by *Bradyrhizobium* and *Rhizobium meliloti*. All *Bradyrhizobium* strains examined produce three cytokinin-active compounds, tentatively identified as ribosylzeatin (ZR), zeatin (Z) and methylthiozeatin (msZ) [11]. No isopentenyladenine derivatives have ever been detected in the dozens of cultures examined, even though they are likely intermediates in the synthesis of Z-type cytokinins. Though yeast extract-containing medium has IP-type cytokinins, these compounds are absent in *Bradyrhizobium* cultures grown in such media.

In contrast, *R. meliloti* 1021 produced two cytokinins which cochromatographed with IPR and IP on Sephadex LH-20 eluted in ethanol. No compounds corresponding to known zeatin derivatives have been detected. When chromatographed on LH-20 eluted with water, the IP fraction again coeluted with authentic IP. However, the IPR fraction did not cochromatograph with the standard under these conditions. This compound has not been identified, but is probably an IP derivative as it reacts with anti-IPR monoclonal antibodies. The unusual cytokinin was not retained on phenylboronate agarose, indicating it is probably not a riboside. Acid hydrolysis under conditions designed to remove attached sugars did not produce IP.

Affinity chromatography on phenylboronate agarose was very effective in purifying cytokinin ribosides. When a cytokinin riboside fraction from an LH-20:H_2O column was examined by HPLC, the UV-absorbing material in the sample obscured any signal due to the cytokinin. However after one pass over the PBA column, the predominant UV-absorbing material in the sample was the cytokinin.

MODIFICATION OF CYTOKININ PRODUCTION

As soybean seed extract has been shown to induce *Bradyrhizobium* nodulation genes, the effect of seed extract on cytokinin production was examined. Cultures were grown in medium containing aqueous soybean seed extract (10%) and were analyzed for cytokinins as described above. The bioassay results are shown in Figure 2. In defined medium, *Bradyrhizobium* 61A68 produced three cytokinins, ZR, Z and msZ. With soybean seed extract, Z disappeared and there was a significant increase in msZ. Cytokinin analysis of control medium containing 10% seed extract showed mostly ZR and a trace of Z, but no msZ. Thus the additional msZ did not come directly from the seed extract. Cytokinins in the seed extract may have served as precursors for cytokinin synthesis. However, adding a similar amount of ZR to defined medium did not affect cytokinin

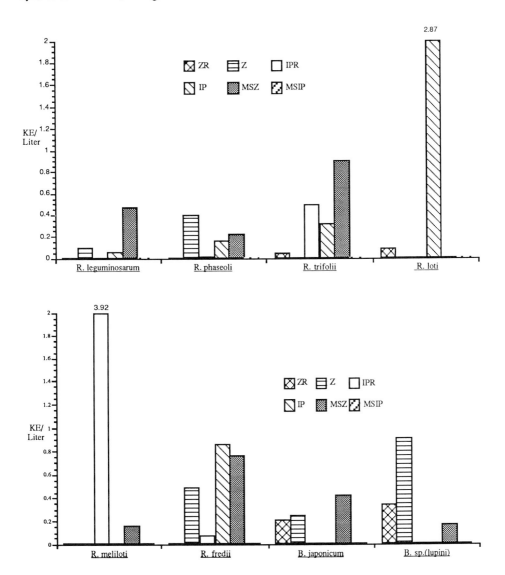

Figure 1. Cytokinin production by rhizobia. Partially purified culture filtrates from 8 type strains of rhizobia were fractionated on Sephadex LH-20 columns. Column fractions were pooled according to the elution positions of cytokinin standards and were tested in the tobacco bioassay. Results for fractions corresponding to 6 common cytokinins are shown. Fraction labels denote cochromatography with indicated cytokinin standards, not necessarily identity. Abbreviations as in text.

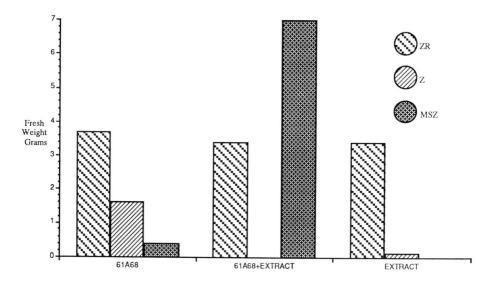

Figure 2. Effect of soybean seed extract on cytokinin synthesis. *B. japonicum* 61A68 was grown with and without 10% soybean seed extract. Medium plus extract served as a control. The resulting culture media were fractionated on Sephadex LH-20 and bioassayed. Data are expressed as the fresh weight of tobacco callus produced by the test sample in the bioassay.

synthesis. In fact, most of the ZR was recovered from the culture medium, suggesting that it was not taken up or metabolized to any significant extent. Attempts to reproduce the effect of seed extract by adding genistein or daidzein gave inconsistent results.

Subsequent experiments have utilized *B. japonicum* ZB977, a derivative strain of 110 containing a *nodY: lacZ* fusion, to allow both *nod* gene induction and cytokinin synthesis to be monitored. Like 110, ZB977 produces mostly msZ and some ZR. The msZ was identified by HPLC retention time, UV scan and reactivity with anti-tZR antibodies. In the presence of 2 μM genistein, the cytokinin activity shifted from the msZ fraction to a later-eluting one. The identity of this compound has not been determined. This shift in activity did not occur when strain 110 was grown under the same conditions.

A shift in cytokinin species was also seen when *R. meliloti* was grown in the presence of 2 μM luteolin (Fig. 3). Uninduced cultures produced primarily a cytokinin-active compound which coeluted with IPR, while induced cultures showed a decrease in this compound and an increase in activity attributable to IP and methylthioisopentenyladenine (msIP). Addition of adenine, a cytokinin precursor, also affected cytokinin synthesis in defined medium. Adenine at 10 mg/L caused a 25% increase in cytokinin activity in uninduced cultures and a two-fold increase when added together with luteolin.

ANALYSIS OF NODULATION MUTANTS

All strains of rhizobia examined have been found to produce cytokinins. Preliminary attempts to demonstrate a role for the Sym plasmid in cytokinin synthesis have been unsuccessful. When cytokinin synthesis was examined in pairs of strains, one of which had been cured of the Sym plasmid, no difference in cytokinin production was seen. Similarly, cytokinin synthesis by Nod-*Bradyrhizobium* transposon mutants in *nod A,B,C* or *D* did not differ significantly from the parental strain when grown under noninducing conditions. The cytokinin profile of *Bradyrhizobium*

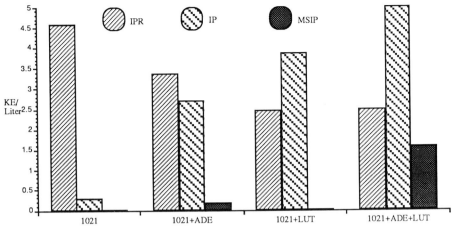

Figure 3. Cytokinin production by *R. meliloti* 1021. Bioassay results are shown for the three cytokinin-active fractions found after LH-20 fractionation. The effect of added 10 mg/l adenine (ade) and 2 μM luteolin (lut), singly and in combination, is also shown.

WAJ 336, a mutant strain with a large deletion including the nodulation genes, also was similar to that of its parental strain 110.

EFFECT OF CYTOKININS ON NODULATION

Preliminary experiments have been carried out to determine whether cytokinins affect nodulation. Inoculated soybean plants treated with 0.5, 1 or 5 μg/L of *trans*-zeatin produced twice as many nodules as the control lacking cytokinin. A *t*-Z concentration of 10 μg/L was inhibitory to nodulation, producing half as many nodules as the control. No nodules were found on cytokinin-treated plants which had not been inoculated with bacteria. The amounts of cytokinin used did not appear to significantly affect root growth, as determined by root length and appearance.

Discussion

All rhizobial strains examined to date produce cytokinins when grown on simple mineral media containing carbohydrate and vitamins. Likewise, all the cultures produce similar amounts of cytokinin, although a cytokinin over-producing strain has been reported [8]. The amount of cytokinin produced by two fast-growing *Parasponia*-nodulating strains, which use a crack entry mode of infection, did not differ significantly from other rhizobia (data not shown). As the diversity of cytokinin-active compounds reported to occur in bacteria is less than that of plants, what appear to be novel bacterial cytokinins in some strains are of particular interest. The significance of the variation in cytokinin production among different species is not known. Variation in cytokinin synthesis also occurs in strains of *Agrobacterium* and plant pathogenic *Pseudomonas* species [1]. In the latter, cytokinin secretion did not correlate with host range or known metabolic markers.

The primarily qualitative effect of plant factors on rhizobial cytokinin synthesis differs from that in agrobacteria, where the effect is quantitative. It may be that the *nod* gene inducers which have been examined do not optimally induce the cytokinin genes. While both soybean seed extract and genistein caused a shift in the cytokinin species produced by bradyrhizobia, the change in cytokinin profile differed. This may be due to strain or medium differences, or the fact that seed extract contained several inducing compounds.

Our limited data concerning the genetics of cytokinin production in rhizobia are consistent with those found for IC3342, an unusual *Rhizobium* which nodulates pigeon pea and which over-produces cytokinin [8]. The loci responsible for cytokinin production in IC3342 are both chromosomal and on the Sym plasmid (*lcr* region), but not at the *nod* loci. Our failure to demonstrate a role for the Sym plasmid may have been caused by reiteration of genes or by the gene(s) remaining silent under uninduced conditions. The first step in cytokinin biosynthesis appears to be a plasmid-coded function in the plant pathogens *Agrobacterium tumefaciens*, *Pseudomonas savastanoi* and *Rhodococcus* (*Corynebacterium*) *fascians*. Interestingly, the *lcr* region showed no DNA homology to known prokaryotic cytokinin genes *ipt*, *tzs* or *ptz* [8]. It remains to be seen whether the *lcr* genes are typical of rhizobia or are unique to this unusual strain. The role of cytokinins in legume nodulation is not known. Enhancement of nodulation by cytokinin was also reported by Yahalom *et al.* [12], who found that benzyladenine (1 nM) significantly increased nodulation above the root tip mark. The elucidation of NodRm-1, a plant-specific signal from *R. meliloti* which elicits root hair deformation and cortical cell division [7], makes it unlikely that cytokinins are the "nodule-inducing principle". It has been proposed for *Agrobacterium* that the *tzs* gene, inducible by plant phenolics, is not necessary for infection but may affect the efficiency of transformation [1,6,13]. Cytokinins may influence host range, susceptibility to infection or tumor initiation [1,6,13]. There is also evidence that *tzs* may be expressed in the tumor, as galls produced by *A. tumefaciens* C58 lacking *tzs* have a different morphology than those produced by C58 [6]. The effect of cytokinins on nodulation and evidence for the production of cytokinin by bacteroids [8], suggest that rhizobial cytokinins may have similar roles.

Acknowledgements

B. japonicum nod gene transposon mutants and ZB977 were provided by Dr. Gary Stacey. *R. meliloti* 1021 and *nod* gene mutants were from Dr. Sharon Long. This research was funded in part by the Feinstone Microbiology Research Fund at MSU and the Himmel Health Foundation.

References

1. Akiyoshi, D. E., D. A. Regier and M. P. Gordon. (1987) Cytokinin production by *Agrobacterium* and *Pseudomonas* spp. J. Bacteriol. 169: 4242-4248.

2. Arora, N., F. Skoog and O. N. Allen. (1959) Kinetin induced pseudonodules on tobacco roots. Am. J. Bot. 46: 610-613.

3. Bauer, W. D., T. Bhuvaneswari, H. E. Calvert, I. J. Law., N. S. A. Malik and S. J. Vesper. (1985) Recognition and infection by slow-growing rhizobia, in H. J. Evans, P. J. Bottomly and W. E. Newton (eds.) Nitrogen Fixation Research Progress, Martinus Nijhoff Publishers, Dordrecht, pp. 247-253.

4. Bhuvaneswari, T. V., B. G. Turgeon and W. D. Bauer. (1980) Early events in the infection of soybean (*Glycine max* L. Merr.) by *Rhizobium japonicum*. Plant Physiol. 66: 1027-1031.

5. Blum, P. H. and B. N. Ames. (1989) Immunochemical identification of a tRNA-independent cytokinin-like compound in *Salmonella typhimurium*. Biochim. Biophys. Acta 1007: 196-202.

6. Castle, L. A. and R. O. Morris. (1988) Investigations into the role of *tzs* in *Agrobacterium tumefaciens* mediated plant transformation. Abstracts, 13th Int. Conf. on Plant Growth Substances, Calgary, Canada.

7. Lerouge, P., P. Roche, C Faucher, F. Maillet, G. Truchet, J. C. Prome and J. Denarie. (1990) Symbiotic host-specificity of *Rhizobium meliloti* is determined by a sulphated and

acylated glucosamine oligosaccharide signal. Nature 344: 781-784.

8. Letham, D. S., R. Zhang, S. Singh, L. M. S. Palni, C. W. Parker, M. N. Upadhyaya and P. J. Dart. (1988) Xylem-translocated cytokinin - metabolism and function. Abstracts, 13th Int. Conf. on Plant Growth Substances, Calgary, Canada.

9. Long, S. R. and J. Cooper. (1988) Overview of symbiosis, in R. Palacios and D.P.S. Verma (eds.), Molecular Genetics of Plant-Microbe Interactions, APS Press, St. Paul, MN, pp. 163-178.

10. Nirunsuksiri, W. and C. Sengupta-Gopalan. (1988) Characterization and regulation of an early nodulin gene in soybeans, in N. T. Keen, T. Kosuge, L.L. Wallings (eds.), Physiology and Biochemistry of Plant Microbial Interactions, American Society of Plant Physiologists, Rockville, MD, pp.171-172.

11. Sturtevant, D. B. and B. J. Taller. (1989). Cytokinin production by *Bradyrhizobium japonicum*. Plant Physiol. 89: 1247-52.

12. Yahalom, E., Y. Okon and A. Dovrat. (1990). Possible mode of action of *Azospirillum brasilense* strain Cd on the root morphology and nodule formation in burr medic (*Medicago polymorpha*). Can. J. Microbiol. 36: 10-14.

13. Zahn, X., D.A. Jones and A. Kerr. (1990) The pTiC58 *tzs* gene promotes high-efficiency root induction by agropine strain 1855 of *Agrobacterium rhizogenes*. Plant Mol. Biol. 14: 785-792

MOLECULAR GENETICS OF THE HYDROGEN UPTAKE SYSTEM OF
RHIZOBIUM LEGUMINOSARUM

T RUIZ-ARGÜESO, E HIDALGO, J MURILLO, L REY, J M PALACIOS
Laboratorio de Microbiología, ETS Ingenieros Agrónomos
Universidad Politécnica
28040 Madrid, Spain.

ABSTRACT. The genetic determinants for H_2-uptake (hup genes) of R. leguminosarum are clustered in a region of about 15 kb of the symbiotic plasmid and are organized in six transcriptional units designated regions hupI to hupVI, all of which are transcribed in the same direction. When DNA restriction fragments containing these hup regions were expressed in E. coli cells under the control of a phage T7 promoter, ten proteins were shown to be specifically expressed by the hup DNA. A 7.3 kb KpnI fragment contaning regions hupI, hupII and part of region hupIII was sequenced. The analysis of the nucleotide sequence of the DNA covering regions hupI/hupII revealed the presence of a putative operon containing the genes for the structural subunits of the hydrogenase and three more open reading frames. A second putative operon containing two ORFs was identified in the DNA corresponding to region hupIII. Both operons have NtrA-binding consensus sequences in the upstream non-coding regions.

1. Introduction

Some strains of <u>Rhizobium</u> and <u>Bradyrhizobium</u> induce in legume nodules the synthesis of a H_2-uptake (Hup) enzyme system that recycle the H_2 generated by the nitrogenase during the nitrogen fixation process. In the <u>B. japonicum</u>-soybean symbiosis, this Hup system has been shown to save energy and provide other biological advantages as well [1].

The first component of the Hup system of <u>B. japonicum</u> is a membrane-bound, [NiFe] hydrogenase which contains two polypeptide subunits of about 35 and 65 kD [1]. The genetic determinants (<u>hup</u> genes) for the two structural subunits of the hydrogenase of <u>B. japonicum</u> lie in a cluster of <u>hup</u>-specific DNA spanning a region of about 16 kb of the chromosomal DNA [2] and have been sequenced [9]. Hydrogenase positive strains of <u>R. leguminosarum</u> contains DNA homologous to <u>B. japonicum</u> <u>hup</u> genes. Based on this homology, a recombinant cosmid, pAL618, containing the entire set of <u>hup</u> genes required for H_2-uptake in pea nodules was isolated from a pLAFR1 gene library of <u>R. leguminosarum</u> strain 128C53 [4].

The organization of <u>R. leguminosarum</u> <u>hup</u> gene cluster in the 20 kb DNA insert of pAL618 was investigated by site-directed transposon mutagenesis and complementation analysis [5]. The <u>hup</u> genes were found to span over 15 kb and to be organized in six transcriptional units designated <u>hup</u>I to <u>hup</u>VI (Fig. 1A) [5]. All the <u>hup</u> regions are

222

H. Hennecke and D. P. S. Verma (eds.),
Advances in Molecular Genetics of Plant-Microbe Interactions, Vol. 1, 222–225.
© 1991 *Kluwer Academic Publishers. Printed in the Netherlands.*

expressed in symbiotic cells but only regions hupV and hupVI are activated in microaerobically grown vegetative cells [8].

We report here preliminary results on the nucleotide analysis of a 7.3 kb KpnI fragment containing regions hupI to hupIII.

2. Results

Several restriction DNA fragments from DNA insert of pAL618 were cloned in both orientations into an expression vector based on a phage T7-RNA polymerase promoter system and the encoded gene products expressed in E. coli cells. Ten proteins synthesized in E. coli were specifically associated to hup DNA from pAL618 (unpublished results). Four proteins of 40, 65, 24 and 28 kD were syntesized by DNA covering regions hupI/hupII. Two proteins were synthesized from DNA of hupIII (17 and 31 kD), hupV (39 and 79 kD), and hupVI (36 and 41 kD). The polypeptide of ca. 65 kD, associated to expression of region hupI, was cross-reactive with antiserum against the large subunit of the B. japonicum hydrogenase [5]. This result suggest that region hupI contains the structural genes of the R. leguminosarum hydrogenase.

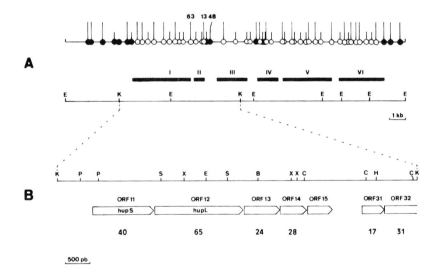

Figure 1. Organization of R. leguminosarum hup genes in the DNA insert of pAL618. A. Transcriptional units (horizontal bars above the restriction map) were defined by transposon mutagenesis and complementation analysis [5]. Vertical lines indicate the positions of Tn5 (long lines) or Tn3-HoHo1 (short lines) transposon insertions. The open or solid circle below each line indicates whether the insertion resulted in a Hup⁻ or a Hup⁺ phenotype, respectively, in symbiosis with peas. B. Gene organization of regions hupI/hupII and hupIII. The open boxes below the restriction map indicate the positions of the ORFs identified by nucleotide sequence analysis. Figures under the boxes indicate the molecular weight of the proteins expressed in E. coli by the corresponding ORF.

To identify the individual genes of each of the hup regions, the entire 15 kb DNA of pAL618 containing the hup cluster is currently being sequenced. The 7.3 kb KpnI fragment comprising regions hupI, hupII and part of hupIII has already been sequenced in both directions. The nucleotide sequence revealed five complete adjacents open reading frames (ORFs) (ORF11 to ORF15) in the DNA covering regions hupI/hupII and two adjacents ORFs (ORF31 and ORF32) in the DNA corresponding to region hupIII; each of the ORFs was preceded by a putative ribosome binding site (Fig. 1B).

The ORF11 (360 aa) and ORF12 (596 aa) are highly similar to the genes coding for the small and large subunits of the B. japonicum hydrogenase [9], both at the nucleotide and amino acid (89 % aa conservation) sequence levels [3]. Based on this homology, the gene products from ORF11 and ORF12 were identified as the small and large subunits of the R. leguminosarum hydrogenase and were designated hupS and hupL, respectively. The non coding region upstream of hupS contains the NtrA-binding consensus sequence (CTGG-N$_8$-TTGCA) at position -11 as regard the transcription initiation site. This sequence is typical of σ^{54}-RNA polymerase-dependent promoters and its presence has also been reported in the promoters of NAD-reducing hydrogenase of A. eutrophus [10] and hydrogenase 3 of E. coli [6].

Downstream of hupL three additional ORFs were identified. ORF13 (239 aa) is separated 13 bases from the stop codon of hupL and ORF15 (163 aa) 11 bases from ORF14 (202 aa). Stop codon of ORF13 and start codon of ORF14 are contiguous. Since hupS and hupL are also separated by only 23 bases, this arrangement predicts an organization of the five ORFs as an operon. Tn5 insertions 63 and 13, which completely suppress the Hup phenotype [5] have been precisely mapped to ORF13 and ORF14, respectively, suggesting that they are likely coding sequences. This is also supported by the observation that ORF13 and ORF14 directed in E. coli the synthesis of proteins with an estimated molecular weight of 24 and 28 kD, respectively. The ORF13 predicted polypeptide is very hydrophobic and has four putative hydrophobic membrane-spanning regions. It is not clear whether ORF15 is a coding sequence since Tn5 48, which has been mapped in this ORF, is associated with a Hup$^+$ phenotype (Fig. 1A). All of these results will be reported in detail elsewhere (manuscript in preparation).

The organization of the putative hupI/hupII operon resembles that of the hydrogenase 1 operon of E. coli which contains four ORFs (hyaC to hyaF) downstream of the genes coding for the structural subunits [7]. The derived amino acid sequence of ORF13 is 46 % homologous to hyaC gene and has similar hydrophobicity profile. The ORF14 derived amino acid sequence has 41 % similarity with hyaD gene. These similarities between two taxonomically unrelated bacteria as R. leguminosarum and E. coli suggests that the presence of additional proteins in the operons containing the hydrogenase structural genes could be a general trait in H$_2$-uptake hydrogenase systems. These proteins may play important roles in hydrogenase activity.

Two adjacent ORFs were identified in region hupIII (Fig. 1B). ORF31 encodes a predicted polypeptide of 149 aa (16.3 kD). ORF32 is not complete in the 7.3 kb KpnI DNA fragment sequenced. It encodes a predicted polypeptide higher than 180 aa. The molecular weight of these polypeptides agree quite well with the proteins synthesized in E. coli cells from DNA of hupIII region (17 and 31 kD). Stop codon of ORF31 overlaps the initiation codon of ORF32 suggesting that both ORFs belong to the same operon. In the non-coding region upstream of ORF31, a NtrA-binding consensus sequence, highly homologous to that found upstream of hupS, was identified 45 bases from the translation start codon. . This result suggest that the R. leguminosarum hup operons may contain a similar type of promoters.

Acknowledgments. This work was supported by grant from the Comision Interministerial de Ciencia y Tecnología (CICYT) PBT87-0029.

3. References

1 Evans, H. J. , Harker, A., Papen, H., Russell, S., Hanus, F. J., and Zuber, M. (1987) 'Physiology, biochemistry and genetics of the uptake hydrogenase in rhizobia', Ann. Rev. Microbiol. 41, 335-361.
2 Haugland, R., Cantrell, M., Beaty, J., Hanus, F., Russell, S., and Evans, H. J. (1984) 'Characterization of Rhizobium japonicum hydrogen uptake genes', J. Bacteriol. 139, 1006-1012.
3 Hidalgo, E., Leyva, A., and Ruiz-Argüeso, T. (1990) 'Nucleotide sequence of the structural genes from Rhizobium leguminosarum', Plant Mol. Biol., in press.
4 Leyva, A., Palacios, J. M., Mozo, T., and Ruiz-Argüeso, T. (1987) 'Cloning and characterization of hydrogen uptake genes from Rhizobium leguminosarum', J. Bacteriol. 169, 4929-4934.
5 Leyva, A., Palacios, J. M., Murillo, J., and Ruiz-Argüeso, T. (1990) 'Genetic organization of the hydrogen uptake (hup) cluster from Rhizobium leguminosarum', J. Bacteriol. 172, 1647-1635.
6 Lutz, S., Böhm, R., Beier, A., and Böck, A. (1990) 'Characterization of divergent NtrA-dependent promoters in the anaerobically expressed gene cluster coding for hydrogenase 3 components of Escherichia coli', Mol. Microbiol. 4, 13-20.
7 Menon, N. K., Robbins, J., Peck, H. D., Chatelus, C. Y., Choi, E., and Przybyla, A. E. (1990) 'Cloning and sequencing of a putative Escherichia coli [NiFe] hydrogenase-1 operon containing six open reading frames', J. Bacteriol. 172, 1969-1977.
8 Palacios, J. M., Murillo, J., Leyva, A., Ditta, G., and Ruiz-Argüeso, T. (1990) 'Differential expression of hydrogen-uptake (hup) genes in vegetative and symbiotic cells of Rhizobium leguminosarum', Mol. Gen. Genet. 221, 363-370.
9 Sayavedra-Soto, L., Powell, G., Evans, H. J., and Morris, R. (1988) 'Nucleotide sequence of the genetic loci encoding subunits of Bradyrhizobium japonicum uptake hydrogenase', Proc. Natl. Acad. Sci. USA 85, 8395-8399.
10 Tran-Betcke, A., Warnecke, U., Böcker, C., Zaborosch, C., and Friedrich, B. (1990) 'Cloning and nucleotide sequences of the genes for the subunits of NAD-reducing hydrogenase of Alcaligenes eutrophus H16', J. Bacteriol. 172, 2920-2929.

β-GLUCURONIDASE (GUS) OPERON FUSIONS AS A TOOL FOR STUDYING PLANT-MICROBE INTERACTIONS

KATE J. WILSON[1,2], KEN E. GILLER[1] AND RICHARD A. JEFFERSON[2]
[1]*Wye College, University of London, Wye, Ashford, Kent, UK;* [2]*Joint Division of the Food and Agriculture Organization of the United Nations and the International Atomic Energy Agency, P.O. Box 100, A-1400, Vienna, Austria.*

ABSTRACT. We are developing the *Escherichia coli gus* operon as a transgenic marker for the detection and study of bacteria of the rhizosphere and phyllosphere. This operon includes genes encoding ß-glucuronidase (GUS) and a glucuronide-specific membrane transporter. We have introduced the entire *gus* operon cloned from a natural isolate of *E. coli* into *Agrobacterium* and *Rhizobium* strains. The operon was expressed in both species. Alfalfa nodules induced by *R. meliloti* strains carrying the *gus* operon stained blue on incubation with the GUS substrate X-Gluc. To enable introduction of GUS into bradyrhizobia, a translational fusion of the Tn5 *aph* gene to *gusA* was constructed. Blue-stained infection threads in root hairs of siratro plants inoculated with a *Bradyrhizobium* strain carrying this fusion were clearly visible after incubation with X-Gluc.

Introduction

Populations of soil bacteria exert a great influence on the growth of plants. However, our present understanding of the role of particular bacterial strains in soil ecology and crop growth is restricted by the difficulty of studying bacterial populations *in situ* in the soil or in tight association with plants. Molecular biology is providing new tools with which to address questions of microbial ecology, most notably the use of marker genes and of techniques for analysing DNA extracted from the spectrum of organisms present in a given microbial community. The latter approach recently provided firm evidence that the microorganisms that we can culture in the laboratory form only a fraction of a given microbial community (1,2), thereby emphasising the need for methods to analyse the behaviour of specific populations of microorganisms *in situ* within a given community, rather than in artificial conditions in a laboratory. To this end, we are developing the *gus* operon as a transgenic marker for use in the detection and analysis of living microorganisms *in situ* in their natural environment.

The *E. coli gus* operon comprises three genes. The first is *gusA*, the structural gene encoding the enzyme ß-glucuronidase (GUS) (3), which is now widely used as a reporter gene in gene fusion analysis in plants (4) and, more recently, in fungi (5) and in bacteria (6). It is followed by *gusB*, encoding a glucuronide-specific permease, and a third gene, *gusC* of unknown function (W-J Liang, T.J. Roscoe and RAJ, unpublished). The glucuronide permease is remarkable in that it actively accumulates an extremely wide variety of glucuronides, including the GUS substrates 4-methylumbelliferyl ß-D-glucuronide, p-nitrophenyl ß-D-glucuronide and 5-bromo-4-chloro-3-indolyl ß-D-glucuronide (X-Gluc)

226

H. Hennecke and D. P. S. Verma (eds.),
Advances in Molecular Genetics of Plant-Microbe Interactions, Vol. 1, 226–229.

(KJW and RAJ unpublished,7). Expression of the operon is specifically induced by glucuronide substrates which act by relieving transcriptional repression by the product of the *gusR* gene which lies immediately upstream from *gusA*.

There are many reasons to choose the *gus* operon as a transgenic marker system for bacteria. There is the total absence of background GUS activity in most bacterial, fungal and plant systems tested, which vastly enhances the sensitivity of detection and enables the spatial relationship of microorganisms to plants to be determined. The enzyme is very stable in diverse biochemical conditions and a wide variety of fluorogenic and colorigenic GUS substrates are available. Most importantly, functional expression of the glucuronide permease in other bacteria will enable accumulation of GUS substrates from low external concentrations into transgenic organisms dispersed in the soil or the rhizosphere, so permitting them to be detected and monitored without the necessity for laboratory culture.

We have recently cloned the entire *gus* operon from non-K12 strains of *E. coli* (in K12 strains the permease does not actively transport glucuronides (8; W-J Liang, P. Henderson and RAJ, unpublished data)). Here we describe the introduction of the *gus* operon into *Rhizobium* and *Agrobacterium* strains and initial experiments demonstrating the potential of GUS and the *gus* operon as a marker for detecting strains of *Rhizobium* and *Bradyrhizobium* in symbiosis with their host plants.

Materials and Methods

Bacterial strains, plasmids, media and genetic manipulations. Rm1021 is wild-type *R. meliloti*; LBA4404 is a disarmed *Agrobacterium tumefaciens*; NC92 is a *Bradyrhizobium* cowpea-group strain (10); pLAFR3 is a broad-host-range cloning vector; pSUP1021 is a Tn5 suicide vector (9); pKW210 is the *gus* operon from a faecal isolate of *E. coli* in pLAFR3. Standard LB or YM media were used. Antibiotic concentrations were standard. X-Gluc was used at 50 μg/ml. Conjugation of plasmids was performed with the mob+ *E. coli* strain S17-1 (9). pKW107 contains an *aph-gusA-rrnB* fusion cloned between the *SmaI* and *BstXI* sites of Tn5 in pSUP1021 (KJW, unpublished).

Growth and analysis of plants. Alfalfa and siratro plants were grown in seed pouches as described (10). For histochemical assays, whole roots were immersed in X-Gluc buffer (50 mM NaPO$_4$, pH 7.0, 1 mM EDTA, 0.1% Triton, 0.1% Sarkosyl, 0.05% SDS, 50 μg/ml XGluc). X-Gluc was from BioSynth AG, Switzerland. Blue-stained nodules were fixed for 48 hours in formalin acetic acid, dehydrated in ethanol, cleared in Histoclear (Agar Aids) and embedded in paraffin wax.

Results and Discussion

THE GUS OPERON IS EXPRESSED IN BOTH *AGROBACTERIUM* AND *RHIZOBIUM*

In initial experiments plasmid pKW210, a pLAFR3 clone carrying the *gus* operon from a faecal isolate of *E. coli* was introduced into *R. meliloti* strain Rm1021 and into *Agrobacterium tumefaciens* strain LBA4404 by conjugation. Transconjugants containing pKW210 formed tiny deep-blue colonies on LB plates containing X-Gluc, a GUS substrate which, on cleavage by GUS forms a deep indigo precipitate. Control transconjugants containing only the pLAFR3 plasmid with no insert did not develop any blue colour. This indicated qualitatively that the cloned *E. coli gus* operon was being expressed, probably from its own promoter. Preliminary quantitative assays indicate that expression levels may be similar to those observed in *E. coli*.

To determine whether expression of the *gus* operon could be used to detect and monitor the infection process, the *R. meliloti* transconjugants were inoculated onto alfalfa plants. At early stages of the infection process (up to 84 hours post-inoculation) curled root hairs were abundant but infection threads were not detected by blue-staining upon incubation in X-Gluc buffer. However, young nodules (nine days post inoculation) stained bright blue and appeared like a string of blue beads down the root.

Alfalfa forms determinate nodules with the symbiotic zones clearly differentiated spatially within the nodule. Therefore, mature, cylindrical nodules (44 days post-inoculation) were assayed to look at the spatial distribution of *gus* expression. In these nodules blue staining was again strong, but was restricted to the tips of the nodules distal to the root. When 6 μm thick sections were examined, blue staining was very clear in individual cells and was localized to the cells of the early symbiotic (meristematic) zone. No staining was observed in the nodule cortex, except for occasional patches of blue on the exterior face, presumably due to clumps of bacteria adhering to the outer surface of the nodule. Nor was there any staining in the cells of the mature symbiotic zone. This localization of the *gus* operon expression indicates either that the plasmid, pKW210, was lost from the mature, nitrogen-fixing bacteroids, or that the promoter from which it is being expressed is not transcribed at this stage of development. This question can easily be resolved by using chromosomally-integrated copies of the operon.

MARKING OF A BRADYRHIZOBIUM STRAIN

We wish to develop this system from the outset to be useful not just in *Rhizobium* and *Agrobacterium* strains which are well characterized genetically, but also in bacteria which are of considerable ecological and economic importance, but which are less easy to manipulate, such as bradyrhizobia. *Bradyrhizobium* sp. (*Arachis*) strain NC92 (like many other bradyrhizobia) has endogenous resistance to tetracyline, prohibiting selection of pKW210 in this strain. To overcome this limitation, and to determine whether higher level expression of *gusA* would enable routine detection of marked bacteria in infection threads, we constructed a Tn*5* derivative containing a translational fusion between the promoter region and 11 N-terminal amino acids of the *aph* gene (encoding the neomycin-resistance determinant of Tn*5*) and *gusA*. This promoter was chosen because it is active in a wide variety of gram-negative bacteria, including bradyrhizobia. GUS+, kan^r NC92 derivatives were obtained following conjugation with S17-1 (pKW107); one such GUS+ derivative was used to inoculate siratro plants. Blue-stained infection threads were readily observed in curled root hairs in one section of the root, while no blue-staining was observed in other areas of the root, including those with abundant root hairs. This observation fits with the observation that zones of the growing root are only transiently susceptible to (brady)rhizobial infection. Mature nodules formed by this strain could also be detected by X-Gluc staining, as with the alfalfa plants described above. Control nodules formed by the parent strain NC92 did not develop any blue colour on incubation in X-Gluc buffer.

FUTURE PROSPECTS

Our goal is to develop the *gus* operon as a marker system that can be routinely used to mark and study free-living, symbiotic and pathogenic bacteria. Developments in progress include manipulation of the glucuronide permease to function efficiently in diverse bacteria, construction of gene fusions that are tightly regulated, yet expressed at high level when induced, manipulation of plasmids and other delivery vehicles to ensure efficient introduction of the operon into the microorganism of choice, as well as the development and optimization of methods for the detection and monitoring of bacteria in soil or in tight association with plants without laboratory culturing. In this paper we indicate the possibilities for the application of this system to greatly facilitate studies of rhizobial infection and competition for nodule occupancy.

Acknowledgements

We thank Fiona Holt for her expert assistance with microscopy. K.E.G. and K.J.W. thank the Leverhulme Trust for generous financial support. R.A.J. is grateful to the S.G. Hughes Anglo-Italian Fund for Itinerant Scientists. K.J.W. and R.A.J thank S.G.H for temporary bench space and design fluid.

References

1) Giovannoni, S.J., Britschgi, T.B., Moyer, C.L. and Field, K.G. (1990). Genetic diversity in Sargasso Sea Bacterioplankton. Nature 345:60-63.
2) Ward, D.M., Weller, R. and Bateson, M.M. (1990). 16SrRNA sequences reveal numerous uncultured microorganisms in a natural community. Nature 345:63-65.
3) Jefferson, R.A., Burgess, S.M. and Hirsh, D (1986) ß-glucuronidase from *Escherichia coli* as a gene fusion marker. Proc. Natl. Acad. Sci. USA. 86:8447-8451.
4) Jefferson, R.A., Kavanagh, T.A and Bevan, M.W. (1987) GUS fusions: ß-glucuronidase as a sensitive and versatile gene fusion marker in higher plants. EMBO J. 6:3901-3907.
5) Roberts, I.N., Oliver, R.P. Punt, P.J. and van den Hondel, C.A.M.J.J.(1989) Expression of the *Escherichia coli* ß-glucuronidase gene in industrial and phytopathogenic filamentous fungi. Curr. Genet.
6) Sharma, S.B. and Signer, E.R. (1990) Temporal and spatial regulation of the symbiotic genes of *Rhizobium meliloti* in planta revealed by transposon Tn5-*gusA*. *Genes Dev.* 4:344-356.
7) Jefferson, R.A. (1989) The GUS reporter-gene system. Nature 342:837-838
8) Stoeber, F. (1961) Etudes des proprietes et de la biosynthese de la glucuronidase et de la glucuronide-permease chez *Escherichia coli*. These de Docteur es Sciences, Paris.
9) Simon, R., O'Connell, M., Labes, M. and Pühler A. (1986). Meth. Enzymol. 118: 640-659.
10) Wilson, K.J., Nambiar, P.T.C., Anjaiah, V. and Ausubel, F.M. (1987) Isolation and characterization of symbiotic mutants of *Bradyrhizobium* sp. (*Arachis*) strain NC92; mutants with host-specific defects in nodulation and nitrogen fixation. J. Bacteriol. 169:2177-2186.

Section III

PLANT-FUNGUS INTERACTIONS

SPECIFICITY OF PLANT-FUNGUS INTERACTIONS: MOLECULAR ASPECTS OF AVIRULENCE GENES

P.J.G.M. DE WIT, J.A.L. VAN KAN, A.F.J.M. VAN DEN ACKERVEKEN
and M.H.A.J. JOOSTEN.
Department of Phytopathology,
Wageningen Agricultural University,
P.O. Box 8025,
6700 EE Wageningen,
The Netherlands

ABSTRACT. A short overview will be presented on genetic and molecular aspects of fungal avirulence genes. The work on *Bremia lactucae* and *Magnaporthe grisea* will be discussed. However, emphasis will be on the state of the art concerning avirulence gene *Avr9* of the fungal pathogen *Cladosporium fulvum*, the causal agent of tomato leaf mould, which has been used as a model to study gene-for-gene relationships in our laboratory. A race-specific peptide elicitor, the putative product of *Avr9*, induces a hypersensitive response (HR) on Cf9 tomato genotypes. Oligonucleotide probes were designed based on the amino acid sequence. With one of these probes a cDNA clone was isolated which encodes the *Avr9* elicitor. The mRNA contains an open reading frame of 63 amino acids, with the sequence of the mature elicitor at the C-terminus. Under certain growth conditions the *Avr9* mRNA can be induced in liquid shake cultures. The *Avr9* gene appears to be a single copy gene which is absent in fungal races which are virulent on tomato Cf9 genotypes.

1. Introduction

The genetic studies by Flor on flax (*Linum usitatissimum*) and its pathogen, the obligate parasite *Melampsora lini*, which eventually led to the gene-for-gene hypothesis (Flor, 1946) strongly influenced hypotheses and investigations of the basis of host-parasite interactions during the last five decades. Since Flor's discovery, gene-for-gene relationships have been reported for many other plant-fungus relationships (Crute, 1985). Geneticists and molecular plant pathologists have studied many interactions for which a gene-for-gene relationship has been proven or assumed. Although much of the research has been focused on the identification of avirulence (A) genes and resistance (R) genes, the development of our understanding has been slow and it is only recently that we are beginning to understand some of the molecular basis of gene-for-gene relationships. This understanding is most advanced for those interactions involving bacteria and their host plants (see contributions by Keen and Staskawicz in this volume). A brief review will be given on the genetic basis of gene-for-gene systems involving plant pathogenic fungi. A few model systems which have been studied rather extensively at the molecular level in recent years will be discussed in more detail with special emphasis on the molecular aspects of avirulence genes.

233

H. Hennecke and D. P. S. Verma (eds.),
Advances in Molecular Genetics of Plant-Microbe Interactions, Vol. 1, 233–241.

2. Genetic analysis of gene-for-gene relationships

The gene-for-gene relationship as discovered by Flor (1946) was proposed as the simplest explanation of the results of studies on the inheritance of virulence in the flax rust fungus *M. lini*. On varieties of flax, *L. usitatissimum* with one gene for resistance to the avirulent parent race, F2 cultures of the fungus segregated into monofactorial ratios while on varieties with 2, 3 or 4 genes for resistance, the F2 cultures segregated into bi-, tri- or tetrafactorial ratios. This suggests that for each gene that conditions a resistance response in the host, there is a corresponding gene in the fungus conditioning avirulence. Each gene in each member of the plant-fungus system can only be identified by its counterpart in the other member of the system. Resistance and avirulence are frequently reported to inherit in a dominant way (Crute, 1985). Prediction of a gene-for-gene relationship for a plant-fungus interaction is easy when the (a)virulence spectrum of a given race can be established by a selfing study of the fungus on differentials of the host carrying single resistance genes. However, many plant pathogenic fungi have no sexual stage, which means that conventional genetic studies cannot be carried out. In this case the Person analysis can be performed (Person, 1959).

2.1. THE INHERITANCE OF RESISTANCE AND AVIRULENCE

The expectation that in gene-for-gene systems resistance and avirulence are dominant has so widely been accepted that recessive resistance and recessive avirulence would be considered as exceptions to this rule. The investigations of dominance relations in plants are based on a small number of examples from a small number of cases where gene-for-gene interactions have been demonstrated (Barret, 1985). The fact that gene-for-gene interactions have mainly been described for cultivated species indicates that the prevalence of dominant resistance could be a consequence of plant breeding. The breeder will detect dominant resistance easier than recessive resistance and consequently overlook or miss potential recessive resistance sources. Heterozygotes which express any level of resistance are lumped together in one category and will be classified as resistant and dominant. The breeding process thus tends to favour the selection of dominant resistance. Indeed, in the literature a number of clear examples of recessive resistance towards rusts and smuts are described (Christ et al., 1987).

Similarly within the pathogen dominant avirulence has been described more frequently than recessive avirulence. However, under natural conditions the concept is not applicable for plant pathogenic Ascomycetes and Fungi Imperfecti as they are haploid. Therefore no dominance relations can be studied under natural conditions. Studying forced parasexual recombination and transformation can shed some light on dominance or epistatic relationships within these organisms. The few cases where real dominance relationships can be investigated are confined almost exclusively to the plant pathogenic Basidiomycetes, and Oomycetes (Christ et al., 1987; Ilott et al., 1989; Spielman et al., 1989).

2.2. CONTROL OF RESISTANCE AND AVIRULENCE BY TWO GENES

In some plant-fungus interactions, involving rusts, smuts but also Oomycetes such as *Bremia lactucae* and *Phytophthora infestans* there have been indications for the involvement of two genes governing resistance towards a certain race and of two genes matching towards one resistance gene (Christ et al., 1987; Ilott et al., 1989; Spielman et al., 1989). Often, however, this apparent involvement of two genes can be explained by non-allelic interactions between the locus for resistance in the host or the locus for virulence or avirulence in the fungus by so-called dominant inhibitor or suppressor genes resulting in non-functioning of resistance genes, virulence or avirulence genes respectively. Sometimes the presence of inhibitor genes is less clear than initially thought because of uncharacterized background in either host or pathogen (Ilott et al., 1989). The establishment of non-allelic modification of virulence and avirulence genes in a number of plant-fungus interactions suggests that there are different levels to control gene-for-gene interactions. These findings have implications for studies directed towards a further understanding of the molecular basis of gene-for-gene relationships. However, besides all the reported apparent exceptions, the gene-for-gene theory still appears to give an adequate genetic description of most differential host-pathogen interactions. Clearly the existence of inhibitor genes does not weaken the gene-for-gene hypothesis.

3. Models Explaining Molecular Aspects Of Gene-for-Gene Interactions

The development of the methodology especially the genetic manipulation and transformation of fungi and plants, has positively influenced progress in research concerning molecular aspects of gene-for-gene interactions. Unfortunately some genetically well defined gene-for-gene systems are often difficult to study at the molecular level as a number of molecular techniques cannot yet be applied to these systems easily. Table 1 gives a number of desired criteria which make plant-fungus systems amenable for studies at the molecular level.

It is clear that none of the plant-fungus systems presently studied meets all these criteria. Some are more important than others, but lack of an efficient transformation system for one or both partners will greatly hamper progress in the long term. Eventually, one has to prove the role of a putative resistance or avirulence gene by transforming it back into the host or pathogen, respectively. In practice, absence of the desired characteristics has limited molecular research to only a few host-pathogen systems. Despite the fact that the genetics of plant-bacterium interactions have not been studied as extensively as plant-fungus interactions, molecular studies on plant pathogenic bacteria have expanded rapidly recently, mainly because bacteria have small genomes and can be transformed efficiently. The availability of molecular techniques made it possible to dissect the bacterial genome of certain pathovars or species and express their genomes part by part in another background with the result of discovering new avirulence genes (Keen and Staskawicz, 1988). Similarly dissection of plants with small genomes such as *Arabidopsis thaliana* will probably generate a number of yet unknown resistance genes to pathogens of the plant and to pathogens in other crop plants (Somerville, 1989).

TABLE 1. Characteristics which make a plant-pathogen system amenable for studies at the molecular level.

1. Host and pathogen are genetically well defined.
2. Host and pathogen can easily be cultured and have a short generation cycle.
3. The sexual cycle of both host and pathogen can easily be studied.
4. Host and pathogen can be transformed in order to introduce new genes or inactivate existent ones.
5. Host and pathogen have a relatively small genome.
6. The existence of putative products of either avirulence or resistance genes.

3.1. *BREMIA LACTUCAE*-LETTUCE

Bremia lactucae is an Oomycete which is genetically very well characterized (Ilott et al., 1987; Ilott et al., 1989; Hulbert et al., 1988). The major determinants of specificity are involved in a gene-for-gene interaction (Judelson and Michelmore, 1989). One of the goals of the research group of Michelmore is to identify and clone avirulence genes of *B. lactucae*. The major disadvantage of this obligate parasite is that until now no DNA-mediated transformation has been described. The genetics of both host and pathogen have been analysed simultaneously (Michelmore et al., 1984; Hulbert et al., 1988; Farrara et al., 1988). Thirteen single dominant genes for avirulence have been characterized in *B. lactucae*. These loci are matched by 13 dominant loci for resistance to downy mildew (*Dm*) in lettuce. There is clear evidence for the existence of one dominant inhibitor of avirulence epistatic to *Avr5/8*. Other inhibitor genes or genes of minor effect may exist but they do not alter the specificity determined by the major avirulence and resistance genes. Very little is known about putative products of avirulence genes conferring an incompatible response in the host cultivars carrying the appropriate resistance genes. The shotgun approach to clone avirulence genes from bacteria as used by Staskawicz et al. (1984) and others cannot be used for *B. lactucae*. Restriction fragment length polymorfism (RFLP) markers that are linked to avirulence genes have been identified (Judelson and Michelmore, 1989; Hulbert et al., 1988). These probes will be used as starting points of chromosome walks to the avirulence genes using a cosmid library. Clones of the fungus containing putative avirulence genes will be transformed into virulent races as soon as a transformation system becomes available. By this approach only a limited number of clones will need to be introduced into *B. lactucae* rather than the large numbers required for the shotgun approach. By using RFLP markers one linkage of 6.5 cM between the RFLP probe *G538* and avirulence gene *Avr6* has been identified. By assuming a constant relationship between physical and genetic distance 6.5 cM equals approximately 160 kb in *B.lactucae*. Each walking step through this region will provide new polymorphic probes which can be used to estimate the proximity to the avirulence gene and the direction of the walk. Also pulsed-field gel electrophoretic separation of chromosomes of *B. lactucae* could facilitate chromosome walking.

However, independent of the approach used to clone an avirulence gene there is need to develop an efficient transformation system for *B. lactucae*. Many of the methods used to transform other fungi cannot be used for *B. lactucae* because of its strict obligate nature and possibly because Oomycetes are quite different from Ascomycetes and Basidiomycetes (Michelmore et al., 1988). To achieve transformation of *B. lactucae*, vectors need to be constructed that will express markers for selection. The DNA must be introduced into intact spores or germlings and the selection must work *in planta*.

3.2. *MAGNAPORTHE GRISEA*-RICE

Magnaporthe grisea is a fungal pathogen which has received much attention in recent years as a model pathogen in molecular plant pathology by several research groups. Cultivars of rice that carry single dominant resistance genes effective against certain races of the pathogen have been developed by Yamada et al. (1976). Genetic studies with the pathogen were previously impossible due to lack of fertile isolates of *M. grisea* that infect rice. Presently genetic studies are performed by crossing a sterile field isolate pathogenic on rice and on weeping lovegrass with a highly fertile strain that is only pathogenic on weeping lovegrass. The progeny that were still pathogenic on rice were backcrossed several times with the sterile rice/weeping lovegrass isolate as recurrent parent (Valent and Chumley, 1989; Leung et al., 1988). By checking the progeny on differentials of rice several avirulence genes were identified (Valent and Chumley, 1989). The identified avirulence genes *avrC039*, *avrM201* and *avrIYM* appeared to be inherited from the parent that is nonpathogenic on rice, as the rice pathogen parent is pathogenic on the three rice cultivars

CO39, M201 and Yashiro-mochi, respectively. A similar result was obtained by Yaegashi and Asaga (1981) who reported that a finger millet pathogen carries an avirulence gene corresponding to the *M. grisea* resistance gene Pi-a. Additional crosses have identified other avirulence genes: *avr2YM* and *Pw12* which appear to be unstable. From strains carrying those avirulence genes spontaneous virulent mutants often appear in standard differential assays.

Whether the identified avirulence genes are dominant or not remains to be seen. *M. grisea* is a haploid organism and presently it is impossible to obtain stable vegetative diploids or heterokaryotic conidia (Crawford et al., 1986). Linkage of RFLPs to avirulence genes will be one of the first steps towards cloning of avirulence genes. In this respect *M. grisea* is easier to handle than *B. lactucae* as the former can be transformed efficiently (Parsons et al., 1987). On the other hand transformation of the host plant rice is more difficult than transformation of lettuce. Transformation of rice is a prerequisite to eventual isolation and characterization of the corresponding resistance genes. There is yet much to be done before the cloning of avirulence genes and the isolation of the corresponding resistance genes in the interactions of *B. lactucae*-lettuce and *M. grisea*-rice become a reality.

3.3. *CLADOSPORIUM FULVUM*-TOMATO

Cladosporium fulvum is a biotrophic fungal parasite which enters tomato leaves through stomata, colonizes the intercellular spaces between mesophyll cells and is confined to the apoplast during the main part of its life cycle. The fungus does not form haustoria and causes no visible damage to the mesophyll cells and their walls. The interaction between *C. fulvum* and tomato is supposed to fit into a gene-for-gene relationship (De Wit and Oliver, 1989). Many genes for resistance have been identified in tomato cultivars. Of these resistance genes, Cf2, Cf4, Cf5, and Cf9 are available in near-isogenic lines of the cultivar Moneymaker. These lines give a clear differential response to the presently known races, resulting either in a compatible or an incompatible interaction. The *C. fulvum*-tomato interaction is an ideal model system to study communication between plant and pathogen, as the interface exchange of molecules is confined to the apoplast, from which washing fluids can be easily obtained (De Wit and Oliver, 1989; De Wit and Spikman, 1982; De Wit et al., 1984; De Wit et al., 1985). Apoplastic fluids of infected leaves can be obtained by *in vacuo* infiltration with water or buffer followed by low speed centrifugation. In the apoplastic fluids of *C. fulvum*-infected leaves, in addition to many other compounds, fungal proteins occur which are constitutively produced or specifically induced by the plant. The fungal proteins can be divided in two catagories. The first are those proteins important for obtaining or establishing basic compatibility (De Wit et al., 1986; Joosten and De Wit, 1988). The second category contains proteins which play a role as inducers of HR in resistant cultivars of tomato, the putative products of avirulence genes. Apoplastic fluids obtained from different compatible race-cultivar interactions have been shown to contain race-specific elicitors which specifically induce HR in tomato cultivars carrying the corresponding genes for resistance (De Wit and Spikman, 1982; De Wit et al., 1984; De Wit et al., 1985; Scholtens-Toma et al., 1988, 1989) Here the state of the art concerning the biochemical and molecular characterization of one of the race-specific elicitors and the cloning of its encoding avirulence gene *Avr9* will be discussed in detail.

3.3.1. *Cloning and Characterization of* Avr9.
The intercellular fluids isolated from different compatible *C. fulvum*-tomato interactions contain race-specific proteinaceous elicitors (De Wit and Spikman, 1982; De Wit et al., 1984; Higgins and De Wit,1985; De Wit et al., 1985;). One of these elicitors, the putative product of avirulence gene *Avr9* which specifically interacts with a product of resistance gene Cf9, has been purified and its amino acid sequence has been determined (Scholtens-Toma and De Wit, 1988). It was isolated from apoplastic fluids of compatible interactions with races carrying *Avr9*. The elicitor was

not detectable in compatible interactions involving race 2.4.5.9, which does not carry *Avr9*. Recently we have found that new Dutch, French and Polish isolates of *C. fulvum* carrying virulence gene a9 (race 2.5.9, 2.4.9.11 and 2.4.5.9.11, respectively) also do not produce the necrosis-inducing peptide indicating that absence of this peptide and virulence towards Cf9 genotypes is strongly associated (Scholtens-Toma et al., 1989). The production of the peptide was strongly induced in the host plant and could not be detected in culture filtrates of the fungus grown under optimal conditions *in vitro*. The peptide could also not be detected in incompatible interactions, presumably because it is only present in very low amounts and it may bind to receptor sites as soon as it is produced.

To clone the *Avr9* gene of *C. fulvum* an approach significantly different from that used to clone bacterial avirulence genes has been followed. As the amino acid sequence of the race-specific elicitor, the putative product of *Avr9*, is known (Scholtens-Toma and De Wit, 1988) the approach from protein to gene was followed rather than the shotgun approach. A degenerated oligonucleotide probe derived from the amino acid sequence was used to screen a cDNA library of a compatible *C. fulvum*-tomato interaction involving a race carrying *Avr9*. The cDNA library was prepared from the interaction at the time that there was a high production of the race-specific elicitor. From this cDNA library one clone was obtained with an insert that contained the entire amino acid sequence of the race-specific elicitor as determined previously (Scholtens-Toma and De Wit, 1988). Analysis of the cDNA clone revealed that the *Avr9* encodes a protein of 63 amino acids, containing the sequence of the mature elicitor at the C-terminus (Van Kan et al., 1990). Southern blot analysis of DNA isolated from races of *C. fulvum* with and without *Avr9* revealed that the cDNA clone hybridized to single bands in various restriction enzyme digests of the races with *Avr9*, i.e. races 0, 2, 4, 5, 2.4, 2.5, 2.4.11, 2.4.5. and 2.4.5.11 respectively, but not to DNA from races which lack *Avr9*, i.e. races 2.5.9, 2.4.5.9, 2.4.9.11 and 2.4.5.9.11, respectively (Van Kan et al., 1990). Apparently the races that are virulent on tomato genotypes that carry resistance gene Cf9 completely lack the *Avr9* gene. This indicates that there is no evidence for a recessive allele of *Avr9* present in races which are virulent on Cf9 genotypes. From these results it may be concluded that *Avr9* is located on an episomal factor, on a B-chromosome or on an unstable part of a chromosome. A genomic clone of *Avr9* has been sequenced and it appeared that the gene contains an intron of 59 bp in the coding region of the gene. The splicing junctions (AT..AG) and consensus sequence TACTAAC are both present in the intron. The region upstream of the transcription start contains a possible TATA-box approximately at position -30 and two repeats further upstream. The importance of these elements in the regulation of expression of the *Avr9* gene will be studied by making deletions and testing for promotor activity. Direct repeats and inverted repeats (RNA) are present downstream of the coding sequence and might play a role in termination of transcription. *Avr9* is the first fungal avirulence gene ever cloned. It will be interesting to know whether the other avirulence genes inducing HR on tomato genotypes carrying resistance genes Cf2, Cf4 and Cf5 can be lost from the *C. fulvum* genomes in a similar way as *Avr9*. If this would be the case then the term recessive allele is misleading and may not be applicable to *C. fulvum*.

By growing the fungus under different conditions in shake cultures it appeared from preliminary results that the *Avr9* mRNA is generally not expressed *in vitro* except under conditions of low nitrate.

3.3.2. *Transformation of Virulent Races of* C.fulvum *with* Avr9. A further proof for having cloned the *Avr9* will be obtained by transforming virulent races of *C. fulvum* such as race 2.5.9, 2.4.5.9, 2.4.9.11. and 2.4.5.9.11 with the clone carrying the *Avr9*. Transformants should become avirulent on tomato genotypes carrying resistance gene Cf9. A transformation system for *C. fulvum* has been established (Oliver et al., 1987; Roberts et al., 1989). As mentioned above the *Avr9* is completely lacking in the virulent races 2.5.9, 2.4.5.9, 2.4.9.11 and 2.4.5.9.11. There are some preliminary indications that the *Avr9* is located on part of a chromosome which can be lost quite easily. A race lacking *Avr9* carries a chromosome which seems to be 500 kb shorter than the corresponding chromosome in the races carrying *Avr9* (Oliver et al., unpublished results). It will be interesting to know whether position effects play a role for the functioning of the *Avr9* once introduced in the transformants. It is also of interest to find out whether the 500 kb fragment on which the *Avr9* gene appears to be located contains other important functions related to avirulence. A disadvantage of *C. fulvum* is the lack of a sexual stage which makes it impossible to perform conventional genetic studies. However, it seems possible to perform parasexual studies with this fungus as reported by Talbot et al. (1988a), who constructed diploids by protoplast fusion. To this end a small collection of selectable mutants has been made using UV and transformation (Talbot et al., 1988a,b). Regeneration frequencies of the double-selected phenotypes were in the range of 10* to 10*. Some stable diploids were obtained, but in most cases spontaneous haploidisation occurred. Preliminary evidence suggests that haploids are recombinant but heterokaryosis has not been ruled out.

A retrotransposon-like element has been detected in the genome of *C. fulvum* (McHale et al., 1989). If this element proves to be a classical retrotransposon, it may help explain the ability of fungal races to mutate in order to overcome plant resistance genes. However, this conclusion is rather preliminary as transposition of this element still has to be proven.

4. Concluding Remarks

Many gene-for-gene relationships have been reported of which some are genetically well characterized. However, only a few of them have become amenable to study at the molecular level. Data of molecular genetic studies on model plant-fungus interactions reported here support the gene-for-gene hypothesis as has been put forward by Flor nearly fifty years ago. Many bacterial avirulence genes and one fungal avirulence gene have been cloned until now, but he first resistance gene has to be cloned yet. Based on the data obtained, the specific elicitor-receptor model is still functional to explain gene-for-gene interactions at the molecular level (Keen, 1982; De Wit, 1987). However, resistance genes and their products need to be identified and their interaction with specific elicitors needs to be studied before the specific elicitor-receptor model can be fully accepted.

5. Acknowledgements:

The authors' research is supported by grants from the EEC (in the framework of the Biotechnology Action Programme; contract no. BAP-0074-NL) and the Foundation for Biological Research (BION), which is subsidized by the Netherlands Organisation for Scientific Research (NWO).

6. References

Barret, J.A. (1985) 'The gene-for-gene hypothesis: parable or paradigm', in D.Rollinson and R.M. Anderson (eds.), Ecology and Genetics of Host-Parasite Interactions.(Linnean Society of London), Academic Press, New York-Sydney-Tokyo, pp. 215-225.

Christ, B.J., Person, C.O. and Pope, D.D. (1987) 'The genetic determination of variation in pathogenicity', In M.S. Wolfe and C.E. Caten (eds.), Populations of Plant Pathogens, Their Dynamics and Genetics. (Britisch Society of Plant Pathology), Blackwell Scientific

240

Publications, Oxford-London-Boston, pp. 21-37.

Crawford, M.S., Chumley, F.G., Weaver, C.G. and Valent, B. (1986) 'Characterization of the heterokaryotic and vegatative diploid phases of Magnaporthe grisea, Genetics 114, 1111-1129.

Crute, I.R. (1985) 'The genetic bases of relationships between microbial parasites and their hosts', in R.S.S. Fraser (ed.), Mechanisms of Resistance to Plant Diseases. Martinus Nijhoff/Dr. W. Junk Publishers, Dordrecht-Boston-Lancaster, pp. 80-142.

De Wit, P.J.G.M. (1987) 'Specificity of active resistance mechanisms in plant-fungus interactions', in G.F. Pegg and P.G. Ayres (eds.), Fungal Infection of Plants, Cambridge University Press, Cambridge-New York-Sydney pp.1-24.

De Wit, P.J.G.M., Buurlage, M.B. and Hammond, K.E. (1986) 'The occurrence of host, pathogen and interaction-specific proteins in the apoplast of *Cladosporium fulvum (syn. Fulvia fulva)* infected tomato-leaves', Physiol. Mol. Plant Pathol. 29, 159-172.

De Wit, P.J.G.M., Hofman, J.E. and Aarts, J.M.M.J.G. (1984) 'Origin of specific elicitors of chlorosis and necrosis occurring in intercellular fluids of compatible interactions of *Cladosporium fulvum (syn. Fulvia fulva)* and tomato', Physiol. Plant Pathol. 24, 17-23.

De Wit, P.J.G.M., Hofman, J.E., Velthuis, G.C.M. and Kuc, J.A. (1985) 'Isolation and characterization of an elicitor of necrosis isolated from intercellular fluids of compatible interactions of *Cladosporium fulvum (syn. Fulvia fulva)* and tomato', Plant Physiol. 77, 642-647.

De Wit, P.J.G.M. and Oliver, R.P. (1989) 'The interaction between *Cladosporium fulvum (syn. Fulvia fulva)* and tomato: a model system in molecular plant pathology', in H. Nevalainen and M. Penttilä (eds.), Molecular Biology of Filamentous Fungi, Found. Biotechn. Industr. Ferment. Res. 6, 227-236.

De Wit, P.J.G.M. and Spikman, G. (1982) 'Evidence for the occurrence of race and cultivar-specific elicitors of necrosis in intercellular fluids of compatible interactions of *Cladosporium fulvum* and tomato', Physiol. Plant Pathol. 21, 1-11.

Farrara, B.T., Illot, T.W. and Michelmore, R.W. (1988) 'Genetic analysis of factors for resistance to downy mildew *(Bremia lactucae)* in lettuce *(Lactuca sativa)*', Plant Pathol. 36, 499-514.

Flor, H.H. (1946) 'Genetics of pathogenicity in *Melampsora lini*', J. Agr. Res.73, 335-357.

Higgins, V.J. and De Wit, P.J.G.M. (1985) 'Use of race and cultivar specific elicitors from intercellular fluids for characterizing races of *Cladosporium fulvum* and resistant tomato cultivars', Phytopathol. 75, 695-699.

Hulbert, S.H., Ilott, T.W., Legg, E.J., Lincoln, S.E., Lander, E.S. and Michelmore, R.W. (1988) 'Genetic analysis of the fungus *Bremia lactucae*, using restriction fragment length polymorfisms', Genetics 120, 947-958.

Ilott, T.W., Hulbert, S.H. and Michelmore, R.W. (1989) 'Genetic analysis for the gene-for-gene interaction between lettuce *(Lactucae sativa)* and *Bremia lactucae*', Phytopathol. 79, 888-897.

Joosten, M.H.A.J. and De Wit, P.J.G.M. (1988) 'Isolation, purification and preliminary characterization of a protein specific for compatible *Cladosporium fulvum (syn. Fulvia fulva)*-tomato interactions', Physiol. Mol. Plant Pathol. 33, 142-253.

Judelson, H.S. and Michelmore R.W. (1989) 'Strategies for cloning avirulence genes from *Bremia lactucae*', in B. Staskawicz , P. Ahlquist and O. Yoder (eds.), Molecular Biology of Plant-Pathogen Interactions. UCLA Symposia on Molecular and Cellular Biology 101, Alan R. Liss Inc., New York, pp.71-85.

Keen, N.T. (1982) 'Specific recognition in gene-for-gene host parasite systems', Adv. Plant Pathol. 1, 35-82.

Keen, N.T. and Staskawicz B.J. (1988) 'Host range determinants in plant pathogens and symbionts', Ann. Rev. Microbiol. 42, 421-440.

Leung, H., Borromeo, E.S., Bernardo, M.A. and Notteghem, J.J. (1988) 'Genetic analysis of virulence in the rice blast fungus Magnaporthe grisea', Phytopathol. 78, 1227-1233.

McHale, M.T., Roberts, I.N., Talbot, N.J. and Oliver, R.P. (1989) 'Expression of reverse transcriptase genes in *Fulvia fulva*, Mol. Plant Microb. Int. 2, 165-168.

Michelmore, R.W., Ilott, T.W., Hulbert, S.H. and Farrara, B. (1988) 'The downey mildews', Adv. Plant Pathol. 6, 53-79.

Michelmore, R.W., Norwood, J.M. Ingram, D.S., Crute, I.R. and Nicholson, P. (1984) 'The inheritance of virulence in *Bremia lactucae* to match resistance factors 3, 5, 6, 8, 9, 10 and 11 in lettuce *(Lactuca sativa)*', Plant Pathol. 13, 301-315.

Oliver, R.P., Roberts, I.N., Harling, R., Kenyon, L., Punt, P.J., Dingemanse, M.A. and Van den Hondel, C.A.M.J.J. (1987) 'Transformation of *Fulvia fulva*, a fungal pathogen of tomato, to hygromycin B resistance', Curr. Genet. 12, 231-233.

Person, C.O. (1959) 'Gene-for-gene relationships in host:parasite systems', Can. J. Bot. 37, 1101-1130.

Parsons, K.A., Chumley, F.G. and Valent, B. (1987) 'Genetic transformation of the fungal pathogen responsible of rice blast disease', Proc. Natl. Acad. Sci. USA 84, 4161-4165.

Roberts, I.N., Oliver, R.P., Punt, P.J. and van den Hondel, C.A.M.J.J. (1989) 'Expression of the E. coli ß-gluruconidase gene in filamentous fungi', Curr. Genet. 15, 177-180.

Scholtens-Toma, I.M.J. and De Wit, P.J.G.M. (1988) 'Purification and primary structure of a necrosis inducing peptide from apoplastic fluids of tomato infected with *Cladosporium fulvum (syn. Fulvia fulva)*', Physiol. Mol. Plant Pathol. 33, 59-67.

Scholtens-Toma. I.M.J., De Wit, G.J.M. and De Wit, P.J.G.M. (1989) 'Characterization of apoplastic fluids isolated from tomato lines inoculated with new races of *Cladosporium fulvum*', Neth. J. Plant Pathol. 95, 161-168.

Somerville, C. (1989) '*Arabidopsis* blooms', Plant Cell 1, 1131-1135.

Spielman, L.J., McMaster, B.J. and Fry, W.E. (1989) 'Dominance and recessiveness at loci for virulence against potato and tomato in *Phytophthora infestans*', Theor. Appl. Genet. 77, 832-838.

Staskawicz, B.J., Dahlbeck, D. and Keen, N.T. (1984) 'Cloned avirulence gene of *Pseudomonas syringae* pv. glycinea determines race-specific incompatibility on *Glycine max* (L.) Merr', Proc. Natl. Acad. Sci. USA 81, 6024-6028.

Talbot, N.J., Coddington, A., Roberts, I.N. and Oliver, R.P. (1988a) 'Diploid construction by protoplast fusion in *Fulvia fulva (syn. Cladosporium fulvum)*: genetic analysis of an imperfect fungal plant pathogen', Curr. Genet. 14, 567-572.

Talbot, N.J., Rawlins, D. and Coddington, A., (1988b) 'A rapid method for ploiding determination in fungal cells', Curr. Genet. 14, 51-52.

Valent, B. and Chumley, F. (1989) 'Genes for cultivar specificity in the rice blast fungus, *Magnaporthe grisea*', in B.J.J. Lugtenberg (ed.), Signal Molecules in Plants and Plant-Microbe Interactions, Springer, Berlin-Heidelberg-New York, pp.415-422.

Van Kan, J.A.L., Van Den Ackerveken, A.F.J.M. and De Wit, P.J.G.M. (1990) 'Cloning and characterization of the avirulence gene *Avr9* of the fungal pathogen *Cladosporium fulvum*', Mol. Plant Microb. Int. in press.

Yaegashi, H. and Asaga, K. (1981) 'Further studies on the inheritance of pathogenicity in crosses of *Pyricularia oryzae* with *Pyricularia* sp. from finger millet', Ann. Phytopath. Soc. Jpn. 47, 677-679.

Yamada, M., Kiyosawa, S., Yamamuchi, T., Hirano, T., Kobayashi, T., Kushibuchi, K. and Watanabe, S. (1976) 'Proposal of a new method for differentiating races of *Pyricularia oryzae* Cavara in Japan', Ann. Phytopathol. Soc. Jpn. 42, 216-219.

MUTUAL TRIGGERING OF GENE EXPRESSION IN PLANT-FUNGUS INTERACTIONS

P. E. Kolattukudy, G. K. Podila, B. A. Sherf, M. A. Bajar and
R. Mohan
The Ohio State University
Ohio State Biotechnology Center
206 Rightmire Hall
1060 Carmack Road
Columbus, Ohio 43210

Abstract

Plant-fungus interactions that ultimately result in disease development or resistance involve mutual triggering of gene expression. Penetration of pathogenic fungi into the plant through the cuticle is achieved by the production of cutinase. The transcription of cutinase gene is induced by the unique cutin monomers. Transformation of Fusarium solani pisi and deletion analysis identified monomer-inducible cutinase promoter activity in a specific 135 bp region. The same 5'- flanking region showed binding to the protein factor required for cutinase transcription activation in isolated fungal nuclei. The protein factor requires phosphorylation for binding and a protein kinase C appears to be involved. Fungal invasion triggers expression of many plant genes including those involved in reinforcing the host cell walls by making them resistant to the fungal hydrolytic enzymes. To achieve this the cell walls are suberized. A highly anionic peroxidase plays a crucial role in this process. The expression of the host gene coding for this enzyme is triggered by the fungal signals in the host that is resistant to the fungal attack but not in the susceptible host. The wound-inducible expression of this peroxidase gene in transgenic plants has been demonstrated.

Introduction

Molecular communications between the fungal spore and the plant begin soon after the two come into contact with each other. In the earliest period the fungal spore senses contact with the plant, probably via the chemical and/or physical features of the plant surface. These interactions trigger germination of the fungal spore and direct penetration into the plant, or in some cases, subsequent differentiation of the infection structure called the appressorium which gives rise to the infection peg that penetrates into the plant. In either case, the fungus must penetrate the outer barriers of plants to infect the plant. The outermost barrier is the cuticle which is composed of a structural component, cutin, consisting of hydroxy and epoxy fatty acids and associated soluble waxes (1).

We shall discuss the recent progress in our understanding of the molecular mechanisms by which the fungus perceives contact with the plant and how the plant-derived signal activates transcription of the fungal cutinase to assist the fungus to breach the cuticular barrier. Once the fungus penetrates through the cuticle and the underlying carbohydrate barriers, the products generated in this process are perceived by the plant as signs of fungal attack. The fungal signals, in turn, trigger expression of plant defense genes. One such defense reaction involves reinforcement of plant cell walls to make them resistant to fungal extracellular enzymes. We shall briefly discuss recent findings that fungal signals trigger the expression of the plant gene encoding a highly anionic peroxidase that is involved in reinforcement of cell walls by suberization; the ability of the plant to respond in this manner confers resistance to fungal disease.

H. Hennecke and D. P. S. Verma (eds.),
Advances in Molecular Genetics of Plant-Microbe Interactions, Vol. 1, 242–249.
© 1991 *Kluwer Academic Publishers. Printed in the Netherlands.*

INDUCTION OF CUTINASE BY CUTIN MONOMERS

Pathogenic fungi penetrate the cuticular barrier using cutinase that is induced by the cutin monomers generated by the small amount of cutinase carried by the spore when it lands on its host (Fig. 1) (1). How the unique monomers of cutin trigger cutinase gene expression in the fungus is not well understood. Nuclear run-off experiments showed that the hydroxy fatty acid monomers of cutin regulated cutinase gene expression at a transcriptional level (2). In isolated fungal nuclei, cutinase gene transcription was activated by a fungal protein factor and cutin monomer (3). This selective activation of cutinase gene transcription was inhibited by novobiocin, a selective inhibitor of transcription initiation, whereas this compound had no effect on the cutinase transcript production in the nuclear run-off experiments with nuclei from already-induced cells. That the transcription of cutinase gene initiated by the cutin monomer and protein factor in the nuclei from uninduced fungal cultures was properly terminated was shown by the observation that the cutinase transcript generated *in vitro* by the nuclei was identical in size to the *in vivo* generated mRNA. Studies with analogues of cutin monomers showed that activation of fungal cutinase gene expression observed in the isolated nuclei required the structural features naturally found in cutin. Thus, this *in vitro* system provides a reliable test system with which to study the mechanism by which the plant signal triggers fungal gene expression.

The role of the protein factor in enhancing cutinase gene transcription is not understood. Fractionation of the protein factor in the presence of low and high ionic strength indicated that low ionic strength caused aggregation, leading to high molecular weight species that were excluded even from Sepharose 2B. At high ionic strength, the protein factor that showed activation of cutinase gene transcription behaved as a 100 kD protein (4). This protein fraction showed binding to a 360 bp segment 5'- flanking region of cutinase gene from F. solani pisi, whereas no binding to the 3'- flanking region was detected. This protein fraction, together with the cutin monomer, caused phosphorylation of a 55 kD protein in the presence of fungal nuclei. It is likely that the protein factor is of nuclear origin and the need for addition of this protein for optimal activation of transcription arose because the protein is lost from the nuclei during isolation of nuclei. If this is so, the nuclear protein that is phosphorylated might represent the transacting factor required to trigger cutinase gene expression.

Figure 1. Schematic representation of how the plant induces cutinase in a fungal spore.

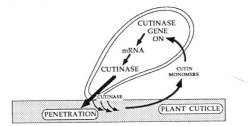

IDENTIFICATION OF CUTINASE PROMOTER INDUCED BY CUTIN MONOMERS

To identify the promoter region of cutinase, the 5'- flanking region was fused 5'- to a phygromycin resistance gene containing no other promoter and this plasmid was introduced into F. solani pisi mycelia by the Li salt method. Formation of stable transformants, selected on the basis of hygromycin resistance, showed that the 5'- flanking region of cutinase had promoter activity (5). Even though deletion of the 5'- end of cutinase gene could be tested as a means of identifying the promoter region, we could not detect inducible cutinase promoter. Therefore, the 5'- flanking region of cutinase gene was placed upstream from a marker gene, chloramphenicol acetyl transferase (CAT), so that promoter strengths could be measured quantitatively. The same plasmid also contained the hygromycin resistance gene driven by a constitutive Cochliobolus promoter (Fig. 2). Such a plasmid containing both features was introduced into the protoplasts of F. solani pisi by electrophoration. Upon selection, on hygromycin-containing medium, 20 stable transformants were obtained per μg of DNA used for electrophoration (A. Bajar, G. K.

244

Podila and P. E. Kolattukudy, manuscript in preparation). Virtually all transformants obtained in the first selection on hygromycin turned out to be mitotically stable. Transformants generated with 1.5 kb 5'- flanking region of cutinase gene showed CAT activity that was inducible by cutin monomers and this induction was repressed by glucose. A 450 bp 5'- flanking region of the cutinase gene also showed inducible expression of CAT and glucose repression. Shortening of the promoter by selected restriction enzymes showed that a 360 bp cutinase promoter conferred inducibility by cutin monomers and glucose repression to CAT, whereas deletion of the next 135 bp of 5'- end abolished inducible CAT expression. (Fig. 2). This 135 bp segment contained one cAMP-responsive element, the only one found in the 3 kb of DNA including cutinase gene and its flanking regions. This segment that was required to confer inducible CAT expression also contained a Sp 1 transcription factor binding element and a nested 11nt direct repeat. Even though functional significance of each of these features has not been demonstrated, the fact that these elements are uniquely present in the segment of DNA that is required to confer inducibility by cutin monomers suggests a functional role. In further support of this conclusion, a cAMP-responsive element, Sp 1 binding site and nested directed repeat were also found in the 5'-flanking region of cutinase gene from Colletotrichum gloeosporioides and C. capsici.

Figure 2. Promoter analysis of cutinase gene from Fusarium solani pisi. CAT activity of Fusarium transformants containing the indicated segments of the cutinase promoter were assayed after induction by the addition of cutin hydrolysate.
(I, induced; U, uninduced)

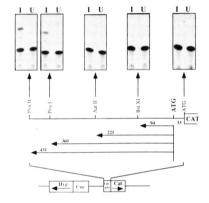

TRANSACTING FACTOR(S) IN CUTINASE INDUCTION

To test whether the transacting factors suspected to be present in the protein factor/nuclear extract can bind to the inducible promoter identified above by transformation, DNA binding studies were performed using gel retardation assays. The 360 bp promoter segment of cutinase showed a retardation band when incubated with the nuclear protein (A. Bajar, G. K. Podila and P. E. Kolattukudy, unpublished results). When an excess of unlabeled 360 bp promoter was included, no band retardation was observed, showing that the nuclear protein bound to the 360 bp DNA fragment. When 135 bp from the 5'- end of this DNA was removed by Pvu 1 cleavage, the resulting 225 bp segment did not bind to the nuclear protein. Thus, the same 135 bp fragment that is essential for conferring inducibility by cutin monomers is also involved in binding the transacting factor(s).

In order to test whether the cutin monomer affects the state of phosphorylation of a nuclear protein and thus influences its interaction with cutinase gene, the effect of cutin monomer was tested on phosphorylation. In fact, cutin monomer addition caused phosphorylation of a 50 kD nuclear protein and the protein factor promoted the phosphorylation of this 50 kD protein. This phosphorylation was inhibited by H7 that preferentially inhibits protein kinase C. Less inhibition was observed with genestein, a selective inhibitor of protein tyrosine kinase. Analysis of the hydrolysis products of the phosphorylated proteins by thin-layer chromatography revealed the presence of phosphoserine and phosphotyrosine. In view of the fact that even in mammalian tissues where the induction of tyrosine phosphorylation is well-known, phosphotyrosine constitutes only a fraction of a percent of the total phosphoamino acids; therefore it is possible that in the fungal system phosphotyrosine level could have been below the limits of detection.

If the phosphorylation of the protein caused by cutin monomer is relevant to regulation of cutinase gene expression, DNA binding and promoter recognition might be affected by the state of phosphorylation of this 50 kD protein. To test this possibility, the nuclear protein fraction was treated with immobilized phosphatase before incubation with DNA. Upon electrophoresis, no retardation band was observed, whereas untreated nuclear protein clearly showed retardation of mobility of the labeled 360 bp promoter. Therefore, it is concluded that phosphorylation of a transacting factor induced by cutin monomer is required for the binding of the factor to the cutinase promoter. Proof that the protein whose phosphorylation is promoted by the cutin monomer is the same as that which binds to the promoter is not yet available.

To test whether protein phosphorylation is relevant to activation of cutinase gene transcription, the effect of protein kinase inhibitors was tested on activation of transcription of cutinase gene in isolated nuclei from $\underline{F}$. solani pisi. When the nuclear preparation was incubated with protein factor and cutin monomer, there was usually a 30 minute lag before linear rates of cutinase gene transcription could be observed (3). Obviously the biochemical events that occur during this early period are critical to activation of cutinase gene transcription. It appears likely that one of these crucial biochemical events involve phosphorylation of transacting factor(s). When a protein kinase inhibitor, H_7, was present during this initial 30 minute period before transcription was initiated by the addition of the necessary reagents including labeled nucleotide, cutinase transcription was severely inhibited. On the other hand, addition of the inhibitor after the initial pre-incubation had little effect on cutinase transcription. Similarly, the presence of genestein during the pre-incubation period caused severe inhibition of cutinase transcription; in this case, presence of genestein only during the transcription period also was inhibitory to cutinase transcription. Presence of phosphotyrosine antibodies during pre-incubation inhibited cutinase transcription, whereas addition of the antibodies after the pre-incubation period had little effect. These results strongly suggest that the early biochemical events that result in linear increase in transcription in the isolated nuclei include phosphorylation of protein(s), possibly at tyrosine residues.

The results thus far obtained strongly suggest that the unique hydroxy acids of cutin cause phosphorylation of a transcription factor, resulting in the binding of the factor to the cutinase promoter, either alone or in combination with other transcription factors (6). Since the molecular weight of the protein factor required for cutinase transcription activation showed a native molecular weight of about 100k and the phosphoprotein showed molecular weight of about 50k by SDS gel electrophoresis, it appears possible that the transacting factor involved in the regulation of cutinase gene transcription by the unique hydroxy acids of cutin is a dimer, as is often found with transcription factors involved in regulation by hydrophobic regulators (6).

PLANT DEFENSE BY CELL WALL REINFORCEMENT

Once the invading fungus gets through the cuticle, it encounters the carbohydrate barrier. Fungal extracellular enzymes are used to get through the barrier. For example, germinating $\underline{F}$. solani pisi spores produced polygalacturonase that reached a maximal level at the time of peak germination. Two polygalacturonases and a pectate lyase were isolated from $\underline{F}$. solani pisi and characterized (Table 1). The lyase was demonstrated to be essential for pathogenesis. Thus, inclusion of antibodies prepared against the lyase in the spore suspension placed on pea stem prevented infection, whereas control serum promoted infection (7). Similar experiments with antibodies prepared against the individual polygalacturonases failed to protect the host (M. Crawford and P. E. Kolattukudy, unpublished). It is possible that when multiple enzymes can break down the carbohydrate barriers, a test for the involvement of individual enzymes would not be conclusive. In any case, the lyase appears to be essential for pathogenesis.

246

Table 1. Pectin Degrading Enzymes from Fusarium solani pisi

Enzyme	Mol. weight	pI	pH Optimum	Type
Anionic Hydrolase	45 kD	5.0	5.5	Endo
Cationic Hydrolase	58 kD	8.0	5.5	Exo
Lyase	26 kD	8.3	8.5	Endo

Mark Crawford and P.E. Kolattukudy, unpublished results.

The oligosaccharides generated by the lyase would be expected to trigger a variety of defensive reactions in the plant. One such reaction involves reinforcement of the walls to prevent further ingress of the fungus into the host and thus limit the damage done by the pathogen. This process could include cross-linking proteins and other components such as phenolics that are present in the wall at the time of fungal attack. This initial process would be followed by deposition of newly synthesized phenolics that are also cross-linked to provide further protection of walls from degradation by fungal extracellular enzymes. Finally, the protection is completed by the suberization of the wall, making the walls virtually impermeable to diffusion of solutes or fungal enzymes. Deposition of phenolics on the wall as a response to fungal infection is often noted. However, such phenolics are seldom characterized chemically and are variously designated as ligmis, suberin, lignin-suberin-like, etc. (8). The cross-linking processes most probably involve peroxidase(s). In fact, the polymerization of the phenolics involved in suberization has been shown to be mediated by a highly anionic peroxidase exclusively located in the cell walls of the suberizing cells (9). The type of superficial examination often used (autofluorescence or simple reactions to release some phenolics, as done with lignin) does not allow proper characterization. It is probable that the composition of the materials deposited in all cases is not the same. Even within the same host-pathogen interaction, the composition of the phenolics on the wall probably changes from the initial period of attack to the formation of the final protective polymer in the wall. Until the materials are chemically analyzed, imprecise use of the terms to describe the phenolics will continue. In one case chemical analysis was recently done and it was found that suberization is a key defense reaction used by a resistant host line. In response to infection by Verticillium albo-atrum, resistant tomato line forms a vascular coating response, whereas near isogenic susceptible tomato line could not respond by erecting the coating that restricts the fungal spread (10). Analysis of the depolymerization products of the vascular coating material by capillary GLC/MS showed that it contained aliphatic components characteristic of suberin (11). Only the infected resistant tomato line showed the deposition of these suberin monomers.

PLANT PEROXIDASE GENE EXPRESSION TRIGGERED BY FUNGAL SIGNALS

If the ability to erect a suberized barrier in response to a vascular fungal attack is a basis of the resistance of the host, the suberization-associated highly anionic peroxidase should preferentially appear in the resistant tomato line upon fungal invasions because this peroxidase is known to be a key regulatory enzyme involved in suberization (12). Upon introduction of V. albo-atrum spores through the petiolar vascular system of excised tomato leaves, vascular coating appeared in the resistant, but not in the susceptible, tomato line. Northern blot analysis with a highly homologous potato peroxidase cDNA showed that the vascular tissue of the resistant line responded to the fungal infection by triggering the expression of the gene for the highly anionic peroxidase (11). The differential ability to express the peroxidase gene in response to fungal attack could also be demonstrated in cell suspension cultures. When low levels (ng quantities) of an elicitor preparation from V. albo-atrum were added to the culture medium, peroxidase transcripts appeared in the cells from the resistant tomato line but the cells from the near isogenic susceptible line could not respond in this manner (Fig. 3) (13). Peroxidase gene expression could be detected in the cells from the resistant line within minutes after the addition of the fungal elicitor.

Figure 3. Autoradiogram of a Northern blot of 10 μg total RNA from susceptible (**A**) and resistant (**B**) tomato cell suspension cultures treated with different concentrations of elicitor from <u>V. albo-atrum</u>. The cells were harvested 15 min and 3h after elicitor treatment and the blots were probed with 32P-labeled peroxidase cDNA.

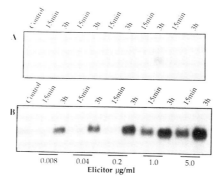

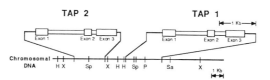

Figure 4. Schematic representation of the tomato anionic peroxidase genes. (H, HindIII; X, XbaI; Sp, SphI; P, PstI; Sa, SalI).

Tomato contains two highly homologous copies of the highly anionic peroxidase gene found in tandem (Fig. 4) (14). When one of these genes with its 5'- flanking region was introduced into tobacco by agrobacterium-mediated transformation, the transgenic tobacco showed wound-inducible expression of the gene in leaves and stems (A. Bajar and P. E. Kolattukudy, manuscript in preparation). When the 5'- flanking region of this gene was fused to β-GUS gene and introduced into tobacco, the transgenic plants expressed β-GUS in a wound-inducible manner (Fig. 5). The expression of β-GUS was limited to the wounded region. This observation is consistent with the finding that suberization is limited to a few cell layers adjacent to the broken cells. When this peroxidase gene was fused to a double construct of cauliflower mosaic virus 35S promoter and introduced into tobacco, constitutive expression of the peroxidase was observed (Table 2). Even when the peroxidase level was 400 times that found in control tobacco, the plants developed normally but matured slowly. No wilting was observed at any time in the transgenic tobacco plants that expressed the peroxidase gene at low to very high levels. This observation is in contrast to that reported with transgenic tobacco plants that over-expressed a different peroxidase gene (15). Peroxidases constitute a class of enzymes and should not be viewed as one enzyme, as some tend to do. The tobacco peroxidase that is thought to be involved in lignification was no more homologous to the suberization-associated tomato peroxidase than to horse radish peroxidase. Obviously some homology should be expected in the regions that code for common elements mechanistically involved in the basic reaction catalyzed by the enzyme. Such regions include the heme-binding region that is highly conserved, as expected. The tobacco enzyme is obviously quite different from that involved in suberization. In fact, tobacco RNA from control plants showed no hybridization with the potato peroxidase cDNA that is highly homologous to the tomato peroxidase gene. Electrophoresis of the proteins from the transgenic tobacco plants that showed high levels of tomato peroxidase transcripts also showed the anionic peroxidase isozyme pattern similar to that found in wound-healing potato and tomato. These isozymes coded by the exogenous gene introduced into the tobacco plants must represent different post translational modifications. In any case, with the availability of cloned peroxidase and transgenic plants that express this gene, it should be possible to elucidate the role that this highly anionic peroxidase plays in the plant defense against pathogenic fungi.

248

Figure 5. **A,** Structure of the tomato anionic peroxidase (TAP-1)-β-GUS fusion gene. A 767 bp DNA fragment of the TAP-1 gene was cloned into the plant reporter vector pBI101.2 and transfered into tobacco via the <u>Agrobacterium</u> mediated leaf-disc method. The DNA fragment of TAP-1 gene contained the 5' promoter region (462 bp) and the DNA sequences that coded for the putative 74 amino acid leader sequence.

B, Time course of induction of β-GUS in a transgenic tobacco plant that expressed the wound-inducible TAP-1 promoter. 5mm square leaf-pieces were cut, allowed to wound-heal over the time period indicated, and assayed for β-GUS activity.

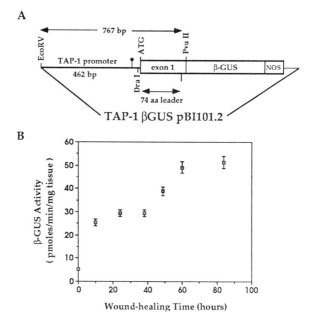

Table 2. Peroxidase Activity Assays of *N. tabaccum* (35 S)$_2$/ TAP-1 Expressors

Transgenic Plants	Specific Activity[*]
B	407
7	240
P	154
A	146
L	127
5	110
100	109
16	99.2
Non-Transformed Controls	
cv. NC2326	1.02
cv. Coker	2.94

* S.A. = OD^{470}/min./mg total protein

Acknowledgements

This work was supported, in part, by grant DCB-8819008 from the National Science Foundation.

References

1. Kolattukudy, P. E. (1985) 'Enzymatic penetration of the plant cuticle by fungal pathogens', Ann. Rev. Phytopathol. 23, 223-250.
2. Kolattukudy, P. E., Podila, G. K., Roberts, E. and Dickman, M. D. (1989) 'Gene expression resulting from the early signals in plant-fungus interaction', in B. Staskawicz, P. Ahlquist and O. Yoder (eds.), Molecular Biology of Plant-Pathogen Interactions, Alan R. Liss, Inc., New York, pp. 87-102.
3. Podila, G. K., Dickman, M. B. and Kolattukudy, P. E. (1988) 'Transcriptional activation of a cutinase gene in isolated fungal nuclei by plant cutin monomers', Science 242, 922-925.
4. Podila, G. K., Dickman, M. B., Rogers, L. M. and Kolattukudy, P. E. (1989) 'Regulation of expression of fungal genes by plant signals', in H. Nevalainen and M. Penttilä (eds.), Molecular Biology of Filamentous Fungi, Foundation for Biotechnical and Industrial Fermentation Research, Helsinki, pp. 217-226.
5. Lin, T. S. and Kolattukudy, P. E. (1978) 'Induction of a biopolyester hydrolase (cutinase) by low levels of cutin monomers in *Fusarium solani* f. sp. *pisi*,' J. Bacteriol. 133, 942-951.
6. Fawell, S. E., Lees, J. A., White, R. and Parker, M. G. (1990) 'Characterization and colocalization of steroid binding and dimerization activities in the mouse estrogen receptor', Cell 60, 953-962.
7. Crawford, M. S. and Kolattukudy, P. E. (1987) 'Pectate lyase from *Fusarium solani* f. sp. *pisi*: purification, characterization, *in vitro* translation of the mRNA, and involvement in pathogenicity', Archives of Biochemistry and Biophysics 258, 196-205.
8. Kolattukudy, P. E. (1980) 'Biopolyester membranes of plants: cutin and suberin', Science 208, 990-1000.
9. Espelie, K. E., Francheschi, V. R. and Kolattukudy, P. E. (1986) 'Immunocytochemical localization and time-course of appearance of an anionic peroxidase associated with suberization in wound-healing potato tuber tissue', Plant Physiol. 81, 487-492.
10. Street, P. F. S., Robb, J. and Ellis, B. E. (1986) 'Secretion of vascular coating components by xylem parenchyma cells of tomatoes infected with *Verticillium albo-atrum*', Protoplasma 132, 1-11.
11. Robb, J., Lee, S-W., Mohan, R. and Kolattukudy, P. E. (1990) 'Vascular coating: suberization of the xylem in tomatoes infected by *Verticillium albo-atrum*' (submitted).
12. Roberts, E., Kutchan, T. and Kolattukudy, P. E. (1988) 'Cloning and sequencing of cDNA for a highly anionic peroxidase from potato and the induction of its mRNA in suberizing potato tubers and tomato fruits', Plant Molec. Biol. 11, 15-26.
13. Mohan, R. and Kolattukudy, P. E. (1990) 'Differential activation of expression of a suberization-associated anionic peroxidase gene in near-isogenic resistant and susceptible tomato lines by elicitors of *Verticillium albo-atrum*', Plant Physiol. 921, 276-280.
14. Roberts, E. and Kolattukudy, P. E. (1989) 'Molecular cloning, nucleotide sequence, and abscisic acid induction of a suberization-associated highly anionic peroxidase', Mol. Gen. Genet. 217, 223-232.
15. Lagrimini, M. L., Bradford, S. and Rothstein, S. (1990) 'Peroxidase-induced wilting in transgenic tobacco plants', Plant Cell 2, 7-18.

FUNGAL SIGNALS INVOLVED IN THE SPECIFICITY OF THE INTERACTION BETWEEN BARLEY AND RHYNCHOSPORIUM SECALIS

W. KNOGGE, M. HAHN, H. LEHNACKERS, E. RÜPPING AND L. WEVELSIEP
Max-Planck-Institut fuer Zuechtungsforschung
Department of Biochemistry
Carl-von-Linne-Weg 10
D-5000 Koeln 30
Federal Republic of Germany

ABSTRACT. The barley-Rhynchosporium secalis pathosystem is being used as an experimental model to identify fungal molecules involved in the expression of race-cultivar specificity. Evidence is presented that a necrosis-inducing peptide may be the product of a fungal avirulence gene.

1. Introduction

The gene-for-gene hypothesis assumes that the basis of race-cultivar specific resistance is the interaction of an avirulence gene of a given pathogenic race with the corresponding resistance gene in the host cultivar [1]. This implies a crucial role for the products of these genes in recognition of the pathogen by the plant, leading to the onset of defense reactions. A functional approach towards the identification of a resistance gene requires, as a first step, the identification and isolation of the respective avirulence gene product, which should determine the specificity of the interaction. Pathogen-induced phenotypic differences specifying susceptibility or resistance of the plant can serve as the experimental starting point to isolate compounds from the pathogen such as specific toxins or elicitors of the plant defense response.

Rhynchosporium secalis (Oudem.) J.J. Davis is a Deuteromycete which causes leaf scald of barley (Hordeum vulgare L.). Studies on the inheritance of scald resistance in barley have indicated its monogenic nature and have defined several resistance loci [2]. In the present work two near-isogenic barley cultivars, "Atlas" (resistance locus Rrs2) and "Atlas 46" (Rrs1 in addition to Rrs2), and R.secalis, race US238.1, which is virulent on "Atlas" and avirulent on "Atlas 46", represent the components of the experimental system.

During development of the fungus on the two barley cultivars no microscopically visible differences occur until 3-4 d p.i., after which fungal growth on the resistant cultivar slows down and finally stops. During this initial recognition phase, there is no direct contact between the fungus and the plant plasmalemma [3]. Resistance must therefore be triggered by diffusible factors crossing the plant cell wall.

250

H. Hennecke and D. P. S. Verma (eds.),
Advances in Molecular Genetics of Plant-Microbe Interactions, Vol. 1, 250–253.
© 1991 Kluwer Academic Publishers. Printed in the Netherlands.

2. Fungal Molecules Involved in the Expression of Plant Susceptibility

2.1. IDENTIFICATION OF NECROSIS-INDUCING PEPTIDES

The most obvious phenotypic difference between the infected barley cultivars is the occurrence of necrotic lesions on cultivar "Atlas" but not on cultivar "Atlas 46". From fungal culture filtrates, a small family of necrosis-inducing peptides (NIPs) was isolated and purified. NIP1 appears to be a homodimer of 7.2 kDa with disulfide-linked subunits of 3.8 kDa and to be less toxic than NIP2 or NIP3. NIP1 and NIP2 (6.8 kDa) are not glycosylated whereas NIP3 is a glycoprotein of 9.2 kDa. After removal of the carbohydrate moiety the peptide (5.4 kDa) retained the toxic activity. Upon infiltration of primary leaves, the toxic activity of the NIPs was not cultivar-specific since all caused necrosis on both cultivars (Table 1). However, when extracts from infected leaves were probed on Western blots with antisera raised against each of the three peptides, NIP3 could be detected only in the susceptible cultivar during late stages of pathogenesis and neither NIP1 nor NIP2 could be found in leaves of either cultivar. In the susceptible cultivar, however, a 45 kDa protein was found to cross-react with NIP1-antisera. Its appearance also correlated with lesion development. This molecule, which has not yet been isolated, may be either a precursor or the native form of NIP1 ("in vivo-NIP1"). Western dot blots of the same plant extracts using antisera raised against fungal cell wall material demonstrated that although fungal biomass increased continuously in susceptible leaves, NIP3 and in vivo-NIP1 occurrence showed a burst correlating with lesion formation. This suggests an involvement of plant compounds in triggering a physiological switch in the fungus which results in the appearance of both proteins. Since the occurrence of NIP3 and in vivo-NIP1 in planta, unlike their toxicity, appears to be cultivar-specific, they may be involved in lesion development in vivo.

2.2. MODE OF ACTION OF NECROSIS-INDUCING PEPTIDES

Some of the symptoms described for R.secalis-infected barley leaves such as membrane permeability changes and ion leakage [4] point to the plant plasmalemma as the putative target site for fungal effector proteins. The influence of NIPs from R.secalis on a key plasmalemma enzyme, ATPase, was therefore analyzed using plasma membrane vesicles isolated from barley primary leaves [5]. While NIP2 had no effect, NIP1 and NIP3 proved to be strong stimulators of the enzyme in vesicle preparations from both cultivars (Table 1). Like necrosis-induction the peptide moiety of NIP3 retained the ability to stimulate ATPase activity.

Photoaffinity labelling studies and affinity chromatography using NIP3 as a ligand revealed that this compound does not bind directly to the ATPase (100 kDa). Two NIP3-binding membrane proteins (~ 66 kDa) were, however, identified. The mode of action of NIP3 (and possibly also of in vivo-NIP1) may, therefore, involve energy depletion of cells via indirect stimulation of the plasmalemma ATPase. Since NIP2 causes necrosis to the same extent as NIP3 without influencing ATPase activity, its mode of action must be different.

3. Fungal Molecules Involved in the Expression of Plant Resistance

3.1. IDENTIFICATION OF A SPECIFIC RESISTANCE RESPONSE OF BARLEY

Resistance-specific reactions of the plant are a necessary prerequisite for the identification of cultivar-specific fungal elicitors. A number of cDNA-clones which hybridize to mRNAs accumulating in wheat upon mildew infection [6] were therefore tested in the barley-R.secalis pathosystem. In Northern blots, the clone WIR3, encoding a peroxidase, hybridized to an mRNA species from infected barley primary leaves. The induction was transient and occurred in both cultivars, showing a maximum at about 24 h p.i. The clone WIR2 yielded a strong signal with RNA from the resistant cultivar but a much weaker signal with RNA from the susceptible cultivar.

The WIR2 cDNA clone was used to isolate a homologous cDNA clone (BIP2) from a barley cDNA library. Nucleotide sequence analysis revealed a potential signal peptide of 20 amino acids and similarity to the thaumatin-like pathogenesis-related (PR) proteins from a number of plants. Furthermore, beginning with amino acid 21, the predicted amino acid sequence of this protein was identical to the N-terminal 27 amino acids of a thaumatin-like PR-protein isolated from mildew-infected barley leaves [7].

3.2 IDENTIFICATION OF A RACE- AND CULTIVAR-SPECIFIC ELICITOR

The accumulation of BIP2 mRNA provided an assay system to screen for fungal elicitors. Testing of various fungal fractions derived from mycelial walls or spore germination fluids, yielded negative results. NIP1, however, induced the accumulation of high levels of BIP2 mRNA in primary leaves of the resistant cultivar "Atlas 46" while only a minor effect was discerned on the susceptible cultivar "Atlas" (Table 1). Since NIP2 and NIP3 did not exhibit similar elicitor activity, it can be concluded that neither cell death nor ATPase stimulation are sufficient for gene activation.

TABLE 1. Effects of R.secalis, race US238.1, and its necrosis-inducing peptides on two near-isogenic barley cultivars

Plant phenotype	Fungus		NIP1		NIP2		NIP3	
	R	S	R	S	R	S	R	S
Induction of necrosis	−	+++	+	+	+++	+++	+++	+++
Stimulation of ATPase	?	?	+++	+++	−	−	+++	+++
Induction of BIP2 mRNA	+++	+	+++	(+)	(+)	−	(+)	−

A comparison of culture filtrates from various fungal races on Western blots demonstrated the presence of NIP2 in all races, NIP3 in most, and NIP1 exclusively in race US238.1. Taken together with the specificity of BIP2 mRNA induction, this indicates that NIP1 may be the product of an avirulence gene of R.secalis. F_2-individuals from crosses between the two barley cultivars are presently being analyzed for cosegregation of

NIP1-induced BIP2 mRNA accumulation and resistance to race US238.1. In addition, other barley cultivars are being tested for a correlation between BIP2 mRNA induction by NIP1 and the presence of resistance locus *Rrs1*.

4. Conclusions

The individual peptides isolated from *in vitro*-cultures of *R.secalis*, race US238.1, may have different roles *in vivo*. While the function of NIP2 remains unclear, NIP3 and *in vivo*-NIP1 appear to be toxins involved in lesion development on the susceptible cultivar. Two lines of evidence imply that NIP1 is an avirulence gene product of *R.secalis*, race 238.1: 1) *in vitro*, NIP1 was detected in culture filtrates only of this particular fungal race; 2) upon treatment of barley leaves with NIP1, the synthesis of a barley PR-protein was induced only in the cultivar carrying the resistance locus *Rrs1*. Genetic studies are underway (see 3.2.) to support this hypothesis.

5. Acknowledgements

We thank C. Buchen and B. Hoss for excellent technical assistance. This work was supported by the Bundesministerium für Forschung und Technologie.

6. References

[1] Flor, H.H. (1956) "The complementary genic system in flax and flax rust", Advan. Genet., 8, 29–54.
[2] Shipton, W.A., Boyd, W.J.R., and Ali, S.M. (1974) "Scald of barley", Rev. Plant Pathol., 53, 839–861.
[3] Lehnackers, H. and Knogge, W. (1990) "Cytological studies on the infection of barley cultivars with known resistance genotypes by *Rhynchosporium secalis*", Can. J. Bot., in press.
[4] Jones, P. and Ayres, P.G. (1972) "The nutrition of the subcuticular mycelium of *Rhynchosporium secalis* (barley leaf blotch): permeability changes induced in the host", Physiol. Plant Pathol., 2, 383–392.
[5] Widell, S., Lundborg, T., and Larsson, C. (1982) "Plasma membranes from oats prepared by partition in an aqueous polymer two-phase system", Plant Physiol., 70, 1429–1435.
[6] Schweizer, P., Hunziker, W., and Mösinger, E. (1989) "cDNA cloning, *in vitro* transcription and partial sequence analysis of mRNAs from winter wheat (*Triticum aestivum* L.) with induced resistance to *Erysiphe graminis* f.sp. *tritici*", Plant Molec. Biol., 12, 643–654.
[7] Bryngelsson, T. and Green, B. (1989) "Characterization of a pathogenesis-related, thaumatin-like protein isolated from barley challenged with an incompatible race of mildew", Physiol. Molec. Plant Pathol., 35, 45–52.

MOLECULAR DETERMINANTS OF PATHOGENESIS IN *USTILAGO MAYDIS*

S. A. Leong, [1,2] Eunice Froeliger,[2] Allen Budde,[2] Baigen Mei,[2] Christophe
Voisard[2] and James Kronstad.[3] USDA/ARS[1] and Department of Plant
Pathology,[2] University of Wisconsin, Madison 53706 U. S. A.;
Biotechnology Laboratory, University of British Columbia, Vancouver,
Canada V6T1W5.[3]

ABSTRACT. Our work is focused on the identification and analysis of genes required for pathogenic growth of *Ustilago maydis*, the cause of corn smut disease. Successful infection of maize by *U. maydis* requires that haploid cells of the opposite mating type fuse to form a dikaryotic mycelium. This parasitic mycelium colonizes host tissues, induces gall development and eventually becomes sporogenous giving rise to diploid teliospores. The two mating type loci *a* and *b* control cell fusion, sexual development and pathogenicity. Our longterm goal is to understand how the products of the *a* and *b* mating type loci function to control these processes. Toward this end, the two alleles of the *a* locus and six alleles of the *b* locus have been cloned and their structure and expression is under study. The six *b* alleles were each found to encode an open reading frame of 410 amino acids. The different *b* alleles showed considerable sequence similarity; the N-proximal portion of the amino acid sequence was more variable between alleles than the C-terminal portion. By contrast, hybridization analysis of the two *a* alleles indicated that they are dissimilar. Gene replacement experiments were conducted to replace the genomic *a2* allele with the cloned *a1* allele to determine whether heterozygosity at *a* is required for dikaryon formation and pathogenicity. These experiments confirmed the results of Banuett and Herskowitz (1989) who demonstrated that diploid cells homozygous at *a* are affected in the production of dikaryotic hyphae in culture but not in pathogenicity. Finally, disruption of *b* alleles in haploid cells gave rise to sterile, nonpathogenic mutants; disruption of a *b1* allele in a pathogenic diploid (*b1/b2*) produced a nonpathogenic mutant. These data indicate that the *b* locus acts in a positive manner to control sexual development and pathogenicity.

We are also conducting a systematic analysis of the siderophore-mediated iron uptake system of *U. maydis* in order to assess its role in pathogenicity, spore physiology, and survival of the pathogen. *U. maydis* produces two cyclic hydroxamate siderophores, ferrichrome and ferrichrome A. Mutants defective in hydroxylation of ornithine, the first committed step in the biosynthesis of the ferrichromes, have been characterized further. Preliminary studies of enzyme activity indicate that this enzyme is a flavoprotein which utilizes NADH or NADPH as a source of electrons. As expected, enzyme activity was absent in a mutant unable to produce δ-N-hydroxyornithine. A genomic DNA clone capable of restoring siderophore production in trans in this mutant was identified. In addition, a genomic clone capable of complementing in trans three mutants deregulated for siderophore production was isolated. Disruption of these sequences in the *U. maydis* genome gave rise to the expected mutant phenotypes indicating that these clones likely encode the structural genes for ornithine-N^5- oxygenase and a regulator of siderophore biogenesis.

H. Hennecke and D. P. S. Verma (eds.),
Advances in Molecular Genetics of Plant-Microbe Interactions, Vol. 1, 254–263.
© 1991 *Kluwer Academic Publishers. Printed in the Netherlands.*

1. Introduction

Ustilago maydis, the causal agent of corn smut disease, is a basidiomycete fungus with a dimorphic growth habit (Christensen, 1963). In the laboratory we commonly work with the haploid, budding yeast phase which can be manipulated in much the same way as *Saccharomyces cereviseae* (Holliday, 1974). The fungus grows on chemically defined media, has a rapid doubling time, and is amenable to genetic manipulation. For example, mutants are readily isolated by treatment with chemical mutagens or UV light and stable diploids can be constructed for complementation analysis and mitotic recombination analysis. *U. maydis* also has a sexual phase; however, the sexual cycle is only completed when the fungus is growing in living host tissue. In fact, the haploid yeast-like cells are not pathogenic, only diploid or dikaryotic cells that result from the mating of haploid cells of compatible mating type cause disease. This dikaryotic mycelium invades the vascular bundle of aerial tissues of the corn plant, a process which triggers new cell division in the bundle sheath and vascular parenchyma. These newly divided cells eventually become filled with black, zygotic teliospores of the fungus to form the gall.

Over the last several years, much progress has been made in the development of the molecular genetic tools that will enable an analysis of this host-parasite interaction at the molecular level. Stable, integrative transformations systems (Wang et al, 1988; Banks and Taylor, 1988) have been developed for the fungus. In addition, a high frequency transformation system based on a replicating vector is available for *U. maydis* (Tsukuda et al, 1988). Protocols for single step gene replacement and gene disruption have been developed (Kronstad et al, 1989; Fotheringham and Holloman, 1989). Numerous genes have been cloned and characterized from the fungus (Bank and Taylor, 1988; Fotheringham and Holloman, 1989; Froeliger and Leong, 1989; Hargreaves and Turner, 1989; Holden et al, 1989a; Holden et al, 1989b; Kronstad and Leong, 1989a; Kronstad et al, 1989b; Kronstad and Leong, 1990; Schultz et al, 1990; Smith and Leong, 1990; Tsukuda et al, 1989; Wang et al, 1989). Finally, molecular karyotype analysis indicates that *U. maydis* has at least 20 yeast-sized chromosomes (Kinscherf and Leong, 1988). Several genes have been located on the karyotype by hybridization analysis (Budde and Leong, 1990).

Using these tools we are investigating two gene systems, genes that control mating and sexual development and genes involved in the high affinity, siderophore-mediated iron acquisition system of *U. maydis*. We wish to understand what the products of the mating type loci encode, how they function to control mating and sexual development and why heterozygosity at the mating type loci is required to activate these processes. We also wish to understand what role siderophores might play in the interaction of this pathogen with its host and the environment.

2. Results and Discussion

2.1 ANALYSIS OF MATING TYPE CONTROL

Two loci control the ability of *U. maydis* to mate, complete the sexual cycle and cause disease. The *a* locus is diallelic and controls fusion of haploid cells carrying different *a* alleles (reviewed in Froeliger and Kronstad, 1990). Heterozygosity at *a* also appears to be required for production of hyphae in culture (Banuett and Herskowitz, 1989); diploid cells homozygous for *a* but heterozygous at *b* are unable to produce mycelium in culture. The *b* locus is multiallelic and controls formation of the dikaryon and subsequent events in sexual development (reviewed in

Froeliger and Kronstad, 1990). While mating and production of dikaryotic hyphae can occur in culture, formation of teliospores requires that the fungus be in contact with the host.

2.1.1. *The a mating type locus*.
An *a2* allele of the *a* locus was cloned by its proximity to the biochemical marker *pan1-1* (Holliday, 1974). A haploid strain (*a1,b1, pan1-1*) was transformed with a cosmid library prepared in pJW42, a cosmid that can self replicate in *U. maydis* (Leong et al, 1990). Prototrophic transformants were tested for their ability to mate with a haploid tester (*a1, b2*). Of those tested, two had acquired the mating type activity associated with the *a2* allele. As expected, the transformants, but not the recipient, were also pathogenic when crossed with the the tester strain.

Subcloning experiments further localized the *a2* mating type activity and *pan1* gene to a 6 kb *Bam*H1 fragment. Southern hybridization analysis of genomic DNA using the 6 kb fragment as a probe indicated that only a single copy of this DNA sequence was present in the genome of the *a2* haploid strain. When used as a probe to genomic DNA of an *a1* haploid strain, a single 10 kb *Bam*H1 fragment was identified. This fragment was cloned and the minimal sequences required to confer mating type activity in each fragment were identified by transformation with subportions of the fragments. A 2 kb region of each *Bam*H1 fragment was found to confer full *a* mating type activity when transformed into cells with the alternate allele.

Comparison of the restriction maps of the two *Bam*H1 fragments revealed a considerable amount of restriction fragment length polymorphism. Of the 6 kb *Bam*H1 fragment, approximately 4 kb was found to be in common with that of the 10 kb fragment as determined by DNA hybridization analysis and restriction mapping. To obtain a more complete picture of the locus, chromosome walking experiments were conducted to obtain the adjacent genomic DNA fragments surrounding the *a2* allele. Comparison of the restriction map of the two alleles is shown in figure 1. An area of approximately 5 kb in the 10 kb *a1* clone and 10 kb in the 14 kb region carrying the *a2* mating type activity were found to be nonhomologous by DNA hybridization (figure 1). Like a number of other mating type loci (Giasson et al 1989; Glass et al, 1988; Herskowitz, 1988), the alternate alleles of the *a* locus of *U. maydis* are dissimilar. Such nonidentical mating type alleles have been termed idiomorphs (Metzenberg and Glass, 1990).

To provide final proof for cloning of the locus, a gene replacement experiment was conducted in which the 10 kb *Bam*H1 fragment carrying the *a1* allele was transformed into haploid cells of *a2* genotype. Transformants were tested for *a1* mating type activity. Replacement of the *a2* allele with the recombinant *a1* allele was confirmed by Southern hybridization and analysis of the meiotic segregation of the *a* and *b* mating type alleles in a test cross. These data along with the proximity of the *a* mating type activity to *pan1* indicate that this region of DNA encodes the *a* mating type locus.

To investigate the role of the *a* locus in control of infectious hyphae, the *a2* locus of a diploid strain (*a1/a2, b2/b1, pan1-1/+, nar1-1+, +/ad1-1, +/me1-2, +/nar1-6*) was replaced by transformation with the 10 kb fragment carrying the *a1* mating type allele. DNA of transformants was screened by Southern hybridization for loss of the *a2* allele. Two transformants were homozygous for *a1*. Analysis of meiotic products obtained from these strains provided genetic proof for homozygosis as only basidiospore segregants of *a1* genotype were found. Interestingly, both of the replacement strains showed reduced capacity to produce infectious hyphae in culture. By contrast, both showed normal ability to cause disease when compared to the parent diploid strain (*a1/a2*). These data substantiate the work of Banuett and Herskowitz (1989), who found similar results with *a* homozygous diploid strains which they constructed by

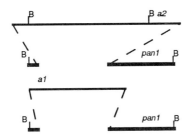

Figure 1. The *a* mating type region of *U. maydis*. Heavy bars indicate regions of sequence similarity and light bars indicate regions of sequence dissimilarity as determined by restriction mapping and Southern hybridization analysis. Restriction site *Bam*H1 is denoted by B.

UV treatment of a diploid strain heterozygous at *a*. Current efforts involve disruption and deletion of the *a* alleles in the genome of *U. maydis* and DNA sequence analysis of the minimal regions required to confer *a* allele activity. We are also investigating the production of pheremones and their control by the *a* locus in the fungus.

2.1.2. The b mating type locus.

We previously reported the cloning of *b1* and *b2* alleles of the *b* locus of *U. maydis* (Kronstad and Leong, 1989a). Both alleles are encoded on 8 kb *Bam*H1 fragments and no evidence was found for additional copies of the alleles in the genome of haploid cells by Southern hybridization at high stringency. To further locate the *b* allele activity, the *b1*-containing fragment was subjected to Tn5 mutagenesis and Tn5-marked fragments were tested for *b1* activity by transformation of a diploid (*b2/b2*). Transformants that were heterozygous at *b* became mycelial on mating agar while transformants receiving the mutated *b* allele remained yeast-like. These data, in concert with subcloning and restriction mapping experiments, localized the *b1* region to a 1.4 kb BglII/*Sal* I fragment.

DNA sequence analysis of the fragment and surrounding region revealed an open reading frame (ORF) of 410 amino acids (Kronstad and Leong, 1989b). A similar ORF was identified at the same location in the *b2* allele clone. Comparison of the predicted amino acid sequences of the two alleles identified a region of variable sequence in the first 160 amino acids while the C-terminal portion of the predicted proteins were nearly identical. DNA sequence analysis of four additional alleles revealed a similar overall structure (Kronstad and Leong, 1990). These data are consistent with those of Schultz et al (1990) who determined the sequence of four *b* alleles. Neither group found any extensive DNA sequence similarities with those available in the databases. However, we identified a short sequence that matches the nucleus localization sequence of several proteins (Kronstad and Leong, 1990), while Shultz et al (1990) found a sequence having identity to the homeodomain of the *Drosophila antennapedia* gene and yeast mating type genes a1 and α2. These features are consistent with the *b* allele product functioning as a regulatory molecule.

To investigate whether the product of the *b* locus acts positively or negatively, the *b1* and *b2* alleles of haploid strains and the *b1* allele of a diploid strain were disrupted. In the former case, the disrupted strains were both sterile and nonpathogenic, either by themselves or when mated to the appropriate haploid tester strain. In the latter case, the disrupted strain, which still had a functional *b2* allele, was itself nonpathogenic and was able to mate only with haploid strains of compatible *b* genotype (Kronstad and Leong, 1990). These data suggest that the *b* locus functions in positive manner either directly or indirectly to regulate genes required for formation of the parasitic mycelium and subsequent events in sexual development. Future work will focus on a mutational analysis of the cloned *b1* allele and will attempt to identify genes that interact with the *b1* allele by isolating second site revertants which are restored in *b* mating type activity.

2.2. HIGH AFFINITY IRON TRANSPORT.

Microorganisms growing in an aerobic environment have evolved an iron gathering system that consists of siderophores, microbial high affinity iron transport compounds, and a membrane bound uptake system designed to recognize these iron chelates (Winkelmann, et al, 1988). Siderophores are recognized as virulence factors for some plant (Expert et al, 1988) and animal pathogens (Payne, 1988). Moreover, siderophores are known to be an iron storage mechanism for a number of fungi (Matzanke, 1987) and are germination factors for spores of others (Charlang et al, 1981; Horowitz et al, 1976). Siderophores clearly play an important role in the ecology of the rhizosphere and the biological control of plant disease (Leong and Expert, 1990; Loper and Buyer, 1990).

We have isolated and characterized two cyclic hydroxamate siderophores, ferrichrome and ferrichrome A, from *U. maydis* (Budde and Leong, 1989). As with other microorganisms, biosynthesis of the siderophores is negatively regulated by the iron concentration of the growth medium. Mutants defective in the biogenesis (Wang et al 1989) or regulated synthesis of siderophores (Leong et al, 1987) were isolated by NTG mutagenesis. Those blocked in siderophore biosynthesis fell into two categories: 1) mutants unable to synthesize ferrichrome and ferrichrome A and 2) mutants unable to synthesize ferrichrome. The deregulated mutants do not respond normally to the iron concentration of the growth medium and produce siderophores constitutively.

2.2.1. *Ornithine-N^5-Oxygenase.* Based on biochemical feeding tests, genetic analysis and enzymatic assay, mutants unable to synthesize either ferrichrome or ferrichrome A were concluded to be defective in the conversion of ornithine to δ–N-hydroxyornithine. This is the first committed step in siderophore biosynthesis in *U. maydis*. Preliminary analysis of enzyme activity was performed using the assay developed for the lysine hydroxylase of *Escherichia coli* (Plattner et al, 1989). The *U. maydis* enzyme appears to be a flavoprotein and to require NADH or NADPH as a source of electrons. Extracts prepared from siderophore-nonproducting mutant UMS023 showed reduced oxygenase activity when compared to those from the wild type. Activity was also shown to be absent in extracts prepared from wild type cells grown in high iron medium.

An 8.1 kb *Hind*III fragment of genomic DNA was previously reported to restore siderophore production in these mutants (Wang et al, 1989). These experiments were performed using integrative transformation assays. When the 8.1 kb fragment was cloned in the replicating vector pCM54 (Tsukuda et al, 1988), no complementation was observed. This indicated that the complete gene was not present on this fragment. A 3.2 kb subclone from one end of the 8.1 kb *Hind*III fragment also restored siderophore production in UMS023 when integrated into the genome (figure 2). Fragments including this region of DNA were isolated from the original complementing cosmid, cloned in pCM54, and tested for their ability to complement mutant

UMS023. A 7 kb *Ssp*I overlapping fragment was able to fully complement the mutant, while the 6.6 kb overlapping *Bgl*II fragment and 4 kb *HindIII/Ssp*I fragment failed to complement This indicated that the entire gene was present on the SspI fragment and that *Bgl*II and *Hind*III sites interrupt the gene.

A gene disruption experiment was conducted to verify the importance of the *Bgl*II site in the 3.2 kb *Ssp*I/*Eco*R1 fragment. A 3 kb fragment carrying the chimeric selectable marker for hygromycin resistance (Wang et al, 1988) was cloned into the *Bgl*II site of the 3.2 kb *Eco*R1/*Ssp*I fragment and the disrupted, linear fragment was transformed into the wild type fungus. Numerous transformants were identified that failed to produce siderophores. Southern hybridization analysis of DNA of several siderophore-nonproducing mutants indicated that the wild type fragment was replaced with the mutant fragment in all cases. These results indicate that the *Bgl*II site interrupts the structural gene for the oxygenase or a biosynthetic regulatory gene that controls expression of the oxygenase. These possibilities will be resolved once we have isolated the oxygenase enzyme and shown that the amino acid sequence of its N-terminus matches the sequence deduced from the cloned DNA.

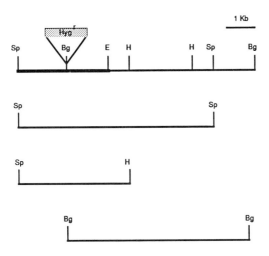

Figure 2. Ornithine N^5-Oxygenase region of *U. maydis*. Restriction sites shown are: *Ssp*I, Sp; *Bgl*II, Bg; and *Hind*III, H.

Current and future experiments involve DNA sequence analysis of the complementing region of DNA and overexpression and purification of the enzyme. We are keen to understand the properties of this enzyme further as it is unique to microorganisms that produce siderophores containing δ-N-hydroxyornithine. The development of inhibitors of this enzyme may provide a rational approach to design of highly targeted antimicrobials which may have application to human, animal and/or plant health. Finally, with disruption mutants in hand, we will be able to assess the role of siderophores in the life cycle of *U. maydis*. Previous work involved the analysis of pathogenesis using chemically induced mutants and suggested that siderophores contribute marginally to pathogenesis (Leong et al, 1988). These studies suffered in that the mutant strains we employed were not completely isogenic with the parent strain and could harbor secondary mutations. Furthermore, we could not control for reversion of the siderophore biosynthetic defect in planta. We are now poised to conduct a more definitive study of siderophores and their contribution to corn smut disease.

2.2.2. *Deregulation of Siderophore Biosynthesis.* Normally siderophore biosynthesis is tightly regulated by the concentration of iron in the growth medium (Bagg and Neilands, 1987). In fact, extracellular ferrichrome and ferrichrome A are not produced in cultures containing 10 uM added iron (Budde and Leong, 1989). Three mutants constitutive for siderophore production were isolated by screening for their ability to inhibit the growth of *Salmonella typhimurium* mutant TA2701 (*ent⁻, TonA⁻*) seeded in an iron replete, complete medium (Leong et al, 1987). The *Ustilago* mutants were confirmed to produce siderophores in liquid culture medium containing 10 uM iron. Genetic analysis of one mutant UMC002 indicated that the regulatory defect is controlled by a single genetic locus and that the mutation is recessive.

Cosmid DNA clones were identified that restored normal regulation of siderophore biosynthesis in UMC002 (Leong et al, 1990). From this cosmid a 5 kb *Bam*H1/*Xba*I fragment cloned in the replicative vector pCM54 was found to encode the entire complementing region. Gene disruption experiments were conducted to further localize the gene. The 5 kb fragment was disrupted at the unique *Xho*I site with a chimeric selectable marker for hygromycin resistance (Wang et al, 1989) (figure 3). Transformation of the wild type with the linearized construct gave rise to many colonies that were constitutive for siderophore production. Preliminary Southern hybridization analysis of DNA from these transformants gave a pattern of hybridization that is consistent with a gene replacement event.

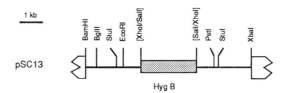

Figure 3. Disruption of the 5 kb *Bam*H1/*Xba*I fragment that complements mutants constitutive for siderophore production.

The nature of the function encoded by this fragment remains to be determined. The current data are compatible with the gene product acting either directly or indirectly to negatively regulate siderophore biogenesis. This gene may encode a protein like the *Escherichia coli fur* gene, which specifies a repressor of siderophore biosynthetic and transport genes (Bagg and Neilands, 1987). It seems unlikely that this gene encodes some function involved with transport of siderophores or iron into cells. For example *exbB* mutants of *E. coli* hyperexcrete enterobactin, its native siderophore, in media containing 100 uM iron (Gutterman, 1973). The product of *exbB* gene is a cytoplasmic membrane protein required for uptake of siderophores and vitamin B12 (Braun et al, 1987). Unlike *exbB* mutants which grow poorly even in the presence of iron, the *U. maydis* constitutive mutants attain similar cell densities as the wild type parent strain under all conditions of growth indicating that transport of siderophores or iron is not limiting in these mutants. Future experiments will involve DNA sequence analysis of the minimal region required for complementation, analysis of trnascriptional regulation of the oxygenase and regulatory genes in wild type and constitutive mutants, and purification and biochemical characterization of the gene product.

3. Acknowledgement.

This work was supported by the U.S.D.A., the Graduate School of the University of Wisconsin, and Public Health Service grant 1 RO1 GM33716 from the National Institutes of Health to S.A.L., a fellowship from the Schwiezencher Nationalfonds to C. V., and an operating grant from the Natural Sciences and Engineering Research Council of Canada to J.W.K.

4. References

Bagg, A. and J. B. Neilands. 1987. Molecular mechanism of regulation of siderophore-mediated iron assimilation. *Microbiol. Rev.* 51: 509-518.

Banuett, F. and I Herskowitz. 1989. Different *a* alleles of *Ustilago maydis* are necessary for maintenance of filamentous growth but not for meiosis. *Proc. Natl. Acad. Sci. U.S.A.* 86:5878-5882.

Banks, G. R. and S. Y. Taylor. 1988. Cloning of the PYR3 gene of *Ustilago maydis* and its use in DNA transformation. *Mol. Cell. Biol.* 8:5417-5424.

Braun et al. 1987. Iron transport systems in *Escherichia coli*. In: Iron Transport in Microbes, Plants and Animals (G. Winkelmann, D. van der Helm and J. B. Neilands, eds.) VCH, Verlagsgesellschaft. pp.35-51

Budde, A. D. and S. A. Leong. 1989. Characterization of siderophores from *Ustilago maydis*. *Mycopathologia* 108:125-133.

Budde, A. D. and S. A. Leong. 1990. *Ustilago maydis*. In: Genetic Maps, Vol. 5 (S. J. O'Brien, ed.) Cold Spring Harbor Laboratory, New York.

Christensen, J. J. 1963. Corn smut caused by *Ustilago maydis*. Amer. Phytopathological Soc. Monograph No. 2.

Fotheringham, S. and W. K. Holloman. 1989. Cloning and disruption of *Ustilago maydis* genes. *Mol. Cell. Biol.* 9:4052-4055.

Froeliger, E.H. and J. W. Kronstad. 1990. Mating and pathogenicity in *Ustilago maydis*. In: Seminars in Developmental Biology: Fungal development and mating interactions (C.A. Raper and D.I. Johnson, eds) Saunders Scientific Publications: London. In press

Froeliger, E. H., and S. A. Leong. 1989. Isolation of mating type genes from *Ustilago maydis*. Genetics and Cellular Biology of Basidiomycetes. University of Toronto. (Abstr.)

Giasson, L et al. 1989. Cloning and comparison of Aα mating-type alleles of the basidiomycete *Schizophyllum commune*. *Mol. Gen. Genet.* 218:72-77.

Glass, N. L et al 1988. DNAs of the two mating type alleles of *Neurospora crassa* are highly dissimilar. *Science* 241:570-573.

Hargreaves, J. A. and G. Turner. 1989. Isolation of the acetyl-CoA Synthase gene from the Corn Smut pathogen. J. Gen. Microbiol. 135:2675-2678.

Herskowitz, I., 1988. Life cycle of the budding yeast *Saccharomyces cerevisiae*. *Microbiol. Rev.* 52:536-553.

Holden, D. W., Kronstad, J. and S. A. Leong. 1989a. Mutation in a heat-regulated hsp70 gene of *Ustilago maydis*. *EMBO J.* 8:1927-1934.

262

Holden, D. W., Spanos, A. and G. R. Banks. 1989b. Nucleotide sequence of the REC1 gene of *Ustilago maydis*. Nucl. Acids Res. 17:10489.

Holliday, R. 1974. *Ustilago maydis*. In: Handbook of Genetics (R. C. King, ed.) Vol. 1, Plenum Press, New York. pp575-595

Kinscherf, T. and S. A. Leong. 1988. Molecular analysis of the karyotype of *Ustilago maydis*. *Chromosoma* 96:427-433.

Kronstad, J. et al. 1989. Isolation of metabolic genes and demonstration of gene disruption in *Ustilago maydis*. *Gene* 79:97-106.

Kronstad, J. W. and S. A. Leong. 1989a. Isolation of two alleles of the *b* locus of *Ustilago maydis*. *Proc. Natl. Acad. Sci. U.S.A.* 86:978-982.

Kronstad, J. W. and S. A. Leong. 1989b. Molecular characterization of the *b* locus of *Ustilago maydis*. *Fungal Genetics Newsletter* 36:26.

Kronstad, J. W. and S. A. Leong. 1990. The *b* mating type locus of *Ustilago maydis* contains variable and constant domains. Genes and Develop. 4:1384-1395.

Leong, S. A. et al. 1987. Molecular strategies for the analysis of the interaction of *Ustilago maydis* and maize. In: Molecular Strategies for Crop Protection (C. Arntzen and C. Ryan, eds.) Alan R. Liss, New York. pp. 95-106

Leong, S. A. et al. 1988. Identification and molecular characterization of genes which control pathogenic growth of *Ustilago maydis* in maize. In: Molecular Genetics of Plant-Microbe Interactions (D. P. Verma and R. Palacios, eds.) APS Press. St.Paul. pp.241-246

Leong, S. A. and D. Expert. 1989. Siderophores in plant-pathogen interactions. In: Plant-Microbe Interactions-A Molecular Genetic Perspective, Vol. 3 (E. Nester and T. Kosuge, eds.) McGraw Hill, New York. pp.62-8

Leong, S. A. et al. 1990. Molecular analysis of pathogenesis in *Ustilago maydis*. In: Molecular Strategies of Pathogens and Host Plants (S. S. Patil, D. Mills, and C. Vance, eds.) Springer-Verlag, New York. In press

Loper, J. and J. Buyer. 1990. Siderophores in relation to biological control of plant disease. MPMI. In Press.

Matzanke, B. F. 1987. Mossbauer Spectroscopy of microbial iron uptake and metabolism. In: Iron Transport in Microbes, Plants and Animals (G. Winkelmann, D. van der Helm, and J. B. Neilands, eds.) VHC, Verlagsgesellschaft. pp.251-284

Metzenberg, R. and N. L. Glass. 1990. Mating type and mating strategies in *Neurospora*. BioEssays 12:53-59.

Payne, S. 1988. Iron and virulence in the family enterobacteriaceae. CRC Microbiol. 16:81-111.

Plattner, H. et al. 1989. Isolation and some properties of lysine N^6-hydrolase from *Escherichia coli* strain EM222. Biol. Metals 2:1-5.

Schultz, B. et al. 1990. The *b* alleles of *U. maydis*, whose combinations program pathogenic development, code for polypeptides containing a homeodomain-related motif. *Cell* 60:295-306.

Smith, T. and S. A. Leong. (1990) Isolation and characterization of a *Ustilago maydis* glyceraldehyde-3-phosphate dehydrogenase encoding gene. *Gene*. In press

Tsukuda, T. et al. 1988. Isolation and characterization of an automonously replicating sequence from *Ustilago maydis*. *Mol. Cell. Biol.* 8:3703-3709.

Tsukuda, T., Bauchwitz, R. and W. K. Holloman. 1989. Isolation of the REC1 gene controlling recombination in *Ustilago maydis*. Gene 85:335-341.

Wang, J., Holden, D. and S. A. Leong. 1988. Gene transfer system for the phytopathogenic fungus *Ustilago maydis*. *Proc. Natl. Acad. Sci. USA* 85:865-869.

Wang, J., Budde, A. and S. A. Leong. 1989. Analysis of ferrichrome biosynthesis in the phytopathogenic fungus *Ustilago maydis*: Cloning of an ornithine-N^5-oxygenase gene. *J. Bacteriol.* 171:2811-2818.

THE *B* LOCUS OF *USTILAGO MAYDIS:* MOLECULAR ANALYSIS OF ALLELE SPECIFICITY

M. DAHL, M. BÖLKER, B. GILLISSEN, F. SCHAUWECKER,
B. SCHROEER and R. KAHMANN
Institut für Genbiologische Forschung Berlin GmbH
Ihnestr.63
D-1000 Berlin 33, FRG

ABSTRACT. *Ustilago maydis* is a dimorphic fungal pathogen of corn. One form is yeast-like and non-pathogenic; the other is filamentous and pathogenic. The two loci *a* and *b* with two and 25 different alleles, respectively, regulate this dimorphism. Any combination of two different alleles at both loci allows cell fusion and triggers pathogenic development. The cloning and analysis of different *b* alleles revealed that they contain a single open reading frame of 410 amino acids with a variable N-terminal region and a highly conserved C-terminal region. The constant region contains a motif characteristic for homeodomain proteins suggesting that combinatorial interactions between *b* polypeptides generate regulatory proteins that determine the developmental program of the fungus. We report here the localization of allele specificity determinants in *b*, the construction of a *b*-null mutant and the cloning of the gene loci *a*1 and *a*2.

Introduction

Ustilago maydis belongs to the large group of plant pathogenic Basidiomycetes known as the smut fungi (for reviews see Christensen, 1963; Banuett and Herskowitz, 1988). In its life cycle two forms can be distinguished: one is unicellular and haploid, divides by budding, forms yeast-like colonies on defined media, and is non-pathogenic; the other form is the filamentous dikaryon, whose growth is dependent on the corn plant and causes tumors on leaves, stems, tassels and ears (Christensen, 1963). This dimorphism is governed by the incompatibility or mating type loci *a* and *b*. Development of the pathogenic dikaryon is initiated after cell fusion of two haploids that differ at both loci (Rowell and DeVay, 1954; Rowell, 1955; Holliday, 1961; Puhalla, 1970; Day et al., 1971).

The *b* locus is quite intriguing because there are estimated to be twenty-five naturally-occurring alleles (Rowell and DeVay, 1954; Puhalla, 1968) which function in all pairwise combinations to initiate filamentous growth and pathogenicity. This program is not initiated if

264

H. Hennecke and D. P. S. Verma (eds.),
Advances in Molecular Genetics of Plant-Microbe Interactions, Vol. 1, 264–271.
© 1991 *Kluwer Academic Publishers. Printed in the Netherlands.*

strains carry the same *b* allele. The *a* locus has only two alleles, *a*1 and *a*2, and different *a* alleles are necessary for cell fusion and together with different *b* alleles, for the development of the filamentous form (Rowell and DeVay, 1954; Holliday, 1961; Puhalla, 1968; Banuett and Herskowitz, 1989). In other Basidiomycetes like *Schizophyllum commune* and *Coprinus cinereus*, fruiting body formation requires the presence of different alleles at two incompatibility loci, both of which have multiple alleles (Raper, 1983; Casselton, 1978). Thus, an understanding of the underlying mechanism for self and nonself recognition is crucial for gaining insight into the molecular basis of sexual development in these fungi. For *U. maydis* it is expected that such studies will also provide important clues to plant-pathogen interactions as completion of the sexual cycle requires the corn plant.

A variety of molecular explanations have been proposed for the mechanism of recognition of identical versus non-identical alleles of the Basidiomycete incompatibility loci (see Ullrich, 1978; Metzenberg, 1990). With molecular approaches now possible, e.g. the cloning and characterization of the incompatibility loci in filamentous fungi we are at the beginning of a new era. In this article we shall focus on recent accomplishments in the analysis of the *b* and *a* mating type loci of *Ustilago maydis* .

Materials and Methods

The following *U.maydis* strains were kindly provided by F.Banuett and are described in Banuett and Herskowitz (1989): FB1 (*a*1*b*1), FB2 (*a*2*b*2), FB6a (*a*2*b*1), FB6b (*a*1*b*2), FBD12-11 (*a*2/*a*2 *b*1/*b*2), FBD11-21 (*a*1/*a*2 *b*1/*b*2). All other techniques concerning mating reactions, DNA transformation and tests for pathogenicity were performed as described (Schulz et al., 1990).

Results and Discussion

CHARACTERIZATION OF *b* ALLELES IN *U.MAYDIS*

Two *b* alleles have been cloned initially by using a gene complementation approach: diploids heterozygous for *a* and homozygous for *b* which grow yeast-like (Fuz⁻ phenotype) were transformed with a gene bank from a strain containing a different *b* allele. Transformants were screened for filamentous growth on nutrient charcoal media (Fuz + phenotype) and subsequently tested for pathogenicity by inoculating corn plants with pure cultures of the respective transformants (Kronstad and Leong, 1989, Schulz et al., 1990). Recloning of the integrated DNA in such transformants and identification of the cosmid clone which caused the Fuz + phenotype, respectively, allowed to confine the complementing activity to a BamHI fragment of 8 kb. Other *b* alleles could be cloned and identified by hybridization to this 8 kb BamHI fragment indicating a high degree of homology between different alleles. The sequence analysis of genomic DNA of four different alleles by our group (Schulz et al., 1990) and of six *b* alleles by Kronstad and Leong (1990) has revealed essentially the same results: the *b* locus contains one open reading frame for a polypeptide of 410 amino acids. The N-terminal 110 amino acids show a high degree of variability between different alleles (about 60% identity) whereas the remaining portion of the open reading frame is conserved to a degree of about 90% in all alleles analyzed. Characterization of a partial cDNA clone for the b4 allele has revealed that the ORF contains a 73 bp intron near the C-terminus, splicing does eliminate the stop codon for the 410 aa ORF and extends it to 472 aa (Schulz et al., 1990). The C-terminus of *b*, however, is not essential for *b* activity as deletions which eliminate the C-terminus up to amino acid position 330 are fully active. On the other hand is the effect of small insertions or deletions within the variable domain quite dramatic, they all abolish *b* activity completely (Schulz et al., 1990). The constant region of the polypeptides encoded by the *b* alleles contains the four highly conserved amino acids WF-N-R (Schulz et al., 1990) found in all eukaryotic homeodomain proteins (see Scott et al., 1989). These residues are part of the putative DNA recognition helix in the helix-turn-helix motif of homeodomain proteins (Quian et al., 1989). This motif is also found in the yeast family of regulatory proteins MATa1 and PHO2 from *Saccharomyces cerevisiae* and mat-Pi from *Schizosaccharomyces pombe* which are related to homeodomain proteins. The *S. cerevisiae* MAT α2 protein has three of these residues. MATa1, MAT α2, and PHO2 are all known to be DNA binding proteins (see Scott et al., 1989). A high degree of similarity is observed in comparisons of a 40 amino acid region containing WF-N-R between the *U. maydis b* polypeptides and these known yeast DNA binding proteins (Schulz et al., 1990) suggesting that the *b* polypeptides function as regulatory proteins. Given the biological observation that the presence of different *b* alleles is necessary for filamentous growth and pathogenicity, we have proposed that the status of the *b* alleles in the cell is monitored by interaction of monomers, resulting in formation of a multimeric regulatory protein that governs sexual development (Schulz et al., 1990). For the function of the *b* polypeptides one can envision two basic scenarios: (1) they might, as monomer or homodimer, act as repressors for genes affecting filamentous growth (*fuz* genes) and tumor development (*tum* genes) in haploid cells. In diploids, with two different *b* polypeptides present, heterodimers could form preferentially which are unable to repress the respective gene set. (2) The alternative model proposes that the monomer or homodimeric species is inactive and that a heterodimeric species activates the *fuz* and *tum* genes in the dikaryon. Neither of these models addresses how the association of *b* polypeptides is governed (through the variable or the constant domains ?) nor how protein function or non-function is determined in the many possible associations of at least 25 different allelic variants. Undoubtedly it will be a challenging task to determine what governs allele specificty and how it evolved.

THE PHENOTYPE OF A *b*-NULL MUTATION

In light of the proposed models for the function of the *b* polypeptides the phenotype of a null mutation in *b* should be most informative: if the repressor model is correct a haploid strain carrying a deletion in *b* should become pathogenic whereas the activator model predicts that such a strain neither by itself nor in combination with another *b* allele should cause symptoms in planta.

We have constructed a b1-null allele on a plasmid (pb1-0-BX) by making an internal deletion in the 8 kb BamHI fragment of *b1* which encompasses sequences 5' to the *b1* coding region up to position -409 (BglII site) as well as the N-terminal 678 bp of the *b1* coding region (XhoI site). This fragment was replaced with a cassette encoding hygromycin resistance. Strain FB1 (*a1b1*) was transformed with pb1-0-BX DNA that had been restricted with BamHI. Transformants were selected for hygromycin resistance and genomic DNA was hybridized with a *b*-specific probe after digestion with BglII and BamHI. The BglII-BamHI fragment in FB1 comprises 6.7 kb, if the b1-0-BX allele has replaced the resident *b1* allele the *b1* specific fragment should be absent and instead a new fragment representing the 10 kb b1-0-BX allele should hybridize. Of 16 clones analyzed one such strain, FB1/b1-0-BX#11 was found. In mating assays with FB2 (*a2b2*) this strain gave no Fuz reaction. In planta we have not observed tumor induction when this strain was inoculated alone or in combination with FB2. The same observations were made by Kronstad and Leong (1990) who have introduced by gene disruption a similarly constructed b1-null allele. These results support models which propose that the interaction of two different b polypeptides generates a novel regulatory species which functions as an activator for genes leading to pathogenic development.

THE SPECIFICITY DOMAIN OF *b*

To gain insight into mechanisms that govern the process of recognition between identical or non-identical *b* polypeptides we have mapped the region which confers allele specificity by generating hybrid alleles. Suitable restriction sites within the coding regions of *b2*, *b3*, and *b4* were used. Hybrid alleles were introduced into haploid strains carrying as resident alleles either *b1*, *b2*, *b3* or *b4* and scored for their Fuz and Tum phenotype. All hybrid alleles were active and displayed a Fuz$^+$ Tum$^+$ phenotype with nonparental *b* alleles. In combination with the parental alleles all hybrids displayed the specificity of one of the parents (Fuz$^-$ Tum$^-$ phenotype in the respective strain): if the N-terminal domain up to amino acid position 56 was derived from *b2* and the C-terminus from *b3* the hybrid had *b3* specificity; if amino acids 1 to 115 were from *b2* and the C-terminal part originated from *b3* the hybrid showed *b2* specificity. This allowed to confine the specificity domain to 60 amino acids (position 56 to 115) within the variable region.

We have recently determined the nucleotide sequence for the variable region of four additional alleles from the ATCC collection: *b6* (K), *b8* (N), *b9* (P) and *b10* (R) after cloning the respective alleles by PCR. Inspection of sequences comprising the specificity domain of the eight alleles now in our collection detects at least 11 amino acid exchanges between any two of these alleles. Does this mean that multiple changes must occur to generate new specificities? And is this one of the reasons why new specificities have never appeared in conventional crosses (Puhalla, 1970)? At present we cannot give a definite answer to this question. However, we have generated one mutant allele of *b3* (a genuine PCR "artifact") which causes a Tum$^+$ phenotype when introduced into a haploid strain carrying *b3*, indicating that it has aquired a specificity which is non-identical with *b3*. Subsequent sequence analysis revealed that this *b3*-mut allele differs from *b3* in only three amino acid positions, two of the mutations fall in the specificity domain. It should be most informative to see if any of the three single mutations alone can cause the change in specificity.

CLONING OF THE *a* MATING TYPE LOCUS

The *a* locus of *U. maydis* exists in two alleles, *a1* and *a2*. From the behavior of diploid lines homozygous at *a* and heterozygous at *b* (Fuz$^-$ Tum$^+$) or heterozygous at *a* and homozygous at *b* (Fuz$^-$ Tum$^-$), respectively, it has been inferred that tumor induction requires different alleles at *b* but not at *a* while filamentous growth on charcoal nutrient media necessitates different alleles at *a* and *b* (Banuett and Herskowitz, 1989). Fuz$^-$ diploids of the above types will show a mating reaction provided the mating partner carries the compatible *a* or *b* allele, respectively, indicating that heterozygosity at *b* or *a* does not block the mating reaction. In fact, diploids, heterozygous at *a* and homozygous at *b* (e.g. *a1/a2 b1/b1*) have become dual maters, they show filamentous growth when mated with a1b2 and a2b2 strains (Banuett and Herskowitz, 1989). For our attempts to identify a clone carrying the a locus we made use of this latter property. We transformed the haploid strain FB2 (a2b2) with a cosmid bank derived from FBD11-21 (a1/a2 b2/b2) strain. About 20.000 hygromycin resistant transformants were screened by replica mating with FB6a (a2b1). FB2 will not mate with Fb6a because both strains carry the same a allele. One transformant, FB2Hyg-a1, however, was detected which gave a mating reaction with FB6a. Subsequent analysis showed that this clone mated not only with FB6a but also with FB1 (a1b1) (Figure 1), indicating that it had acquired a dual mater phenotype.

Figure 1: Dual mating phenotype of strain FB2Hyg-a1. Saturated cultures were cross-streaked against each other on charcoal nutrient media. Top horizontal line is FB1(a1b1), bottom horizontal line is FB6a (a2b1). Vertical lines from left to right are FB2(a2b2), FB2Hyg-a1, FB2Hyg-a1, FB6b(a1b2). The white fuzziness develops if two compatible strains fuse and form dikaryotic filaments. The FB2Hyg-a1 transformant is able to form dikaryotic filaments with both tester strains.

By recloning parts of the integrated cosmid in this transformant and using this DNA as probe for screening the cosmid library we identified the cosmid carrying the *a*1 locus. After subcloning *a*1 was initially located on a 10 kb BamHI fragment and subsequently on a 3.9 kb EcoRI fragment. When introduced in strain FB2 (*a2b2*) all transformants acquired the dual mater phenotype. As further test for the successful cloning of *a*1 the plasmid containing the 10 kb BamHI fragment was introduced into the Fuz⁻ strain FBD12-11 (*a2/a2 b1/b2*). All trans-formants displayed a Fuz⁺ phenotype indistinguishable from diploid strains heterozygous at *a* and *b*. Different parts of the 10 kb BamHI fragment were then used as hybridization probes for southern blots from strains carrying either *a*1 or *a*2. The 3.9kb EcoRI fragment did not hybridize with DNA from the *a*2 strain, indicating the absence of *a*1 homologous sequences in *a*2 strains. A probe located adjacent to the *a*1 EcoRI fragment revealed a restriction length polymorphism between *a*1 and *a*2, indicating that this fragment might encompass the border between sequences homologous at *a*1 and *a*2 and sequences different at *a*1 and *a*2. This fragment was used to screen a phage λ library from an *a2* strain. By mapping a series of lambda clones which hybridized to this probe and by subcloning fragments we were able to assign *a2* activity to a 6 kb BamHI fragment. In comparing restriction maps of *a*1 and *a*2 it became apparent that about 4.5 kb of the *a*1 locus had no counterpart in *a*2, whereas a stretch of about 8 kb encompassing *a*2 was absent from *a*1 strains. This situation is reminiscent of the A and a mating type loci of *Neurospora crassa* where the term idiomorph has been introduced to denote sequences which occupy the same locus but share no apparent homology (Metzenberg and Glass, 1990). We will use the term idiomorph for the *a* mating type locus of *Ustilago maydis* in all further discussions. By subcloning restriction fragments we were able to locate *a*1 activity on a 1.1 kb fragment close to the left border of the *a*1 idiomorph.The *a*2 activity was shown to reside in the rightmost 2 kb of the *a*2 idiomorph. We have sequenced the entire *a*1 idiomorph and the region of *a*2 which is sufficient to provide *a*2 function. A comparison of border sequences revealed complete identity followed by partial identical sequences for a stretch of about 150 bp, beyond that sequences showed no similarity. This situation is characteristic for both borders, there is no indication of repeated sequence elements at or near the junctions. The genomic sequences of *a*1 and *a*2 do not allow us to pinpoint to an open reading frame which could be responsible for the activity, the isolation of cDNA clones appears necessary to establish that *a*1 and *a*2 code for polypeptides. Such experiments are in progress.

PERSPECTIVE

With the molecular cloning and characterization of the *a* and *b* mating type loci of *U.maydis* we have the tools for approaching a molecular understanding of how these loci control the developmental program of this fungus. It is amazing to see how different the *a* and *b* loci are organized: for *a*1 and *a*2 there appears to be no sequence similarity, both activities reside on opposite ends of idiomorphs differing in length. It is unknown how these sequences evolved, how they are maintained and which function they serve besides specifying the *a*1 and *a*2 activities. For the multiallelic *b* locus, on the other hand, the sequence analysis has revealed true alleles, genes which share a high degree of identity in some parts and some striking variability   in others. It was by no means surprising that allele specificity resides in the most variable N-terminal domain. If the *b* polypeptides are regulatory proteins, as the presence of the homeodomain-related motif suggests, the puzzling question of recognition and discrimination between at least 25 identical and at least 300 non-identical polypeptide combinations remains unanswered. Clues are undoubtedly going to come from mutant alleles with new, constitutive, relaxed or altered specificity for which we have presented one first example. In other Basidiomycete and Ascomycete fungi the molecular analysis of mating type loci is progressing at high speed (see Metzenberg, 1990) and it should be exciting to see if similar concepts are used to control such basic processes of life as sexual reproduction. Only recently homology between a mating type gene from *S. pombe* and a gene from the sex-determining region on the Y chromosome has been reported (Sinclair et al., 1990). Furthermore, with the key regulatory repertoire of *a* and *b* available we can now proceed to identify target genes and establish how the products of such genes promote sexual development and pathogenicity.

ACKNOWLEDGEMENT

We thank Flora Banuett for strains, stimulating discussions and encouragement. This research was supported by a grant from the German Minister of Science and Technology to R.K.

REFERENCES

Banuett, F. and Herskowitz, I. (1988). Ustilago maydis, smut of maize. In Genetics of Plant Pathogenic Fungi, Advances in Plant Pathology, G.S. Sidhu, ed. (Academic Press, London), vol.6, pp.427-455.

Banuett, F. and Herskowitz,I. (1989). Different a alleles of Ustilago maydis are necessary for maintenance of filamentous growth but not for meiosis. Proc. Natl. Acad. Sci. USA, 86, 5878-5882.

Casselton, L.A. (1978). Dikaryon formation in higher basidiomycetes. In The Filamentous Fungi, J.E.Smith and D.R.Berry, eds (Arnold, London), vol.3, pp.275-297.

Christensen, J.J. (1963). Corn smut caused by *Ustilago maydis*. Amer. Phytopathol. Soc. Monogr. No.2.

Day, P.R., Anagnostakis, S.L., and Puhalla, J.E. (1971). Pathogenicity resulting from mutation at the *b* locus of *Ustilago maydis*. Proc. Natl. Acad. Sci. USA, 68, 533-535.

Holliday, R. (1961). Induced mitotic crossing-over in *Ustilago maydis*. Genet. Res., 2, 231-248.

Kronstad, J.W. and Leong, S.A. (1989). Isolation of two alleles of the *b* locus of *Ustilago maydis*. Proc. Natl. Acad. Sci. USA 86, 978-982.

Kronstad, J.W. and Leong, S.A. (1990). The *b* mating-type locus of *Ustilago maydis* contains variable and constant regions. Genes & Development, 4, 1384-1395.

Metzenberg, R.L. (1990). The role of similarity and difference in fungal mating. Genetics, 125, 457-462.

Metzenberg, R.L. and Glass, N.L. (1990). Mating type and mating strategies in *Neurospora* BioEssays, 12, 53-59.

Puhalla, J.E. (1968). Compatibility reactions on solid medium and interstrain inhibition in *Ustilago maydis*. Genetics 60, 461-474.

Puhalla, J.E. (1970). Genetic studies on the incompatibility locus of *Ustilago maydis*. Genet. Res., 16, 229-232.

Quian, Y.Q., Billeter, M., Otting, G., Müller, M., Gehring, W.J., and Wüthrich, K. (1989). The structure of the Antennapedia homeodomain determined by NMR spectroscopy in solution: comparison with prokaryotic repressors. Cell 59, 573-580.

Raper, C.A. (1983) Controls for development and differentiation of the dikaryon in

Basidiomycetes. In Secondary Metabolism and Differentiation in Fungi, J.W. Bennett and A.Ciegler, eds. (Marcel Dekker, Inc., New York, Basel), pp. 195-238.

Rowell, J.B. and DeVay, J.E. (1954). Genetics of Ustilago zeae in relation to basic problems of its pathogenicity. Phytopathology, 44, 356-362.

Rowell, J.B. (1955). Functional role of compatibility factors and an in vitro test for sexual compatibility with haploid lines of *Ustilago zeae*. Phytopathology, 45, 370-374.

Schulz, B., Banuett, F., Dahl, M., Schlesinger, R., Schäfer, W., Martin, T., Herskowitz, I., and Kahmann, R. (1990).The *b* alleles of *U.maydis*, whose combinations program pathogenic development, code for polypeptides containing a homeodomain-related motif. Cell, 60, 295-306.

Scott, M.P., Tamkun, J.W., and Hartzell III, G.W. (1989). The structure and function of the homeodomain. Biochim. Biophys. Acta, 989, 25-48.

Sinclair, A.H., Berta, P., Palmer, M.S., Hawkins, J.R., Griffiths, B.L., Smith, M.J., Foster, J.W., Frischauf, A.-M., Lovell-Badge, R., and Goodfellow, P.N. (1990). A gene from the human sex-determining region encodes a protein with homology to a conserved DNA-binding motiv. Nature, 346, 240-244.

Ullrich, R.C. (1978). On the regulation of gene expression: incompatibility in Schizophyllum. Genetics, 88, 709-722.

AN *IN PLANTA* INDUCED GENE OF *PHYTOPHTHORA INFESTANS* CODES FOR UBIQUITIN

C.M.J. PIETERSE, E.P. RISSEEUW and L.C. DAVIDSE
Department of Phytopathology
Agricultural University of Wageningen
P.O. Box 8025
6700 EE Wageningen
The Netherlands

ABSTRACT. An *in planta* induced gene of *Phytophthora infestans* (the causal organism of potato late blight) was isolated from a genomic library by differential hybridization using labelled cDNA derived from poly(A)$^+$ RNA of *P. infestans* grown *in vitro* and labelled cDNA made from potato-*P. infestans* interaction poly(A)$^+$ RNA as probes. Sequence analysis showed that the gene codes for ubiquitin, a highly conserved protein which plays an important role in several cellular processes. The structure of the polyubiquitin gene is consistent with the stucture of other known polyubiquitin genes. Northern and Southern blot analyses revealed that the polyubiquitin gene is a member of a multigene family of which all genes show induced expression *in planta*.

1. Introduction

Potato late blight caused by the fungus *Phytophthora infestans* (Mont.) de Bary (Oomycetes) is one of the most important diseases of potato. Leaves and tubers of susceptible cultivars become readily infected by this pathogen. The fungus spreads rapidly through the plant tissue causing a destructive necrosis.

It can be assumed that the establishment of a pathogenic relation between the potato plant and *P. infestans* involves mutual interference in cellular processes of each partner. Defence responses of host tissue being colonized are relatively well studied [1-3], but nothing is known about the molecular basis underlying pathogenicity of the fungus. In order to gain more insight in the molecular processes involved we are studying the gene expression of *P. infestans* during pathogenesis on potato by differential screening of a genomic library of *P. infestans* DNA. Several differentially hybridizing clones containing putative *in planta* induced genes of *P. infestans* were isolated. The coding region of one of these genes was mapped on the lambda clone and completely sequenced. The sequence appeared to code for ubiquitin, a highly conserved protein in eukaryotic organisms which is reported to be involved in several important cellular processes such as intracellular protein breakdown, maintainance of chromatin structure, regulation of gene expression and modification of cell surface receptors (for reviews see [4-6]). The significance of the *in planta* induced expression of this gene will be discussed.

2. Differential screening of the genomic library of *P. infestans* DNA

Differential screening of the genomic library with fungal and interaction cDNA is performed under non-saturating conditions. Under these conditions the intensity of each obtained signal corresponds with the abundancy of a particular cDNA in the probe. A stronger signal obtained

272

H. Hennecke and D. P. S. Verma (eds.),
Advances in Molecular Genetics of Plant-Microbe Interactions, Vol. 1, 272–275.
© 1991 *Kluwer Academic Publishers. Printed in the Netherlands.*

after hybridization with interaction cDNA compared to the signal obtained after hybridization with fungal cDNA indicates a higher abundancy of those cDNA's which are complementary to the DNA in the hybridyzing genomic lambda clone. Since the filters are hybridized with equal amounts of labeled cDNA the procedure even underestimates the difference in abundancy because the quantity of fungal derived cDNA present in the interaction cDNA probe is much less than in the fungal cDNA probe.

The genomic library of *P. infestans* DNA in phage lambda EMBL3 was differentially screened using equal amounts (10^7 cpm) of ^{32}P-labeled cDNA made from poly(A)$^+$ RNA of the fungus grown *in vitro* and ^{32}P-labeled cDNA derived from interaction poly(A)$^+$ RNA as probes. Several clones giving a relatively strong signal after hybridization with labeled interaction cDNA and a relatively weak or no signal after hybridization with labeled cDNA of the fungus grown *in vitro* were isolated. One of these clones was purified after a second round of hybridization. Southern analysis of various restriction fragments of the isolated lambda clone with interaction cDNA as probe revealed a strongly hybridizing *Sst*I-fragment of 3.2 kb in length. This fragment containing a putative *in planta* induced gene was subcloned in pTZ19U and a map of restriction sites was constructed from comparisons of single and multiple restriction enzyme digests of the obtained plasmid pUB-S. The approximate location of the coding region of the differentially expressed gene (closed bar in Fig. 1a) was determined by Southern blot analysis using interaction cDNA as probe.

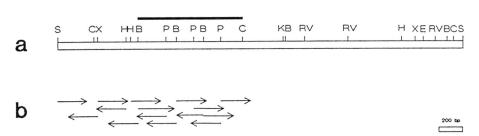

Figure 1. a) Restriction map of the 3.2 kb *Sst*I insert of pUB-S. The closed bar indicates the approximate location of the coding region of the differentially expressed gene as determined by Southern hybridization using labeled interaction cDNA as probe. b) The sequencing strategy is indicated by arrows. B=*Bgl*II; C=*Cla*I; E=*Eco*RI; H=*Hind*III; K=*Kpn*I; P=*Pvu*II; S=*Sst*I; X=*Xho*I; RV=*Eco*RV.

3. Characterization of the *in planta* induced gene of *P. infestans*

Several fragments of the 3.2 kb *Sst*I-insert of pUB-S were subcloned and the sequence of 1632 nt was determined by dideoxy sequencing on both strands of various overlapping clones (Pieterse *et al*, submitted). The sequencing strategy is summarized in Fig. 1b. An open-reading frame of 687 nt containing three almost identical 228 nt repeats was found in the 1632 nt sequence. Comparison of the 687 nt sequence with the sequence databank revealed that the sequence codes for polyubiquitin. The coding region encodes three ubiquitin units with a length of 76 amino acids in a head-to-tail arrangement followed by an extra asparagine residue at the carboxy-terminal end (Fig. 2). Although the ubiquitin-coding repeats within the gene differ up to 8 out of 228 bases, they code for identical amino acid sequences. Only 5 amino acids of the 76 amino acid ubiquitin sequence differ from the sequence of human, chicken and yeast ubiquitin.

274

```
  1    MQIFVKTLTG KTITLDVEPS DSIDNVKQKI QDKEGIPPDQ QRLIFAGKQL

 51    EDGRTLSDYN IQKESTLHLV LRLRGGMQIF VKTLTGKTIT LDVEPSDSID

101    NVKQKIQDKE GIPPDQQRLI FAGKQLEDGR TLSDYNIQKE STLHLVLRLR

151    GGMQIFVKTL TGKTITLDVE PSDSIDNVKQ KIQDKEGIPP DQQRLIFAGK

201    QLEDGRTLSD YNIQKESTLH LVLRLRGGN*
```

Figure 2. Predicted amino acid sequence of the cloned ubiquitin gene of *P. infestans*. Arrows indicate the start of the first, second and third ubiquitin encoding repeat.

4. *In planta* induced expression of *P. infestans* ubiquitin genes

Northern blot analysis of RNA isolated from the *in vitro* grown mycelium, using the two 228 bp *Pvu*II-fragments from the ubiquitin coding region of pUB-S as probe, shows five bands of approximately 640, 870, 1100, 1650 and 2150 nucleotides in length. The size differences between the transcripts are such that each transcript appears to contain a different number of ubiquitin repeats (2, 3, 4, 6 and 8 respectively). This indicates that the genome of *P. infestans* contains multiple copies of ubiquitin encoding genes of different lengths. Southern blot analysis of digested genomic DNA in which the ubiquitin encoding 228 bp *Pvu*II-fragments of pUB-S were used as a probe, confirmed that the identified ubiquitin gene belongs to a multigene family.

The differential expression of the ubiquitin genes was confirmed by Northern blot analysis of equal amounts of interaction and fungal poly(A)$^+$ RNA. RNA isolated from colonized leaf tissue and RNA isolated from the fungus grown *in vitro* give rise to comparable signals upon hybridization with the ubiquitin coding region. Although the amount of fungal RNA as a percentage of the total RNA obtained from the interaction is difficult to measure, it most certainly is very low (5-10%). Taking the strong under representation of fungal RNA in the interaction RNA mixture in consideration, a several fold higher expression of all the ubiquitin encoding genes during growth of the fungus *in planta* could be deduced from the Northern blot.

5. Discussion

Ubiquitin is one of the most highly conserved proteins known to date (for reviews see [4-6]). The 76 amino acid protein occurs in all eukaryotic cells, either free or covalently attached to proteins in the cytosol, plasma membrane or to chromosomal histones. Ubiquitin has been shown to play a key role in several important cellular processes such as the selective degradation of intra-cellular proteins, maintainance of chromatin structure, regulation of gene expression and modification of cell surface receptors. Genes encoding ubiquitin have been characterized from a variety of organisms such as yeast [7], human [8], chicken [9]. In each case, ubiquitin is encoded by one or more polyubiquitin genes which consist of direct repeats of the 76 amino acid coding units. The last repeat at the 3' end of the polyubiquitin gene is usually followed by an extra amino acid residue which is not conserved among different species. Although the unique structure of polyubiquitin genes has been conserved in evolution, considerable variation exists in the number of repeats within each polyubiquitin gene and the number of polyubiquitin encoding loci in the genome.

Induced expression of polyubiquitin genes by heatshock or other types of stress has been observed in a number of organisms [9, 10]. This facilitates an increased production of ubiquitin monomers for the ubiquitin-mediated degradation of abnormal proteins which arise during stress. The induced expression of ubiquitin genes in *P. infestans* during colonization of potato leaves may reflect the highly-active metabolic state of the mycelium in the host tissue. During exponential growth *in vitro* however, the mycelium is also in a highly-active metabolic state but the ubiquitin encoding genes show no induced expression. This implies that the host environment specifically induces the expression of the ubiquitin genes. In view of the regulatory function of ubiquitin in gene expression it is tempting to speculate that induction of the *P. infestans* polyubiquitin genes may lead to expression of genes involved in pathogenicity.

6. References

1. Coolbear, T. and Threlfall, D.R. (1985) 'A comparison of the sites of phytoalexin accumulation and of biosynthetic activity in potato tuber tissue inoculated with biotic elicitors', Phytochemistry 24, 2219-2224.
2. Kombrink, E., Schröder, M. and Hahlbrock, K. (1988) 'Several "pathogenesis-related" proteins in potato are 1,3-ß-glucanase and chitinase', Proc. Natl. Acad. Sci. USA 85, 782-786.
3. Cuypers, B., Schmelzer, E. and Hahlbrock, K. (1988) '*In situ* localization of rapidly accumulated phenylalanine ammonia-lyase mRNA around penetration sites of *Phytophthora infestans* in potato leaves', Molecular Plant-Microbe Interactions 1, 157-160.
4. Monia, B.P., Ecker, D.J. and Crooke, S.T. (1990) 'New perspectives on the structure and function of ubiquitin', Bio/technology 8, 209-215.
5. Finley, D. and Varshavsky, A. (1985) 'The ubiquitin system: Functions and mechanisms', Trends Biochem. Sci. 10, 343-347.
6. Hershko, A. (1988) 'Ubiquitin-mediated protein degradation', J. Biol. Chem. 263, 15237-15240.
7. Ozkaynak, E., Finley, D., Solomon, M-J. and Varshavsky, A. (1987) 'The yeast ubiquitin genes: A family of natural gene fusions', EMBO J. 6, 1429-1439.
8. Wiborg, O., Pedersen, M.S., Wind, A., Berglund, L.E., Marcker, K.A. and Vuust, J. (1985) 'The human ubiquitin gene family: Some genes contain multiple directly repeated ubiquitin coding sequences', EMBO J. 4: 755-759.
9. Bond, U. and Schlesinger, M.J. (1985) 'Ubiquitin is a heat shock protein in chicken embryo fibroblasts', Mol. Cell Biol. 5, 949-959.
10. Muller-Taubenberger, A., Hagmann, J., Noegel, A. and Gerisch, G. (1988) 'Ubiquitin gene expression in *Dictyostelium* is induced by heat and cold shock, cadmium, and inhibitors of protein synthesis', J. Cell Sci. 90, 51-58.

STRATEGIES FOR THE CLONING OF GENES IN TOMATO FOR RESISTANCE TO *FULVIA FULVA*

M. DICKINSON, D. JONES, C. THOMAS, K. HARRISON, J. ENGLISH,
G. BISHOP, S. SCOFIELD, K. HAMMOND-KOSACK and J.D.G. JONES,
Sainsbury Laboratory, John Innes Centre for Plant Science Research,
Colney Lane, Norwich, NR4 7UH, England.

ABSTRACT We are developing two alternative strategies to clone one or more of the *Cf2*, *Cf4*, *Cf5*, *Cf9* or *Cf11* genes of tomato (*Lycopersicon esculentum*) that confer resistance to *Fulvia fulva* (syn. *Cladosporium fulvum*). The first approach is tagging with the maize transposon *Ac*. Since *Ac* preferentially transposes to closely linked sites in both maize and tobacco, close linkage should improve tagging efficiency. We are therefore mapping RFLP mapping T-DNAs carrying *Ac* which have been transformed into tomato using as probes inverse polymerase chain reaction (IPCR) amplified tomato sequences adjacent to the T-DNA ends. Similarly, we are mapping the *Cf* genes on the RFLP map, so that we can identify for further analysis those transformants with T-DNAs closely linked to *Cf* genes. The second strategy involves the development of a technique of subtractive cDNA cloning, in which cDNA from a Cf0 plant (plant with no detectable resistance genes) is used to subtract cDNA which is common to a near-isogenic Cf2 plant, leaving *Cf2* specific cDNA which can be cloned, and used as a probe for differential screening.

Introduction

Ever since plant breeders recognized the existence of resistance genes in plants to their fungal pathogens, they have been using them to control diseases in major crops. However, despite their economic importance, the isolation and characterisation of these genes has not been achieved. Genetic studies indicate a gene for gene relationship in which the product of an avirulence gene in the pathogen is recognized by a corresponding product of a resistance gene in the plant. This recognition results in the induction of an active defence mechanism culminating in pathogen death. If there is no recognition, the pathogen escapes detection and a successful infection occurs.

The system we have chosen for study is the interaction between tomato (*Lycopersicon esculentum*) and its pathogen *Fulvia fulva* (leaf mould). The major advantages of this system are: 1. Tomato is readily amenable to transformation; 2. *Ac* has been shown to be active in tomato (Yoder et al., 1988); 3. the tomato genome has been extensively mapped, and comprehensive classical and restriction fragment length polymorphism (RFLP) maps are available. The approximate locations of several resistance genes (*Cf* genes) to *F. fulva* are known on the genetic map, and we are currently positioning five of these on the RFLP map.

This paper details the strategies we have been developing to clone a *Cf* gene. The tagging strategy, using the maize transposable element *Ac*, takes advantage of the fact that *Ac* preferentially transposes to closely linked sites (Greenblatt, 1984; Jones et al., in press), so the genetic locations of the T-DNAs carrying *Ac*, and the *Cf* genes need to be well defined. The subtractive cDNA strategy assumes that the resistance gene transcripts are present at the time that plant material is harvested and that there are no homologous transcripts in near-isogenic tomato

276

H. Hennecke and D. P. S. Verma (eds.),
Advances in Molecular Genetics of Plant-Microbe Interactions, Vol. 1, 276–279.
© 1991 *Kluwer Academic Publishers. Printed in the Netherlands.*

lines lacking the resistance genes (Cf0 plants). We justify this assumption on the basis that most of the *Cf* genes have been introgressed into tomato from wild relatives. Based on these assumptions, it should in principle be possible to subtract cDNA made to mRNA from a Cf2 plant with cDNA made to mRNA from a near-isogenic Cf0 plant, and remove all the common cDNA leaving the *Cf2* specific cDNA which can be cloned.

Methods

TRANSFORMATION WITH T-DNAS CARRYING *Ac*, IPCR, CLONING AND MAPPING OF FLANKING DNA

T-DNAs carrying *Ac* between the promotor and the coding region of various excision markers have been transformed into tomato. Ideally, the excision markers should allow visual distinction between somatic and germinal excison of *Ac* reflecting the transmission of pre- and post- excision alleles of the T-DNA. Methods for IPCR and cloning of tomato sequences flanking T-DNAs were essentially as described by Ochman et al. (1988). The cloned sequences have been mapped on the tomato RFLP map by using them as probes to blots of DNA from *L.esculentum* x *L.pennellii* F2 plants segregating RFLP's and using the Mapmaker program (both kindly made available by S. Tanksley, Cornell University, Ithaca, NY and E. Lander, MIT).

MAPPING OF THE *Cf* GENES ON THE RFLP MAP.

L.esculentum stocks carying the various *Cf* genes have been crossed to L. *pennellii* and then test-crossed to an *L. esculentum* stock with no detectable *Cf* gene (Cf0). *Cf* gene segregation has been scored either by inoculation with race 0 of *F. fulva*, or by injection with inter-cellular fluid from the compatible Cf0/race 0 interaction (containing all the *Avr* gene products) and scoring for a necrosis/chlorosis reaction indicating an interaction between the *Avr* product and the resistance gene product (De Wit et al., 1985).

SUBTRACTIVE cDNA CLONING.

Poly-A+ mRNA was prepared from young leaves of Cf0 and Cf2 plants grown under identical conditions. Double-stranded (ds)cDNA was made to both mRNAs by priming the first strand with oligo-dT and using RNAse H and DNA polymerase I for the synthesis of the second strand. One set of primer adaptors was ligated to the ends of the Cf0 cDNA and a second set of primer adaptors containing an internal *EcoRI* site was ligated onto the ends of Cf2 cDNA. The primer adaptors consist of two oligonucleotides, one a 24-mer and the other a 14-mer complimentary to the 3' end of the 24-mer. These ligate onto both ends of the cDNA and the 24-mer is then used as the primer in the PCR amplification of the cDNA. Following amplification, the Cf0 cDNA was biotinylated with photobiotin (Forster et al., 1985). 10 µg of biotinylated Cf0 cDNA was then mixed with 0.5 µg unbiotinylated Cf2 cDNA in a 5 µl volume of 1M NaCl, 10mM EPPS pH 8.0, 1mM EDTA and overlaid with mineral oil. The mixture was boiled for 1 min to denature the cDNA and incubated at 65C for 24-48 hours. Biotinylated hybrids were then removed by the method described by Straus and Ausubel (1990), and the cDNA was then subtracted up to 5 more times with excess biotinylated Cf0 cDNA. Following the final subtraction, the cDNA was amplified using the 24-mer unique to the Cf2 cDNA, then cut with *EcoRI*, and cloned into lambda gt 11.

Results

MAPPING OF T-DNAS CARRYING *Ac* ON THE RFLP MAP.

Using the technique of IPCR, 8 T-DNAs carrying *Ac* have been mapped on the tomato RFLP map, and a number of others are currently being mapped. One of these, 1515Q LB2, has been mapped to a location close to the *Cf4* gene on chromosome 1, and we are currently determining whether there is linkage between this T-DNA insertion and *Cf4*.

MAPPING OF THE *Cf* GENES.

Cf genes have only been located on the classical map. Since there is little data on correspondence between this and the RFLP map, we are positioning the *Cf* genes on the latter using probes donated by S. Tanksley. Using this technique, the *Cf2* gene has been mapped relative to RFLP markers on chromosome 6. The same technique indicates that the *Cf9* gene may not be located on chromosome 10 as was previousely reported (Kanwar et al., 1980).

SUBTRACTIVE cDNA CLONING.

As a control experiment to determine the efficiency of the subtractive cDNA cloning technique, cDNA was synthesised from Cf0 and a transformant 601D expressing the neomycin phosphotransferase (*npt II*) gene. Following the addition of primer adaptors and PCR amplification, the 601D cDNA was subtracted with biotinylated Cf0 cDNA. Libraries were constructed in lambda gt 11 of 601D cDNA, and 601D cDNA after 2 and 4 cycles of subtraction. Plaque lifts of these libraries were probed with the *npt II* gene to determine the relative levels of *npt II* cDNA in each library. The results indicate that this increased from 1 in 10,000 plaques in the unsubtracted library, to 1 in 1,000 after 2 cycles of subtraction, and 1 in 500 after 4 cycles.

Cf 2 SUBTRACTION.

One observation from the subtractive cDNA technique is that the size population of cDNA following subtraction and PCR amplification tends to be much smaller than that of the original cDNA, generally being in the 200 - 500bp size range. The reasons for this are unclear, but a consequence of this is that the technique may be selecting for the 3' ends of mRNA. To prevent this, one modification to the technique is to cut a proportion of the initial Cf2 cDNA with *Alu I* before the addition of the primer adaptors. This ensures that the starting population of cDNAs to be subtracted in the 200-500 bp size range is more representative of the whole cDNA. Using this technique, a library of Cf2 cDNA following 6 cycles of subtraction has been constructed. Duplicate plaque lifts of this library have been probed with Cf0 and subtracted Cf2 cDNA to identify those clones which appear to be present only in the subtracted Cf2. Six clones have been identified for further analysis, and these are being used as probes to RFLP blots to determine their location on the RFLP map, and whether they segregate with the *Cf2* resistance gene.

Discussion

This paper has decribed the alternative strategies we have been developing to clone resistance genes from tomato. Using IPCR to clone tomato DNA flanking T-DNA insertions, we have mapped the locations of 8 T-DNAs carrying *Ac*. Similarly we have mapped the *Cf2* gene on the RFLP map and are currently mapping the *Cf5* and *Cf9* genes. When a transformed plant is identified whose T-DNA is linked to a *Cf* gene, a plant will be bred homozygous for the pre-excision T-DNA and the *Cf* gene. A tagging experiment will then be conducted by crossing to Cf0, selecting progeny carrying a germinal excision of *Ac*, if they can be identified, and screening them for inactivation of the *Cf* gene due to insertion of *Ac*. This transposon tagging strategy has an additional usage in that a proportion of the T-DNA insertions will inevitably be closely linked to other potentially interesting genes including genes for resistance to other pathogens. These transformants will be useful tools for the cloning of these other genes.

The alternative strategy of subtractive cDNA cloning has been developed to take advantage of the availability of near-isogenic lines of tomato with and without the *Cf* genes introgressed from wild relatives of tomato. Providing the predictions we have made about resistance genes are valid, in principle it should be possible to clone these genes directly by subtractive techniques. Even if this is not the case and there are transcripts with regions of homology to the *Cf* genes in Cf0 plants, it may still be possible to clone parts of the cDNAs which are unique by for example cutting the cDNA with a restriction enzyme before subtraction The results for the *npt II* control experiment indicate that the strategy of subtractive cDNA cloning using PCR amplification of cDNA and biotinylation is a succesful strategy for the enrichment of *npt II* clones, and we are now in a positon to test the clones obtained from the Cf2 subtraction, to determine if any are closely linked to the *Cf2* gene. As an alternative strategy, we are using the genomic subtraction techniques developed by Straus and Ausubel (1990). This approach should enable us to obtain clones from the introgressed region of DNA spanning the *Cf2* gene which have no homologous sequences in the Cf0 genotype. If this is the case, these will be used to fine map this introgressed region and walk to the *Cf2* gene.

As with the transposon tagging strategy, the subtractive cloning techniques will also be useful for cloning other genes involved in pathogenesis. The cloning of such genes will enable us to study the interaction between the gene products of resistance and avirulence genes, and to understand the mechanism by which a plant protects itself from fungal attack.

References

De Wit, P.J.G., Hofman, A.E., Velthuis, G.C.M. and Kuc, J.A. (1985). Isolation and characterisation of an elicitor of necrosis isolated from intercellular fluids of compatible interactions of *Cladosporium fulvum* (Syn. *Fulvia fulva*) and tomato, Plant Physiology 77, 642-647.

Forster, A.C., McInnes, J.L., Skingle, D.C. and Symons, R.H. (1985). Non-radioactive hybridisation probes prepared by the chemical labelling of DNA and RNA with a novel reagent, photobiotin, Nucleic Acids Research 13, 745-761.

Greenblatt, I.M. (1984). A chromosome replication pattern deduced from pericarp phenotypes resulting from movements of the transposable element *Modulator* in maize, Genetics 108, 471-485.

Jones, J.D.G., Cowland, F., Lim, E., Ralston, E. and Dooner, H. (In press). Preferential transposition of the maize element *Activator* (*Ac*) to linked chromosomal locations in tobacco, The Plant Cell.

Kanwar, J.S., Kerr, E.A. and Harney, P.M. (1980). Linkage of *Cf1* to *Cf11* genes for resistance to tomato leaf mould, *Cladosporium fulvum* Cke., Tomato Genetics Co-operative Report 30, 20-22.

Ochman, H., Gerber, A.S. and Hartl, D.L. (1988). Genetic applications of an inverse polymerase chain reaction, Genetics 120, 621-623.

Straus, D. and Ausubel, F.M. (1990). Genomic subtractions for cloning DNA corresponding to deletion mutations, Proceedings of the National Academy of Sciences, USA 87, 1889-1893.

Yoder, J.I., Palys, J., Alpert, K. and Lassner, M. (1988). *Ac* transposition in transgenic tomato plants, Molecular and General Genetics 213, 291-296.

DOWNY MILDEW OF *ARABIDOPSIS THALIANA* CAUSED BY *PERONOSPORA PARASITICA*: A MODEL SYSTEM FOR THE INVESTIGATION OF THE MOLECULAR BIOLOGY OF HOST-PATHOGEN INTERACTIONS

A. J. SLUSARENKO & BRIGITTE MAUCH-MANI
Institute for Plant Biology
Zollikerstr. 107
CH-8008 Zuerich, Switzerland

ABSTRACT. The results of crosses between resistant and susceptible parents are reported and a strategy to clone and characterise genes associated with resistance in *Arabidopsis* is described. The use of mutants to elucidate the putative signal transduction pathway from perception of the pathogen to activation of defence genes is also discussed. Preliminary results of experiments to characterise *Arabidopsis* defence mechanisms are reported, for example the production of phytoalexins, hydrolytic and oxidative enzymes, and callose encasement of haustoria.

1. Introduction

1.1 THE PATHOSYSTEM

A naturally occurring infection of *Arabidopsis* by *Peronospora parasitica* was described recently (Koch & Slusarenko 1990a). Resistant ecotypes of *Arabidopsis* (such as RLD, Columbia) respond to attempted penetration by the fungus with a typical hypersensitive response. In contrast, susceptible ecotypes (such as Weiningen, Landsberg erecta, Kelsterbach 4) are quickly colonised by intercellular hyphae producing intracellular haustoria. Sexual and asexual reproduction occurs by some 6 days after inoculation. Since infected leaves contain oospores, the fungus is presumed to be either **homothallic**, or a mixture of two mating types which we have not yet been able to separate.

It is worthwhile mentioning that in the absence of further isolates of the pathogen which show a reciprocal check type of interaction with different host ecotypes (*i.e.* host A susceptible to isolate 1 but resistant to isolate 2, host B resistant to isolate 1 but susceptible to isolate 2), one can only speculate that one is working with **vertical resistance** in a classical gene-for-gene interaction. It is possible that the different host ecotypes simply show different degrees of **race non-specific (horizontal) resistance** (*i.e.* quantitative resistance to all races/isolates of a pathogen rather than qualitative resistance to some). However, although it can be absolute, horizontal resistance is not usually associated with a hypersensitive reaction, and, in addition, race/cultivar specialisation has been reported for *Peronospora parasitica* isolates from other crucifers *e.g.* broccoli and cabbage (Natti *et al.* 1967) and oilseed rape (Lucas *et al.* 1988). It is hoped that more pathogen isolates will come to light which enable compliance with the reciprocal check criterion; especially since horizontal resistance, contrary to popular belief, can be determined by one or a few loci and need not be multigenically determined (Vanderplanck 1978).

In order to gain more information on the genetic control of resistance of *Arabidopsis* to *P. parasitica* crossing programmes between resistant and susceptible *Arabidopsis* ecotypes have been

280

H. Hennecke and D. P. S. Verma (eds.),
Advances in Molecular Genetics of Plant-Microbe Interactions, Vol. 1, 280–283.
© 1991 *Kluwer Academic Publishers. Printed in the Netherlands.*

initiated. One hundred F1 plants from a Weiningen x RLD cross were all resistant to *Peronospora* infection. Since Weiningen was the female parent in the crosses this suggests that resistance is not a cytoplasmically inherited trait. The F1 plants were allowed to self and, of 59 F2 plants scored so far, only 2 were susceptible. This would fit with two independently segregating resistance loci where one would expect a ratio of 15:1, resistant:susceptible ($X^2 = 0.82$, P = >0.05). However, there are a number of other possible explanations which fit these data. For example, it is also possible that more than two resistance genes are present and that there is linkage between some of them, this would tend to increase the proportion of parental types in the progeny. Thus, it would appear that our *P. parasitica* isolate might be a complex race with at least two avirulence loci. Backcrosses between the F1 and Weiningen have been set up in order to confirm whether two loci are involved (expected ratio 3:1, resistant:susceptible). Backcrosses between randomly selected F2 individuals and Weiningen have also been performed to try and separate the two postulated resistance loci.

Parallel crosses between Columbia and Landsberg erecta are also underway since these are the ecotypes of choice to use for subsequent analysis and cloning of putative resistance loci.

2. Strategy to isolate classical resistance genes and genes important for defence

2.1 CLASSICAL RESISTANCE GENES

The theoretical aspects are quite straightforward and, since we are only at the beginning and have no results yet, we shall not go into great detail. *Arabidopsis*, with its small genome, low amount of repetitive DNA and transformability by *Agrobacterium*, is an ideal candidate for cloning genetic loci characterised only by their phenotype. We intend to follow co-segregation of the resistance trait with mapped RFLP markers, first localising the resistance locus to a particular chromosome, then determining its position by further fine mapping and eventually isolating the relevant clones by chromosome walking. Transformation of susceptible *Arabidopsis* strains, followed by pathogenicity testing, should identify clones which need further analysis. It is hoped that problems due to variation of the genetic background will not arise.

2.2 MUTANTS

We are also screening EMS-generated M2 seed for mutations which lead to susceptibility. Theoretically, these might have lesions in any of three important components of the interaction:
   a) classical resistance genes
   b) genes involved in the signal transduction pathway
   c) genes essential for a major resistance mechanism.
In a host with multiple resistance genes, such as RLD, one might expect mutants in class a) to be less likely than classes b) or c). Mutants in the signal transduction pathway would be particularly interesting. We have been kindly provided with M4 seed from Columbia susceptible to the plant pathogenic bacterium *Xanthomonas campestris* pv. *campestris* isolate 1067 (Mike Daniels this volume). It seems unlikely that the same classical resistance gene functions in the recognition of both pathogens, so if the M4 individuals prove susceptible to our *P. parasitica* isolate (Columbia is usually resistant), then the chances are high that the lesion falls into class b) or c).

2.3 THE PATHOGENICITY ASSAY

All the above requires the screening of large numbers of plants for resistance or susceptibly to *P. parasitica* and the need for a rapid, reliable, and preferably not too destructive, pathogenicity assay. We have developed a detached leaf assay in which single leaves are inoculated with droplets of spore suspension and incubated overnight in repli-dishes. Trypan blue stain and then destain can be added sequentially to the wells and the leaves scored by examination under the microscope. Resistant leaves show hypersensitive cell-necrosis and susceptible leaves show successful hyphal penetration and formation of haustoria in epidermal cells.

3. How is *Arabidopsis* resistant to infection?

Plants respond to pathogen attack with a number of well documented putative defence responses, *e.g.* synthesis and accumulation of phytoalexins, induction of hydrolytic enzymes such as chitinase and beta-1,3-glucanase and, in the case of biotrophic pathogens such as *P. parasitica*, it is conceivable that hypersensitive cell death prevents establishment of the necessary nutritional relationship between host and pathogen and thus leads to resistance.

3.1 PHYTOALEXINS

Several sulphur- and nitrogen-containing compounds have been reported as phytoalexins from various members of the Cruciferae (Rouxel *et al.* 1989 and references cited therein), but up to now none from *A. thaliana*. Elicitor treatment of suspension-cultured cells of *A. thaliana* was demonstrated to increase the activity of a number of enzymes involved in phenylpropanoid metabolism; *i.e.* phenylalanine ammonia lyase (PAL), 4-coumarate:CoA ligase (4CL) and caffeic acid O-methyl transferase (CMT) (Davis & Ausubel 1989). Steady state levels of PAL and 4CL mRNA were also shown to increase transiently, both with similar kinetics. These results suggest that phenylpropanoid biosynthesis may be an important component of the defence response of *Arabidopsis*. Since the authors found no induction of chalcone synthase they suggest that flavonoid derivatives were not produced as defence substances. Since CMT and peroxidase activities were induced it was suggested that lignin-like compounds might be involved in the defence response.

TLC of ethanolic extracts of infected and non-infected plants showed the induction, on infection, of a long-wave-UV-fluorescing compound which showed antibacterial activity in a plate overlay assay (Slusarenko *et al.* 1989), (Fig. 1a). The substance has an R_f of 0.33 in chloroform:methanol (95:5), fluoresces violet-purple at 360 nm (no change on perfusion with ammonia vapour), is weakly absorbing at 254 nm and has absorbtion maxima (in ethanol) at 217, 273 and 316 nm (Fig. 1b), on addition of alkali there appeared to be decomposition. The substance has not yet been characterised further or definitively identified which makes quantification difficult. The substance accumulates in both compatible and incompatible combinations, but as yet we have no detailed time-course data for comparison of relative rates of production or localisation in resistant or susceptible reactions.

a

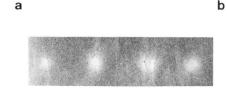

b

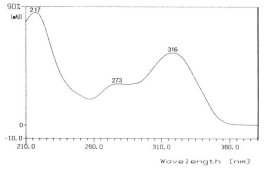

Fig.1. a) Pale spots indicate inhibition of bacterial growth in a TLC overlay assay.
b) UV absorbtion spectrum of the active substance shown in a)

3.2 HYDROLYTIC ENZYMES AND STRUCTURAL DEFENCE RESPONSES

Davis & Ausubel (1989) showed elicitor induction of beta-1,3-glucanase in *Arabidopsis* cell suspension cultures and ethylene induction of chitinase gene activity has been reported (Samac *et al.* 1990). We have been able to demonstrate chitinase activity in *P. parasitica* infected tissues of *Arabidopsis* (K. Croft unpublished results) and are currently investigating relative induction kinetics in compatible and incompatible combinations.

In the compatible combination, encasements around haustoria of *P. parasitica* were frequently observed (Koch & Slusarenko 1990a). Specific staining has shown that these contain callose. Peroxidase activity has been detected in infected tissues (K. Croft unpublished results), and in addition to other possible functions, might indicate that lignification is also occurring, perhaps in haustorial encasements and elsewhere. Studies to detect and localise lignin deposition are in progress.

4. Concluding remarks

There is great potential for using *Arabidopsis* as a model system to characterise molecular and biochemical events in plant-fungal interactions. Certainly, *Arabidopsis* is host to several pathogenic fungi, both biotrophic, e.g. *P. parasitica*, *Erysiphe cruciferarum*, *Plasmodiophora brassicae*, and non-biotrophic, e.g. *Botrytis cinerea* and *Rhizoctonia solani* (Koch & Slusarenko 1990b), and the next few years will no doubt bring additions to this list. Characterisation of the biochemical basis of resistance mechanisms is only a prelude to using *Arabidopsis* where it is most powerful - for molecular analysis. Thus, isolation of genes involved in all facets of the host-pathogen interaction (see section 2.2) will enable the role of any given gene in resistance to be assessed; for example by carrying out mutant complementation studies, over-expression in transgenic plants and down-regulation of candidate genes by antisense technology.

Davis, K.R. & Ausubel, F.M. (1989) "Characterization of elicitor-induced defence responses in suspension-cultured cells of *Arabidopsis*", Molecular Plant-Microbe Interactions 2, 363-368.

Koch, E. & Slusarenko, A.J. (1990a) "*Arabidopsis* is susceptible to infection by a downy mildew fungus", The Plant Cell 2, 437-445.

Koch, E. & Slusarenko, A.J. (1990b) "Fungal pathogens of *Arabidopsis* (L.) Heyhn.", Botanica Helvetica 100 No. 2 (in press).

Lucas, J.A., Crute, I.R., Sherriff, C. & Gordon, P.L. (1988) "The identification of a gene for race-specific resistance to *Peronospora parasitica* (downy mildew) in *Brassica napus* var. *oleifera* (oilseed rape)", Plant Pathology 37, 538-545.

Natti, J.J., Dickson, M.K. & Atkin, D.D. (1967) "Resistance of *Brassica oleracea* varieties to downy mildew", Phytopathology 57, 144-147.

Rouxel, T., Sarniguet, A., Kollmann, A. & Bousquet, J-F. (1989) "Accumulation of a phytoalexin in *Brassica* spp. in relation to a hypersensitive resistance to *Leptosphaeria maculans*", Physiological & Molecular Plant Pathology 34, 507-517.

Samac, D.A., Hironaka, C.M., Yallaly, P.E. & Shah, D.M. (1990) "Isolation and characterisation of the genes encoding basic and acidic chitinase in *Arabidopsis thaliana*", Plant Physiology 93, 907-914.

Slusarenko, A.J., Longland, A.C. & Whitehead, I.M. (1989) A convenient, sensitive and rapid assay for antibacterial activity of phytoalexins", Botanica Helvetica 99, 203-207.

Vanderplanck, J.E. (1978) "Genetic and molecular basis of plant pathogenesis", Advanced series in agricultural sciences No. 6, Springer-Verlag.

HOST-PATHOGEN INTERACTIONS IN THE SYSTEM *ARACHIS HYPOGAEA-CERCOSPORA ARACHIDICOLA*

A.J. BUCHALA, S. ROULIN, J.L. COQUOZ and H. MEIER
Institut de Biologie végétale et de Phytochimie
Université de Fribourg
CH-1700 Fribourg
Switzerland

ABSTRACT. With the aim of producing transgenic peanut plants, resistant to the early leafspot disease, the reaction of various peanut genotypes upon infection with the pathogen *Cercospora arachidicola* has been examined. Besides the production of stilbene-type phytoalexins, suspension-cultured cells of a susceptible genotype did not react typically when examined for responses to infection. Only slight differences were observed when cells derived from a hypersensitive genotype were examined. Submerged cultures of the pathogen, grown on peanut cell walls, produced various carbohydrase activities capable of degrading plant cell wall polysaccharides including β-glucans. The apoplastic fluid from infected plant tissue also contained increased carbohydrase activities which could degrade peanut cell walls but no significant difference was observed between susceptible and resistant genotypes.

Introduction

The peanut plant (*Arachis hypogaea* L.) is commercially important and cultivated on over 18×10^6 ha. Despite improved cultural practices, the use of pesticides and the development of cultivars resistant to specific pathogens, disease still results in the annual loss of many millions of dollars. Early (*Cercospora arachidicola* Hori) and late (*Cercosporidium personatum*) leaf spot disease, which cause defoliation and yield loss, are the major pathogens in the USA. The use of classical fungicides is undesirable and cultivars with increased resistance usually have decreased yield and delayed maturity. A better understanding of the the interaction between the plant and the pathogen would clearly help in devising improved protection.

AIMS OF THE PRESENT STUDY

Many biochemical plant/pathogen interactions are initiated at the carbohydrate level. Polysaccharides of the pathogen cell wall are recognised by the plant, which produces enzymes that can degrade the pathogen cell wall and produce oligosaccharides (β-glucan fragments) which in turn trigger other defence mechanisms, *e.g.* the production of phytoalexins or characteristic proteins. In other cases, enzymes from the pathogen produce

H. Hennecke and D. P. S. Verma (eds.),
Advances in Molecular Genetics of Plant-Microbe Interactions, Vol. 1, 284–287.
© 1991 *Kluwer Academic Publishers. Printed in the Netherlands.*

oligosaccharides (oligouronides), which derive from the plant cell wall and which have a similar effect. The delay between inoculation and the appearance of the first symptoms *in vivo* is long and not regular. However in general, plant cells cultured in suspension react more rapidly.

The aim of the present study is to characterise the cell walls of the plant and the pathogen, the induction of carbohydrase activity upon their interaction, and to study the effect of oligosaccharides derived from the cell walls on peanut cells cultured in suspension.

Results and discussion

Morphology.
Peanut plants, genotype Flory Giant susceptible to the pathogen, and genotype P. I. 109839, described as resistant [1], were grown and infected, on the lower leaf surface, with a spore suspension of *C. arachidicola*, in a growth chamber at 25 °C and 80-100% humidity. The typical symptoms appeared about 3 weeks later. Anatomically, the lesion *per se* consisted of dead and collapsed cells exhibiting coagulation of their cytoplasmic contents, disappearance of chloroplasts and starch grains, and ultimately accumulation of dark brown material (phenolic compounds). Damage to the protoplasts was more obvious than damage to the cell walls. The pathogen entered the plant either through the stomata or directly through the epidermis cells, and was localised mainly intracellularly. No haustoria were observed.

Cell wall composition of the plant and the pathogen.
The cell wall polysaccharides of peanut leaves were examined and found to be typical of leguminous plants. The hydrolysate of purified cell walls contained 70% glucose, 8% arabinose, 5% galactose, 5% xylose, 4% rhamnose, 3% mannose and 2% fucose as well as uronic acids. Fractionation of the cell walls gave material rich in pectic substances (24%) and xyloglucan; $(1\rightarrow3)$-β-glucan could also be identified.

A hydrolysate of purified cell walls from *C. arachidicola* contained 60% glucose, 20% mannose, 20% galactose and traces of aminosugars. Most of the glucose could be shown to derive from a $(1\rightarrow3)$-β-glucan, containing some $(1\rightarrow6)$-linkages, by specific enzymic hydrolysis with an exo-$(1\rightarrow3)$-β-glucanase from Basidiomycete QM-806. The cell wall thus contains a β-glucan similar to that found in *Phytophthora* spp. A galactomannan is also probably present.

Infection of suspension-cultured cells.
Suspension cultures were established for the susceptible genotype and for a genotype (P.I. 276233) claimed to exhibit hypersensitivity [1]. Inoculation of suspension cultures of the susceptible genotype gave rise to the production of stilbene-type phytoalexins within 48h (resveratrol) [2]. Phytoalexins were strongly induced not only upon addition of spores and cell walls of the pathogen, but also upon addition of cell wall from *Phytophthora megasperma*, the soya bean pathogen (see Fig. 1).

The situation seems also to be analogous to the interaction between *P. megasperma* and soya bean, where infection induces the secretion of β-glucanase into the plant cell wall.

The enzyme hydrolyses the β-glucan of the fungal cell wall to produce certain oligosaccharides which in turn elicit phytoalexin synthesis in the plant [3].

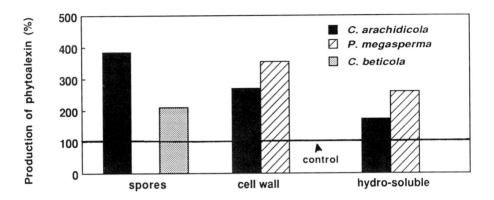

Fig.1: Production of phytoalexin (% of control) in suspension-cultured cells. Suspension cultures were infected by direct addition of the fungus or of cell wall fragments to the culture medium at the end of logarithmic growth. Control cultures were inoculated with water. 48h after inoculation, phytoalexins were extracted with methylacetate and detected by fluorometry at 410 nm.

Enzyme activities in the culture medium of the pathogen.
C. arachidicola was grown in submerged culture on a KNO_3 medium containing sucrose (1% or 0.2%) with or without starch-free cell walls (0.25%) from suspension-cultured groundnut cells. Although the production of biomass was always best on 1% sucrose, several carbohydrase activities were significantly higher in the culture medium containing cell walls and little (0.2%) or no sucrose. Exo- and endo-$(1{\rightarrow}3)$-β-glucanases, invertase, aryl α- and β-glucosidase, α- and β-galactosidase, and β-xylosidase were always found in the culture medium. α-Amylase and amyloglucosidase were only found in cultures on cell walls (± 0.2% sucrose). The presence of the latter two activities correlates with the disappearance of starch in the leaf cell walls. Low levels of pectin-degrading enzymes, e.g. polygalacturonase or pectin lyase, were also found but no additional activity was observed upon addition of peanut cell walls. This may be taken to support the idea that pectinases are not important in the initial phases of infection and that the plant has not developed a defense mechanism triggered by oligouronides. Hemicellulase activities could not be detected with simple assays. However, using radio-labelled groundnut cell walls as substrate, it could be shown that, upon addition of cell walls, higher carbohydrase activities were present in the culture medium (Fig. 2) and that the liberated oligosaccharides contained mannose, glucose and galactose.

Enzyme activities in the apoplast of infected leaf tissue.
Intercellular fluids (IF) from infected leaves of the susceptible and the resistant genotypes were collected at intervals during 3 weeks after infection. Leaves were infiltrated, under vacuum, with 0.1 M phosphate buffer (pH 6.0) and centrifuged in a syringe placed in a

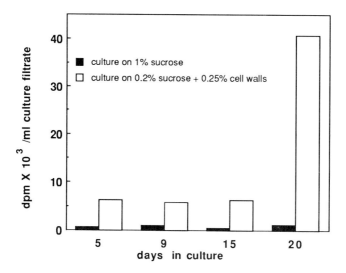

Fig. 2: Carbohydrase activity in the culture medium of *C. Arachidicola* - radioactivity released from labelled peanut cell walls

tube. IF were collected at the bottom of the tube and incubated, for 20h, with a suspension of radio-labelled groundnut cell walls. The radioactivity liberated in the supernatant was measured but no significant global differences were observed between the two genotypes. This does not exclude the possibility that different carbohydrase activities were responsible for the liberation of cell wall polysaccharide fragments.

CONCLUSIONS

None of the above-mentioned reactions to infection with the pathogen occur more intensely or more rapidly in the resistant or hypersensitive genotypes examined. Other biochemical responses, e.g. peroxidase activity and synthesis of hydroxyproline-rich glycoprotein will be examined in the near future.

REFERENCES

1 Melouk, H.A. and Banks, D.J. (1982) "A method of screening peanut genotypes for resistance to *Cercospora* leafspot", Peanut Science **5**, 112-114.
2 Rolfs, C., Fritzmeier, K.H. and Kindl, H. (1981) "Cultured cells of *Arachis hypogaea* susceptible to induction of stilbene synthase (resveratrol-forming)", Plant Cell Reports **1**, 83-85.
3 Ebel, J. and Grisebach, H. (1988) "Defense strategies of soybean against the fungus *Phytophthora megasperma* f. sp. *glycinea*: a molecular analysis", Trends in Biochemical Science **13**, 23-27.

Section IV

HOST PLANT RESPONSE TO MICROBIAL INFECTION

GENESIS OF ROOT NODULES AND FUNCTION OF NODULINS

D.P.S. VERMA, G.-H. MIAO, C.-I. CHEON & H. SUZUKI
Department of Molecular Genetics and Biotechnology Center
The Ohio State University
Columbus, Ohio 43210-1002
USA

ABSTRACT. A highly coordinated activity of many bacterial and plant genes gives rise to an organ capable of housing *Rhizobium* and supporting the metabolic demands of symbiotic nitrogen fixation. Several host genes encoding nodule-specific proteins (nodulins) have been isolated and characterized from many legumes. Most of the early nodulin genes identified to date appear to encode structural proteins while the late nodulins mainly participate in the metabolic functions of nodules. The nodulin genes are derived from other plant genes encoding structural and metabolic proteins as they have similar functions or sequence homologies. However, the nodulin genes have come under the nodule developmental control in order to adapt to this unique organ. We have demonstrated that endocytosis release of *Rhizobium* is not necessary for the development of an organized nodule structure, suggesting that certain bacterial signals trigger the nodule developmental program prior to the release of bacteria inside the "infected" cells. Recent studies suggest that "physiological internalization" of the peribacteroid membrane compartment, housing the bacteria, is essential for the establishment of the symbiotic state. One of the nodulins, nodulin-26, is shown to play a critical role in this process. This nodulin forms an ion channel in the peribacteroid membrane. Symbiotically fixed nitrogen seems to be entering the host cell by diffusion, as all cells in the infection zone, delimited by the endodermis, have a uniform concentration of ammonia. The latter was determined by an ammonia-sensitive glutamine synthetase promoter linked with a reporter gene. Conversion of amides into ureides is essential in tropical legumes and several nodulins, including nodulin-35, are involved in this pathway.

1. Introduction

Development of the capacity in legume plants to harbor soil bacteria (*Rhizobium*, *Bradyrhizobium* and *Azorhizobium* sps.) in symbiotic association has provided this group of plants the ability to be autotropic for fixed nitrogen and thus grow in many diverse habitats. This ability manifested in the production of protein rich grains providing much needed nutrition to the seedling during germination period. Since the effective associations were mutually advantageous to both organisms, effectiveness has been selected over ineffectiveness during evolution and the process has thus been optimized. That a coevolution of both genomes has taken place, is evident from many mutations in bacteria and plant conferring similar phenotypes (see Verma and Stanley, 1989).

Root nodule, an organ *sui generis*, is a product of interaction of bacterial and plant genes that initiate a developmental program forming this unique structure. It has been shown that many bacterial mutants (Long, 1989) and chemicals, such as triiodobenzoic acid (Hirsch *et al.*, 1989) can form nodule-like structures (pseudonodules). This suggests that hypertrophy (localized proliferation of tissue) and the endocytotic processes

291

H. Hennecke and D. P. S. Verma (eds.),
Advances in Molecular Genetics of Plant-Microbe Interactions, Vol. 1, 291–299.

are separate events and are manifested by different signals (Morrison and Verma, 1987). Since cellular proliferation is required for endocytosis, cell division commences ahead of the infection thread (Dudley et al., 1987) which demarks the potential receptive zone for the endocytosis of bacteria. Since different degrees of efficiency exist in this association, it suggests that many plant genes have been recruited to perform specialized functions in nodules and only optimum expression of these genes gives rise to a fully effective state.

2. Early nodulins and nodule morphogenesis

A number of plant genes are induced soon after the contact of host root with rhizobia. The invasion of bacteria is not essential for the expression of some of the early nodulin genes (see Verma et al., 1988a). Since these genes can also express during hypertrophic growth caused by auxin transport inhibitors (Hirsch et al., 1989), it suggests that root nodules evolved from such outgrowths. The nodule primordia originates from the inner cortex of root in response to a signal (hormone gradient) produced during the infection process. Many of the early nodulins resemble hydroxy proline-rich cell wall proteins, eg. ENOD2 which is expressed in the inner cortex. Based on its location, it has been suggested to play a role in the diffusion of oxygen across this tissue (van de Wiel et al., 1990). A protein similar to ENOD2 is found to be expressed in the cell wall of axes tissue of germinating soybean (Averyhart-Fullard et al., 1988). Appearance of specific cell wall proteins may mark the boundaries of the tissue and help differentiation in nodule.

A cascade of expression of early nodulin gene follows during nodule development and several early nodulin genes have been found to express specifically in the infected cells, eg. ENOD3, ENOD5 and ENOD14 along with leghemoglobin (Scheres, 1990). The indeterminate nodules carrying apical meristem resemble lateral roots. Many nodules formed on non legume plants resemble lateral roots. The "stem nodules" on Sesbania are originated from lenticels primordia capable of giving rise to lateral roots. Thus, the primary difference between a nodule primordia and lateral roots seems to be that Rhizobium is able to create a new primordia in the inner cortex which behaves as a lateral root primordia and the cells derived from such primordia are able to be infected by Rhizobium. Persistence of meristem, which may in turn be controlled by hormone gradient, gives rise to an indeterminate or a determinate nodule. A unique organization of endodermis dividing the nodule cortical tissue into inner and outer cortex, makes root nodule a special organ since other plant organs are devoid of such structures (van de Wiel et al., 1990).

3. Late nodulins and nodule metabolism

As nodule grows, bacteria inside the host cell proliferate deriving both their carbon and nitrogen requirements from the plant. This creates nitrogen limiting conditions which may be responsible for triggering nif genes in Rhizobium. Concomitantly, the rapid demand for oxidative metabolism causes hypoxic conditions in the infected cells. Appearance of leghemoglobin inside the infected cells helps in maintaining the flux of oxygen. It has been demonstrated that leghemoglobin facilitates respiration in mitochondria (Suganuma et al., 1987) and may thus be considered as a "defense" molecule against hypoxia. However, hypoxia alone does not induce leghemoglobin.

A majority of the nodulins produced late in the infection process, but prior to the commencement of nitrogen fixation, have various metabolic roles in the nodule. These have been studied in detail in soybean (see Verma *et al.*, 1986; Verma and Delauney, 1988b). For example: Nodulin-100 constitutes sucrose synthetase, a tetrameric enzyme, activity of which is very high in nodules (Thummler and Verma, 1987). This enzyme helps maintain the carbon flow in nodules and its activity is apparently controlled by the availability of free heme. The primary assimilation of reduced nitrogen, transported into the host cell from bacteroids, takes place by the host glutamine synthetase. This enzyme occurs in nodule-specific forms in some legumes (Forde *et al.*, 1989), while in others, it has come under the control of ammonia (see Hirel *et al.*, 1987; also below). While the amides can be directly transported in temperate legumes, they must be converted to ureides in tropical legume nodules. Nodulin-35 is a subunit of nodule-specific uricase, an enzyme involved in ureide metabolism.

A high level of CO_2 fixation by phospoenolpyruvate carboxylase has been observed in alfalfa nodules (Miller *et al.*, 1987); however, it is not known whether this activity is encoded by a nodule-specific gene. This localized fixation of CO_2 may help to meet some of the carbon requirements in this energy demanding tissue.

While several nodulins are involved in carbon/nitrogen metabolism in nodules, the function of many others is still unknown. Six nodulins encoded by abundant transcripts, produced prior to the commencement of nitrogen fixation, have been identified in soybean (see Verma and Delauney, 1988b). These proteins share sequence homologies at both amino and carboxy termini, suggesting a common evolutionary origin from an ancestral gene. Presence of a putative signal sequence on the amino terminus of some of these proteins indicate that they may be membrane or extracellular proteins (Jacobs *et al.*, 1987). Thus, a number of early and late nodulins appear to have originated from existing genes as shown in Figure 1.

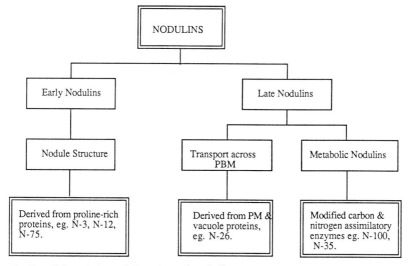

Figure 1: Possible origin and function of nodulin genes
PM, Plasma membrane; PBM, peribacteroid membrane; N-3, N-12, N-75, early nodulins (see van de Wiel *et al.*, 1990); N-26 (Nodulin-26), an integral peribacteroid membrane protein (Fortin *et al.*, 1987); N-35, N-100 are nodulins representing uricase and sucrose synthetase (Nguyen *et al.*, 1985; Thummler and Verma, 1987).

3.1. PERIBACTEROID MEMBRANE NODULINS

In addition to the nodulins involved in enzymatic functions, several major nodulins are integral membrane proteins of the peribacteroid membrane (PBM). Of these, nodulin-26, appears to be an ion channel, which may facilitate transport of specific metabolites across the PBM. This protein has homology with a number of other membrane proteins (see Verma *et al.*, 1990), each of which may have acquired different specificities during evolution. In *Escherichia coli*, a homolog of this nodulin-26 acts as a glycerol facilitator, while in animals, nodulin-26 resembles an integral membrane protein (MIP-26) in the eye lens membrane (Gorin *et al*, 1984), and a protein involved in neuro differentiation in *Drosophila* (Rao *et al.*, 1990). Recently, a tonoplast protein (Johnson *et al.*, 1990) and a turgor regulated protein (Guerrero *et al.*, 1990) have also been found to have homology with nodulin-26. These studies suggest that an ancestral protein involved in transport of a specific group of compounds (polyols) has been recruited in one of the nodule functions. Studies on the transport of specific metabolites mediated by nodulin-26 will shed light on the physiology and metabolism of this unique organ.

3.1.1. *Targeting of PBM Nodulins*. *De novo* formation of the subcellular compartment housing the bacteria demands a massive synthesis of new membrane (PBM) which is estimated to be twenty times more than plasma membrane in a normal cell. The PBM contains many nodulins which must be specifically targeted to this membrane and not to the plasma membrane from which the PBM is derived. However, comparison of putative sequences that may act as signals for this targeting did not yield any consensus (Jacobs *et al.*, 1987). Moreover, the two PBM nodulins analyzed in detail use different strategies to reach to the PBM, nodulin-24 carries a cleavable amino terminus signal while the nodulin-26 has a non-cleavable internal signal. The nodulin-24 does not contain any membrane-spanning region and seems to be located on the surface of the PBM facing the bacteroid (Fortin *et al.*, 1987). This has been recently confirmed by *in vitro* co-translation and processing data (C.-I. Cheon and D.P.S. Verma, unpublished). The nodulin-26 spans the membrane six times with its amino and carboxy ends facing the cytoplasm (G.-H. Miao and D. P. S. Verma, unpublished) and it is phosphorylated towards its carboxy end (Weaver *et al.*, 1990). A single glycosylation site that is co-translationally glycosylated faces the bacteroidal surface. It is possible that some specific lipids and/or glycan may be involved in targeting this protein to the PBM. Since nodulin-26 has homology with a tonoplast membrane protein (Johnson *et al.*, 1990), the PBM compartment may behave like a vacuoles.

3.2. REGULATION OF NITROGEN ASSIMILATION IN NODULES

3.2.1. *Glutamine Synthetase*. This is a key enzyme in root nodules assimilating ammonia from nitrogen fixation. A nodule-specific GS isoform has been reported in some legumes where the expression of GS is coordinated with root nodule development (Forde *et al.*, 1989), however, in the case of soybean, a GS gene has been identified which responds directly to the flux of ammonia entering the host cell from bacteroids. Isolation of this gene followed by transformation of a legume (*Lotus corniculatus*) and a non-legume (tobacco) plant revealed that ammonia inducibility may be specific to the legume background (Miao *et al.*, 1990).

Fusion of the soybean GS promoter with a reporter gene (GUS) and its transfer to *L. corniculatus* showed that GUS activity is present uniformly in the ammonia-treated root tissue, while without ammonia treatment the GUS activity is mainly localized in root apices and vascular tissue. The ammonia-specific induction of GS gene in transgenic *L.*

corniculatus takes four to six hours after ammonia application, and reaches a maximum after about 12 hours. These results along with the earlier data (Hirel *et al.*, 1987), clearly demonstrate that the regulation of soybean cytosolic GS by ammonia occurs at the transcription level and is not due to the increased stability of GS transcripts.

Subcellular localization of the GUS activity in young nodules formed on transgenic *L. corniculatus* by *Rhizobium loti* showed that the GUS activity can be detected in emerging nodules (see also Forde *et al.*, 1989) and incipient nodules prior to any morphological demarcation on the root surface. In mature nodules, infection zone containing both infected and uninfected cells, and the cells in the inner cortex had most of the GUS activity (Figure 2). The strongest activity was found in the infected cells. The outer cortex, including the suberized endodermis layer, showed no activity. The activity was also present in the vascular bundles located in the inner cortex. The entire infection zone, delimited by the endodermis (which is known to be impermeable to many gases), may encounter a flux of NH_4^+ which is diffused throughout this tissue, consequently inducing GUS. Since a rapid flux of ammonia occurs from symbiotic nitrogen fixation in root nodules and which must be metabolized to avoid any toxic effects, some of the GS genes may have come directly under the control of ammonia in certain legume plants. The pattern of GS gene expression, however, may change depending upon the nature of the nodules, ie., ureide- or amide-producers. Transfer of this gene to soybean or other tropical legumes such as *Vigna* may help to answer this question. We have recently been able to transform *Vigna aconitifolia* using *Agrobacterium rhizogenes* (H. Suzuki and D.P.S. Verma, unpublished). It is possible that the expression of GS in soybean (a ureide-producing plant) may not be the same as observed in the *Lotus* (an amide-producing plant).

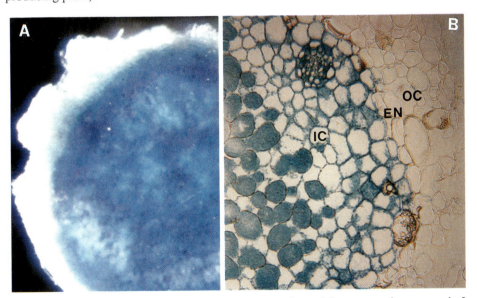

Figure 2: Localization of GUS activity driven by soybean GS promoter in transgenic *L. corniculatus* nodules. EN, endodermis; IC, inner cortex; OC, outer cortex. A, surface view of a cut nodule; B, a thin section after staining for GUS activity (see Miao *et al.*, 1990).

3.2.2. _Expression of uricase (nodulin-35) gene._ Uricase (EC 1.7.3.3) is a key enzyme involved in ureide production in tropical legumes, and the enzyme present in nodules is different from that present in uninfected young seedlings (Bergmann _et al._, 1983). It catalyzes the conversion of uric acid to allantoin, a reaction carried out in the peroxisomes of uninfected cells of nodules. Nodulin-35 is a subunit of uricase-II and is a major protein in root nodules. It is primarily localized in the peroxisomes of uninfected cells of the infection zone (Nguyen _et al._, 1985), including cells in the inner cortex. Since the ureide biosynthetic pathway is compartmentalized between infected and uninfected cells it appears that an intermediate metabolite of this pathway is involved in the induction of the nodulin-35 gene. However, nodulin-35 gene has been observed to be induced prior to ureide production in nodules (Nguyen, _et al._, 1985).

A nodulin-35 (N-35) cDNA was isolated from a soybean (_Glycine max_ L. var. Prize) nodule cDNA expression library using a previously isolated partial cDNA clone. The N-35 cDNA was expressed in _Escherichia coli_ driven by the _lacZ_ promoter and was found to be functionally active. The uricase activity was detected in the cytoplasmic fraction of _E.coli_ with the same pH optimum and apparent K_m values as that in the nodules. The size of the N-35 polypeptide expressed in _E. coli_ is identical to that present in soybean nodules and is assembled into a tetrameric holoenzyme with the same native molecular weight of uricase. Thus, the presence of peroxisomes does not appear to be essential for the proper assembly of the uricase holoenzyme. These data also indicated that post-translational modifications or membrane transport are not essential either for the assembly or for the activity of uricase. We constructed an N-35 derivative devoid of its 34 carboxy terminal amino acids. Although purified N-35 is a very stable enzyme, this truncated protein seems to be unstable in _E. coli_. The instability of the truncated product and the lack of uricase activity with this construct, suggests the importance of the carboxy terminus for the function of this enzyme, which is also involved inperoxisome targeting (Gould _et al._, 1987).

In order to determine the subcellular location of N-35 in _E. coli_, periplasmic proteins were prepared by cold osmotic shock. Most of the activity was found to be located in the cytoplasmic fraction of the cells containing the expression plasmid. Attempts made to express soybean nodulin-35 cDNA in tobacco under the control of the CaMV 35S promoter failed to show any uricase activity. While this protein is functional in _E.coli_ , it is not active and did not accumulate in any significant amount in tobacco cells. In the absence of peroxisomes the N-35 in the plant cytoplasm may be inactive or degraded.

We were not able to induce expression of soybean nodulin-35 gene in tobacco, indicating that there are some specific requirements for nodule factors for the induction of this gene. However, when introduced into yeast (Verma _et al._, 1988a) this gene functions constitutively, suggesting that a repressor mechanism may be operative in non-nodule tissue to control the expression of this gene. Since most of the nodulin gene promoters function only in the legume background, transformation of a ureide-producing legume such as soybean and _Vigna_ (see above) with the nodulin-35 gene may allow dissection of the _cis_-acting elements and _trans_-acting factors as well as identification of metabolites necessary for the induction of this gene in ureide-producing root nodules.

3.3. CARBON METABOLISM IN NODULES

3.3.1. _Regulation of sucrose synthetase (nodulin-100) gene._ Sucrose is the primary photosynthate translocated from the shoot to nodules. It is catabolized by sucrose synthetase, a tetrameric enzyme of nodulin-100 subunits. We have demonstrated (Thummler and Verma, 1987) that the activity of this enzyme may be regulated by the availability of heme. Free heme binds to the holoenzyme and dissociates it into subunits loosing all activity. Such a mechanism may prevent bacteria from becoming pathogenic on the host plant (Verma, 1989). During effective symbiosis there is not free heme in the nodule and its synthesis by rhizobia increases during hypoxia. Also, hypoxia induces sucrose synthetase gene expression.

4. Perspectives

The fact that most of the nodulin genes are induced prior to and independent of nitrogen fixation in nodules, suggests that the nodule developmental program is initiated in response to certain signals from bacteria in anticipation for the availability of reduced nitrogen beneficial for plant growth. In the event that this association becomes ineffective various feedback controls have evolved to restrict the growth of the invaded bacteria. For example, the activity of sucrose synthetase, an enzyme controlling carbon flow to the nodules, may be restricted by heme which becomes dissociated from leghemoglobin in a non-symbiotic state (Thummler and Verma, 1987). Similarly, the host defense mechanism may be evoked leading to the synthesis of phytoalexins (Werner *et al.*, 1985). The unique environment of nodules which is low in oxygen and high in osmoticum (Delauney and Verma, 1990) imposes specific pressure on the cellular metabolism. Since the PBM is the primary interface between the bacteria and the plant cell, it holds the key for symbiosis. Our understanding of the mechanism by which the subcellular compartment housing the bacteria is formed is very limited and much less is known about the transport of specific metabolites to and from bacteria during the symbiotic state. The presence of a highly conserved protein (nodulin-26) in this membrane suggests a need for transport of very specific group of ions across the PBM. A detailed understanding of nodulin genes will help answer how legumes have recruited pre-existing genes to help evolve symbiotic interactions with *Rhizobium*.

5. Acknowledgements

This work was supported by the NSF grants DCB-8819399 and DCB-8904101. We wish to thank Bertrand Hirel for the analysis of GS genes, Robert Ridge for histochemistry and Angela Kalb for preparation of the manuscript.

REFERENCES

Averyhart-Fullard, V., Datta, K. and Marcus, A. (1988), 'A hydroxproline rich protein in the soybean cell wall', Proc. Natl. Acad. Sci. USA 85, 1082-1085.
Bergmann, H., Preddie, E. and Verma, D.P.S. (1983), 'Nodulin-35: A subunit of specific uricase (uricase II) induced and localized in the uninfected cells of soybean nodules', EMBO J. 2, 2333-2339.
Delauney, A.J. and Verma, D.P.S. (1990), 'A soybean gene encoding Δ^1-pyrroline-5-carboxylate reductase was isolated by functional complementation in *Escherichia coli* and is found to be osmoregulated', Mol. Gen. Gen. 221, 299-305.
Dudley, M., Thomas, E., Jacobs, W. and Long, S.R. (1987), 'Microscopic studies of cell divisions induced in alfalfa roots by *Rhizobium meliloti*', Planta 171, 289-301.
Forde, B.G., Day, H.M., Turton, J.F., Shen, W.-J., Cullimore, J.V., and Oliver, J.E. (1989), 'Two glutamine synthetase genes from *Phaseolus vulgaris* L. display

contrasting developmental and spatial patterns of expression in transgenic *Lotus corniculatus* plants', Plant Cell 1, 391-401.

Fortin, M.G., Morrison, N.A. and Verma, D.P.S. (1987), 'Nodulin-26, a peribacteroid membrane nodulin is expressed independently of the development of the peribacteroid compartment', Nucl.Acids Res. 15, 813-824.

Gorin, M.B., Yancey S.B., Cline, J., Revel, J-P and Horwitz, J. (1984), 'The major intrinsic protein (MIP) of the bovine lens fiber membrane: Characterization and structure based on cDNA cloning', Cell 39, 49-59.

Gould, S. J., Keller G.-A. and Subramani, S. (1987) Identification of a peroxisomal targeting signal at the carboxy terminus of Firefly luciferase', J. Cell Biol. 105, 2923-2931.

Guerrero, F.D., Jones, J.T. and Mullet, J.E. (1990), 'Turgor-responsive gene transcription and RNA levels increase rapidly when pea shoots are wilted. Sequence and expression of three inducible genes', Plant Mol. Biol. 15, 11-26.

Hirel, B., Bouet, C., King, B., Layzell, D., Jacobs, F., and Verma, D.P.S. (1987), 'Glutamine synthetase genes are regulated by ammonia provided externally or by symbiotic nitrogen fixation', EMBO J. 5, 1167-1171.

Hirsch, A.M., Bhuvaneswari, T.V., Torrey, J.G. and Bisseling, T. (1989), 'Early nodulin genes are induced in alfalfa root outgrowths elicited by auxin transport inhibitors', Proc. Natl. Acad. Sci. USA 86, 1244-1248.

Jacobs, F.A., Zhang, M., Fortin, M.G. and Verma, D.P.S. (1987), 'Several nodulins of soybean share structural domains but differ in their subcellular location', Nucl. Acids Res. 15, 1271-1280.

Johnson, K.D., Hofte, H. and Chrispeels, M.J. (1990), 'An intransic tonoplast protein of protein storage vacuoles in seeds is structurally related to a bacterial solute transporter (GlpF)', The Plant Cell 2, 525-532.

Long, S.R. (1989), '*Rhizobium*-legume nodulation: Life together in the underground', Cell 56, 203-214.

Miao, G.-H., Hirel, B., Marsolier, M.C., Ridge, R.W. and Verma, D.P.S. (1990), 'Legume-specific ammonia-regulated expression of soybean genes encoding cytosolic glutamine synthetase', (submitted).

Miller, S.S., Boylan, K.L.M. and Vance, C.P. (1987), 'Alfalfa root nodule carbon dioxide fixation. III Immunological studies of nodule phosphoenolpyruvate carboxylase', Plant Physiol. 84, 501-508.

Morrison, N. and Verma, D.P.S. (1987), 'A block in the endocytosis of *Rhizobium* allows cellular differentiation in nodules but affects the expression of some peribacteroid membrane nodulins', Plant Mol. Bio. 9, 185-196.

Nguyen, T., Zelechowska, M., Foster, V., Bergmann, H., and Verma, D.P.S. (1985), 'Primary structure of the soybean nodulin-35 gene encoding uricase II localized in the peroxisomes of uninfected cells of nodules', Proc. Natl. Acad. Sci. USA 82, 5040-5044.

Rao, Y., Jan, L.Y. and Jan, Y.N. (1990), 'Similarity of the product of the *Drosophila* neurogenic gene 'big brain' to trans membrane channel proteins', Nature 345, 163-167.

Scheres, B. (1990) 'Early Nodulins in root nodule development', Ph.D. thesis, University of Wageningen, The Netherlands.

Suganuma, N.M., Kitou and Yamamoto, Y. (1987), 'Carbon metabolism in relation to cellular organization of soybean root nodules and respiration of mitochrondria aided by leghemoglobin', Plant Cell Physiol. 28, 113-122.

Thummler, F. and Verma, D.P.S. (1987), 'Nodulin 100 of soybean is the subunit of sucrose synthase regulated by the availability of free heme in nodules', J. Biol. Chem. 262, 14730-14736.

van de Weil, C., Scheres, B., Franssen, H., Van Lierop, M-J., Van Lammeren, A., Van Kammen, A. and Bisseling, T. (1990), 'The early nodulin transcript ENOD2 is located in the nodule parenchyma (inner cortex) of pea and soybean root nodules', EMBO J. 9, 1-7.

Verma, D.P.S. (1989), 'Plant genes involved in carbon and nitrogen assimilation in root nodules', in J.E. Poulton, J.T. Romeo, E.E. Conn (eds.), Recent Advances in Phytochemistry, Vol. 23: Plant Nitrogen Metabolism, Plenum Publishing Corp, New York, pp 43-63.

Verma, D.P.S., Delauney, A.J., Guida, M., Hirel, B., Schafer, R. and Koh, S. (1988a), 'Control of expression of nodulin genes', in: R. Palacios and D.P.S. Verma (eds.), Molecular Genetics of Plant-Microbe Interactions, APS press, Minnesota, pp 315-320.

Verma, D.P.S. and Delauney, A.J. (1988b), 'Root nodule symbiosis: nodulins and nodulin genes', D.P.S. Verma and R. Goldberg (eds.), In: Plant Gene Research - Temporal and Spatial Regulation of Plant Gene Expression, Springer Verlag Vein Publishers, New York, 10, 169-199.

Verma, D.P.S., Fortin, M.G., Stanley, J., Mauro, V., Purohit, S., and Morrison, N. (1986), 'Nodulins and nodulin genes of Glycine max', Plant Mol. Biol. 7, 51-61.

Verma, D.P.S. and Stanley, J. (1989), 'The legume-Rhizobium equation: A coevolution of two genomes', in: C.H. Stirton & J.L. Zarucchi (eds.), Advances in Legume Biology, Monogr. Syst. Bot. Missouri Bot. Gard. 29, 545-557.

Verma, D.P.S., Miao, G.-H., Joshi, C.P., Cheon, C.-I. and Delauney, A. (1990), 'Internalization of Rhizobium by plant cells: Targeting and role of peribacteroid membrane nodulins', in: Plant Molecular Biology, 1990, R.G. Herrmann and B.A. Larkins (eds.), Plenum Press.

Weaver, C. D., Stacey, G. and Roberts, D.M. (1990) 'Phosphorylation of nodulin-26 by a Ca dependent protein kinase', Proceedings from the 8th International Congress on Nitrogen Fixation, Knoxville, Tenn. USA, G-18.

Werner, D., Mellor, R.B., Hahn, M.G. and Grisebach, H. (1985), 'Soybean root response to symbiotic infection. Glyceollin I accumulation in an ineffective type of soybean nodules with an early loss of the peribacteroid membrane', Z. Naturforsch 40c, 179-181.

EARLY NODULINS IN PEA AND SOYBEAN NODULE DEVELOPMENT

Ton Bisseling, Henk Franssen, Francine Govers, Beatrix Horvath, Marja Moerman, Ben Scheres, Clemens van de Wiel, Wei-Cai Yang.
Department of Molecular Biology, Agricultural University, Dreijenlaan 3, 6703 HA Wageningen, The Netherlands.

During the last few years we have isolated early nodulin clones of both pea and soybean. In this chapter we will summarize the characteristics of these clones (Table 1). Furthermore we will describe how we started to use early nodulin clones to isolate *Rhizobium* signal molecules.

TABLE 1.

clone	characteristics	site of expression in the nodule
pPsENOD2	repeating pentapeptides, Pro-rich, cell wall protein ?	nodule parenchyma
pPsENOD3	metal binding protein ?	early symbiotic zone
pPsENOD5	amino acid sequence resembles arabinsgalactan proteins.	invasion and early symbiotic zone in infected roots; in cells containing infection thread
pPsENOD12	repeating pentapeptides, Pro-rich cell wall protein ?	invasion zone, in infected roots; in cells containing infection thread and in growth of infection thread tip
pPsENOD14	homologous to PsENOD3	early symbiotic zone
pGmENOD2	homologous to PsENOD2	nodule parenchyma
pGmENOD13	homologous to GmENOD2	nodule parenchyma
pGmENOD40	-	pericycle of vascular bundle and uninfected cells
pGmENOD55	Pro and Ser-rich	?

300

H. Hennecke and D. P. S. Verma (eds.),
Advances in Molecular Genetics of Plant-Microbe Interactions, Vol. 1, 300–303.
© 1991 *Kluwer Academic Publishers. Printed in the Netherlands.*

pPsENOD: *Pisum sativum* early nodulin cDNA clone.
pGmENOD: *Glycine max* early nodulin cDNA clone.
Nodule parenchyma: "nodule inner cortex" (2).

PsENOD2, GmENOD2, GmENOD13

The soybean early nodulin clone pGmENOD2 is the first early nodulin clone we isolated (1). At the moment ENOD2 clones have been isolated from several legumes including pea (pPsENOD2) (2). Furthermore the soybean early nodulin clone pGm ENOD13, that is homologous to GmENOD12 (3). The ENOD2/ENOD13 early nodulins are proline-rich proteins, composed of two repeating pentapeptides containing two proline residues each. Based on this sequence we postulated that these early nodulins are cell wall proteins.

By *in situ* hybridization it was shown that the ENOD2/GmENOD13 genes are specifically expressed in the nodule parenchyma ("inner cortex") (2). The nodule parenchyma is the tissue that forms the major oxygen barrier in root nodules (5). This oxygen barrier is caused by the specific morphology of the nodule parenchyma cells, by which only small intercellular spaces are present in this tissue. We postulated that the putative ENOD2/GmENOD13 cell wall proteins contribute to this specific morphology of the nodule parenchyma cells (2).

PsENOD3/PsENOD14

Both pPsENOD3 and pPsENOD14 encode small nodulins with a molecular weight of about 6 kD (4). These two early nodulins are 55 % homologous and contain as most striking characteristic a cluster of four cysteins arranged in such a way that a metal ion can be bound. The PsENOD3 and PsENOD14 genes are expressed in the infected cells of the central tissue of the pea nodule. The PsENOD3/PsENOD14 genes are transcribed at maximal level in the early symbiotic zone. In older cells of the symbiotic zone the PsENOD3/PsENOD14 mRNA concentration decreases (4).

PsENOD5

PsENOD5 is a protein with hydrophobic regions at both the C- and N-terminus, the latter possibly forming a signal peptide. This early nodulin also has a proline rich domain, which inaddition has a high content of glycine, alanine and serine residues (4). The amino acid composition of this domain is reminiscent of that of arabinogalactan proteins.

The PsENOD5 gene is first expressed in pea root hairs that become infected by *Rhizobium*. In the infected pea root cortex the expression of this early nodulin gene is restricted to the cells containing the infection thread tip. Based on these data we concluded that PsENOD5 has a role in the infection process. Because some arabinogalactan proteins are known to be a component of the plasma membrane, PsENOD5 may be part of the plasma membrane of the infection thread. In pea nodules the PsENOD5 gene is expressed in the infected cells of the invasion zone, the zone where infection thread growth takes place, but the highest level of PsENOD5 transcript is found in the early symbiotic zone (4). In this latter zone infection threads have stopped to grow, bacteria proliferate and an active membrane synthesizing apparatus is keeping up with bacterial proliferation to surround bacteria with peribacteroid membranes. The relatively high level of PsENOD5 mRNA in the early symbiotic zone suggests that this early nodulin may not only be part of the plasma membrane of the infection thread, but also of the peribacteroid membrane.

PsENOD12

Like ENOD2, PsENOD12 is composed of two repeating pentapeptides containing two prolines each, suggesting that this early nodulin also is a cell wall protein (7). Expression of the PsENOD12 gene is observed in root hairs and root cortical cells containing a growing infection thread. In contrast to the PsENOD5 gene, this early nodulin gene is also expressed in cells several layers in front of the growing infection thread (7). In these root cortex cells morphological changes occur - including secondary cell wall formation - preceding penetration by an infection thread occur. The putative cell wall protein PsENOD12 may thus be part of the additional cell wall formed in the root cortex cells that prepare for infection thread passage. In addition PsENOD12 may be a component of the infection thread itself. In pea root nodules the expression of the PsENOD12 gene is restricted to the invasion zone, the zone where active infection thread growth occurs. This is consistent with a role of PsENOD12 in the infection process.

GmENOD40 and GmENOD55

GmENOD40 shares no significant homology with any previously characterized protein. Consequently its sequence did not provide a clue about its function. In soybean nodules the highest level of GmENOD40 transcript is found in the pericycle of the vascular bundle, and low level of this mRNA is found in the uninfected cells. In soybean nodules the uninfected cells form a network by which transport of compounds might occur and the pericycle of the vascular bundle is probably involved in intracellular transport of nutrients from and to the vascular bundle (8). Therefore we expect that GmENOD40 has a function in intracellular transport processes.

GmENOD55 is a proline rich protein as well. In this protein the prolines are confined to an internal domain of 32 amino acids in which proline and serine residues occur in an alternating. GmENOD55 mRNA has not yet been localized in soybean nodules. Therefore it is not possible to postulate its function.

Early nodulins as tools to identify signals involved in bacterium- plant interaction

The multistep nature of root nodule formation indicates that at several stages specific signals from *Rhizobium* are involved in inducing the next step in nodule development. To identify *Rhizobium* signal molecules we will use assays that involve the induction of an early nodulin gene that marks a specific step of nodule development. At the moment we have developed two molecular assays to identify bacterial signal molecules which elicit the infection process . These assays involve the induction of the PsENOD12 and PsENOD5 genes in pea root hairs. Interestingly, application of sterile culture filtrates of *R.leguminosarum* bv. *viciae* to pea seedlings elicits both PsENOD12 and PsENOD5 gene expression. In collaboration with J.Dénarié and coworkers (Toulouse) we are characterizing the compounds which trigget the expression of these two early nodulin genes.

References

1 Franssen, H.J., Nap, J.P., Gloudemans, T., Stiekema, W., Van Dam, H., Govers, F., Louwerse, J., Van Kammen, A. and Bisseling, T. (1987) Proc Natl Acad Sci USA 84, 4495-4499.

2 Van de Wiel, C., Scheres, B., Franssen, H., Van Lierop, M.J., Van Lammeren, A.,Van Kammen, A. and Bisseling, T. (1990) EMBO J 9, 1-7.

3 Franssen, H.J., Scheres, B., Van de Wiel, C. and Bisseling, T. (1988) Molecular Genetics of Plant-Microbe Interactions, Palaciosl, R. and Verma, D.P.S. (eds), APS Press, St. Paul, pp 321-326.

4 Scheres, B., Van Engelen, F., Van der Knaap, E., Van de Wiel, C., Van Kammen, A. and Bisseling, T. (1990) The Plant Cell, in press.

5 Witty, J.F., Minchin, F.R., Skot, L. and Sheehy, J.E. (1986) Oxford Surveys of Plant and Cellular Biology 3, 275-315.

6 Knox, J.P., Day, S. and Roberts, K. (1989) Development 106, 47.

7 Scheres, B., Van de Wiel, C., Zalensky, A., Horvath, B., Spaink, H., Van Eck, H., Zwartkruis, F., Wolters, A.M., Gloudemans, T., Van Kammen, A. and Bisseling, T. (1990) Cell 60, 281-294.

8 Gunning, B.E.S., Pate, J.S., Minchin, F.R. and Marks, J. (1974) Symp Soc Exp Biol 28, 87-126

DIFFERENT MODES OF REGULATION INVOLVED IN NODULIN GENE EXPRESSION IN SOYBEAN

Champa Sengupta-Gopalan, Herve Gambliel, Inez Feder, Hannes Richter and Stephen Temple
Dept. of Agron. and Hort./Mol. Biol. Program, New Mexico State University, Las Cruces, NM 88003, USA.

INTRODUCTION
The root nodule is a specialized plant organ in which the resident bacteria belonging to the genus <u>Rhizobium</u>/<u>Bradyrhizobium</u>, fix N_2 while the plant provides the ideal environment for the bacteria to fix N_2. The fixed N_2 is assimilated by host encoded enzymes. Several unique host genes are specifically expressed in the nodule and these gene products, nodulins, probably have specific functions in the nodule. To date, all studies appear to show that nodulin gene expression is regulated primarily at the transcriptional level and induction of transcription precedes the onset of nitrogenase activity. In this chapter, we will focus on the regulatory role of specific AT rich DNA elements, distributed in the 5' and 3' flanking regions of a representative soybean nodulin gene. This chapter will also focus on the regulation of glutamine synthetase (GS) genes and demonstrate that regulation of nodulin gene expression can be controlled at a step other than transcription. Nodule-specific GS genes, like other nodulin genes appears to be developmentally regulated independent of N_2-fixation (5), while the activation of nodule-specific GS enzyme appears to require the products of active N_2-fixation.

RESULTS AND DISCUSSION
Characterization of AT rich DNA elements in nodulin genes: In order to understand how nodulin gene expression is regulated, we have undertaken to analyze the 'cis' acting and 'trans' -activating factors that regulate a representative member of a nodulin gene family. The representative gene encodes for nodulin-20 and analysis of genomic clones containing this gene have shown that the N-20 gene is flanked by a truncated version of the same gene (N-20t). Sequence analysis in and around the N-20 and N-20t genes showed that besides the coding region, a high degree of sequence similarity is maintained in the 5' upstream and 3' downstream regions between the two genes (Fig. 1). The conservation in the 5' and 3' flanking regions of these two genes would suggest involvement of some sort of functional constraints on these regions.

H. Hennecke and D. P. S. Verma (eds.),
Advances in Molecular Genetics of Plant-Microbe Interactions, Vol. 1, 304–309.

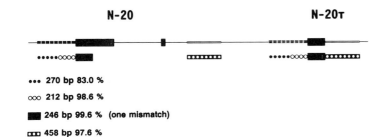

N-20   **N-20T**

••• 270 bp 83.0 %
ooo 212 bp 98.6 %
▉ 246 bp 99.6 % (one mismatch)
ⅲ 458 bp 97.6 %

Fig. 1 Graphic representation of the sequence similarity between N-20 and N-20t. The solid box represents the coding sequence

Short stretches of AT domains in the promoter of the leghemoglobin Lbc3 gene have been reported to be necessary for efficient gene expression in transgenic lotus and interactions have been described between these AT rich domains and nodule specific trans-acting factors (4). The N-20 and N-20t genes show AT rich domains similar to those present in the Lbc_3 gene promoter distributed in the upstream and downstream regions of the N-20 and N-20t genes. In an attempt to identify trans-acting factors that may be involved in the regulation of transcription, we scanned the 5' distal regions of the N-20 gene by gel mobility shift assays. Nuclear factors from 12 day old nodule extracts bind a 245 bp BglII/EcoRV fragment of the N-20 gene that contains two of these dAdT elements. Binding is characterized by a pattern of multiple interactions with the probe. At least 5 complexes can be detected at different extract concentration. Lower extract concentrations revealed two well defined complexes termed I & II as well as complex IV, which is more diffuse and appears to contain two main components in addition to some nonspecific components which cause smearing (see Fig. 2, Panel D). In addition, a faint and a strong signal were observed for complexes III and V at higher extract concentrations but were not characterized any further. Additionally, none of the retarded bands are detected with the N-20 cDNA, (Fig. 2B). Mapping of the N-20/N-20t tandem show that in vitro binding is found both upstream and downstream and within the intron of the N-20/N-20t genes (Fig. 2A). In all cases, binding is localized within small fragments which contain dAdT stretches, homologeous to the binding elements in the gene for Lbc_3.

Complexes formed from fragments within the promoter and 3' untranslated region of both N-20 and N-20t can be titrated with each other respectively (Fig. 2D), but not by a linearized bluescribe plasmid, indicating that they are specific for these regions. In addition, dAdT elements from the promoter regions of the Phytohemagglutinin-L (Fig 2D) and β-phaseolin (data not shown) genes from bean also act as strong competitors. This suggests that it is not a particular dAdT sequence that is being recognized but rather a

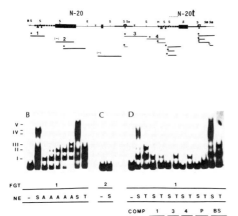

Fig. 2 Mapping of dAdT binding domains on the N-20/N-20t tandem genes. Panel A: Black boxes represent exons. The two promoters are marked by discontinuous lines. Triangles represent homology with a minimum of 12 bp and 87% sequence identity with binding site 1 (lower) or 2 (upper) of the leghemoglobin Lbc₃ (4). Fragments that were tested by the gel retardation assay are shown. ● indicates that the fragment is able to make complex I and II. ★ shows that complex IV is also made. (-) indicates that no binding occurs with these fragments under standard assay conditions. Restriction sites are shown with abbreviations. Panel B: Representative mobility-shift assays with a 245 bp Bgl11/EcoRV N-20 promoter fragment and nuclear extracts (NE) from 12 day old soybean nodules obtained in various fashions: 0.42M NaCl preparation (S), increasing amounts of a 0.1 sulfuric acid extract (A), and a 2% TCA soluble nuclear fraction (T). Panel C: Use of fragment 2, a cDNA fragment, as probe. Panel D: Competition studies using fragments containing a portion of the 3' direct repeat element (3), the NOD-20 promoter (1), NOD-20t promoter (4), phytohemagglutinin dAdT promoter domains (P), and linearized bluescribe plasmid (BS)

particular structural feature of the DNA containing these regions. Poly(dA-dT) is also an effective competitor. In keeping with Jacobsen et al., (3) we have found that the protein(s) responsible for the formation of complex I and II are soluble in .42M NaCl and 2% TCA and therefore meet the operational criteria for high mobility group proteins (Fig. 2B). We have compared the mobility of the above complexes formed by extracts from 12 day old soybean nodules with those obtained using nuclear extracts from alfalfa nodules, soybean leaves and bean seeds (Fig. 3). The results indicate that leaf and nodule tissue are able to form similar complex patterns with these probes. In particular, the mobility of complex I and II appear similar and the same is seen with bean seed extracts. Alfalfa nodule nuclear extracts, however, form complexes of slightly lower mobility than formed with both bean seed and soybean nodules. Thus, there appears to be species specific and organ specific differences as well as similarities in the pattern of retardation. Preliminary results in our laboratory based on southwestern studies indicate that different proteins are responsible for this binding in soybean leaf and nodules.

Our results suggest that multiple protein factors interact with the AT-rich elements in a very complex manner involving interaction between DNA and protein, as well as between different proteins in association with DNA. This complex interaction probably causes local conformational changes in the chromatin. The presence of AT-rich elements in the 5' and 3' flanking regions would suggest they probably define the transcriptional unit.

Fig. 3 Comparison of DNA:protein complexes formed between a promoter fragment and nuclear extracts from different organs of soybean (G)bean (P) and alfalfa (M). Arrows indicate presence of complexes not detectable in the picture.

PROBE 1

NE - NOD LEAF SEED NOD

PLANT G G P M

Regulation of glutamine synthetase (GS) activity during nodule development: Ammonia produced by nitrogen fixation is assimilated in the plant fraction of the nodule via the combined activities of glutamine synthetase (GS) and glutamate synthetase (GOGAT). GS, an octameric enzyme, is encoded by members of a small gene family and expression of the different members is regulated in an organ specific manner (1). We had reported earlier that during root nodule development in soybean, novel forms of GS genes not expressed in the roots, are switched on in the nodule. These GS genes, in common with other nodulin genes, appeared to be turned on independent of the onset of nitrogenase activity (5). However, work from Verma's lab showed the absence of any nodule specific form of GS in soybean and furthermore demonstrated ammonia mediated enhancement in the level of GS transcripts in uninfected roots (2).

To clarify this ambiguity, we have extended our investigation on nodule/root GS genes and GS enzymes. In Fig. 4, total soluble protein from uninfected roots, effective nodules and ineffective nodules defective either in structural nitrogenase (Bj702) or in bacterial release (Bj2101), were analyzed after fractionation for both GS activity and for immunoreaction to GS antibody. A 2D gel profile of the immunoreactive proteins would suggest the presence of about 3 or 4 nodule-specific GS polypeptides, which are present in the nodules formed by Bj110 and Bj702 but not by Bj2101. Immuno-precipitation of _in vitro_ translation products with GS antibody showed that these nodule-specific GS proteins are indeed novel gene products (5). Native gel electrophoresis followed by staining for GS activity (panel B) showed activity in a band common to roots and all

308

the different nodules (GS_{r1}) and a slower migrating broad band defined as GS_n only in the case of the effective nodules. Native gel electrophoresis followed by Western blotting (Fig. 4 panel C) showed that the roots and nodules formed by Bj2101 have identical profiles with two immunoreactive bands (GS_{r1} and GS_{r2}). Nodules formed by Bj702 showed four major immunoreactive bands ($GS_{n1,2,3,4}$) in addition to GS_{r1} and GS_{r2} while effective nodules formed by Bj110, showed GS_{r1}

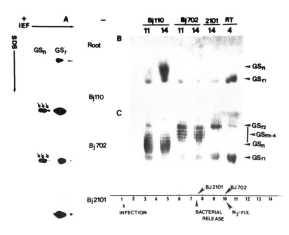

Fig. 4 Identification of nodule-specific GS enzyme. A: 2-D SDS - PAGE Western analysis of the GS polypeptides of nodules formed by different strains of B. japonicum and uninfected roots; B: Fractionation of root and nodule proteins by native gel electrophoresis followed by staining for GS transferase activity; C: Native PAGE Western analysis of GS proteins. The defect in the mutant strains used in this study is indicated. The designation of the GS proteins is described in the text. The numbers on the top indicate days after inoculation.

lower levels of $GS_{n1,2,3,4}$ and immunoreactive material co-migrating with GS_n. Taken together, our results indicate that there are nodule-specific GS genes and that they are indeed turned on independent of the onset of nitrogenase activity. However, activation of the nodule-specific GS (GS_n) enzyme is dependent on N_2 fixation. As seen in Fig. 5A, analysis of GS activity during root nodule development shows a biphasic pattern of increase in GS activity, the first increase coincides with the onset of N_2-fixation and the second increase accompanies an increase in the Lb protein. Immunodetection of the GS proteins on the native gel showed a gradual decrease in the level of GS_{r2} starting around day 9 and a gradual loss in $GS_{n1,2,3,4}$ starting day 12 in the effective nodules. Coincident with the disappearance of the GS_{r2} and $GS_{n1,2,3,4}$ bands, there appears to be a comparable increase in the total GS activity and activity in the GS_n band. Ineffective nodules formed by Bj702, showed no increase in GS activity over the level in uninfected roots. Furthermore, unlike the effective nodules, nodules formed by Bj702 showed no loss in GS_{r2} and $GS_{n1,2,3,4}$ and did not show activity in GS_n band. Some activity could be detected in the region of $GS_{n1,2,3,4}$ in nodules formed by Bj702.

Since our results suggest that activation of the nodule specific GS enzyme in soybeans requires active N_2-fixation, we tested the effect of various N_2 sources on the activation of GS in nodules normally defective in nitrogenase activity (Bj702). Our results showed no stimulation in GS activity with either NH_4^+ or NO_3^- but showed 40 to 50% stimulation in specific activity and the appearance of activity in the nodule specific form of GS with glutamine and glutamate as localized on activity gel. Our preliminary results suggest that probably a product of the GS catalyzed reaction itself is required for activation of GS in the nodules and the activation involves some sort of processing. Work is in progress to understand the mechanism of activation of nodule specific GS and to characterize the genes encoding for the nodule-specific form of GS.

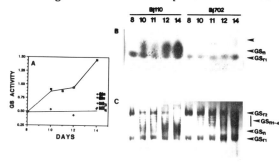

Fig. 5 Analysis of GS activity in developing nodules formed by Bj110 and Bj702. A: Total GS activity (μmol GHA/mg protein/min) at different times following infection (■ Bj110, ● Bj702). The level of total GS activity in nodules formed by Bj702 following 24 hr treatment with 10mm glutamine, glutamate, KNO_3 or NH_4NO_3 is indicated by an asterick. B: Native PAGE of soluble protein from developing nodules formed by Bj110 and Bj702, followed by staining for GS activity. C: Native PAGE western analysis of GS proteins in developing nodules. The numbers indicate days following infection.

REFERENCES

1. Forde, B.G. and Cullimore, J.V. (1989) 'The molecular biology of glutamine synthetase in higher plants', In: Miflin B.J. (ed), Oxford survey of plant molecular and Cell Biology. Oxford Univ. Press.

2. Hirel, B., Bouet, C., King, B., Layzell, D., Jacobs, F. and Verma, D.P.S. (1987) 'Glutamine synthetase genes are regulated by ammonia provided externally or by symbiotic nitrogen fixation', EMBO J 5:1167-1171.

3. Jacobson, K., Laursen N.B., Jensen, E.O., Marcker, A., Poulsen, C. and Marcker, K.A. (1990). 'HMG1-like proteins from leaf and nodule nuclei interact with different AT motifs in soybean nodulin promoters', The Plant Cell, 2:85-94.

4. Jensen, E.O., Marcker, K.A., Schell, J. and de Bruijin, F.J. (1988) 'Interaction of a nodule specific, trans-acting factor with distinct DNA elements in the soybean leghaemoglobin lbc$_3$ 5' upstream region', EMBO J. 7:1265-1271.

5. Sengupta-Gopalan, C. and Pitas, J.W. (1986) 'Expression of nodule-specific glutamine synthetase genes during nodule development in soybeans' Plant Mol. Biol. 7:189-199.

REGULATION OF NODULE-EXPRESSED SOYBEAN GENES

ERIK ØSTERGAARD JENSEN
Laboratory of Gene Expression,
Department of Molecular Biology & Plant Physiology
University of Aarhus,
DK-8000 Aarhus C
Denmark

ABSTRACT. The qualitative and quantitative contributions of four separate *cis*-acting DNA elements controlling the root nodule specific soybean leghemoglobin *lbc₃* gene were analysed in transgenic *Lotus corniculatus* plants. The strong positive element SPE (-1363,-1219; -1090, -947) responsible for high level expression was demonstrated to be an organ regulated element. Deletion of the downstream qualitative organ specific element OSE (-139, -102) resulted in a low expression level. Similarly, regulatory elements were defined in the promoter regions of the nodulin N23, *lba* and Enod2 promoter regions. Three different organ specific nuclear proteins recognizing AT-rich DNA sequences were identified in soybean. One protein (NAT2) is present in mature nodules, another protein (NAT1) is detected in roots and nodules, and a third protein (LAT1) is only observed in leaves. NAT2 recognizes a weak positive element in the *lbc₃* promoter. Biochemical analysis and DNA binding studies of LAT1 and NAT1 proteins indicate a relationship to high mobility group proteins. Purification and identification of the LAT1 and NAT1 activities revealed peptides within a molecular weights size range of 21 to 23 kD.

Introduction

Legume root nodule formation is a result of temporally controlled developmental expression of specific genes residing in the genomes of the host plant and the infecting bacteria . The host genes involved are coding for a group of proteins, which is denoted nodulins. The nodulin genes are divided into two classes, which based on their expression-pattern are termed early and late nodulin genes. Early nodulin genes like Enod2 are activated shortly after infection probably due to rhizobial microsymbiont signal molecules excreted by the bacteria, while late nodulin genes like leghemoglobin and N23 seem to require additional factors for induction (Govers et al. 1986; Scheres et al. 1990). In an approach to study the regulatory circuits directing expression of nodulin genes we have used a *Lotus corniculatus* transformation regeneration system (Petit et al. 1987; Stougaard et al. 1987; Hansen et al. 1989) to define *cis*-acting regulatory DNA elements on both early and late nodulin promoters. Simultaneously, we also studied nuclear protein interactions with late nodulin promoters. By the use of gel-retardation assay and footprinting studies we have identified and purified several factors which interact with AT-rich DNA sequences (Jensen et al. 1988, Jacobsen et al. 1990).

H. Hennecke and D. P. S. Verma (eds.),
Advances in Molecular Genetics of Plant-Microbe Interactions, Vol. 1, 310–316.
© 1991 *Kluwer Academic Publishers. Printed in the Netherlands.*

Identification of Regulatory DNA Elements

Identification of regulatory DNA elements required for regulated expression of soybean nodulin genes relied on a transformation-regeneration system, which we established previously for *Lotus corniculatus* (Birds-foot trefoil). Expression of the heterologous soybean gene promoters in the *Lotus* symbiotic system, indicated that the molecular mechanisms regulating early and late nodulin gene expression are conserved among various legumes.

We have studied expression of reporter genes in transgenic *Lotus* plants employing deletion-mutated promoters, derived from the soybean nodulin genes *lbc3* , *lba*, N23 and Enod2, and we also employed CaMV-35S based chimeric promoters. Combination of data from this expression-analysis resolved the position of several cis-regulatory DNA-elements. Strong positive "enhancer"- like elements, negatively regulating elements, as well as elements determining organ specific expression were identified.

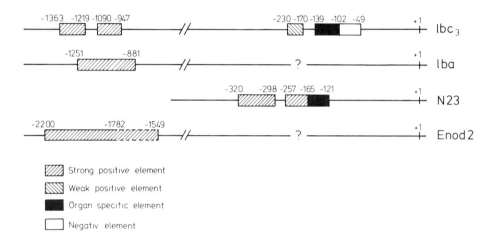

Figure 1. Schematic representation of the regulatory elements in the promoter regions of the early nodulin Enod2 and the late nodulins *lba, lbc₃* and N23.

THE LBC₃ PROMOTER

A 2 kb 5´region of *lbc₃* 5´3´-CAT was initially analysed in a classical deletion analysis (Stougaard et al. 1987). Two upstream quantitative promoter components were defined by this approach. A strong positive element (SPE) composed of two smaller elements located between positions −1363,-1219 and −1090, -947, and a weak positive element (WPE) located between -230,-170. Two qualitative elements were characterised after reactivation of silent deletions using the "constitutive" CaMV 35S enhancer. An organ-specific element (OSE) containing putative nodulin consensus sequences (AAAGAT; CTCTT) , and a negative element (NE) were found overlapping the basic promoter (Fig. 1). The qualitative and quantitative contributions of these elements were also analysed using internal deletions removing one or more of the defined regulatory regions (Stougaard et al. 1990). Low levels of expression resulting from removal of the OSE element indicated that this element also acts as a quantitative element possibly interacting with the upstream SPE. The WPE element previously found responsible for 2 % of full level expression could on the other hand be removed without effecting organ specificity and with only minor effect on the expression level. Nodule specific expression of deletions with only the TATA box region and strong positive element present indicates the presence of specific control sequences on the putative enhancer (SPE) itself. Full level expression seems to depend on the strong positive element, the organ specific element as well as basic promoter elements.

The regulatory capacity of the region carrying the OSE and NE elements was also investigated in hybrid promoter studies using the " neutral constitutively" expressed CaMV 35S promoter. Insertion of the complete -139, -35 region carrying the organ specific and the negative element of the lbc_3 gene promoter into the -90 position of the 35S promoter changed the expression of the 35S promoter towards a higher level in nodules. This effect was even more pronounced when a dimer of the -139,-35 region was used. Low level expression was only detected in roots from a few plants and no activity was detected in leaves (Lauridsen et al. 1990).

Strong preference for nodule expression was therefore conferred on the 35S promoter by the OSE and NE elements acting together. Insertion of the NE element alone (-102,-35) seems to lower expression from the 35S in roots, leaves and nodules. The OSE alone does not confer any nodule specificity on the 35S promoter, but might elevate expression. Presence of both OSE and NE seems therefore required to give preferential expression of the 35S promoter in root nodules.

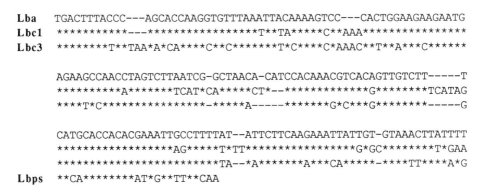

Figure 2. Alignment of DNA sequences from the strong positive elements of the *lba* (-1013, -861) and lbc_3 (-1342, -1168) promoters. Included is also a promoter fragment from the lbc_1 promoter and a sequence from the pea leghemoglobin promoter (-862,-837, Lbps, Nap 1988). Asterisks indicate identity to the *lba* sequence. Dashes indicate gaps in the alignment.

THE N23 PROMOTER

The importance of the lbc_3 OSE region suggests that the consensus sequences 5'AAAGAT and 5'CTCTT are required for efficient and nodule-regulated expression of nodulin genes. By 5' deletion analysis we have located two elements between positions –344 to -293 and -247 to -165, respectively, in the N23 promoter. Removal of the distal element reduced the expression level to approximately 20%, while removal of both elements reduced the level to below detection (Jørgensen et al. 1988). 35S-enhanced expression of the inactive –165 deletion lacking the positive elements resulted in a expression level of 20% of the full level in nodules. However, reactivation of deletion -121 resulted in very low level of expression (2% of full activity). Thus the region between -165 to -121, which contain one of the consensus sequences (5'CTCTT), is important for the organ specific expression (Stougaard et al. 1990). The immediate upstream regions are therefore necessary for high-level nodule-regulated expression of both the lbc_3 and the N23 gene promoters.

OTHER NODULIN PROMOTERS.

Strong positive regulatory elements have also been identified in the lba promoter in the region from -1251 to -881 (She, unpublished) and in the early nodulin promoter Enod2 from -2200 to -1782 (Lauridsen et al. 1990) .In both instances, no residual expression could be detected when deleting these elements. Alignment of the DNA sequences present in the strong elements of the lbc_3 and lba promoters revealed a high degree of similarity. Homologous sequences were also found on an lbc_1 promoter fragment and in the pea leghemoglobin promoter (Fig. 2).

Proteins Interacting with Late Nodulin Promoters

We have studied the interaction of chromatin-derived proteins from soybean-organs (nodules, roots, leaves and seedlings) with late nodulin gene promoters. By the use of radio-labelled DNA-fragments in gel-electrophoretic DNA-mobility shift assays (mobility shift) and footprinting studies, we have identified a number of factors which interact with AT-rich sequences in the lbc_3 and N23 upstream regions.

THE NODULE SPECIFIC NUCLEAR FACTER NAT2

First, a nodule specific DNA-binding protein was identified (NAT2). Two AT-rich DNA sequences in the weak positive element (-230 to -170) of the $lbc3$ gene have been delimited (Jensen et al. 1988), but several other bindingsites are located between the SPE and the WPE elements (Lauridsen et al. 1990). Point-mutation analysis of these binding sites revealed that even single basepair substitutions in the substrate sequence strongly affects binding affinity. Thus, the substrate recognition is highly sequence dependent.

The NAT2 factor is barely detectable in roots and the activity increases in the root nodules until about day 10-12 after infection with *Rhizobium*. Interestingly, many of the late nodulin genes are activated at this stage of nodule development. After appearance of the NAT2 activity, it is persistent throughout the nodule development.

DNA-mobility shifts applied to the NAT2 factor suggests a high order of complexity, which is persistent throughout a set of chromatographic steps aimed at obtaining this abundant nodule-specific factor in purified form. Our current purification scheme employs ammonium-sulphate precipitation, heating to 60°C for 10 minutes and sequential chromatography on DEAE-cellulose, Heparin-Sepharose, CM-Sepharose and Sephacryl S-300. We have obtained more than 5,000-fold purification at the protein-level. SDS-PAGE and silver-staining reveal less than 20 polypeptides in this fraction. After qualitative DNA-affinity chromatography, we observe dominant species of 40 kD, 34 kD, 28 kD and 23 kD. Protein/DNA-UV-crosslinking, employing ^{32}P-labelled, Br-dU substituted probe with less purified fractions, reveal a dominant 28 kD polypeptide after SDS-PAGE and autoradiography. This information is in agreement with the properties defined by the Sephacryl S-300 column.

314

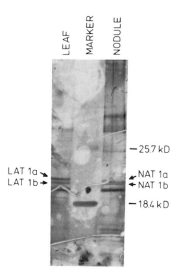

Figure 3. SDS-PAGE of purified leaf and nodule nuclear proteins containing LAT1 and NAT1 binding activities, respectively.

CHARACTERIZATION AND PURIFICATION OF THE LAT1 AND NAT1 PROTEINS

Protein-DNA interaction studies has also been applied to fragments from the N23 promoter (Jacobsen et al. 1990). One of the fragments formed a complex with the NAT2 factor. Two other DNA binding proteins, NAT1 and LAT1, were recognised and their cognate binding sites were identified by DNase I footprinting. The root and nodule factor NAT1 cannot be detected in leaf nuclear extracts, whereas LAT1 was detected and identified from leaf nuclear extracts. These two proteins behave in similar ways in mobility shift assays. The NAT1 binding site overlaps that of the LAT1 factor, but NAT1 requires additional sequences for binding. In the N23 promoter the LAT1 protein binds the same same sequence as NAT2, however, binding affinity is less affected by the pointmutations. NAT1 and LAT1 recognize several DNA sequences in the lbc_3 promoter.

The NAT1 and LAT1 proteins were extracted from chromatin by 0.35 M NaCl and they remain soluble in 2% TCA. This behaviour suggested a relationship with the high mobility group proteins (HMG-proteins). Purification of the LAT1 binding activity revealed the presence of two polypeptides with molecular weights of 21 and 23 kD, respectively. The two LAT1 peptides have identical N-terminal sequences, which bear strong resemblance to a wheat embryo HMG-protein (Spiker 1988). Two peptides with molecular weights of 21 and 22 kD, respectively, co-purified with the NAT1 activity (Fig. 3). Thus the molecular weights, mobility in gel-retardation assay and biochemical properties of LAT1 and NAT2 indicate a close relationship, despite their different binding properties (Table I). The biological effects of the NAT1 and LAT1 factors remain to be studied. However, HMG-proteins was found associated to transcriptional active chromatin in other biological systems (Solomon et al. 1986). Thus, we believe that both LAT1 and NAT1 proteins may have analogous functions, at the level of chromatin conformation, in the organs from which they are derived.

TABLE I. Comparison of the biochemical and DNA binding properties of
LAT1, NAT1 and NAT2

	Tissue specificity	Heat stable	Soluble in 2% TCA	Molecular weight	Recognizes a NAT2 binding site 1 (*)	Mobility of DNA-protein complex
LAT1	leaf	yes	yes	21 & 23 kD	yes	high, like NAT1
NAT1	root & nodules	yes	yes	21 & 22 kD	no	high
NAT2	nodules	no	no	27-28 kD	yes	low

*) See Jacobsen et al. 1990

Conclusion

Several *cis*-elements and *trans*-acting factors have been identified in the soybean leghemoglobin *lbc₃* and N23 promoter regions. So far no plant encoded nuclear proteins interacting with the organ specific element have been identified. Thus a correlation between *in vivo* effects of DNA regulatory elements and factor binding have only been established for the weak regulatory element in the *lbc₃* promoter and the nodule specific NAT2 protein. An internal deletion removing the weak positive element and the two NAT2 binding sites resulted in only minor reduction of the nodule specific expression (Stougaard et al. 1990). However, the presence of other NAT2 binding sites in the *lbc₃* promoter region between the SPE and WPE elements may then compensate the loss of sites at the WPE.

The function of the NAT1 and LAT1 HMG proteins still remains unknown, but in analogy to mammalian systems, they may be involved in chromatin conformation.

Acknowledgements

We would like to thank the Danish Biotechnology Programme for the financial support of this project.

References

Christensen, T, Sandal, N.N., Stougaard, J.and Marcker, K.A.(1989) 5' flanking sequence of the soybean leghemoglobin *lbc₃* gene. Nucleic Acids Res.17: 4383.
Govers, F., Moerman, M., Downie, J.A., Hooykaas, P., Franssen, F.J., Louwerse, J., Van Kammen, A., and Bisseling, T.(1986) Rhizobium *nod* genes are involved in inducing an

early nodulin gene. Nature, 323: 564-566.

Hansen, J., Jørgensen J-E., Stougaard, J., and Marcker, K.A. (1989) Hairy roots - a short cut to transgenic root nodules. Plant Cell Rep., 8: 12-15.

Jacobsen, K., Laursen N.B., Jensen E.Ø., Marcker, A., Poulsen, C., and Marcker, K.A.(1990) HMG I like proteins from leaf and nodule nuclei interact with different AT motifs in soybean nodulin promoters. The Plant Cell, 2: 85-94.

Jensen, E.Ø., Marcker, K.A., Schell, J., and de Bruijn, J.(1988) Interaction of a nodule specific, trans-acting factor with distinct DNA elements in the soybean leghemoglobin lbc_3 5′upstream region. EMBO J., 7: 1265-1271.

Jørgensen, J-E., Stougaard J., Marcker, A., and Marcker, K.A.(1988) Root nodule specific gene regulation: Analysis of the soybean nodulin N23 gene promoter in heterologous symbiotic systems. Nucleic Acids Res. 16: 39-50.

Lauridsen, P., Sandal, N., Kühle, A., Marcker, K.A. and Stougaard, J. (1990) Regulation of nodule specific genes. In 6th Nato Advenced Study Institute of Plant Molecular Biology.

Nap; J.-P.(1988) Nodulins in root nodule development. PhD– thesis, Wageningen, Netherland.

Petit, A., Stougaard J., Kühle, A., Marcker, K.A., and Tempe′, J.(1987) Transformation and regeneration of the legume Lotus corniculatus: A system for molecular studies of symbiotic nitrogen fixation. Mol. Gen. Genet. 207: 245-250.

Scheres, B., Van De Wiel, C., Zalanski, A., Horvath, B., Spaink, H., Van Eck, H., Zwartkruis, F., Wolters, A-M., Gloudemans, T., Van Kammen, A., and Bisseling, T.(1990) The ENOD12 product is involved in the infection process during the pea-Rhizobium interaction. Cell 60::: 281-294.

Solomon, M.J., Strauss, F and Varshavsky, A. (1986) A mammalian high mobility group protein recognizes any strech of six A-T base pairs in duplex DNA. Proc. Natl. Acad. Sci. USA 83: 1276-1280

Spiker, S. (1988) Histone variants and high mobility group non-histone chromosomal proteins of higher plants: Their potential for forming a chromatinn structure that is either poised for transcription or transcriptionally inert. Physiol. Plant. /75: 200.213.

Stougaard, J., Sandal, N.N., Grøn, A., Kühle, A., and Marcker, K.A. (1987) 5′analysis of the soybean leghemoglobin lbc_3 gene: Regulatory elements required for promoter activity and organ specificity. EMBO J., 6: 3565-3569.

Stougaard, J., Jørgensen, J-E., Christensen, T., Kühle, A., and Marcker, K.A. (1990) Interdependence and nodule specificity of cis-acting regulatory elements in the soybean leghemoglobin lbc_3 and N23 promoters. Molec. Gen. Genet. 220: 353-360.

PATTERNS OF NODULE DEVELOPMENT AND NODULIN GENE EXPRESSION IN ALFALFA AND AFGHANISTAN PEA

A.M. HIRSCH, B. BOCHENEK, M. LÖBLER, H.I. MCKHANN, A. REDDY,
H.-H. LI, M. ONG, AND J. WONG
Department of Biology
405 Hilgard Avenue
University of California
Los Angeles, CA 90024-1606 U.S.A.

ABSTRACT. Several investigators have shown a correlation between the morphological events observed during nodule development and the patterns of nodulin gene expression in nodules arrested at different stages of development. We have studied the expression of the early nodulins ENOD2 and ENOD12, and the late nodulin leghemo globin in nitrogen-fixing alfalfa and Afghanistan pea nodules and in *Rhizobium meliloti* mutant-induced alfalfa nodules. Our goal is to use nodulins as molecular markers to track the stages of nodule development. We have also examined the responses of alfalfa and Afghanistan pea to treatment with *N*-1-(naphthyl)phthalamic acid (NPA), a known auxin transport inhibitor. NPA-induced alfalfa nodules differ from wild type *R. meliloti*-induced nodules in their morphology and pattern of gene expression. Only MsENOD2 and Nms30 transcripts are detected; MsENOD12 and leghemoglobin are not expressed. However, the NPA-induced nodule-like structures developed on Afghanistan pea roots contain transcripts for both PsENOD2 and PsENOD12; late nodulin genes are not expressed. We have also determined that certain flavonoids induce alfalfa roots to develop nodule-like structures, which contain transcripts for ENOD2. These flavonoid-induced nodules more closely resemble wild type *R. meliloti* nodules in structure than the nodules induced by *R. meliloti exo* mutants or by NPA.

1. Introduction.

The development of leguminous nitrogen-fixing root nodules has been studied extensively, particularly from a descriptive point of view (see references in Hirsch et al., 1983). This descriptive framework is extremely useful for defining the stages of the interaction between the host legume and the *Rhizobium* that infects it.

Alfalfa root nodule development begins after specific recognition of *R. meliloti*, followed by root hair curling and the formation of an infection thread. Anticlinal cell divisions in the inner cortical cells of the root (Dudley et al., 1987) lead to the formation of a nodule primordium, which is soon invaded by an infection thread that started in an infected root hair. Rhizobia are released from branches of the infection thread into the central cells of the nodule primordium. At the distal (apical) end of the primordium, a persistent nodule meristem, consisting of small, densely cytoplasmic cells, is organized. The nodule meristem gives rise to the central cells of the nodule, to the nodule parenchyma, and to the nodule cortex. A nodule endodermis separates the nodule cortex from the nodule parenchyma and central cell zone, in which both infected and uninfected nodule cells are found.

H. Hennecke and D. P. S. Verma (eds.),
Advances in Molecular Genetics of Plant-Microbe Interactions, Vol. 1, 317–324.
© 1991 *Kluwer Academic Publishers. Printed in the Netherlands.*

1.1. NODULE DEVELOPMENT CAN BE CONSIDERED A LINEAR PATHWAY.

Root nodule development can be considered a linear pathway, extending from time zero -- the point of inoculation, to completion -- the formation of a fully functional nitrogen-fixing root nodule. Both members of the symbiotic partnership differentiate specialized cells, bacteroid or nodule cell, in response to signals received. Consequently, each partner expresses genes that are expressed only in the symbiotic state. The products of these genes can be used as markers to track stages of nodule development. Furthermore, *Rhizobium* mutants, which arrest nodule formation at specific points, can be used to dissect the nodule development pathway.

1.2. NODULINS CAN BE USED AS MOLECULAR MARKERS TO TRACK THE PLANT'S RESPONSE.

Nodulins are defined as proteins that are present in nodules and not in roots (Van Kammen, 1984). For convenience, nodulins have been divided into early and late nodulins (Nap and Bisseling, 1990). We have utilized in our investigations the early nodulins MsENOD2 (Dickstein et al. 1988), MsENOD12 (Löbler and Hirsch, ms. in prep.), and the late nodulin leghemoglobin (Dunn et al., 1988) to study two of the at least three events in nodule development that appear to require plant-microbe signalling (Norris et al., 1988):

1.2.1. *Nodule Formation. Rhizobium nod* genes are involved in signalling root hair deformation and the initiation of mitoses in cells of the root cortex, cells which normally would remain developmentally quiescent (see review by Long, 1989). Characteristic of this stage is the expression of the plant gene, ENOD2. Originally described by Franssen et al. (1987) for soybean, ENOD2 appears to be conserved among a broad range of legumes (Dickstein et al., 1988; Van de Wiel, 1990a).

1.2.2. *Rhizobium release from the infection thread and the differentiation of the central zone containing infected and uninfected host cells*. The signalling mechanism between host and *Rhizobium* at this stage of development so far is unknown. Some evidence suggests that intact rhizobial exopolysaccharide is required (Norris et al., 1988). A number of late nodulins are induced upon bacterial release (see review by Nap and Bisseling, 1990). A molecular marker that can be used to "tag" this developmental stage is leghemoglobin (Lb). Transcripts of Lb have been found, albeit at reduced levels, in mutant-induced nodules that are blocked at the stage after release of rhizobia from infection threads, but before the onset of nitrogen fixation (Morrison and Verma, 1987; Norris et al., 1988). We are also examining the expression of MsENOD12, an early nodulin gene in alfalfa, for which we have recently isolated a cDNA clone (Löbler and Hirsch, ms. in prep.).

1.2.3. *Induction of some of the enzymes of the ammonia assimilation pathway prior to nitrogen-fixing activity*. This stage in alfalfa appears to be correlated with the presence of elongated, fully differentiated bacteroids (Dunn et al., 1988; Norris et al., 1988). The lowered oxygen tension of the nodule which triggers the expression of the *R. meliloti nif*A gene may result in the induction of mRNAs for some of these enzymes.

We have concentrated on the first two developmental stages and also on three nodule-specific genes (MsENOD2, MsENOD12, and Lb) that can be used as markers to track patterns of gene expression in nodules of alfalfa and Afghanistan pea. We have examined the effects on nodule development and nodulin gene expression by either inoculating roots with *Rhizobium* mutants or by treating them with compounds that function as auxin transport inhibitors. We have chosen Afghanistan pea because it is the only pea that we have tested so far that responds, by making nodule-like structures on its roots, to the application of NPA.

These studies provide a framework for examining and characterizing plant mutants that are blocked in nodule formation.

2. Materials and Methods

2.1. PLANTS AND GROWTH CONDITIONS.

Alfalfa (*Medicago sativa* L. cv. Iroquois) and *Pisum sativum* cv. Afghanistan plants were grown as described by Norris et al. (1988). The plants were treated with the ATIs as described by Hirsch et al. (1989).

2.2. RNA ANALYSIS AND *IN SITU* HYBRIDIZATION METHODS.

A2ENOD2 was kindly provided by Dr. Rebecca Dickstein; PsENOD2 and PsENOD12 were generous gifts of Dr. Ton Bisseling. RNA transfer blots were made and probed as described by Van de Wiel et al. (1990b). Tissue was fixed and prepared for *in situ* hybridization by either using radioactively labeled RNA antisense and sense probes or digoxigenin-labeled probes (Boehringer Mannheim) as described by Van de Wiel (1990a, 1990b) and by Bochenek and Hirsch (1991).

3. Results

3.1. NODULIN EXPRESSION IN NORMAL NITROGEN-FIXING NODULES.

3.1.1. *Alfalfa Nodules.* In contrast to pea or soybean ENOD2 (Van de Wiel et al., 1990a), MsENOD2 is expressed early in alfalfa nodule development, as early as 24 hours after inoculation with *R. meliloti* (Löbler and Hirsch, ms. in prep.). In mature nitrogen-fixing nodules, MsENOD2 is expressed in nodule parenchyma cells (Van de Wiel et al., 1990b). A similar pattern of ENOD2 expression has already been described for pea and soybean by Van de Wiel et al. (1990a).

We have isolated a cDNA clone containing an open reading frame of 330 bp for MsENOD12, using PsENOD12 (Scheres et al., 1990) as a probe. The MsENOD12 mRNA is approximately 650 nucleotides long. MsENOD12, like PsENOD12, codes for a proline-rich protein on the basis of deduced amino acid sequence of the cDNA. The proline-rich region consists of 10 repeats of the consensus motif PPIYx (proline-proline-isoleucine-tyrosine-x). The amino terminal signal peptide of the alfalfa ENOD12 cDNA is 92% identical to that of PsENOD12, but the rest of the protein, including the proline-rich domain exhibits only 32% amino acid sequence identity. In pea, ENOD12 is expressed in the early invasion zone of the nodule (Scheres et al., 1990), while in wild type *R. meliloti*-induced alfalfa nodules, transcripts that hybridize to MsENOD12 antisense RNA are detected in the nodule endodermis and in the early invasion zone. Also, unlike PsENOD12 which is expressed in pea root hairs as early as 24 hours after inoculation with *R. leguminosarum*, transcripts hybridizing to MsENOD12 are not detected on RNA transfer blots until 10 days after inoculation of alfalfa roots with *R. meliloti*. Lb transcripts are detected on RNA transfer blots several days after ENOD12 mRNAs are observed (Löbler and Hirsch, ms. in prep.). *In situ* hybridizations show that Lb transcripts are located in the distal region of zone III (Reddy et al., ms. in prep.).

The localization of ENOD2, ENOD12, and Lb transcripts in wild type *R. meliloti* induced nodules is summarized in Fig. 1.

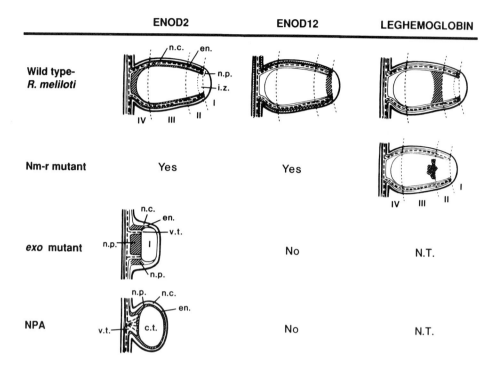

Fig. 1. Patterns of nodulin gene expression in alfalfa. The shaded region indicates the *in situ* localization of the mRNA listed above. Abbreviations: n.c., nodule cortex; v.t., vascular tissue; en., endodermis; n.p., nodule parenchyma; c.t., central tissue; i.z., invasion zone; I, zone I, the meristem; II, zone II, early symbiotic or infection thread zone; III, zone III, late symbiotic or infected cell zone; IV, zone IV, senescent zone.

3.1.2. *Afghanistan pea nodules.* ENOD2 transcripts in *R. leguminosarum* bv. *viciae* TOM-induced nodules are in the same location, nodule parenchyma, as they are in cultivar "Rondo" (Scheres et al., ms. in prep). The location of PsENOD12 transcripts in mature Afghanistan pea nodules by *in situ* hybridization was ambiguous. At times, only the early invasion zone, just behind the apical meristem was labelled and in other instances, both the early invasion zone and the cells of the nodule endodermis were labeled.

The localization of PsENOD2 and PsENOD12 transcripts in nitrogen-fixing Afghanistan pea nodules is presented in Fig. 2.

3.2. EFFECTS OF *RHIZOBIUM MELILOTI* MUTANTS ON NODULIN-GENE EXPRESSION.

3.2.1. *Nm^r R. meliloti Mutants Elicit the Formation of Infected, but Ineffective Nodules.* We have isolated a spontaneous Nm^r mutant of *R. meliloti* that elicits the formation of white, ineffective nodules on alfalfa. Infection threads are formed and proliferate throughout the nodule. In the majority of nodules examined, bacteria are released from the infection threads into the central tissue, but these bacteria do not differentiate into elongate bacteroids (Reddy et al., ms. in prep.). By RNA transfer blot analysis, we have found that the early nodulins, MsENOD2 and MsENOD12, are expressed. *In situ* hybridization studies show that Lb is also expressed and that transcripts are detected in the distal end of zone III just as they are in wild type *R. meliloti*-induced nodules (Fig. 1). However, the infected region of zone III is very limited and frequently consists of few host expanded cells. The released Nm [r] mutant rhizobia degenerate before nitrogen fixation ensues. Few elongate bacteroids are observed in nodules initiated by *R. meliloti* Nm[r] mutants.

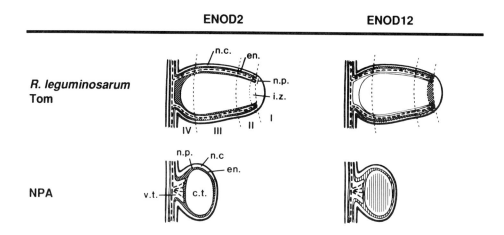

Fig. 2. Patterns of nodulin gene expression in Afghanistan pea. Abbreviations are the same as those in Figure 1.

3.2.2 Exopolysaccharide (exo) Mutants of R. meliloti Induce "Empty" Nodules. As previously described, *exo* mutant-induced nodules are completely free of bacteria; infection threads abort in the outermost cells of the nodule (Finan et al., 1985). The early nodulin MsENOD2 is expressed, but MsENOD12 is not (Löbler and Hirsch, ms. in prep.). However, *in situ* hybridization studies disclose what appears to be a different pattern of ENOD2 gene expression in nodules elicited by *exo* mutants compared to those induced by wild type *R. meliloti*. ENOD2 transcripts are confined to the base of the *R. meliloti exo* mutant-induced nodules, while they extend along the periphery as well as along the base of nitrogen-fixing nodules (Van de Wiel et al., 1990b) (Fig. 1). This apparent difference in pattern of gene expression is most likely related to the lack of sustained meristematic activity in *R. meliloti exo* mutant-induced nodules (Fig. 3b). Several centers of meristematic activity develop and appear to be initiated following divisions of cells from the pericycle as well as from the inner cortex (Fig. 3a) (Yang, Signer, and Hirsch, ms. in prep.).

3.3. AUXIN TRANSPORT INHIBITORS INDUCE NODULE-LIKE STRUCTURES.

Previous studies have shown that NPA and TIBA (2,3,5-triiodobenzoic acid), compounds known to function as auxin transport inhibitors, elicit the formation of nodule-like structures on alfalfa roots (Hirsch et al., 1989). MsENOD2 is expressed in these nodule-like structures, and in a tissue-specific mannner (Van de Wiel et al., 1990b). MsENOD12, on the other hand, is not expressed, nor is Lb (Löbler and Hirsch, ms. in prep.).

322

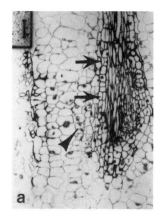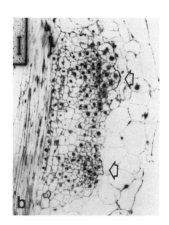

Fig. 3. Longtiudinal sections of alfalfa roots with developing nodules induced by *exo* mutant *R. meliloti*. a) Divisions in the pericycle (arrows) as well as in the cortex (arrowheads). b) Two adjacent centers of meristematic activity within the root cortex (open arrows). Bar = 100 μm.

3.3.1. *NPA Induces Nodule-like Structures on the Roots of Afghanistan Pea*. Of the five different cultivars of pea tested, including "Little Marvel", "Alaska", "Sparkle", and "Rondo", only cv. "Afghanistan" forms nodule-like structures in response to NPA (Scheres et al., ms. in prep). These nodule-like structures are similar in structure to those elicited on alfalfa roots. In contrast to the NPA-elicited structures on alfalfa roots which contain only MsENOD2, those nodules formed on Afghanistan pea contain transcripts for both PsENOD2 and PsENOD12.

Like the NPA-elicited nodule-like structures on alfalfa, PsENOD2 is expressed in the nodule parenchyma. At this time, we cannot unequivocally state the site of PsENOD12 expression in NPA-induced nodule-like structures of Afghanistan pea. Transcripts hybridizing to PsENOD12 have been found in the central tissue as well as in the nodule parenehyma (Fig. 2).

3.3.2. *Flavonoids Stimulate the Formation of Nodule-like Structures on Alfalfa*. We have discovered that compounds that function as natural auxin transport inhibitors (Jacobs and Rubery, 1988) and also as inducers of *Rhizobium nod* genes (see references in Long, 1989), namely, certain flavonoids, elicit the formation of nodule-like structures on alfalfa roots (McKhann et al., ms. in prep.). Quercetin or luteolin at 10^{-5} M elicit the formation of nodule-like structures on approximately 10% of the treated alfalfa plants. In addition, alfalfa seed exudate, which contains a large number of different flavonoids, induces nodules on 10-15% of the alfalfa plants. These nodules are frequently large and multi-lobed and have peripheral vascular bundles. They closely resemble the spontaneous alfalfa nodules described by Truchet et al. (1989). In contrast to spontaneous nodule formation, however, seed exudate-, quercetin- or luteolin-induced nodulation is apparently not repressed by the addition of 20 mM potassium nitrate (McKhann et al., ms. in prep.). The flavonoid-induced nodules contain transcripts for MsENOD2. However, we do not yet know if the transcripts are localized in a tissue-specific manner.

4. Discussion.

This analysis was undertaken to provide molecular "guideposts" for analyzing the stages of nodule development in alfalfa and Afghanistan pea. In alfalfa, MsENOD2 gene expression is associated with nodule morphogenesis and because of its early expression (24 hours after inoculation *with R. meliloti*), it is likely that MsENOD2 is expressed during some of the earliest stages of nodule formation. On the other hand, MsENOD12, which like MsENOD2 is a proline-rich protein, is expressed later in alfalfa nodule development; most likely, after nodules become infected by *R. meliloti*. However, MsENOD12 expression in alfalfa nodules contrasts with the situation in pea where PsENOD12 is expressed very early (Scheres et al., 1990). Moreover, we find that there is a difference in the location of gene expression when PsENOD12 and MsENOD12 are compared. Transcripts that hybridize to antisense ENOD12 RNA are detected mainly in the nodule endodermis in mature nitrogen-fixing nodules of alfalfa and Afghanistan pea, but also in the early invasion zone. We assume that a major transcript of 650 nucleotides is the one detected in the endodermis of alfalfa nodules and possibly another cross-hybridizing transcript is expressed in the invasion zone. Lb expression also appears to correlate with bacteria release from infection threads. In the Nm ^{r} *R. meliloti* mutant-induced nodules, bacteria, although released, do not elongate. However, Lb is expressed in the "correct" location in these nodules, i.e., in a region which is equivalent to the distal end of zone III.

Our studies have shown that alfalfa nodule formation arrests at certain defined points with respect to nodulin gene expression. For example, *exo* mutant rhizobia and NPA appear to block at the same step of nodule morphogenesis, while Nm ^{r} *R. meliloti* arrest nodule formation in a similar way to the *lps* mutants described earlier (Norris et al., 1988). This lack of finetuning of the events resulting in nodulin gene expression makes it difficult to deduce the exact sequence of interaction between the two symbiotic partners during the formation of a nodule. The analysis of defined plant mutants may offer a way to overcome this hurdle.

5. Acknowledgements.

We acknowledge the generosity of Agway, Inc. of Syracuse, N.Y. for alfalfa seeds and Drs. T.A. Lie and Tom LaRue for seeds of "Afghanistan" pea. We thank our colleagues in Wageningen, The Netherlands for their encouragement and help with various aspects of this research. We also thank Margaret Kowalczyk for the artwork. This research was supported by National Science Foundation grant DCB-8703297.

6. References.

Bochenek, B., and Hirsch, A.M. (1991) ' *In situ* hybridization of nodulin mRNAs in root nodules using non-radioactive probes', Plant Mol. Rep. Submitted for publication.

Dickstein, R., Bisseling, T., Reinhold, V.N., and Ausubel, F.M. (1988) 'Expression of nodule-specific genes in alfalfa root nodules blocked at an early stage of development', Genes Develop. 2, 677-687.

Dudley, M.E., Jacobs, T.W., and Long, S.R. (1987) 'Microscopic studies of cell divisions induced in alfalfa roots by *Rhizobium meliloti*', Planta 171, 289-301.

Dunn, K.R., Dickstein, R., Feinbaum, R., Burnett, B.K., Peterman, T.K., Thoidis, G., Goodman, H.M., and Ausubel, F.M. (1988) 'Developmental regulation of nodule-specific genes in alfalfa root nodules', Mol. Plant-Microbe Interact. 1, 66-75.

Finan, T.M., Hirsch, A.M., Leigh, J.A., Johansen, E., Kuldau, G.A., Deegan, S., Walker, G.C., and Signer, E.R. (1985) 'Symbiotic mutants of *Rhizobium meliloti* that uncouple plant from bacterial differentiation', Cell 40, 869-877.

Franssen, H.J., Nap, J.-P., Gloudemans, T., Stikema, W., Van Dam, H., Govers, F., Louwerse, J., Van Kammen, A., and Bisseling. T. (1987). 'Characterization of cDNA for nodulin-75 of soybean: a gene product involved in early stages of root nodule development', Proc. Natl. Acad. Sci. USA.' 84, 4495-4499.

Hirsch, A.M., Bang, M., and Ausubel, F.M. (1983) 'Ultrastructural analysis of ineffective alfalfa nodules formed by *nif*::Tn5 mutants of *Rhizobium meliloti*', J. Bacteriol. 155, 367-380.

Hirsch, A.M., Bhuvaneswari, T.V., Torrey, J.G., and Bisseling, T. (1989) 'Early nodulin genes are induced in alfalfa root outgrowths elicited by auxin transport inhibitors', Proc. Natl. Acad. Sci. (USA) 86, 1244-1248.

Long, S.R. (1989) '*Rhizobium*-legume nodulation: life together in the underground', Cell 56, 203-214.

Morrison, N., and Verma, D.P.S. (1987) 'A block in the endocytosis of *Rhizobium* allows cellular differentiation in ndoules but affects the expression of some peribacteroid membrane nodulins', Plant Mol. Biol. 9, 185-196.

Nap, J.-P. and Bisseling, T. (1990) 'Nodulin function and nodulin gene regulation in root nodule development,' in P.M. Gresshof (ed.), Molecular Biology of Symbiotic Nitrogen Fixation, CRC Press, Boca Raton, F, pp. 181-229.

Norris, J.H., Macol, L.A., and Hirsch, A.M. (1988) 'Nodulin gene expression in effective alfalfa nodules and in nodules arrested at three different stages of development', Plant Physiol. 88, 321-328.

Scheres, B., Van de Wiel, C., Zalensky, A., Horvath, B., Spaink, H., Van Eck, H., Zwartkruis, F., Wolters, A.-M., Gloudemans, T., Van Kammen, A., and Bisseling, T. (1990) 'The ENOD12 gene product is involved in the infection process during the pea-*Rhizobium* interaction', Cell 60, 281-294.

Truchet, G., Barker, D.G., Camut, S., de Billy, F., Vasse, J., and Huguet, T. (1989) 'Alfalfa nodulation in the absence of *Rhizobium*', Molec. Gen. Genet. 219, 65-68.

Van de Wiel, C., Scheres, B., Franssen, H., van Lierop, M.-J., Van Lammeren, A., Van Kammen, A., and Bisseling, T. (1990a) 'The early nodulin transcript ENOD2 is located in the nodule parenchyma (inner cortex) of pea and soybean root nodules', EMBO J. 9, 1-7.

Van de Wiel, C., Norris, J. H., Bochenek, B., Dickstein, R., Bisseling, T. and Hirsch, A.M. (1990b) 'Nodulin gene expression and ENOD2 localization in effective, nitrogen-fixing and ineffective, bacteria-free nodules of alfalfa', Plant Cell. In press.

Van Kammen, A. (1984) 'Suggested nomenclature for plant genes involved in nodulation and symbiosis', Plant Mol. Rep. 2, 43-45.

ENDOCYTOSIS AND THE DEVELOPMENT OF SYMBIOSOMES IN THE PEA-*RHIZOBIUM* SYMBIOSIS

S. PEROTTO, A.L. RAE, E.L. KANNENBERG and N.J.BREWIN
John Innes Institute,
John Innes Centre for Plant Science Research,
Colney Lane, Norwich, NR4 7UH, U.K.

ABSTRACT. In order to investigate the structure and function of the peribacteroid membrane in plant-microbe surface interactions, rat monoclonal antibodies have been used to identify the major components of this membrane and to investigate their cytological distribution. Most of the antigenic determinants that have been identified are carbohydrate in nature, as judged by periodate sensitivity and sugar inhibition of the antibody-antigen binding reaction. These carbohydrate antigens are components of macromolecules that are either protease-sensitive (i.e. glycoproteins) or protease-insensitive (glycolipids), the latter group being extractable in organic solvents. Immunolocalisation studies of longitudinal sections of pea nodule tissue suggest that the expression of these glycoconjugate antigens (collectively termed the 'glycocalyx') is developmentally regulated. Enhanced expression in infected nodule tissue is correlated with the onset of endocytosis and the development of peribacteroid membranes.

1. Introduction

Rhizobium leguminosarum biovar *viciae* induces nitrogen-fixing nodules on the roots of pea and vetch seedlings. Bacteria gain access to the cortical tissues of the root through intercellular and intracellular tunnels termed infection threads (VandenBosch et al., 1989). Subsequently, bacteria from infection threads extrude into unwalled infection droplets (Figure 1) and are then released into the cytoplasm, being engulfed by the naked plant cell plasma membrane (Robertson et al., 1985). The internalization of *Rhizobium* represents a rare example of endocytosis in plants cells, where the presence of a rigid cell wall normally prevents the access of large particles to the plasma membrane. The nature of endocytosis in plants is not understood. After endocytosis, bacteria continue to divide and progressively develop into endosymbiotic nitrogen-fixing bacteroids. Every time the intracellular bacterium divides, the enveloping plant-derived peribacteroid membrane also divides concomitantly, but the basis for this synchronous division is not understood. In pea and clover nodule cells, the bacteroids are individually enclosed by plant membrane (Robertson et al., 1985), forming an organelle-like structure termed the symbiosome (Mellor, 1989; Mellor & Werner, 1987). Examination of ultra-thin nodule sections by electron microscopy has often suggested the presence of a physical interaction between the peribacteroid membrane and the bacteroid surface (Robertson et al., 1985), and *in vitro* studies have also suggested that some form of interaction may be possible (Bradley et al., 1986). In order to investigate the molecular basis for surface interactions between the peribacteroid membrane (*pbm*) and the endosymbiotic bacteroids, it is first necessary to identify the main components of both the plant and the bacterial membranes.

325

H. Hennecke and D. P. S. Verma (eds.),
Advances in Molecular Genetics of Plant-Microbe Interactions, Vol. 1, 325–330.
© 1991 *Kluwer Academic Publishers. Printed in the Netherlands.*

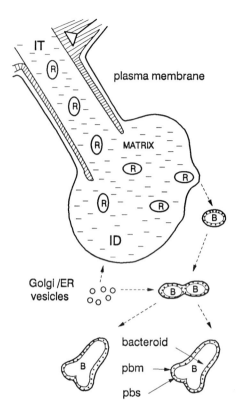

Figure 1. Endocytosis and the development of symbiosomes.

Monoclonal antibodies have proved to be very useful as specific molecular probes for the analysis of cell-cell surface interactions. In this laboratory, we are currently using monoclonal antibodies to investigate four different aspects of tissue invasion by *Rhizobium*: the nature of the plant matrix glycoprotein localised in infection threads (VandenBosch et al, 1989); structure-function relationships in *Rhizobium* lipopolysaccharide (Kannenberg et al, 1989); an immunological analysis of the peribacteroid membrane (Brewin et al, 1985, Bradley et al, 1988); and the nature of plant-derived acid hydrolases in the peribacteroid space (M.F.LeGal, unpublished results). In this particular presentation, we will focus on the immunochemical analysis of the peribacteroid membrane and its developmental role in symbiosis (Figure 1).

2. Results and Discussion

2.1 COMPARISON OF THE PERIBACTEROID AND PLASMA MEMBRANES

The relatedness of the peribacteroid membrane with the plant cell plasma membrane has been emphasised by the recent observation that a monoclonal antibody which was first isolated as

Table 1. Monoclonal antibodies reacting with antigens on the peribacteroid membrane.

Class	Examples	Antigen characters
I	MAC64, MAC207	Membrane glycoproteins
II	MAC206, MAC255, MAC268	Membrane glycoproteins and glycolipids
III	MAC254, MAC266	Membrane and soluble glycoproteins
IV	MAC275	Membrane protein (75K)

reacting with peribacteroid membrane (Bradley et al., 1988) has subsequently been shown to recognise a plasma membrane glycoprotein antigen that is a common feature of all angiosperms so far tested (Pennell et al., 1989; Pennell & Roberts, 1990). In the case of the plasma membrane, it has been suggested that oligosaccharide side-chains of membrane bound macromolecules project into the primary plant cell wall (Roberts, 1990), taking the form of a 'glycocalyx' somewhat analogous to the membrane-associated glycoproteins and glycolipids that ensheath animal cells.

A possible function of the plant cell glycocalyx might be to anchor other components of the plant cell wall to the underlying plasma membrane (Knox et al., 1989; Roberts, 1990). In this respect, the peribacteroid membrane is exceptional because it is not associated with plant cell wall components such as pectins, xyloglucans and cellulose (VandenBosch et al., 1989; Rae, Bonfante-Fasolo, unpublished results). However, the peribacteroid membrane is in close physical contact with the bacterial cell wall of the intracellular endosymbiotic bacteroids, and it will be interesting to investigate whether components of the plant glycocalyx are involved in any form of surface interaction with the bacterial surface.

2.2 BIOCHEMICAL ANALYSIS

The monoclonal antibodies listed in Table 1 have identified a range of epitopes associated with various components of the peribacteroid membrane. The carbohydrate nature of many of these epitopes was suggested by disruption of the antibody-antigen binding reaction by periodate oxidation or by competitive inhibition with sugars (Table 2). The evidence from sugar-inhibition studies indicated that in many cases derivatives of L-arabinose and/or D-galactose are components of these epitopes, as has been shown for many other extracellular plant glycoproteins (Fincher et al., 1983; Anderson et al., 1984) and membrane-associated plant glycoproteins (Norman et al., 1990; Pennell et al., 1989). The carbohydrate epitopes that we have identified on the *pbm* are frequently covalently attached to protease-sensitive macromolecules (glycoproteins). However, in the case of Class II antibodies (Table 1), the carbohydrate epitopes are associated with fast-migrating protease-insensitive antigens which are extractable in organic solvents and may represent glycolipids. Some of the epitopes recognised by Class II antibodies (e.g. MAC 268) appeared to be present on both glycolipid and glycoprotein components, as has been shown for other oligosaccharides present in glycoconjugates of pea membranes (Hayashi & Maclachlan, 1984).

2.3. ORIENTATION OF ANTIGENS IN THE PERIBACTEROID MEMBRANE

It was shown by immunofluorescence staining that the antigens associated with the *pbm* were not exposed on the cytoplasmic surface of membrane-enclosed bacteroids (symbiosomes) isolated by fractionation of nodule homogenates. However, when the *pbm* envelope was

Table 2. ELISA assays showing percentage inhibition of antigen-antibody binding in a hapten competition assay with 100 mM sugar solutions

ANTIBODY				SUGAR			
Class	MAC No	L-Ara	D-Mel	D-Gal	GlcA	GalA	L-Rha
I	64	45	-	50	-	-	-
	207	40	-	-	90	-	-
II	255	100	100	60	-	-	40
	267	70	70	-	-	-	-
	268	80	80	-	-	-	-
III	266	-	40	-	-	60	-

ruptured, the membrane antigens became accessible to antibody, indicating that only the luminal face of the *pbm* carries these antigens. This face is topologically equivalent to the external face of the plasma membrane and to the internal face of Golgi-derived vesicles which are the probable precursors of membrane components for both *pbm* and plasma membrane (Brewin et al., 1985).

2.4. TISSUE DISTRIBUTION OF ANTIGENS

If particular components of the plant membrane glycocalyx are involved in plant-microbial surface recognition, they might be expected to be more abundant on the peribacteroid membrane, relative to other plant membranes. Certain glycolipid antigens recognised by Class II antibodies were found to be more abundant in the *pbm* from isolated bacteroids relative to root membrane material (e.g. MAC 206). These differences were also consistent with the results of *in situ* studies of antigen distribution within longitudinal sections of pea nodules. The expression of the glycolipid antigens recognised by MAC 255 and all other Class II antibodies was very weak in the region of the nodule meristem but strongly enhanced in the tissue zones associated with endocytosis and the development of peribacteroid membranes. Glycoproteins recognised by Class I antibodies (e.g. MAC 207) appeared to be strongly represented in the membranes of this early infection zone but diminished in the mature symbiotic zone, whereas the converse was true for antigens recognised by Class III antibodies (e.g. MAC 266).

2.5. POSSIBLE PLANT-MICROBIAL CELL SURFACE INTERACTIONS

Although it shares many features in common with the glycocalyx of the plasma membrane, the glycocalyx associated with the peribacteroid membrane is functionally unique, in that it is not associated with an adjacent plant cell surface but rather with the surface of endosymbiotic bacteroids. It has recently been shown that the glycocalyx of animal cells is commonly involved in physical interactions with the surface of pathogenic bacteria (Karlsson, 1989), and glycolipid components are particularly important in this respect. Similarly, some of the glycocalyx antigens associated with the peribacteroid membrane may have an important role in the plant-microbial surface interaction leading to endocytosis and the development of symbiosomes. This could be tested using the same techniques that have been applied to the analysis of the animal glycocalyx in cell-pathogen interactions (Karlsson, 1989). The monoclonal antibodies described here could be used as molecular probes to identify the components involved and to monitor the process of association.

3. Acknowledgements

S.P. was supported by a sectoral training grant from the European Commission. E.L.K. acknowledges funding from Deutsche Forshcungsgemeinschaft.

4. References

-Anderson, M.A., Sandrin, M.S., and Clarke, A.E. (1984) 'A high proportion of hybridomas raised to a plant extract secrete antibody to arabinose or galactose', Plant Physiol. 75, 1013-1016.
-Bradley, D.J., Butcher, G.W., Galfré, G., Wood, E., and Brewin, N.J. (1986) 'Physical association between the peribacteroid membrane and lipopolysaccharide from the bacteroid outer membrane in *Rhizobium*-infected pea root nodule cells'. J. Cell Sci. 85, 47-61.
-Bradley, D.J., Wood, E.A., Larkins, A.P., Galfré, G., Butcher, G.W., and Brewin, N.J. (1988) 'Isolation of monoclonal antibodies reacting with peribacteroid membranes and other components of pea root nodules containing *Rhizobium leguminosarum*', Planta 173, 149-160.
-Brewin, N.J., Robertson, J.G., Wood, E.A., Wells, B., Larkins, A.P., Galfré, G., and Butcher, G.W. (1985) 'Monoclonal antibodies to antigens in the peribacteroid membrane from *Rhizobium*-induced root nodules of pea cross-react with plasma membranes and Golgi bodies', EMBO J. 4, 605-611.
-Fincher, G.B., Stone, B.A., and Clarke, A.E. (1983) 'Arabinogalactan proteins: structure, biosynthesis and function', Ann. Rev. Plant Physiol. 34, 47-70.
-Hayashi, T. and Maclachlan, G. (1984) 'Glycolipids and glycoproteins formed from UDP-galactose by pea membranes', Phytochemistry 23, 487-497.
-Kannenberg, E.L., and Brewin, N.J. (1989) 'Expression of a cell surface antigen from *Rhizobium leguminosarum* is regulated by oxygen and pH', J. Bacteriol. 171, 4543-4548.
-Karlsson, K-A. (1989) 'Animal glycosphingolipids as membrane attachment sites for bacteria', Annu. Rev. Biochem. 58, 309-350.
-Knox, J.P., Day, S., and Roberts, K. (1989) 'A set of cell surface glycoproteins forms an early marker of cell position, but not cell type, in the root apical meristem of *Daucus carota* L', Development 106, 47-56.
-Mellor, R.B., and Werner, D. (1987) 'Peribacteroid membrane biogenesis in mature legume nodules', Symbiosis 3, 75-100.
-Mellor, R.B. (1989) 'Bacteroids in the *Rhizobium*/legume symbiosis inhabit a plant internal lytic compartment:implications for other microbial endosymbiosis', J. Exp. Botany 40, 831-839.
-Norman, P.M., Kjellbom, P., Bradley, D.J., and Lamb, C.J. (1990) 'Immunoaffinity purification and biochemical characterization of plasma membrane arabino-galactan-rich glycoproteins of *Nicotiana glutinosa*', Planta 181, 365-373.
-Pennell, R.I., and Roberts, K. (1990) 'Sexual development in pea is presaged by a switch in arabinogalactan protein expression', Nature 344, 547-549.
-Pennell, R.I., Knox, J.P., Scofield, G.N., Selvendran, R.R., and Roberts, K. (1989) 'A family of abundant plasma membrane-associated glycoproteins related to the arabinogalactan proteins is unique to flowering plants', J. Cell Biol. 108, 1967-1977.
-Roberts, K. (1989) 'The plant extracellular matrix', Current Opinion in Cell Biology 1, 1020-1027.
-Robertson, J.G., Wells, B., Brewin, N.J., Wood, E.A., Knight, C.D., and Downie, J.A. (1985)

'The legume-*Rhizobium* symbiosis: a cell surface interaction', J. Cell Sci. Suppl. 2, 317-331.

-VandenBosch, K.A., Bradley, D.J., Knox, J.P., Perotto, S., Butcher, G.W., and Brewin, N.J. (1989). 'Common components of the infection thread matrix and the intercellular space identified by immunochemical analysis of pea nodules and uninfected roots', EMBO J. 8, 335-342.

PLANT GENETIC CONTROL OF NODULATION IN LEGUMES

PETER M. GRESSHOFF, DEBORAH LANDAU-ELLIS, ROEL FUNKE, LUIS SAYAVEDRA-SOTO AND GUSTAVO CAETANO-ANOLLES.
Plant Molecular Genetics and Center for Legume Research
Institute of Agriculture
The University of Tennessee, Knoxville, TN 37901-1071, USA.

ABSTRACT. This article summarizes recent findings with our supernodulation (*nts*) and nonnodulation (*nod*) mutants of soybean. The nonnodulation mutant nod139 (unable to curl root hairs and to develop subepidermal cell divisions) does not induce the autoregulation response that controls nodulation. In contrast, nonnodulation mutant nod49 (also unable to curl root hairs, but able to induce some cortical cell divisions) induces autoregulation as effectively as the wild type. Since we perceive autoregulation of nodulation to be an arrest of nodule development, we postulate that cell division associated events signal the shoot of the plant of their activity and that the shoot responds by restricting their further development. We discovered a second nodulation control mechanism, recognized by nodule excision, which acts independently of, and after the autoregulation mechanism, and is different in soybean and alfalfa. Using RFLP mapping we were able to position the *nts* locus on linkage group F. We are using chromosome walking strategies to clarify the relation between the genetic and physical distance between closely linked loci. Nodules formed in the absence of *Rhizobium* (Nar) share functional and structural properties with *Rhizobium*-induced nodules. Alfalfa plants with 100% Nar progeny were selected. Observations like the above suggest that the plant controls the entire nodulation ontogeny and that the bacterium has evolved to activate the process in restrictive genotypes.

1. Introduction:

1.1. BACKGROUND AND UPDATE:

The analysis of the role of the plant in the nodulation process was benefited by the isolation and characterization of symbiotic plant mutants. The subject has been reviewed (see Gresshoff and Delves, 1986; Rolfe and Gresshoff, 1988), so here we summarize our advances that have occurred since the last meeting in 1988.

Other groups have also used induced mutagenesis to isolate symbiotically altered plant mutants of similar physiological characteristics to ours (Gremaud and Harper (1988); Buzzell et al (1990)). For example, Harper's group isolated three hypernodulating and one nonnodulating mutant of soybean cultivar Williams. Buzzell et al (1990) isolated supernodulation mutants of soybean variety Elgin, and point towards the involvement of possibly two genetic loci in supernodulation control, contrary to the conclusions of Delves et al (1988), who showed single gene inheritance in a number of supernodulation isolates.

Genetical analysis showed that all *nts* isolates in Bragg were part of the same complementation group, although they had phenotypic differences. Inheritance was mendelian recessive (Delves et al, 1988).

Anatomical studies by Mathews et al. (1989a) demonstrated that *nts* plants convert more early nodulation cell clusters to mature nodules. This supported the concept that the *nts*

H. Hennecke and D. P. S. Verma (eds.),
Advances in Molecular Genetics of Plant-Microbe Interactions, Vol. 1, 331–335.
© 1991 *Kluwer Academic Publishers. Printed in the Netherlands.*

mutation was either an alteration of the autoregulation loop, or an escape from the functions of that regulatory circuit. We know that the shoot of the plant is directly involved with the phenotype (Delves et al, 1986; Olsson et al, 1988). Mathews et al (1989b), and in more detail Sutherland et al (1990), used *nod* gene-*lacZ* fusions to quantify the inducing activity of supernodulation and nonnodulation mutants of Bragg and found that there is no significant difference in inoculated and uninoculated plants during the 12 day test period.

Gresshoff et al (1988) suggested that the supernodulation phenotype was correlated with changes in the ABA level of the soybean leaf. We have been unable to confirm this result in recent studies (D. Eskew, unpubl. data). Likewise we have been unable to confirm the absence of a leaf protein in nodulated *nts* plants (Sayavedra-Soto et al, 1991).

2. Results and Discussion:

2.1. AUTOREGULATION REQUIRES THE COTRICAL CELL DIVISION STAGE:

Mathews et al (1990) showed that the nonnodulation phenotype epistatically suppresses the supernodulation phenotype. Reciprocal grafting demonstrated that the nodulation phenotype of nod139 and nod49 was root-controlled. Using approach grafts between wild-type, supernodulation, or nonnodulation plants, and time delaying the inoculation on the spacially separated root systems, we showed that autoregulation in the shoot was initiated only if the root system was able to induce cortical cell divisions (Caetano-Anollés and Gresshoff, 1990). In contrast, nod139, which did not form subepidermal cell divisions, and was controlled by a separate gene (Mathews et al., 1989c), was unable to control nodulation. However, its shoot was able to do so, if grafted to a cell division competent root (Mathews, 1989), implying that events closely related to cortical cell divisions systemically signal the shoot. To test the importance of cell divisions we found that the excision of lateral root tips caused an increase in nodule number per plant in wild type but not in the supernodulation mutant (Caetano-Anollés et al, 1991).

2.2. NODULE EXCISION DEMONSTRATES ANOTHER CONTROL MECHANISM:

Cytological analysis of inoculated wild-type plants revealed the abundance of prenodule structures and cell division foci in the root cortex of soybean. We tested if these structures are irreversible arrested or were able to develop into nodules by surgically removing nodules from the tap root of wild-type and *nts* plants, superinfecting with an antibiotic resistant *Bradyrhizobium* strain, and monitoring the reappearance of nodules (Caetano-Anollés et al, 1991). The same region of the tap root that gave rise to the original burst of nodules also formed new nodules, which harboured the original inoculum, suggesting that original infection foci, arrested by the initial set of nodules, developed into nodules. This phenomenon also occurred in *nts* plants, indicating that this mechanism was different from the autoregulation response. Initially formed nodules developed a 'suppressive field' around themselves. This suppression was released, if the original nodules were removed.

Similar experiments with alfalfa plants showed a lower frequency of prenodule structures in the inner root cortex of alfalfa when compared with soybean (Caetano-Anollés and Gresshoff, 1991). Removal of nodules resulted in renewed nodulation from the growing root tip region of the entire plant, but not from the region of original nodule emergence. Alfalfa has a different nodule control mechanism, functioning earlier than that of soybean. Whether this is a reflection of a difference of the temperate legume versus tropical legume or an annual versus perennial plant is unclear. It indicates that the autoregulation response is not the only mechanism controlling nodule number, and that the relative contribution of different mechanisms may be different within legumes.

2.3. RFLP MAPPING OF SOYBEAN NODULATION GENES:

Soybean has a relatively narrow genetic basis. For this reason DNA probes were isolated to detect restriction fragment length polymorphisms between *G. max* and *G. soja*, an ancestral soybean, still capable of fertile hybridization with modern cultivars.

The RFLP map generated by the researchers at USDA-ARS (Iowa State University) comprises about 230 markers and covers about 2500 centimorgans (Randy Shoemaker, pers. comm.). The soybean haploid genome size is estimated to be 1.8×10^9 bp (determined by reassociation kinetics) or 1×10^9 bp (by Feulgen microdensitometry). The pachytene chromosome map reveals 20 distinct chromosomes with about 40% heterochromatic regions (mainly centromeric). This gives 395-750 kb DNA per cM.

We have completed a crossing program of our *nts* and *nod* soybean lines with *G. soja*, and have obtained F1 hybrids and F2 segregants (Landau-Ellis et al, 1990). The F2 plants (segregating 62 wild-type and 20 Nts) were tested for supernodulation to provide homozygous recessives. Genomic DNA was probed with diagnostic probes representing the different linkage groups of the RFLP map. We found close linkage with probe pA36, which gave about 10% recombination with the *nts* locus. This places the soybean *nts* gene 10 cM from the pA36 cluster on linkage group F (Fig. 1). Experiments delineating the closer linkage to flanking markers like pA130 and pA132 as well as pA703 as well as the nonnodulation loci are in progress.

Figure 1: RFLP linkage group F of *Glycine max*.(R. Shoemaker, unpubl. data)

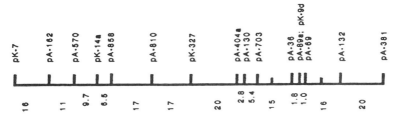

Using pulse field gel electrophoresis we attempted to resolve the molecular make-up of this small region of the soybean genome. Preliminary data from double probing of PFGE gels (Funke et al, 1991) with probes pA69 and pK9d suggests that they, being 1 cM apart by recombination, are positioned on the same 100 kb fragment. We have constructed a partial linking library and jumping library in lambda phage, and have detected clones with homology to pA130. We expect to construct a map of overlapping clones in the pA36 cluster, thereby providing information on the correlation between the genetic distances and the amount of DNA between RFLP markers. With the possible isolation of additional *nts*-linked RFLP clones, we may be able to consider a chromosome walking strategy to isolate the gene, for which at present we know only its phenotype and not its gene product.

2.4. SPONTANEOUS NODULATION OF ALFALFA:

Truchet et al (1988) reported that alfalfa plants were able to form nodule-like structures in the absence of *Rhizobium* (Nar). His further studies and ours have illustrated that these structures are indeed 'genuine' nodules, rather than abnormal lateral root outgrowths. Presumably they occur in many legumes, but were discarded over the years as accidentally contaminated plants. In alfalfa we found that spontaneous nodules appeared in the same region of the root that normally gave rise to tap root nodulation. Their appearance was delayed, resulting in their suppression in inoculated roots by *Rhizobium*-induced nodules. Like Truchet et al (1989) we found that nitrate suppresses the formation of spontaneous nodules. The ontogeny of these nodules was identical to that of *Rhizobium*-induced nodules, starting with cell divisions in the inner cortex and not the pericycle, as those leading to the formation of lateral roots (Joshi et al, 1991). Spontaneous nodules elicit the autoregulation response of nodulation and are themselves subject to autoregulation (Caetano-Anollés and Gresshoff, 1991). Truchet et al. (1989) showed that the nodulin Enod2 is expressed. Spontaneous nodules contain different cell types just like the *Rhizobium*-induced nodules, some of which were enlarged and contained numerous starch deposits (Joshi et al, 1991).

Inoculation of alfalfa plants with *nodH⁻ Rhizobium meliloti* resulted in a significant increase in the frequency of nodulated plants, compared to sham inoculations (Caetano-Anollés, unpubl. data). Reisolation of bacteria from the nodule showed the stability of the original non-nodulation phenotype.

Instead of the normal 3-4% of the total plant population of cultivar Vernal with spontaneous nodules, inoculation with *nodH* bacteria resulted in 17% nodulated plants. Inoculation with *nodABC⁻* mutant gave no such increase, showing that a precursor of the NodRm1 signal (Lerouge et al, 1990) affected the efficiency of the plant response. Our data suggest a caution about reporting nodulation phenotypes based on a small percentage of nodulated plants.

Spontaneous nodulation has not been reported (yet) in other legumes. Has agricultural development selected against the phenotype? Do we see spontaneous nodulation because of the heterogeneous nature of the alfalfa population? If so, would a heterozygous clover or primitive *Glycine* population show the same phenotype?

When Nar⁺ plants were self-fertilized, we isolated progeny in which 100% of the plants formed spontaneous nodules. The mode of inheritance indicated that a single dominant genetic element controlled Nar in the tetraploid alfalfa. Nodule numbers per plant and the positioning of the nodules remained the same, possibly due to a strong genetic control over the ability of an alfalfa plant to form spontaneous nodules. Most features of a nodule are not under the inductive control of *Rhizobium*, but represent an internal developmental program, whose expression is presumably optimized in some homozygous recessive plant genotypes by *Rhizobium*. The preferential deposition of starch in some nodule cells suggests that the ancestral nodule may have functioned as a carbon-storage organ. *Rhizobium* possibly acquired the ability to induce these structures at high frequencies in plant genotypes selected against the formation of these structures. By recognizing this we may be able to find the nodule formation ability in non-legumes, but still with a carbon storage rather than nitrogen fixation role.

3. References:

Buzzell, R.I., Buttery, B.R., and Ablett, G. (1990) Supernodulation mutants in Elgin 87 soybean. in P.M. Gresshoff, G. Stacey, L.E. Roth and W.E. Newton (eds.) Nitrogen Fixation: Achievements and Objectives. Chapman Hall, New York. p. 726.
Caetano-Anollés, G., and Gresshoff, P.M. (1990) Early induction of feedback regulatory responses governing nodulation in soybean (*G. max* (L) Merr.) Plant Science, in press.
Caetano-Anollés, G., and Gresshoff, P.M. (1991) Feedback regulation of nodule formation in alfalfa. mature nodules control infection initiation. Plant Physiol., submitted.
Caetano-Anollés, G., Paparozzi, E. T., and Gresshoff, P.M.(1991) Mature nodules and root tips control nodulation in soybean. J. Plant Physiol., in press.
Caetano-Anollés, G., Joshi, P.A., and Gresshoff, P.M. (1990) Spontaneous nodules induce feedback suppression of nodulation in alfalfa. Planta, in press.
Gresshoff, P.M., Krotzky, A., Mathews, A., Day, D.A., Schuller, K.A., Olsson, J.E., Delves, A.C., and Carroll, B.J. (1988) Suppression of symbiotic supernodulation symptoms in soybeans. J. Plant Physiol. 132, 417-423.

Delves, A.C., Carroll, B. J., and Gresshoff, P.M. (1988) Genetic analysis and complementation studies on a number of mutant supernodulating soybean lines. J. Genetics 67, 1-8.

Delves, A.C., Mathews, A., Day, D.A., Carter, A.S., Carroll, B.J. and Gresshoff, P.M. (1986) Regulation of the soybean-*Rhizobium* symbiosis by shoot and root factors. Plant Physiol. 82, 588-590.

Funke, R., Sayavedra-Soto, L.A. and Gresshoff, P.M. (1991) Physical mapping of the supernodulation (*nts*) region of the soybean genome using pulse field gel electrophoresis. Mol. Gen. Genet., submitted.

Gremaud, M.F. and Harper, J.E. (1988) Selection and initial characterization of partially nitrate tolerant nodulation mutants of soybean. Plant Physiol. 89, 169-173.

Gresshoff, P.M. and Delves, A.C. (1986) Plant genetic approaches to symbiotic nodulation and nitrogen fixation in legumes. in A.D. Blonstein and P.J. King (eds.): Plant Gene Research III. A genetical approach to plant biochemistry, pp. 159-206. Springer Verlag, Wien.

Joshi, P.A., Caetano-Anollés, G., Graham, E., and Gresshoff, P.M. (1991) Ontogeny and ultrastructure of spontaneous nodules in alfalfa (*Medicago sativum*) Protoplasma, submitted.

Landau-Ellis, D., Shoemaker, R. and Gresshoff, P.M. (1990) Molecular mapping of the soybean supernodulation locus to RFLP linkage group F. Nature, submitted.

Lerouge, P., Roche, P., Faucher, C., Maillet, F., Truchet, G., Promé, Je., and Dénarié, J. (1990) Symbiotic host specificity of *Rhizobium meliloti* is determined by a sulphated and acylated glucosamine oligosaccharide signal. Nature 344, 781-784.

Mathews, A. 1989. PhD thesis, ANU, Canberra, Australia.

Mathews, A., Carroll, B.J., and Gresshoff, P.M. (1990) The genetic interaction between nonnodulation and supernodulation in soybean: an example of developmental epistasis. Theor. Appl. Genetics 79, 125-130.

Mathews, A., Carroll, B.J., Gresshoff, P.M. (1989a) Development of *Bradyrhizobium* infections in supernodulating and non-nodulating mutants of soybean (*G. max* (L.) Merrill). Protoplasma 150, 40-47.

Mathews, A., Kosslak, R.M., Sengupta-Gopalan, C. Appelbaum, E.R., Carroll, B.J. and Gresshoff, P.M. (1989b) Biological characterization of root exudates and extracts from nonnodulating and supernodulating soybean mutants. MPMI 2, 283-290.

Mathews, A., Carroll, B.J. and Gresshoff, P.M. (1989c) A new recessive gene conditioning non-nodulation in soybean. J. Hered. 80, 357-360..

Olsson, J.E., Nakao, P., Bohlool, B.B. and Gresshoff, P.M. (1989) Lack of systemic suppression of nodulation in split root systems of supernodulating soybean (*Glycine max* (L.) Merr.) mutants. Plant Physiol. 90, 1347-1352.

Rolfe, B.G. and Gresshoff, P.M. (1988) Genetic analysis of legume nodule initiation. Ann. Rev. Plant Physiol. Plant Mol. Biol. 39, 297-319.

Sayavedra-Soto, L.A., Angermüller, S.A., Prabdu, R., and Gresshoff, P.M. (1991) Polypeptide patterns in leaves of *Glycine max* (L) Merr. cv. Bragg and its supernodulation mutant during nodulation. The New Biologist, submitted.

Sutherland, T.D., Bassam, B.J., Schuller, L.J., Gresshoff, P.M. (1990) Early nodulation signals in wild type and symbiotic mutants of soybean (*Glycine max* (L.) Merr. cv. Bragg). MPMI 3,122-128.

Truchet, G., Vasse, J., Odorico, R., de Billy, F., Camut, S., and Huguet, T. (1988) Discrimination between nodules and root-derived structures: Does alfalfa nodulate spontaneously? in R. Palacios, and D.P.S. Verma (eds) : Molec. Genet. of Plant-Microbe Interactions. pp. 179-180.

Truchet, G., Barker, D. G., Camut, S., de Billy, F., Vasse, J., and Huguet, T. (1989) Alfalfa nodulation in the absence of *Rhizobium*. Mol. Gen. Genet. 219, 65-68.

GENETIC AND CELLULAR ANALYSIS OF RESISTANCE TO VESICULAR ARBUSCULAR (VA) MYCORRHIZAL FUNGI IN PEA MUTANTS

V. GIANINAZZI-PEARSON, S. GIANINAZZI, J.P. GUILLEMIN,
A. TROUVELOT and G. DUC
Station de Génétique et Amélioration des Plantes
INRA
BV 1540
21034 Dijon Cédex
France

ABSTRACT. Screening of nodulation mutants of *Pisum sativum* has yielded mutants showing resistance to VA fungi (termed myc⁻). Most of these have aborted infections (myc⁻$^{(1)}$ phenotype) which are characterised by host cell reactions recalling those in certain pathogen infections. Mutants affected in later steps of mycorrhiza development (myc⁻$^{(2)}$ phenotype), with blocking of arbuscule formation, were less frequent. The myc⁻$^{(1)}$ character is recessive, segregates monogenically and occurs on at least five different, independently mutated loci, indicating that VA endomycorrhiza formation is under multiple gene control. Expression of the myc⁻$^{(1)}$ character is indissociable from that of nod⁻ in mutants and appears to be the result of pleiotropic effects of single genes. In late mutants where arbuscule formation is blocked, nodules develop but are inefficient (nod⁺,fix⁻). Coincidences between myc and nod characters may reflect common mechanisms in plant control over some step(s) in endomycorrhiza and nodule symbioses. Since chemical mutagenesis generally causes loss of gene function, inactivation of symbiosis-specific susceptibility genes in the mutants could affect production of signal molecules and somehow lead to stronger expression of plant resistance mechanisms to the symbionts.

1. Introduction

Vesicular arbuscular (VA) endomycorrhizas represent the most widespread type of fungal infection in plant roots. With evolution most terrestrial plants have become susceptible to VA endomycorrhizal fungi, obligate biotrophs infecting over 80% of plant species, including many cultivated plants. The multiple events leading to the formation of VA endomycorrhizas involve complex interactions between host and fungal cells which must be determined by the genomes of both symbionts (Gianinazzi-Pearson and Gianinazzi, 1989). The ubiquitous nature of VA endomycorrhizas implies the existence of fungal virulence genes with a broad spectrum of action and of common plant genes to control the infection. The analysis of genetic determinants involved in the formation and function of VA endomycorrhizas has been hampered by the fact that VA fungi cannot be grown in pure culture and by the lack of genetically defined host variants which are

336

H. Hennecke and D. P. S. Verma (eds.),
Advances in Molecular Genetics of Plant-Microbe Interactions, Vol. 1, 336–342.

deficient for mycorrhizal formation. Non mycorrhiza-forming species do exist among not more than 5% of plant species. The reason for this non host resistance is not known, but experiments with lupin, the only legume which does not form VA endomycorrhiza, show that both root exudate components and mobile shoot factors may be involved (Gianinazzi-Pearson and Gianinazzi, 1989 ; Gianinazzi-Pearson et al., 1989). In order to gain insight into the genetic control of the plant over the symbiosis and to better analyse how VA endomycorrhizal systems function, we have previously looked amongst nodulation mutants in the host species, pea and fababean, for mutants that show a phenotypic deficiency for VA endomycorrhiza formation (Duc et al., 1989). In the present paper we report additional mutants in another pea line, together with further investigations on the genetical and cellular aspects of resistance to VA endomycorrhizal fungi in such mutants.

2. Material and methods

66 independent nodulation mutants (screened against a multistrain Rhizobium inoculum) from separate chemical (ethyl methane sulphonate) mutagenesis programs on two pea (Pisum sativum L.) lines, cv. Frisson (Duc and Messager, 1989) and cv. Finale (Engvild, 1987), were tested for their ability to develop VA endomycorrhizas. The mutants, named P and DK respectively, and their isogenic parent lines were inoculated with Rhizobium and a mixed inoculum of Glomus intraradices and Glomus mosseae in pots containing a clay-loam soil. After 6 weeks' growth, root systems were examined for nodulation and for endomycorrhizal infection structures in the light microscope after trypan blue staining (Philipps & Hayman, 1970) and in the electron microscope after double fixation and embedding (Jacquelinet-Jeanmougin et al., 1987).

3. Results and discussion

Both parent lines had a myc$^+$ phenotype with formation of typical VA endomycorrhizas. Average values for infection parameters were : frequency of infection units (F)=75-79%, colonisation intensity of the root cortex (M)=25-43%, and arbuscule frequency (A)=13-32%, estimated by the method of Trouvelot et al. (1986) (table 1).

Two types of endomycorrhiza mutant phenotypes were found (figure 1). The most common phenotype, termed myc$^{-(1)}$, was found in 21 out of 45 non-nodulating (nod$^-$) mutants from the two pea lines and the infection was blocked at an early stage immediately after the formation of appressoria. Amongst 21 mutants forming inefficient, white nodules (nod$^+$,fix$^-$), two out of 11 DK mutants showed inhibition of endomycorrhizal development at a later stage, characterised by intercellular hyphal development but lack of arbuscule formation in the root parenchyma cortex (phenotype myc$^{-(2)}$). All the remaining mutants had a myc+ phenotype.

Parameters for infection were lower in all myc$^-$ mutants as compared to the parent lines (table 1). When a myc$^{-(1)}$ mutant was grown in the same pot as a parent line, there was no change in the phenotype. Arbuscules were not formed and colonisation of the root cortex (M value) remained low, but the frequency (F) of infection units (limited to penetration points in the mutants) increased to 74%, suggesting that the roots of parent lines produced a fungal growth stimulus which was lacking in the root exudates of mutants.

338

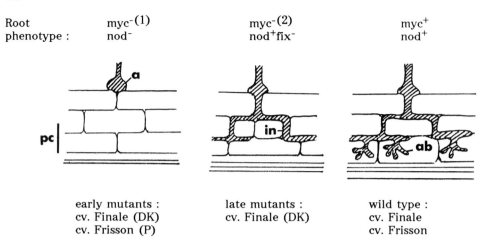

| Root phenotype : | myc⁻(1) nod⁻ | myc⁻(2) nod⁺fix⁻ | myc⁺ nod⁺ |

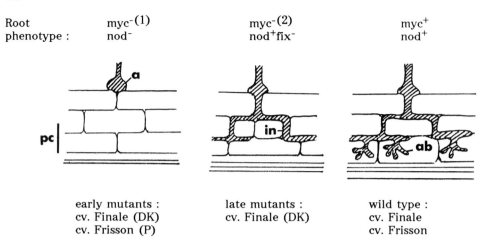

Root phenotype : $\text{myc}^{-(1)}$ nod⁻ $\text{myc}^{-(2)}$ nod⁺fix⁻ myc^+ nod⁺

early mutants :
cv. Finale (DK)
cv. Frisson (P)

late mutants :
cv. Finale (DK)

wild type :
cv. Finale
cv. Frisson

Figure 1 : Diagrammatic representation of endomycorrhizal infections in parent lines and mutants of pea (a, appressorium; in, intercellular hyphae; ab, arbuscules; pc, parenchymal cortex).

Table 1 : Parameters of endomycorrhizal infection in parent lines and mutants of pea

Infection parameters :	F%	M%	A%
cv-Frisson	79	25	13
mutants myc⁻(1) of cv-Frisson (P)	40	0.6	0
cv-Finale	75	43	32
mutants myc⁻(1) of cv-Finale (DK)	56	0.6	0
mutants myc⁻(2) of cv-Finale (DK)	58	1.7	0

Ultrastructural observations of normal endomycorrhizal infections in the parent lines of pea showed that host reaction to external hyphae at penetration points was limited to an occasional slight and localised thickening of the epidermal cell wall which did not impede fungal infection (figure 2A).

Very different ultrastructural modifications were associated with the blockage of fungal development in myc⁻(1) mutants. The presence of a fungal hypha at the junction of outer root cells induced the formation of a thick, two-layered deposit on the inner face of the adjacent host wall (figure 2B,C). The structure of this material, which was different from the normal host wall (figure

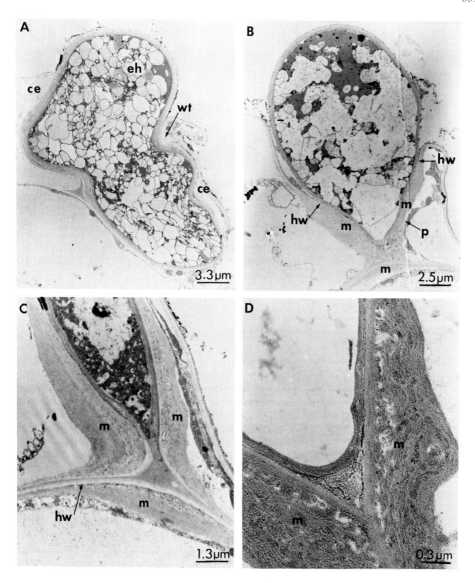

Figure 2 : (A) Ultrastructural details of infection points of a VA fungus on roots of pea cv-Frisson. External hypha (eh) penetrating between two epidermal cells (ec), which show slight wall thickenings (wt) close to the fungus.

(B,C,D) Ultrastructural details of infection points of a VA fungus on roots of a myc⁻(1) mutant of pea cv-Frisson. (B) Thick deposit of material (m) between the plasmalemma (p) and wall (hw) of three outer root cells adjacent to a penetrating hypha. (C) Detail of (B) showing the two-layered structure of the deposited material (m). (D) PATAg-positive reaction of deposited material (m).

2C), was particularly evident after the PATAg reaction (Thiéry, 1967) for 1-4 polysaccharides (figure 2D), and it resembled callose/lignin-containing wall appositions reported to be formed in resistance reactions to certain pathogenic infections (Bracker & Littlefield, 1971) or in incompatible Frankia/*Alnus* cellular interactions (Berry and McCully, 1990).

Analysis of the myc⁻ character in the F1 generation from diallel crosses between the myc⁻ mutants revealed that they can be classed into 5 groups of complementation, suggesting that at least five different, independently mutated loci are involved (table 2), and indicating that endomycorrhiza formation is under multiple gene control. Analysis of the F1 and F2 generations from crosses made with parent lines showed that the myc⁻$^{(1)}$ character, like the nod⁻ character, is recessive and segregates monogenically (table 3). It is highly probable that the expression of myc⁻ and nod⁻ characters is the result of pleiotropic effects of single genes because of (1) the low probability of several double mutation events following mutagenesis, (2) the identity of complementation groups for both nod⁻ and myc⁻$^{(1)}$ characters, and (3) the absence of recombinant nod⁻,myc⁺ or nod⁺,myc⁻ plants in the 272 F2 plants so far observed (table 3). In grafting experiments both myc⁻ and nod⁻ characters were controlled by the rootstock and not the scion (table 4), which lends further support for a close link between the determinants of the myc⁻ and nod⁻ characters.

4. Conclusion

The discovery of genetically defined myc⁻ mutants opens the possibility of identifying plant genes involved in VA endomycorrhizal infections, and of determining their products and their role in the symbiosis. Furthermore, comparative physiological studies of the infection process in mutants blocked at different morphological changes of mycorrhiza formation should provide information concerning those fungal structures (intercellular hyphae, arbuscules) that are essential to the functional symbiosis. The constant coincidence of the myc⁻ and nod⁻ characters in the mutants, together with the similar sensitivity of *Rhizobium* and VA fungi to some flavonoids (Gianinazzi-Pearson *et al.*, 1990), suggest that common mechanisms may control some early infection event(s) in both types of symbiosis.

Since chemical mutagenesis generally causes loss of gene function, inactivation of symbiosis-specific susceptibility genes by the induced mutations could be affecting synthesis of signal molecules which in myc⁺ plants somehow control activation of resistance genes, so that only very weak plant defence reactions normally occur to VA endomycorrhizal infections (Gianinazzi and Gianinazzi-Pearson, 1990). The lack of, or reduction in, such molecules in myc⁻ plants could lead to the observed stronger expression of plant resistance mechanisms to the VA fungal symbionts. Recent studies have shown, however, that high temperatures which break down resistance reactions to pathogens in leaves (Gianinazzi, 1984) do not break down the mutant resistance in roots to VA fungi (unpublished data), suggesting that different mechanisms may be involved.

5. References

Berry, A.M. and McCully, M.E. (1990) 'Callose-containing deposits in relation to root-hair infections of *Alnus rubra* by *Frankia*', Can. J. Bot. 68, 798-802.

Table 2 : Results of diallel crosses realised between the myc$^{-(1)}$, nod$^-$ mutants revealing 5 groups of complementation (A, C, P, Q & R).

Groups	Mutants
A	P1, P2, P3, P53, DK6 & DK9.
C	P4, P6, P55, DK2, DK7, DK16 & DK20.
P	DK10, DK13, DK19 & DK25.
Q	DK12.
R	DK51.

Table 3 : Phenotype of F1 and F2 generations from crosses between pea cv-Frisson and myc$^{-(1)}$, nod$^-$ mutants.

Crosses	Number of plants		% Phenotype myc$^+$, nod$^+$	myc$^{-(1)}$, nod$^-$
cv-Frisson x locus a (mutants P1, P2 & P3)	-- 153	F1 F2	100 70	0 30
cv-Frisson x locus c (mutants P4, P55 & P6)	-- 119	F1 F2	100 73.5	0 26.5

Table 4 : Effect of rootstock and scion grafting between mutants and cv-Frisson on root phenotype for mycorrhiza infection and nodulation.

Grafting scion/rootstock	Root phenotype	
mutant P3/cv-Frisson	myc$^+$	nod$^+$
mutant P55/cv-Frisson	myc$^+$	nod$^+$
cv-Frisson/mutant P3	myc$^{-(1)}$	nod$^-$
cv-Frisson/mutant P55	myc$^{-(1)}$	nod$^-$

Bracker, C.E. and Littlefield, L.J. (1971) Fungal Pathogenicity and the Plant's Response, Eds. R.J.W.Byrde & C.V.Cutting, Academic Press, London, New-York, 159-317.

Duc, G. and Messager, A. (1989) 'Mutagenesis of pea (*Pisum sativum*) and the isolation of mutants for nodulation and nitrogen fixation', Plant Science 60, 207-213.

Duc, G., Trouvelot, A., Gianinazzi-Pearson, V., and Gianinazzi, S. (1989) 'First report of non-mycorrhizal plant mutants (myc-) obtained in pea (*Pisum sativum L.*) and fababean (*Vicia faba L.*)', Plant Science 60, 215-222.

Engvild, K.C. (1987) 'Nodulation and nitrogen fixation mutants of pea, *Pisum sativum L.*', Theor. Appl. Genet. 74, 711-713.

Gianinazzi, S. (1984) 'Genetic and Molecular Aspects of Resistance Induced by Infections or Chemicals', in T. Kosuge and E.W. Nester (eds.), Plant Microbe Interactions, volume 1, MacMillan Publishing Company, New York, pp. 321-342.

Gianinazzi, S and Gianinazzi-Pearson, V (1990) 'Cellular interactions in vesicular-arbuscular (VA) endomycorrhizae : the host's point of view', in P. Nardon, V. Gianinazzi-Pearson, A.M. Grenier, L. Margulis and D.C. Smith (eds.), Endocytobiology IV, INRA-Press, Paris, pp. 83-90.

Gianinazzi-Pearson, V. and Gianinazzi, S. (1989) 'Cellular and genetical aspects of interactions between hosts and fungal symbionts in mycorrhizae', Genome 31, 336-341.

Gianinazzi-Pearson, V., Branzanti, B. and Gianinazzi, S. (1989) '*In vitro* enhancementof spore germination and early hyphal growth of a vesicular-arbuscular mycorrhizal fungus by host root exudates and plant flavonoids', Symbiosis 7, 243-255.

Jacquelinet-Jeanmougin, S., Gianinazzi-Pearson, V., and Gianinazzi, S. (1987) 'Endomycorrhizas in the Gentianaceae. II. Ultrastructural aspects of symbiont relationships in *Gentiana lutea L.*', Symbiosis 3, 269-286.

Philipps, J.M. and Hayman, D.S. (1970) 'Improved procedure for clearing roots and staining parasitic and vesicular-arbuscular fungi for rapid assessment of infection', Trans. Br. Mycol. Soc. 55, 158-161.

Thiéry, J.P. (1967) 'Mise en evidence de polysaccharides sur coupes fines en microscopie électronique', Journal de Microscopie 6, 987-1018.

Trouvelot, A., Kough, J.L. and Gianinazzi-Pearson, V. (1986) 'Mesure du taux de mycorhization VA d'un système radiculaire. Recherche de méthodes d'estimation ayant une signification fonctionnelle', in V. Gianinazzi-Pearson and S. Gianinazzi (eds.), Physiological and Genetical Aspects of Mycorrhizae, INRA-Press, Paris, pp. 217-221.

AGROBACTERIUM RHIZOGENES T-DNA GENES AND SENSITIVITY OF PLANT PROTOPLASTS TO AUXINS

C. MAUREL, H. BARBIER-BRYGOO, J. BREVET, *A. SPENA, J. TEMPE
and J. GUERN
Institut des Sciences Végétales
CNRS
F-91198 Gif-sur-Yvette Cedex
France

Max-Planck-Institut für Züchtungsforschung, D-5000 Köln 30, FRG

ABSTRACT. The sensitivity to auxin of mesophyll protoplasts isolated from *Agrobacterium rhizogenes* transformed tobacco plants was shown to be higher than that of untransformed ones, by studying the action of auxin on the protoplast transmembrane electrical potential difference. This membrane response allowed us to show that single T-DNA genes, namely the *rolA, B* and *C* genes, were also able to confer an increased sensitivity to auxin on transformed protoplasts, *rolB* being the most powerful with a 10,000-fold increase. Plants transformed with *rolB* were thus chosen as a model system for further studies and the regulation of the *rolB* promoter in tobacco was studied by using the ß-glucuronidase (GUS) reporter gene. Tissue-specific expression of our *rolB:GUS* chimeric gene in root meristems and vascular tissues was strongly modified by the addition of exogenous auxin. Furthermore, the slight GUS activity detected in mesophyll protoplasts isolated from *rolB:GUS* plants could be increased 20 to 100 times by adding auxin. These results suggest that auxin plays a central role in the regulation of the *rolB* promoter in tobacco. Consequently, the sensitivity to auxin of *rolB*-transformed protoplasts could be modulated by auxin itself, in good correlation with the activation of the *rolB* gene by auxin. These interactions between the auxin signal and the *rolB* gene might explain some aspects of their cooperative effects for root induction on leaf fragments.

1. Introduction

The soil bacterium *Agrobacterium rhizogenes* induces on dicotyledonous plants, at the site of infection, the neoplastic proliferation of highly branched, hairy roots. The basis of this disease is the expression of genes carried on DNA sequences (T-DNA) transferred from a bacterial plasmid (pRi) into the plant cell (Zambryski et al., 1989). The hairy root transformation has been used in plant gene transfer technology, notably for the ease by which transformed cells can be selected and regenerated to whole plants (Tempé and Casse-Delbart, 1989). This disease also provides a system to study plant development and might bring clues on the mechanism of root differentiation.

Genetic analyses have pointed out that the auxin biosynthetic genes carried by the T-DNA of some *A. rhizogenes* strains are not necessary for hairy root proliferation (Vilaine and Casse-Delbart, 1987). This proliferation results mainly from the action of other T-DNA genes, named *rolA, rolB* and *rolC*. As single genes, they modify plant development and, in the case of *rolA* and *rolB*, they can induce transformed roots on inoculated tissues (Cardarelli et al., 1987b; Oono et al., 1987; Spena et al., 1987; Vilaine et al., 1987; Schmülling et al., 1988; Sinkar et al., 1988).

343

H. Hennecke and D. P. S. Verma (eds.),
Advances in Molecular Genetics of Plant-Microbe Interactions, Vol. 1, 343–351.
© 1991 *Kluwer Academic Publishers. Printed in the Netherlands.*

The physiological bases of the hairy root disease are poorly understood. The recent finding that transformed cells exhibit an increased sensitivity to auxin has opened new perspectives to understanding the morphogenetic effects of the Ri T-DNA genes (Shen et al., 1988; Spano et al., 1988). However, evidence is still needed to justify the assumption that the increased sensitivity to auxin of transformed cells might be a prerequisite for root induction.

In this paper we have first investigated the gene(s) responsible for the increased sensitivity to auxin of transformed tobacco. Every single *rolA, B* or *C* gene was found able to increase the sensitivity to auxin of mesophyll protoplasts, *rolB* being the most powerful with increases up to 100,000-fold. Plants transformed with *rolB* were thus chosen as a model system for further studies. In these plants, auxin was found to regulate the *rolB* promoter and, consequently controlled the sensitivity to auxin of transformed protoplasts. These interactions between the auxin signal and the *rolB* gene could explain some aspects of their cooperative action during root induction.

2. Materials and Methods

2.1. TRANSGENIC PLANTS

The *Nicotiana tabacum* plants transgenic for subfragments of the pRiA4 T_L-DNA were those described by Spena et al. (1987) and Schmülling et al. (1988). The construction of a gene fusion between the *rolB* promoter and the GUS coding sequence was described by Maurel et al. (1990). Briefly, a *HpaI-BamHI* fragment containing 1177 bp upstream from the start codon of the *rolB* gene and the first 4 bp of its coding sequence was isolated from pRi1855 and cloned in a pBI101 vector (Jefferson et al., 1987). A second construct in which a functional *rolB* gene and a *rolB*:GUS fusion were cloned in tandem was obtained by placing, beside the GUS fusion described above, a DNA fragment encompassing the *rolB* gene with 1177 bp 5' and 646 bp 3' non-coding sequences (Maurel et al., in preparation). These constructions were introduced into tobacco by *Agrobacterium* in leaf disk inoculations and T-DNA structures were verified by Southern analysis.

2.2. MESOPHYLL PROTOPLAST ISOLATION

Leaf tissues were digested in To medium (Caboche, 1980) over 15 h or 1.5 h periods in the presence of 0.1% cellulase R10 (Yakult), 0.02% macerozyme R10 (Yakult) and 0.05% driselase (Sigma) or of 0.5% cellulase R10 and 0.5 % macerozyme R10, respectively. Usually the digestion mixture contained $1.5x10^{-5}$M naphthaleneacetic acid (NAA) and $5x10^{-6}$M benzylaminopurine. After two washings in 0.3 M KCl, 5mM $CaCl_2$, 1 mM MES pH 5.7, protoplasts were resuspended in To medium with no auxin.

2.3. MEASUREMENT OF THE TRANSMEMBRANE ELECTRICAL POTENTIAL DIFFERENCE (Em) OF PROTOPLASTS

Em measurements were performed on freshly isolated protoplasts by the microelectrode technique as described by Ephritikhine et al. (1987). Auxin effects were studied by performing, for each NAA concentration tested, 15-20 individual measurements on an aliquot of the protoplast stock solution and mean Em values were calculated. Variations from the reference value in the absence of auxin were reported as ΔEm.

2.4. ß-GLUCURONIDASE ASSAYS

Histochemical and fluorimetric assays were performed according to Jefferson et al. (1987).

2.5. ROOT INDUCTION ON CULTURED LEAF FRAGMENTS

Small disks (2 mm in diameter) were punched out from normal or transformed leaves, carefully avoiding areas with primary and secondary ribs. The disks were cultured on a Monnier's medium modified as previously described (Maurel et al., 1990) in the presence of the NAA concentrations indicated and roots were counted after 10-14 days.

3. Results

3.1. SINGLE *rol* GENES INCREASE THE SENSITIVITY TO AUXIN OF TOBACCO MESOPHYLL PROTOPLASTS

Auxin induces variations of the transmembrane potential difference of isolated protoplasts (Ephritikhine et al., 1987). The dose-dependence of these effects can be used to characterize the sensitivity of the protoplasts to the hormone. Figure 1 shows a dose-response curve obtained on normal tobacco mesophyll protoplasts, with a maximal hyperpolarization at 3×10^{-7} M naphthaleneacetic acid (NAA). When protoplasts isolated from tobacco plants transformed by the pRiA4 T-DNA were used, a similar dose-response curve was obtained with respect to the shape and the amplitude of the response (Fig. 1). However, the maximal hyperpolarization induced by 10^{-8} M NAA reflected a 30-fold increased sensivity to auxin.

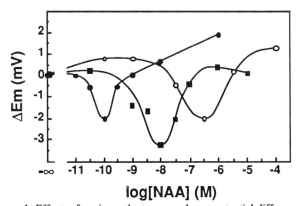

Figure 1: Effects of auxin on the transmembrane potential difference (Em) of mesophyll protoplasts, normal (O), transformed by the whole T-DNA of pRiA4 (■)(clone RSIII) or the *rolB* gene of pRiA4 (●)(clone B1100,2B). ΔEm represents Em variations from the reference value in the absence of auxin. Each point corresponds to a mean Em value calculated from 15-20 individual measurements. Standard errors did not exceed 0.4 mV.

The membrane response of protoplasts was further used to investigate the T-DNA gene(s) responsible for the increased sensitivity to auxin. A T-DNA subfragment carrying the three genes *rolA, B* and *C* is able to induce the full hairy root syndrome in tobacco (Spena et al., 1987; Cardarelli et al., 1987b). Protoplasts transgenic for this fragment exhibited a 1,000-fold increased sensitivity to auxin (Maurel et al., submitted). The individual effects of the *rol* genes, *A, B* or *C*, were then tested. For each gene, the response to auxin of protoplasts isolated from two independently transformed clones was studied in several experiments. Each single *rol* gene confered an increased sensivity to auxin upon transformed protoplasts. However, each clearly exhibited distinct efficiencies: *rolB*, the most powerful, gave increases in sensitivity up to 10,000-fold (see Fig. 1) whereas *rolC*, the least efficient, gave a 10-fold increase (Maurel et al., submitted).

Since *rolB* has also been reported as the most powerful gene for root induction in leaf disk inoculation experiments (Spena et al., 1987), further studies focused on this gene and its effects in plants.

3.2. AUXIN REGULATES THE *rolB* PROMOTER IN TRANSGENIC TOBACCO

The regulation of the *rolB* promoter in tobacco was studied by using the ß-glucuronidase (GUS) reporter gene (Jefferson et al., 1987). The *rolB*:GUS translational fusion we used contains 5' regulatory sequences of the *rolB* gene, up to -1177 bp, and the first 4 bp of its coding sequence (Maurel et al., 1990).

Histochemical analysis of *rolB*:GUS transformed plants revealed, as previously reported by Schmülling et al. (1989), a tissue-specific expression of the *rolB* promoter. In aerial parts, GUS activity was restricted to vascular tissues, identified as internal and external phloem in stem sections. In roots, GUS activity was strongly expressed in apical cap and pith meristems, as well as in lateral root primordia (Schmülling et al.,1989; Maurel et al., 1990).

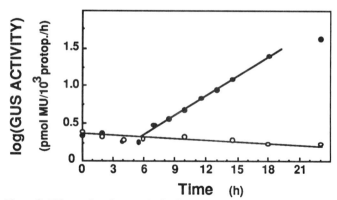

Figure 2: Effects of auxin on *rolB*:GUS expression in protoplasts. Protoplasts were isolated by overnight digestion, in the absence of phytohormone, of leaf tissue from transgenic plants containing the *rolB*:GUS fusion (clone BGUS11). They were then cultured in the absence (O) or in the presence (●) of 10^{-5} M NAA. GUS activity was estimated by fluorimetric measurement of methyl umbelliferone (MU) production from methyl umbelliferyl glucuronide at times indicated.

Regulation by the *rolB* promoter was also investigated in mesophyll protoplasts. When prepared in the absence of phytohormone, mesophyll protoplasts of *rolB*:GUS plants exhibited a low GUS activity, 1-5 times above the background activity of untransformed protoplasts. Figure 2 presents a time course study of GUS activity in isolated protoplasts cultured in the presence or in the absence of auxin. Exponential increase of GUS activity, beginning 5-6 hours after auxin addition, led to a 25-fold stimulation of GUS activity after 24 h incubation. Among the different phytohormones tested (auxin, abscisic acid, gibberellin, cytokinin), only auxin was able to induce such a stimulation. Moreover, this auxin-modulated GUS activity was not found when the GUS gene was under control of the 35S Cauliflower Mosaic Virus (CaMV) promoter (Maurel et al., 1990).

The effects of exogenous auxin on *rolB*:GUS expression in tissues were also tested. A 24 h incubation of excised roots or leaves in the presence of 10^{-5} M NAA induced GUS expression in previously silent tissues, in root cortex and in mesophyll. These results further suggest that auxin plays a central role in the regulation of the *rolB* promoter in tobacco (Maurel et al., 1990).

3.3. AUXIN CONTROLS THE SENSITIVITY TO AUXIN OF *rolB*-TRANSFORMED PROTOPLASTS

In order to combine both physiological and genetical approaches, a functional *rolB* gene with its own regulatory sequences and a *rolB:GUS* fusion as described above were cloned in tandem. Tobacco plants transgenic for this *B-B*:GUS construct allow simultaneous analysis of *rolB* promoter expression (as detected by GUS activity) and *rolB* physiological effects (as studied by auxin responses on protoplasts or organs).

The functionality of the *rolB* gene was determined by electrophysiology. Maximal hyperpolarization of *B-B*:GUS protoplasts was induced by 3×10^{-11} M NAA, reflecting a sensitivity to auxin 100,000 times higher than that of untransformed protoplasts (maximal hyperpolarization at 3×10^{-6} M NAA). On the other hand, *B-B*:GUS plants displayed, as previously described in *rolB*:GUS plants, a highly regulated expression of the *rolB* promoter. For instance, the presence of 10^{-5} M NAA during overnight digestion of *B-B*:GUS leaf tissues led to a level of GUS activity in protoplasts 6 times higher than that in auxin-deprived protoplasts.

One should recall that during protoplast isolation for physiological studies auxin was included in the digestion mixture at a concentration of 1.5×10^{-5} M NAA, which is optimal for *rolB* promoter expression. In order to test whether *rolB* gene expression level could influence the sensitivity to auxin of protoplasts, NAA was omitted during their preparation. *B-B*:GUS protoplasts obtained in this condition exhibited a low sensitivity to auxin with a maximal hyperpolarization at 3×10^{-6} M NAA. Experiments with intermediate auxin concentrations during protoplast preparation confirmed that the sensitivity to auxin of *B-B*:GUS protoplasts was highly correlated to the *rolB* promoter expression level (Maurel et al., in preparation).

In *B-B*:GUS protoplasts, GUS activity started to increase 6 h after auxin addition. Digestion of normal leaf tissues in the presence of auxin over periods of 1.5 h or 15h led to protoplast suspensions showing similar sensitivities to auxin. On the other hand, the sensitivity of *B-B*:GUS protoplasts, 30 times higher than the control when protoplasts were prepared in the presence of auxin over a 1.5 h period, was increased up to 100,000 times the control when digestion times were increased to 15 h (Maurel et al., in preparation). Such a rise in the sensitivity to auxin is consistent with the 'late' activation of the *rolB* promoter by auxin.

These results show that the sensitivity to auxin of *rolB*-transformed protoplasts can be modulated by auxin itself. Such effects are well correlated to the auxin-dependent expression of the *rolB* gene.

3.4. AUXIN CONTROLS THE *rolB*-DEPENDENT ROOT INDUCTION ON LEAF FRAGMENTS

Auxin and the *rolB* gene strongly interact at the cellular level; possible interactions during root induction were investigated in *B-B*:GUS transformed tobaccos. Small disks were excised from the lamina of normal and transformed tobacco leaves, carefully avoiding primary and secondary veins. The disks were then cultured *in vitro* in the presence of auxin at concentrations ranging from 0 to 10^{-4} M NAA. Normal and *rolB*-transformed disks reacted to auxin by exhibiting similar increases in diameter and peripheral callus formation. However, they strongly differed in their rhizogenic response. Disks from normal leaves were unable to differentiate roots, at any auxin concentration tested (Fig. 3). *rolB*-transformed disks exhibited a slight spontaneous rhizogenesis in the absence of auxin and root formation was strongly stimulated by auxin with an optimal concentration of 10^{-6} M NAA (Fig. 3). Highly rhizogenic areas were located on the periphery of disks in which auxin strongly stimulated *rolB* expression, several days before root appearence (Maurel et al, in preparation).

These experiments show that auxin controlled the *rolB*-dependent rhizogenesis on leaf fragments. Such a control might involve the early activation of the *rolB* gene by auxin.

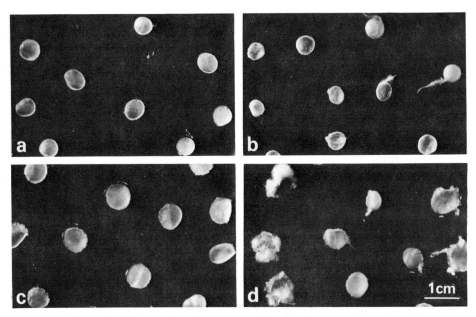

Figure 3: Effects of auxin on root induction in normal (a, c) and
rolB-transformed (b, d) leaf explants. Excised leaf disks were cultured *in vitro*
for two weeks in the absence (a, b) or in the presence (c, d) of 10^{-6} M NAA.

4. Discussion

Rhizogenic effects of auxin have been well known for a long time. The morphogenetic action of *A. rhizogenes* has thus been attributed to auxin-type effects of the Ri T-DNA genes, though their mode of action was unknown. Several authors (Cardarelli et al., 1987a; Vilaine and Casse-Delbart, 1987) have shown that the T_R-DNA of *A. rhizogenes* agropine type strains, which codes for auxin biosynthesis enzymes, was not necessary for hairy root proliferation. It was therefore suggested that the increased sensitivity to auxin of Ri-transformed materials could be responsible for their high rhizogenic potential (Shen et al., 1988; Spano et al., 1988).

In this work we have used the electrical response of protoplasts to auxin to investigate the Ri T-DNA gene(s) responsible for the increased sensitivity to auxin in tobacco. Each of the three *rol* genes *A*, *B* or *C* was able to increase the sensitivity to auxin of mesophyll protoplasts by factors up to 100,000 in the case of *rolB*. The *rolC* gene led to a 10-fold increase and was the least efficient of the *rol* genes. These results indicate a major role of the *rolB* gene, as already described for root induction, and led us to choose this gene as a model.

The importance of *rol* gene expression levels for their effects on auxin responses was first suggested by the reinforced effects of *rolC* on protoplasts when this gene was placed under control of the strong 35S CaMV promoter. The sensitivity to auxin of 35S:*rolC*-transformed protoplasts, though still lower than that of *rolB*-transformed protoplasts, was raised 200 times over the sensitivity of normal protoplasts (Maurel et al., submitted). With respect to *rolB*, we found, as previously reported by Schmülling et al. (1989), that its promoter is regulated in a tissue-specific manner in tobacco. Moreover its expression in tissues or protoplasts could be modulated by the addition of exogenous auxin (Maurel et al., 1990). Such a regulation leads to tremendous effects on the auxin response of *rolB*-transformed protoplasts: activation of the *rolB* gene by auxin led to increases in the sensitivity to auxin of 3,000-fold at least.

Although the studies with protoplasts suggest that the auxin response of *rolB*-transformed cells can be controlled by auxin itself, the precise mode of action of *rolB* remains poorly understood. Sequence analysis did not reveal any obvious function for the *rolB* gene product. Barbier-Brygoo et al. (1990) suggested that differences in sensitivity to auxin observed on protoplasts, normal or transformed by the whole Ri T-DNA, could reflect a differential expression of auxin receptors at the plasmalemma surface. This proposal is based on experiments in which functional auxin receptors were inactivated using antibodies raised to an auxin-binding protein from corn coleoptile. Immunoinactivation of the auxin receptors was estimated by the specific action of the antibodies on the electrical response of protoplasts to auxin . Experiments are in progress to see whether the single *rolB* gene induces similar effects.

In this work, we found that the different efficiencies of the *rol* genes in increasing the sensitivity to auxin of protoplasts were consistent with their respective abilities to induce roots on leaf fragments (Spena et al., 1987). A similar correlation between sensitivity to auxin of protoplasts and rooting capacity has already been described in the case of a non-rooting tobacco mutant (Ephritikhine et al., 1987). These correlations support our assumption that a high sensitivity of plant cells to auxin could be a major determinant for root differentiation. Such a question was directly asked in our study of root induction in *rolB*-transformed plants. The use of small leaf disks with no major rib allowed us to consider *rolB*-dependent rhizogenesis only. Control of this root induction by auxin is consistent with the observations of Capone et al. (1989) on carrot disks. Histological and kinetic data show that this control might involve the early activation of the *rolB* gene by auxin. It is tempting to speculate that, at this stage, the increased sensitivity to auxin conferred by *rolB* might direct the transformed cell into root organogenesis. However the multiple action of auxin in this system, in stimulating cell division, inducing the *rolB* gene and possibly acting on cells whose responsiveness is modified by *rolB* effects is difficult to interprete. Expression of *rolB* under control of an inducer with no auxin activity would be useful to reduce the complexity of this system. However, this complexity certainly illustrates the fine interactions which exist between *A. rhizogenes* T-DNA genes and the transformed plant cell.

Acknowledgements. We thank F. Paquereau for technical assistance in root induction experiments and Dr Stephen K. Farrand for help with the manuscript. C. Maurel was recipient of a doctoral grant from the I.N.R.A..This work was supported by funds from the C.N.R.S. (UPR0040), I.N.R.A. and E.E.C. (BAP-0015-F).

References

Barbier-Brygoo H, Guern J, Ephritikhine G, Shen WH, Maurel C, and Klämbt D (1989) The sensitivity of plant protoplasts to auxin: modulation of receptors at the plasmalemma. *In* Plant Gene Transfer, C Lamb and R Beachy (eds), UCLA symposia on Molecular and Cellular Biology, New Series, Vol 129, Alan R. Liss, Inc, New York, NY.

Caboche M (1980) Nutritional requirements of protoplast-derived, haploid tobacco cells grown at low cell densities in liquid medium. Planta, 149, 7-18.

Capone I, Cardarelli M, Trovato M, and Costantino P (1989) Upstream non-coding region which confers polar expression to Ri plasmid root inducing gene *rolB*. Mol Gen Genet, 216, 239-244.

Cardarelli M, Spano L, Mariotti D, Pomponi M, Mauro ML, Van Sluys MA, and Costantino P (1987a) The role of auxin in hairy root induction. Mol Gen Genet, 208, 457-463.

Cardarelli M, Mariotti D, Pomponi M, Spano L, Capone I, and Costantino P (1987b) *Agrobacterium rhizogenes* T-DNA genes capable of inducing hairy root phenotype. Mol Gen Genet, 209, 475-480.

Ephritikhine G, Barbier-Brygoo H, Muller JF, and Guern J (1987) Auxin effect on the transmembrane potential difference of wild-type and mutant tobacco protoplasts exhibiting a differential sensitivity to auxin. Plant Physiol, 83, 801-804.

Jefferson RA, Kavanagh TA, and Bevan MW (1987) GUS fusions: ß-glucuronidase as a sensitive and versatile gene fusion marker in higher plants. EMBO J , 6, 3901-3907.

Maurel C, Brevet J, Barbier-Brygoo H, Guern J, and Tempé J (1990) Auxin regulates the promoter of root inducing *rolB* gene of *Agrobacterium rhizogenes* in transgenic tobacco. Mol Gen Genet, 223, 58-64.

Oono Y, Handa T, Kanaya K, and Uchimiya H (1987) The T_L-DNA gene of Ri plasmids responsible for dwarfness of tobacco plants. Jpn J Genet, 62, 501-505.

Schmülling T, Schell J, and Spena A (1988) Single genes from *Agrobacterium rhizogenes* influence plant development. EMBO J, 7, 2621-2629.

Schmülling T, Schell J, and Spena A (1989) Promoters of the *rolA, B,* and *C* genes of *Agrobacterium rhizogenes* are differentially regulated in transgenic plants. Plant Cell, 1, 665-670.

Shen WH, Petit A, Guern J, and Tempé J (1988) Hairy roots are more sensitive to auxin than normal roots. Proc Natl Acad Sci USA, 85, 3417-3421.

Sinkar VP, Pythoud F, White FF, Nester EW, and Gordon MP (1988) *rolA* locus of the Ri plasmid directs developmental abnormalities in transgenic tobacco plants. Genes Dev, 2, 688-697.

Spano L, Mariotti D, Cardarelli M, Branca C, and Costantino P (1988) Morphogenesis and auxin sensitivity of transgenic tobacco with different complements of Ri T-DNA. Plant Physiol, 87, 479-483.

Spena A, Schmülling T, Koncz C, and Schell J (1987) Independent and synergistic activity of the *rolA*, *B* and *C* loci in stimulating abnormal growth in plants. EMBO J, 6, 3891-3899.

Tempé J, and Casse-Delbart F (1989) Plant gene vectors and genetic transformation: *Agrobacterium* Ri plasmids. *In* Cell culture and somatic cell genetics of plants: The molecular biology of nuclear genes, Vol 6, pp 23-49, Academic Press Inc., San Diego, California.

Vilaine F, and Casse-Delbart F (1987) Independent induction of transformed roots by the T_L and T_R regions of the Ri plasmid of agropine type *Agrobacterium rhizogenes* . Mol Gen Genet, 206, 17-23.

Vilaine F, Charbonnier C, and Casse-Delbart F (1987) Further insight concerning the T_L region of the Ri plasmid of *Agrobacterium rhizogenes* strain A4: Transfer of a 1.9 kb fragment is sufficient to induce transformed roots on tobacco leaf fragments. Mol Gen Genet, 210, 411-415.

Zambryski P, Tempé J, and Schell J (1989) Transfer and function of T-DNA genes from *Agrobacterium* Ti and Ri plasmids in plants. Cell, 56, 193-201.

CHIMAERAS AND TRANSGENIC PLANT MOSAICS: A NEW TOOL IN PLANT BIOLOGY

A. SPENA and S.C. SCHULZE
MPI für Züchtungsforschung
Carl von Linné Weg 10
5000 Köln 30
FRG

ABSTRACT. Chimaeras are genetic mosaics composed of intermixed genetically different tissues. Transposable elements can be used to activate gene expression in clonal populations of cells and consequently to generate genetic mosaics. To achieve this a gene of interest (e.g. morphogenetic genes from plant pathogens) is split by a transposable element in such a way that excision events will reactivate gene expression in transgenic plants. Excision events taking place late during plant organ development will generate extra-apical mosaics composed of intermixed tissues with wild-type and mutant phenotype. In particular, the use of a transposon-split *35sipt* gene to raise plants transgenic for a cytokinin synthetising gene, whose constitutive expression is not compatible with plant regeneration is presented. Moreover we have constructed a chimaeric gene composed of the 35s RNA promoter of cauliflower mosaic virus and the indoleacetic-acid lysine synthetase of *Pseudomonas savastanoi* and defined the morphological alterations established by *iaal* gene expression in transgenic plants.

1. Introduction

At the beginning of the sixteenth century, Rabelais asking "Can a chimaera, swinging in the void, swallow second intentions ?" was condemning chimaeras to mean the vain or the foolishly fancy (cited in Borges, 1974). However chimaeras are pregnant of second intentions as shown, in biology, with the use of genetic mosaics which have made perceivable what is otherwise deceitful.

Plant chimaeras were first reported in the middle of the 17th century and attracted human being attention for their bizarre appearance. They appeared to combine characteristics of two distinct varieties like the citrus fruit Bizzaria, composed of intermixed orange and lemon parts discovered in the Panciatichi garden in Florence and described by Ferrarius in 1646, or the black and white striped grape fruits reported by Gallesio in its Pomonia Italiana (1834).

For long time chimaeric fruits and plants appeared to satisfy simply the human wish to join features that in the real world are separated, in

H. Hennecke and D. P. S. Verma (eds.),
Advances in Molecular Genetics of Plant-Microbe Interactions, Vol. 1, 352–356.
© 1991 *Kluwer Academic Publishers. Printed in the Netherlands.*

other words to make possible what is looking impossible. Later, when biologists became aware that the study of abnormalities could help our understanding of normal developmental processes (Masters, 1869), chimaeras started to be built and used with the intention to address biological problems.

Chimaerism is the "leitmotiv" of our experimental approach. The first type of chimaerism, common to animal and plant genetic engineering, is at the level of genetic information, that is in the construction of genes consisting of DNA fragments isolated from the genome of different organisms. In this regard, we have used chimaeric genes containing the coding regions of the *rol* genes of *Agrobacterium rhizogenes* to establish developmental alterations in transgenic plants (Spena et al., 1987; Schmülling et al., 1988), and here we present the use of a chimaeric gene containing the coding region of the *iaal* gene from *Pseudomonas savastanoi* to test whether expression of indole-acetic acid lysine synthetase causes morphological alterations in transgenic plants.

The second type of chimaerism is in the use of a transposable element to activate gene expression in clonal populations of cells and consequently to generate, by plant genetic engineering, genetic mosaics. This second type of chimaerism is perceivable at the organism level as a contrast of different phenotypes within the same organ (Spena et al., 1989) and it has been used to evaluate whether a morphogenetic gene affects chlorophyll pigmentation and leaf morphology in a cell autonomous way. Here we present data investigating the use of genetic switches based on transposable elements and morphogenetic genes to circumvent regeneration problems (Spena, 1990). In this perspective we have used the *ipt* gene of *Agrobacterium tumefaciens* and shown that while its expression under the control of the *35S* RNA promoter of cauliflower mosaic virus is not compatible with plant regeneration (Smigocki and Owens, 1988), plants can be raised when the *35sipt* gene is splitted by a transposable element.

2. Materials and Methods

Standard techniques were used for the construction of recombinant DNA plasmids (Maniatis et al., 1982). Constructs were delivered to plant cells using the pPCVOO2 binary vector (Koncz and Schell, 1986), and transgenic plants were raised by following a modification of the leaf disc transformation procedure of Horsch et al. (1984).

3. Results and Discussion

3.1. EXPRESSION OF THE INDOLEACETIC ACID-LYSINE SYNTHETASE OF *PSEUDOMONAS SYRINGAE* SUBSP. *SAVASTANOI* CAUSES MORPHOLOGICAL ALTERATIONS IN PLANTS.

The *iaal* gene of *Pseudomonas syringae* susp. *savastanoi* encodes an indoleacetic acid lysine synthetase able to conjugate indoleacetic acid with lysine (Glass and Kosuge, 1985; Roberto et al., 1990).

A chimaeric gene consisting of the *iaal* coding region under the control of the 35S RNA promoter of cauliflower mosaic virus was constructed and used to raise transgenic tobacco plants. Tobacco plants transgenic for the *35siaal* chimaeric gene dysplay morphological and developmental alterations suggestive of an altered auxin metabolism. Their major phenotypic alteration is an increased leaf nastic curvature (i.e. epinasty). Other alterations caused by *iaal* gene expression are reduction in root growth and increased side shoot formation. Transgenic tobacco plants transmit the forementioned alterations to progeny as a dominant mendelian gene cosegregating with the kanamycin resistance marker. Similar alterations are caused by *iaal* gene expression also in potato plants (manuscript in preparation). In conclusion the use of the *35Siaal* chimaeric gene has allowed us to show that expression of the *iaal* gene from *Pseudomonas savastanoi* causes, directly or indirectly, developmental alterations in transgenic plants.

3.2. TRANSPOSON-SPLIT GENE FOR CYTOKININ SYNTHESIS, AND ITS USE TO CIRCUMVENT REGENERATION PROBLEMS CAUSES BY CONSTITUTIVE EXPRESSION OF A MORPHOGENETIC GENE.

Smigocki and Owens (1988) have delivered a chimaeric *35sipt* gene, composed of the 35s RNA promoter from cauliflower mosaic virus and the coding region of the *ipt* gene of *Agrobacterium tumefaciens*, to *Nicotiana tabacum*, *rustica* and *plumbaginifolia*. Expression of the *35sipt* gene caused formation of shoots exhibiting loss of apical dominance which were not able to root even on medium supplemented with 1-naphtaleneacetic acid. These effects are most likely due to the (up to 137-fold) increase in cytokinin concentration caused by the expression of the *ipt* gene when expressed under the control of the *35s* RNA promoter of cauliflower mosaic virus.

To circumvent the regeneration problems due to the inhibitory effect of high doses of cytokinin on root formation, we have split the *35sipt* gene with the *Ac* transposable element of maize (Behrens et al., 1984). Insertion of the *Ac* transposable element in the untranslated leader sequence of the *35sipt* gene prevents expression. In fact, *ipt* gene expression will take place only after transposon excision and reconstruction of the *35sipt* gene. Excision events taking place late during organ development will generate plant mosaics, composed of clonal cellular populations expressing the *ipt* gene, and neighbouring not expressing cells.

8 out of 80 shoots induced from leaf-discs transformed with the 35s-Ac-ipt construct, were able to root and to grow in soil. They show alterations in growth, development and physiology indicative of an increased cytokinin activity (manuscript in preparation). In conclusion, by using a transposon splitted gene construct we have raised plant mosaics transgenic for the *35sipt* gene. In other words we have made possible, what was looking not possible. These hormomal chimaeras are currently used to investigate physiological and developmental effects of cytokinins.

Extraapical plant chimaeras could be also used to evaluate whether expression of a gene of interest affects cell-cell and plant-pathogen interactions. For example, mosaic leaves composed of intermixed tissues expressing and not expressing the gene of interest could be used to evaluate whether a paticular gene action influences sensitivity or resistance to pathogens.

4. Acknowledgements

We thank the late Tsune Kosuge and Frank Roberto for plasmid pLG87 and for having communicated us data prior publication.

5. References

Behrens, U., Fedoroff, N., Laird, A., Müller-Neumann, M., Starlinger, P.and Yoder, J. (1984). Cloning of the Zea mays controlling element Ac from the wx-m7 allele. Mol. Gen. Genet. 194, 346-347.

Borges, J.L. (1974). The book of imaginary beings, Penguin, London.

Ferrarius. (1646) Hesperides sive de Malorum Aureorum Cultura et Usu. Rome

Gallesio, G. (1834). Pomona italiana. Capurro, Pisa.

Glass, N.L. and Kosuge, T.. (1986). Cloning of the gene for indoleacetic acid-lysine synthetase from Pseudomonas syringae subsp. savastanoi . J. Bacteriology. 166, 598-603.

Horsch, R., Fraley, R., Rogers, S., Sanders, P., Lloyd, A. and Hoffmann, W. (1984). Inheritance of functional foreign genes in plants. Science 223, 496-498.

Koncz, C. and Schell, J. (1986). The promoter of T$_L$-DNA gene 5 controls the tissue-specific expression of chimaeric genes carried by a novel type of Agrobacterium binary vector. Mol. Gen. Genet. 204, 383-396.

Maniatis, T., Fritsch, E.F. and Sambrook, J. (1982). Molecular cloning: a laboratory manual, Cold Spring Harbor Laboratory, Cold Spring Harbor, NY.

Masters, M.T. (1869). Vegetable Teratology. Hardwicke, London.

Roberto, F.F., Klee, H., White, F., Nordeen, R. and Kosuge, T.. (1990). Expression and fine structure of the gene encoding indole-3-acetyl-L-lysine synthetase from Pseudomonas savastanoi. Proc. Natl. Acad. Sci. USA 87, 1-6.

Schmülling, T., Schell, J. and Spena, A.. (1988). Single genes from Agrobacterium rhizogenes influence plant development. EMBO J. 7, 2621-2629.

Smigocki, A.C. and Owens, L.D.. (1988) Cytokinin gene fused with a strong promoter ehances shoot organogenesis and zeatin levels in transformed plant cells. Proc. Natl. Sci. USA 85, 5131-5135.

Spena, A., Schmülling, T., Koncz, C. and Schell, J.. (1987). Independent and synergistic activity of the rolA, B and C loci in stimulating abnormal growth in plants. EMBO J. 6, 3891-3899.

Spena, A., Aalen, R.B. and Schülze, S.C.. (1989). Cell-autonomous behavior of the *rolC* gene of *Agrobacterium rhizogenes* during leaf development: A visual assay for transposon excision in transgenic plants. The Plant Cell 1, 1157-1164.

Spena, A.. (1990). Unstable liaisons: the use of transposons in plant genetic engineering. T.I.G. 6, 76-77.

IDENTIFICATION OF SIGNAL TRANSDUCTION PATHWAYS LEADING TO THE EXPRESSION OF *ARABIDOPSIS THALIANA* DEFENSE GENES

F.M. AUSUBEL[1], K.R. DAVIS[2], E.J. SCHOTT[1], X. DONG[1], M. MINDRINOS[1]

[1]*Department of Molecular Biology*
Massachusetts General Hospital
Boston, MA 02114

[2]*Department of Plant Biology and the Biotechnology Center*
Ohio State University
Columbus, OH 43210-1002

ABSTRACT. We have developed a model system that utilizes the defense response of *Arabidopsis thaliana* to pathogenic pseudomonads to study the signal transduction pathways leading to the activation of plant defense genes. We have identified *Pseudomonas syringae* strains that are either virulent or elicit a visible hypersensitive response when infiltrated into *A. thaliana* leaves and we have cloned two avirulence *(avr)* genes from an avirulent strain. We have also cloned an *A. thaliana* gene encoding phenylalanine ammonia lyase (PAL) as well as three adjacent genes that encode β-1,3-glucanases (BG1, BG2, and BG3). Our overall strategy is to devise simplified genetic screens in which we monitor the induction of the PAL and BG promoters in response to cloned *avr* genes. We have shown that inoculation of *A. thaliana* leaves with virulent strains leads to the gradual accumulation of mRNA corresponding to BG1, BG2, and BG3 but not to the accumulation of mRNA corresponding to PAL. In contrast, inoculation with an avirulent strain leads to transient accumulation of PAL mRNA but has only a modest effect on the levels of BG1, BG2, and BG3 mRNA. Significantly, virulent strains carrying cloned *avr* genes mimic the avirulent strain in that they strongly elicit PAL mRNA but not β-1,3-glucanase mRNA accumulation.

1. Introduction

Plants mount a highly coordinated defensive response to invasion by phytopathogenic microorganisms (see Lamb et al., 1989; Dixon and Lamb, 1990 for recent reviews). However, in the absence of mutant plants that are defective in defense responses, it has been difficult to demonstrate a significant role for any particular defense-response gene in conferring resistance. Moreover, little is known about the molecular genetic mechanisms involved in the signal transduction pathways leading to the activation of defense-related genes. To facilitate the molecular genetic analysis of the defense response, we have cooperated with Brian

H. Hennecke and D. P. S. Verma (eds.),
Advances in Molecular Genetics of Plant-Microbe Interactions, Vol. 1, 357–364.
© 1991 *Kluwer Academic Publishers. Printed in the Netherlands.*

Staskawicz's laboratory at the University of California to develop a model pathogenesis system that utilizes *A. thaliana* as the host for infection by a variety of *Pseudomonas* strains and species (Davis et al., 1989; Schott et al., 1990). Other laboratories are also investigating the use of *A. thaliana* as a host for bacterial pathogens (Simpson and Johnson, 1990; M. Daniels, personal communication; J. Dangl, personal communication).

    A. thaliana offers several advantages as an experimental system for molecular genetic analysis in comparison to plants used previously to study disease resistance. These advantages include a fast generation time and very small seeds (25 µg/seed) that facilitate the isolation and genetic analysis of mutants (Redei, 1975; Koornneef et al., 1983; Meyerowitz, 1989), and a very small genome (100 Mb) that facilitates the cloning of genes that correspond to mutant phenotypes (Chang et al., 1988; Nam et al., 1989).

2. Identification of *A. thaliana* Pathogens

To identify bacterial pathogens that would infect *A. thaliana*, we screened a large number of *Pseudomonas* strains that are pathogens of plants in the family *Cruciferae* (Schott et al., 1990). Several strains that we characterized in some detail are listed in Table 1. These *Pseudomonas* strains display a variety of phenotypes when infiltrated into *A. thaliana* leaves. The pathogenic or virulent strains, 4326 and 5034, elicit the appearance of a water-soaked lesion and multiply four to six logs *in planta*. In contrast, the avirulent strains, 83-1 and 1065, elicit a hypersensitive response (HR) and only multiply one or two logs *in planta*.

Table 1. *Pseudomonas* Pathogens of *A. thaliana*

SPECIES	PATHOVAR	STRAIN	HOST	PHENOTYPE
P. syringae	*maculicola*	4326 (B70)	radish	Vir
P. syringae	*maculicola*	5034	cabbage	Vir
P. syringae	*tomato*	1065	tomato	Avr
P. cichorii	?	83-1	?	Avr

We used a variety of criteria to classify the *Pseudomonas* strains into appropriate species and pathovar categories. These criteria included host range, lipid composition, metabolic characteristics, and DNA/DNA hybridization. In general, the strains classified as *P. syringae* pv. *tomato* and *P. syringae* pv. *maculicola* are similar to each other but quite distinct from strain 83-1 which was classified as *P. cichorii*.

3. Cloning *P. syringae avr* Genes

We adopted two strategies to identify *avr* genes in the avirulent *P. syringae* pv. *tomato* strain 1065. First, following a strategy used by a variety of labs to clone *avr* genes (Staskawicz et al., 1984), we conjugated a cosmid library of 1065 DNA constructed in the broad host range vector pLAFR3 into the virulent strain 4326 and then screened individual transconjugants for ones that elicited a visible HR on *A. thaliana* ecotypes Columbia and Nossen. Among 350 transconjugants tested, one elicited a strong HR, and the cosmid in this strain was shown to contain an *avr* gene that was delimited to a 3.1 kb region and which reduced the *in planta* growth of strain 4326 by approximately two logs. This same *avr* gene has been cloned in Brian Staskawicz's laboratory and is referred to as *avrRpt2* (B. Staskawicz, personal communication).

The second strategy for identifying 1065 *avr* genes was based on the genomic subtraction technique developed in our laboratory (Straus and Ausubel, 1990). In the case of several *avr* genes identified in other laboratories, it has been found that some virulent strains completely lack DNA sequences homologous to the *avr* genes (Keen and Staskawicz, 1988). We therefore used the genomic subtraction technique to isolate sequences present in strain 1065, an avirulent strain, but absent in 5034, a virulent strain. When these sequences were used to probe a library of strain 1065 DNA in pLAFR3, 243 clones out of 2400 clones tested were found to hybridize. Among these 243 clones, we identified two unique clones that carried *avr* genes by conjugating these clones into *P. syringae* pv. *maculicola* strain 4326 and then testing the transconjugants for their ability to elicit an HR response. One of these *avr* genes is the same as the one identified by conjugating the entire 1065 library into 4326 (*avrRpt2*). The additional *avr* gene identified by genomic subtraction will be described in detail in a separate publication (M. Mindrinos and F.M. Ausubel, in preparation).

When transferred to the virulent strain 4326, both *avr* genes caused a reduction in *in planta* growth of about two logs in the *A. thaliana* ecotype Columbia. Interestingly, however, preliminary results suggest that another *A. thaliana* ecotype, Bensheim, only appears to recognize one of the two cloned *avr* genes. This result indicates that *A. thaliana* is most likely responding to a specific *avr* gene through the action of a single or a limited number of "resistance" genes. Brian Staskawicz's laboratory has also identified *A. thaliana* ecotypes that fail to respond to particular cloned *avr* genes (B. Staskawicz, personal communication).

4. Cloning *A. thaliana* PAL and β-1,3-glucanase Genes

Our overall strategy in developing the *A. thaliana* pathogenesis model is to utilize a molecular genetic approach to define and characterize the signal transduction pathway(s) leading to the expression of plant defense genes. To help achieve this goal, we have chosen two well-characterized plant defense responses to study at the molecular level, the synthesis of the hydrolase β-1,3-glucanase, which may inhibit pathogens by degrading their cell walls, and the synthesis of phenylalanine ammonia lyase (PAL), a key enzyme required for the synthesis of phenylpropanoid phytoalexins (Lamb et al., 1989; Collinge and Slusarenko, 1987; Hahlbrock and Scheel, 1989). Our immediate objective is to establish simplified genetic screens in which we can monitor the induction of a single defense gene promoter (PAL or β-1,3,-glucanase) in response to a bacterial signal specified by a single *Pseudomonas* *avr* gene.

To facilitate the monitoring of the defense response, we have cloned an *A. thaliana* gene that encodes PAL (Davis et al., 1989) and three *A. thaliana* genes that encode β-1,3-glucanases. The *A. thaliana* PAL and β-1,3-glucanase genes were cloned by screening a library of *A. thaliana* DNA (Landsberg erecta ecotype) constructed in λFIX (Voytas and Ausubel, 1988) with heterologous cDNA probes from bean (pPAL5; Edwards et al., 1985) and tobacco (pGL43; Mohnen et al., 1985) respectively. In the case of PAL, it appears that the gene we have cloned is part of a small family of four to five poorly cross-hybridizing genes. On the basis of partial DNA sequence data, the cloned *A. thaliana* PAL gene is 85% homologous at the nucleotide level to the bean PAL gene (Cramer et al., 1989).

In the case of the β-1,3-glucanase genes, restriction mapping, Southern blotting analysis, and DNA sequence analysis showed that *A. thaliana* contains three adjacent β-1,3-glucanase genes, BG1, BG2, and BG3, that are clustered within a 12 kb region in a tandem array and that are transcribed in the same direction. BG1, BG2, and BG3 are 60-70% homologous at the amino acid level to the tobacco β-1,3-glucanase gene used as the hybridization probe (Mohnen et al., 1985). There do not appear to be any other β-1,3-glucanase genes in the *A. thaliana* genome.

5. Induction of *A. thaliana* Defense Genes

In a previous publication, we showed that both PAL and β-1,3,-glucanase mRNA transiently accumulated in *Arabidopsis* tissue culture cells that had been treated with the bacterial pectin-degrading enzyme α-1,4-endopolygalacturonic acid lyase (PGA lyase) (Davis and Ausubel, 1989). As is the case in other tissue culture systems, PAL mRNA was induced rapidly, reaching maximum levels in three hours, whereas β-1,3,-glucanase mRNA reached a maximum level 12 to 18 hours after elicitor treatment. These experiments established that *A. thaliana* responds to elicitors in a similar fashion to other plants that had been studied previously (Edwards et al., 1985; Fritzemeier et al., 1987; Habereder et al., 1989).

Subsequently, we have examined the activation of the PAL and β-1,3-glucanase genes in *A. thaliana* leaves following infiltration with virulent and avirulent *P. syringae* strains. Infiltration with a strongly avirulent strain such as *P. syringae* pv. *tomato* 1065 results in the rapid accumulation and then decline of phenylalanine ammonia lyase (PAL) mRNA, whereas infiltration with most of the strongly virulent strains such as *P. syringae* pv. *maculicola* 4326 has only a limited effect on PAL induction. The opposite results are obtained when β-1,3-glucanase mRNA is examined. Following inoculation with the virulent strain 4326, mRNA corresponding to BG1, BG2, and BG3 gradually accumulates 10 to 20 fold over the course of 48 hours, whereas inoculation with the avirulent strain 1065 has only a modest effect on the levels of BG1, BG2, and BG3 mRNA. These results are summarized in Figure 1.

Significantly, the virulent strain 4326 carrying the cloned *avr* gene, *avrRpt2*, on plasmid pMMXR1, behaved similarly to the avirulent strain 1065 with respect to activation of the PAL and β-1,3-glucanase genes. That is, 4326/pMMXR1 activated the expression of PAL mRNA with the same kinetics as strain 1065 but compared to 4326 had only a modest effect on the activation of the β-1,3-glucanase genes (data not shown).

6. Discussion

The data summarized in Figure 1 indicate that the PAL and β-1,3-glucanase genes respond to different types of signals and are most likely activated by different signal transduction pathways. The fact that PAL mRNA accumulates rapidly in response to infection by avirulent strains suggests that the PAL gene is responding directly to a signal generated by bacterial *avr* genes and that this plant response is most likely mediated by *A. thaliana* "resistance" genes. Similar conclusions have been reached by other investigators concerning the activation of genes involved in phenylpropanoid biosynthesis during the defense response (Bell et al., 1986;

Dhawale et al., 1989; Habereder et al., 1989; Jahnen and Hahlbrock, 1988). In contrast to PAL mRNA, mRNA corresponding to the three *A. thaliana* β-1,3-glucanase genes only accumulates significantly during the extended course of a virulent infection, suggesting that the β-1,3-glucanase genes are activated by a signal transduction pathway that involves a signal that is not specific for a particular pathogen and that requires significant bacterial multiplication *in planta*. The kinetics that we observed for β-1,3-glucanase induction have also been observed in other plants with various pathogen or elicitor treatments (Kombrink and Hahlbrock, 1986; Kauffman et al., 1987; Kombrink et al., 1988; Vogeli-Lange et al., 1988; Benhamou et al., 1989; Wade et al., 1989; Memelink et al., 1990).

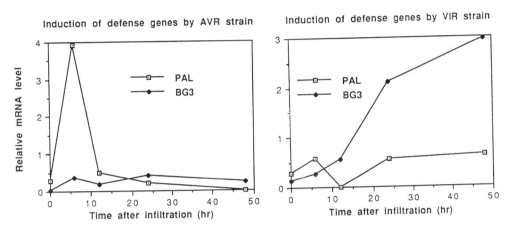

Figure 1. Accumulation kinetics of mRNA corresponding to PAL and BG genes in *A. thaliana* leaves infiltrated with virulent and avirulent *P. syringae* strains. The undersides of *A. thaliana* ecotype Col-0 leaves were infiltrated with 10 µl of an appropriate dilution of a bacterial suspension in 10 mM MgCl$_2$ using a 1 ml syringe with no needle. The virulent strain was *P. syringae* pv. *maculicola* 4326 and the avirulent strain was *P. syringae* pv. *tomato* 1065. At the indicated times, 10-20 leaves were collected and quickly frozen in liquid nitrogen and RNA was extracted using a modification of the phenol-sodium dodecyl sulfate extraction, LiCl precipitation procedure (Ausubel et al., 1990). RNA blots were prepared using 5 µg of RNA per lane as described (Davis and Ausubel, 1989) and then hybridized with radiolabelled gene-specific probes for PAL, BG1, BG2, and BG3. The data shown in this figure were obtained by quantifying the Northern blot data with a Betascope 603 blot analyzer.

Determining the significance of each of the various plant defense responses during pathogen attack has been hampered by the lack of suitable plant mutants that are defective for specific responses. By establishing that *A. thaliana* defense genes respond to genetically well-defined bacterial signals and thereby laying the foundation for the isolation of *A. thaliana* mutations that affect the defense response, we believe that we have overcome some of the problems inherent in other models available for the study of plant-pathogen interactions.

Our overall strategy to isolate *A. thaliana* mutations in defense responses is to construct transgenic *A. thaliana* plants that carry PAL-GUS or BG-GUS fusions and then to screen mutagenized transgenic plants for ones that express the GUS reporter gene (Jefferson et al., 1987) aberrantly following infiltration with virulent or avirulent *P. syringae* strains. Because the PAL and BG genes appear to be activated by different signal transduction pathways, these screens should yield at least two different categories of mutations. In the case of screening for mutants that fail to express the PAL-GUS fusion in response to an avirulent strain, one category of mutations should correspond to *A. thaliana* resistance genes.

7. Acknowledgements

We thank Brian Staskawicz for supplying strains, plasmids, and data prior to publication. This work was supported by a grant from Hoechst A.G. to Massachusetts General Hospital and funds provided by Ohio State University.

8. References

Ausubel, F.M., R. Brent, R.E. Kingston, D.D. Moore, J.G. Seidman, J.A. Smith, and K. Struhl. (1990) Current Protocols in Molecular Biology. Greene Publishing Associates/ Wiley-Interscience, New York.

Bell, J.N., T.B. Ryder, V.P.M. Wingate, J.A. Bailey, and C.J. Lamb. (1986) Differential accumulation of plant defense gene transcripts in a compatible and an incompatible plant-pathogen interaction. Mol. Cell. Biol. 6, 1615-1623.

Benhamou, N., J. Grenier, A. Asselin, and M. Legrand. (1989) Immunogold localization of β-1,3-glucanases in two plants infected by vascular wilt fungi. The Plant Cell 1, 1209-1221.

Chang, C., J.L. Bowman, A.W. DeJohn, E.S. Lander, and E.M. Meyerowitz. (1988) Restriction fragment length polymorphism map for *Arabidopsis thaliana*. Proc. Natl. Acad. Sci. USA. 85, 6856-6860.

Collinge, D.B., and A.J. Slusarenko. (1987) Plant gene expression in response to pathogens. Plant Mol. Biol. 9, 389-410.

Cramer, C.L., K. Edwards, M. Dion, X. Liang, S.L. Dildine, et al. (1989) Phenylalanine ammonia-lyase gene organization and structure. Plant Molec. Biol. 12, 367-383.

Davis, K.R., and F.M. Ausubel. (1989) Characterization of elicitor-induced defense responses in suspension-cultured cells of *Arabidopsis*. Mol. Plant Microbe Interact. 2, 363-368.

Davis, K.R., E. Schott, X. Dong, and F.M. Ausubel. (1989) *Arabidopsis thaliana* as a model system for studying plant-pathogen interactions. In: Signal Molecules in Plants and Plant-Microbe Interactions (B.J.J. Lugtenberg, ed.) Springer-Verlag, Berlin, pp. 99-106.

Dhawale, S., G. Souciet, and D.N. Kuhn. (1989) Increase of chalcone synthase mRNA in pathogen-inoculated soybeans with race-specific resistance is different in leaves and roots. Plant Physiol. 91, 911-916.

Dixon, R., and C. Lamb. (1990) Molecular communication in interactions between plants and microbial pathogens. Annu. Rev. Plant Physiol. Plant Mol. Biol. 41, 339-367.

Edwards, K., C.L. Cramer, G.P. Bolwell, R.A. Dixon, W. Schuch, and C.J. Lamb. (1985) Rapid transient induction of phenylalanine ammonia-lyase mRNA in elicitor-treated bean cells. Proc. Natl. Acad. Sci. USA 82, 6731-6735.

Fritzemeier, K.-H., C. Cretin, E. Kombrink, F. Rohwer, J. Taylor, D. Scheel, and K. Hahlbrock. (1987) Transient induction of phenylalanine ammonia-lyase and 4-coumarate:CoA ligase mRNAs in potato leaves infected with virulent or avirulent races of *Phytophthora infestans*. Plant Physiol. 85, 34-41.

Habereder, H., G. Schroder, and J. Ebel. (1989) Rapid induction of phenylalanine ammonia-lyase and chalcone synthase mRNAs during fungus infection of soybean (*Glycine max* L.) roots or elicitor treatment of soybean cell cultures at the onset of phytoalexin synthesis. Planta 177, 58-65.

Hahlbrock, K., and D. Scheel. (1989) Physiology and molecular biology of phenylpropanoid metabolism. Annu Rev. Plant Physiol. Plant Mol. Biol. 40, 347-369.

Jahnen, W., and K. Hahlbrock. (1988) Cellular localization of non-host resistance reactions of parsley (*Petroselinum crispum*) to fungal infection. Planta 173, 197-204.

Jefferson R.A., T.A. Kavanagh, and M.W. Bevan. (1987) GUS fusions: Beta-glucuronidase as a sensitive and versatile gene fusion marker in higher plants. EMBO J. 6, 3901-3907

Kauffman, S., M. Legrand, P. Geoffroy, and B. Fritig. (1987) Biological function of "pathogenesis-related" proteins: four PR proteins of tobacco have 1,3-β-glucanase activity. EMBO J. 6, 3209-3212.

Keen, N. and B.J. Staskawicz. (1988) Host range determinants in plant pathogens and symbionts. Annu. Rev. Microbiol. 42, 421-440.

Kombrink, E. and K. Hahlbrock. (1986) Responses of cultured parsley cells to elicitors from phytopathogenic fungi. Plant Physiol. 81, 216-221.

Kombrink, E., M. Schroder, and K. Hahlbrock. (1988) Several "pathogenesis-related" proteins in potato are 1,3-β-glucanases and chitinases. Proc. Natl. Acad. Sci. USA 85, 782-786.

Koornneef, M., J. van Eden, C.J. Hanhart, P. Stam, F.J. Braaksma, and W.J. Feenstra. (1983) Linkage map of *Arabidopsis thaliana*. J. Hered. 74, 265-272.

Lamb, C.J., M.A. Lawton, M. Dron, and R.A. Dixon. (1989) Signals and transduction mechanisms for activation of plant defenses against microbial attack. Cell 56, 215-224.

Memelink, J., H.J.M. Linthorst, R.A. Schilperoort, and J.H.C. Hoge. (1990) Tobacco genes encoding acidic and basic isoforms of pathogenesis-related proteins display different expression patterns. Plant Mol. Biol. 14, 119-126.

Meyerowitz, E.M. (1989) *Arabidopsis*, a useful weed. Cell 56, 263-269.

Mohnen, D., H. Shinshi, G. Felix, and F. Meins, Jr. (1985) Hormonal regulation of β-1,3-glucanase messenger RNA levels in cultured tobacco tissues. EMBO J. 4, 1631-1635.

Nam, H.-G., J. Giraudat, B. den Boer, F. Moonan, W.D.B. Loos, B.M. Hauge, and H.M. Goodman. (1989) Restriction fragment length polymorphism linkage map of *Arabidopsis thaliana*. The Plant Cell 1, 699-705.

Redei, G.P. (1975) *Arabidopsis* as a genetic tool. Annu. Rev. Genet. 9, 111-127.

Schott, E.J., K.R. Davis, X. Dong, M. Mindrinos, P. Guevara, and F.M. Ausubel. (1990) *Pseudomonas syringae* infection of *Arabidopsis thaliana* as a model system for studying plant-bacterial interactions. In: *Pseudomonas*: Biotransformation, Pathogenesis, and Evolving Biotechnology (S. Silver, A.M. Chakrabarty, B. Iglewski, and S. Kaplan, eds.) American Society for Microbiology, Washington, pp. 82-90.

Simpson, R.B. and L.J. Johnson. (1990) *Arabidopsis thaliana* as a host for *Xanthomonas campestris* pv. *campestris*. Mol. Plant Microbe Interact. 3, 233-237.

Staskawicz, B.J., D. Dahlbeck, and N. Keen. (1984) Cloned avirulence gene of *Pseudomonas syringae* pv. *glycinae* determines race-specific incompatibility on *Glycine max*. Proc. Natl. Acad. Sci. USA 81, 6024-6028.

Straus, D., and F.M. Ausubel. (1990) Genomic subtraction for cloning DNA corresponding to deletion mutations. Proc. Natl. Acad. Sci. USA **87,** 1889-1893.

Vogeli-Lange, R., A. Hansen-Gehri, T. Boller, and F. Meins, Jr. (1988) Induction of the defense-related glucanohydrolases, β-1,3-glucanase and chitinase, by tobacco mosaic virus infection of tobacco leaves. Plant Sci. 54, 171-176.

Voytas, D.V., and F.M. Ausubel. (1988) A copia-like retrotransposon family in *Arabidopsis thaliana*. Nature 336, 242-244.

Wade, A.A., M. de Tapia, L. Didierjean, and G. Burkard. (1989) Biological function of bean pathogenesis-related (PR 3 and PR 4) proteins. Plant Sci. 63, 121-130.

LOCAL AND SYSTEMIC GENE ACTIVATION FOLLOWING THE HYPERSENSITIVE RESPONSE OF PLANTS TO PATHOGENS

K. HAHLBROCK, N. ARABATZIS, M. BECKER-ANDRÉ, F. GARCIA-GARCIA,
P. GROSS, H.-J. JOOS, H.KELLER, E. KOMBRINK, D. SCHEEL,
M. SCHRÖDER
Max-Planck-Institut für Züchtungsforschung
Department of Biochemistry
Carl-von-Linné-Weg 10
D - 5000 Köln 30, FRG

Major parts of this contribution have recently been published in the form of original papers, review articles or symposium proceedings, and some new results are contained in manuscripts submitted for publication. Rather than reviewing our data again, we are therefore providing here an extended abstract with references for further reading.

Our present state of knowledge concerning the temporal and spatial patterns of the activation of plant defense responses to pathogens is based on work by several groups on a variety of experimental systems. For the purpose of this overview, we are concentrating mainly on one selected system. Where appropriate, however, results obtained with this and other systems are compared to draw some general conclusions.

As an economically important and scientifically feasible system, we are investigating the molecular mechanisms of defense-related gene activation in compatible and incompatible interactions of potato (Solanum tuberosum cv. Datura), carrying resistance gene R1, with appropriate races of the late-blight fungus Phytophthora infestans. The first microscopically visible responses of potato leaves to attempted fungal penetration are callose deposition and the accumulation of UV-fluorescing phenolics in the walls of affected plant cells. Rapid death of these cells (hypersensitive response) gives rise to the formation of small necrotic spots which are surrounded by a sharply localized area of transient activation of defense-related genes (Cuypers and Hahlbrock, 1988; Cuypers et al., 1988).

Among the strongly activated genes are those encoding two key enzymes of phenylpropanoid metabolism, phenylalanine ammonia-lyase (PAL) and 4-coumarate:CoA ligase (4CL) (Fritzemeier et al., 1987), and a member of one class of 'pathogenesis-related' proteins (PR1) (Taylor et al., 1990). In situ mRNA hybridization has revealed not only the rapid and local activation of these genes around fungal infection sites, but also complex, differential expression patterns in various organs and cell types during the development of uninfected plants (N. Arabatzis and F. Garcia-Garcia, unpublished results). The stimulation of phenylpropanoid metabolism is associated with the accumulation of soluble and wall-bound phenolic amides (Scheel et al., 1989). In addition, several

365

H. Hennecke and D. P. S. Verma (eds.),
Advances in Molecular Genetics of Plant-Microbe Interactions, Vol. 1, 365–366.
© 1991 Kluwer Academic Publishers. Printed in the Netherlands.

other genes, e. g. those encoding a different class of PR proteins, including 1,3-ß-glucanase and chitinase, are systemically activated throughout infected leaves (Hahlbrock et al., 1989). Some striking similarities
as well as a few important differences between compatible and incompatible plant-fungus interactions have prompted us to explore new experimental approaches to questions concerning the signal-transduction chain
from fungal elicitors to plant gene activation (Hahlbrock et al., 1990;
Scheel et al., 1990).

Cuypers, B. and Hahlbrock, K. (1988) 'Immunohistochemical studies
    of compatible and incompatible interactions of potato leaves with
    Phytophthora infestans and of the nonhost response to Phytophthora
    megasperma', Can. J. Bot. 66,700-705.
Cuypers, B., Schmelzer, E., and Hahlbrock, K. (1988) 'In situ loca-
    lization of rapidly accumulated phenylalanine ammonia-lyase mRNA
    around penetration sites of Phytophthora infestans in potato
    leaves', Mol. Plant Microbe Interactions 4, 157-160.
Fritzemeier, K.-H., Cretin, D., Kombrink, E., Rohwer, F., Taylor, J.,
    Scheel, D., and Hahlbrock, K. (1987) 'Transient induction of phe-
    nylalanine ammonia-lyase and 4-coumarate:CoA ligase mRNAs in
    potato leaves infected with virulent or avirulent races of
    Phytophthora infestans', Plant Physiol. 85, 34-41.
Hahlbrock, K., Arabatzis, N., Becker-André, M., Joos, H.-J., Kombrink,
    E., Schröder, M., Strittmatter, G., and Taylor, J. (1989) 'Local
    and systemic gene activation in fungus-infected potato leaves',
    in B. J. J. Lugtenberg (ed.), Signal Molecules in Plants and Plant-
    Microbe Interactions, NATO ASI Series H 36, pp. 242-249.
Hahlbrock, K., Groß, P., Colling, C., and Scheel, D. (1990) 'Molec-
    ular basis of plant defense responses to fungal infection', in
    R. G. Herrmann (ed.), Plant Molecular Biology, NATO ASI Series.
Scheel, D., Colling, C., Keller, H., Parker, J. E., Schulte, W., and
    Hahlbrock, K. (1989) 'Studies on elicitor recognition and signal
    transduction in host and non-host plant/fungus interactions', in
    B. J. J. Lugtenberg (ed.), Signal Molecules in Plants and Plant-
    Microbe Interactions, NATO ASI Series H 36, pp. 212-218.
Scheel, D., Colling, C., Hedrich, R., Kawalleck, P., Parker, J. E.,
    Sacks, W. R., Somssich, I. E., and Hahlbrock, K. (1990) 'Signals
    in plant defense gene activation', this book.
Taylor, J. L., Fritzemeier, K.-H., Häusser, I., Kombrink, E., Rohwer,
    F., Schröder, M., Strittmatter, G., and Hahlbrock, K. (1990)
    'Structural analysis and activation by fungal infection of a gene
    encoding a pathogenesis-related protein in potato', Mol. Plant-
    Microbe Interactions 3, 72-77.

PROPERTIES OF PLANT DEFENSE GENE PROMOTERS

C.J. LAMB and R.A. DIXON*
Plant Biology Laboratory
Salk Institute for Biological Studies
10010 North Torrey Pines Road
La Jolla, California 92138

*Plant Biology Division
Samuel Roberts Noble Foundation
P.O. Box 2180
Ardmore, Oklahoma 73402

ABSTRACT. Elaboration of active responses to mechanical damage, elicitors or microbial attack involves transcriptional activation of batteries of defense genes. Data from recent functional analysis of the promoters of the rice RCH10 gene, which encodes a basic endochitinase, and the bean CHS8 and CHS15 genes, which encode the enzyme chalcone synthase involved in isoflavonoid phytoalexin and flavonoid pigment synthesis are described. These studies indicate that defense gene promoters respond to highly conserved signal pathways operative in both monocots and dicots. Conservation of regulation follows biological function such that the bean CHS promoters are elicitor-inducible in tobacco protoplasts and transgenic tobacco plants, even though in tobacco CHS has no role in microbial defense and the endogenous CHS genes are not induced by elicitor or microbial attack. For many defense gene promoters, responses to environmental stress stimuli are superimposed upon complex developmental patterns of expression. The RCH10 and CHS15 promoters are highly compact, and in the latter case detailed functional analysis suggests that developmental expression in petals and root tips, as well as stress induction in leaves, utilize common or overlapping cis-elements. The mechanistic implications for flexible integration of environmental responses within plastic developmental programs are briefly considered.

367

H. Hennecke and D. P. S. Verma (eds.),
Advances in Molecular Genetics of Plant-Microbe Interactions, Vol. 1, 367–372.
© 1991 Kluwer Academic Publishers. Printed in the Netherlands.

1. INTRODUCTION

Plants elaborate a number of inducible defenses following mechanical damage or microbial attack (Lamb *et al.*, 1989). These responses include:

(a) Synthesis of phytoalexins;

(b) Reinforcement of cell walls by deposition of callose, lignin, and related wall-bound phenolics, and accumulation of hydroxyproline-rich glycoproteins;

(c) Production of proteinase inhibitors and lytic enzymes such as chitinase and glucanase, which attack microbial cell walls.

These defenses can also be induced by glycan and (glyco)protein elicitors from fungal cell walls, or metabolites such as arachidonic acid and glutathione (Dixon and Lamb, 1990). With the exception of callose production, induction of these defenses by elicitor, wounding, or infection involves transcriptional activation of the corresponding defense genes as part of a massive switch in the pattern of host gene expression (Lamb *et al.*, 1989; Dixon and Harrison, 1990). These observations focus attention on the mechanisms underlying activation of defense genes in the transduction of stress signals to initiate the induction of active defense responses. In the present paper we review the properties of plant defense gene promoters which have emerged from recent studies of gene fusions in transgenic plants.

2. DEFENSE GENE PROMOTERS RESPOND TO CONSERVED SIGNALS

Phytoalexins are chemically distinct in different plant taxa. For example, in solanaceous plants phytoalexins are isoprenoid derivatives, whereas in many legumes, phytoalexins are isoflavonoid derivatives synthesized from phenylalanine via the phenylpropanoid pathway (Dixon *et al.*, 1983). Chalcone synthase (CHS) catalyzes the first and key regulatory step in the branch pathway of phenylpropanoid biosynthesis specific for the production of flavonoid pigments and UV-protectants, which are ubiquitous in higher plants, as well as the isoflavonoid phytoalexins characteristic of legumes. In bean CHS is encoded by a family of several differentially regulated genes and we have characterized the promoters of two of these genes, designated CHS8 and CHS15.

Analysis of the expression of CHS15 promoter - chloramphenicol acetyltransferase (CAT) gene fusions in electroporated soybean protoplasts showed that the promoter was responsive to fungal elicitor and glutathione (Dron *et al.*, 1988). The kinetics and dose response for induction of the gene fusion in protoplasts closely resembles the response of endogenous CHS genes in cell suspensions. Interestingly, the CHS15-CAT gene fusion was also induced by fungal elicitor in electroporated tobacco protoplasts, even though CHS has no metabolic role in the phytoalexin defense response in this species, and endogenous tobacco CHS genes do not respond to biological stress signals. Thus the CHS15 promoter responds to an elicitor signal pathway conserved between bean and tobacco.

Further evidence of such signal conservation has come from studies of the expression of CHS8-glucuronidase (GUS) gene fusions in transgenic tobacco plants (Doerner *et al.*, 1990; Schmid *et al.*, 1990; Stermer *et al.*, 1990). Thus, wounding of leaves of transgenic plants containing a CHS8-GUS gene fusion causes the local accumulation of GUS activity after 24 to 48 hr. Addition of fungal elicitor to the wound site causes a more marked and rapid local induction of the gene fusion, and elicitor induction can be observed in the absence of concomitant wounding, by infiltration of elicitor into intact leaves or treatment of 10-day-old-seedlings (Doerner *et al.*, 1990). Moreover, the CHS8-GUS gene fusion is also induceed following inoculation of leaves with *Pseudomonas syringae* pv. *syringae*. GUS activity is observed not only in tissue immediately surrounding the hypersensitive lesion, but also in tissue at a distance of several centimeters from the lesion. These data imply that the bean CHS promoter can respond to both the initial signal pathway locally activated in tobacco following elicitor treatment or perception of microbial attack, and also signals for intercellular transmission to distant tissue.

Elicitor signal pathways are also conserved between monocots and dicots. Thus, transcripts of a rice gene, RCH10, encoding a basic endochitinase accumulate in cell suspension cultures following treatment with a fungal elicitor (Z. Qun and C.J. Lamb, unpublished data). Analysis of transgenic tobacco plants containing a RCH10-GUS gene fusion showed that wounding of leaf tissue caused a 5- to 10-fold increase in GUS activity over low basal levels in unwounded leaves. Simultaneous addition of fungal elicitor further induced the gene fusion by 2- to 3-fold and also resulted in a more rapid appearance of GUS activity.

3. DEVELOPMENTAL REGULATION OF DEFENSE GENE PROMOTERS

CHS mRNA and enzyme levels are highly regulated during development associated with the synthesis of flavonoid pigments. Analysis of the regulation of the CHS8-GUS and CHS15-GUS gene fusions in transgenic tobacco plants shows that these individual CHS promoters are not only responsive to diverse environmental stimuli such as light, wounding, elicitor and infection, but also contain the *cis*-elements required to establish tissue- and cell-type-specific accumulation of flavonoids during development.

For example, the CHS8 promoter is highly active in the root apical meristem and in petals, exclusively in those cells of the inner epidermis that accumulate anthocyanins (Schmid *et al.*, 1990). Although the CHS8-GUS gene fusion is only weakly expressed in other floral organs, mature leaves and stems, the early stages of seedling development are characterized by an apparent wound induction of the promoter in the endosperm and strong expression in the immature root. The latter becomes progressively localized to the apical meristem and perivascular tissue at the root-hypocotyl junction. The promoter becomes active during lateral root formation in both the new root and damaged tissue of the main root. The gene fusion is also expressed in greening cotyledons and primary leaves, but not in the shoot apical meristem. Light modulates expression in the cotyledons and the root-shoot junction, but has no effect on other aspects of the developmental program.

Moreover, while the developmental program is characterized by exquisite tissue- and cell-type-specific patterns of expression (eg. the inner epidermis of petals), the response of leaves to wounding or elicitor involves induction in mesophyll and vascular tissues as well as in epidermis (Schmid *et al.*, 1990). The CHS8 promoter exhibits a precise developmental program of expression, and yet is able to respond flexibly to diverse environmental stimuli. Thus the CHS8 promoter contains *cis*-elements responsive to externally applied signals that override or circumvent regulatory elements that establish the strict tissue- and cell-type specificity exhibited in other contexts. Activation in all tissues adjacent to a local wound may be crucial for effective protection against microbial attack after mechanical damage.

A second example of stress inducibility of a defense gene promoter superimposed on a complex developmental program of expression is provided by analysis of the rice RCH10 chitinase gene promoter (Q. Zhu and C.J. Lamb, unpublished). In addition to wound and elicitor induction of the RCH10-GUS gene fusion in leaves of transgenic tobacco, the gene fusion is expressed to high levels in roots and moderate levels in stems during normal development. The promoter is active in the root tip, in vascular tissue and in the epidermis. Strikingly, the promoter is also highly active in the stigma, style and ovary, as well as the anther, but is only weakly expressed in sepals, carpels and petals. It is not known whether expression in the male and female floral organs reflects a defense function for chitinase, perhaps because protection of these organs is crucial for succesful completion of the life-cycle, or whether chitinase has a different function in flowers, eg. generation of signal molecules.

4. DEFENSE GENE PROMOTERS CAN BE HIGHLY COMPACT

The complex pattern of functional activity exhibited by a number of the plant defense gene promoters examined to-date raises the question of how flexible responsiveness to stress stimuli is superimposed upon precise developmental programs of expression. We have attempted to address this question by dissection of the functional architecture of the CHS15 promoter in transgenic tobacco plants (J. Kooter, O. Faktor, R.A. Dixon and C.J. Lamb, unpublished data). Previous studies have shown that 5' flanking sequences to -130 were sufficient for elicitor induction in electroporated soybean protoplasts, whereas deletion to -72 abolished this response. A detailed series of 5' deletions shows that sequences to -84 are sufficient to confer expression in the pigmented regions of petals and in root tips, but that deletion to -72 abolishes these modes of expression. Interestingly, elicitor induction in leaf tissue can likewise be observed with the -84 deletion but not the -72 deletion. Analysis of internal deletions confirms the critical role of the proximal region of the promoter in CHS15 expression, and we can conclude that the CHS15 promoter is highly compact.

The emerging data from this functional analysis in transgenic plants suggests that apparently diverse modes of expression in petals and root tips during normal development, and in leaves in response to stress, may utilize common or overlapping *cis*-elements. An important corollary is that integration of the signal pathways for petal, root and stress induction may occur prior to the promoter, and that the different expression modes are dependent, at least in part, on a common *trans*-factor(s).

The rice RCH10 chitinase promoter has not yet been analyzed in such detail, but recent data indicate that this promoter is also very compact. Thus deletion to -160 does not affect either the kinetics or level of induction in wounded or elicitor-treated leaves, or the developmental expression in roots and floral organs.

The compact organization of these promoters may be related to the requirement for flexible integration of environmental responses within developmental programs, and further mechanistic analysis of the functional architecture of defense gene promoters and the *trans*-factors, which represent terminal elements of the underlying signal transduction pathways, should give insights into the molecular basis of the striking developmental plasticity exhibited by higher plants.

5. ACKNOWLEDGEMENTS

The research described in this paper was supported by grants to C.J.L. from the Samuel Roberts Noble Foundation, the Rockefeller Foundation, the Hermann Frasch Foundation and the U.S. Department of Agriculture, CRGO Grant # 89-37263-4690.

6. REFERENCES

Dixon, R.A. and Harrison, M.J. (1990) Activation, structure and organization of genes involved in microbial defense in plants. *Adv. Genet.* **28**, 165-234.

Dixon, R.A. and Lamb, C.J. (1990) Molecular communication in interactions between plants and microbial pathogens. *Annu. Rev. Plant Physiol. Plant Mol. Biol.* **40**, 347-364.

Dixon, R.A., Dey, P.M. and Lamb, C.J. (1983) Phytoalexins: Enzymology and molecular biology. *Adv. Enzymol. Relat. Areas Mol. Biol.* **55**, 1-135.

Doerner, P.W., Stermer, B., Schmid, J., Dixon, R.A. and Lamb, C.J. (1990) Plant defense gene promoter-reporter gene fusions in transgenic plants: Tools for identification of novel inducers. *Biotechnology* **8**, 845-848.

Dron, M., Clouse, S.D., Lawton, M.A., Dixon, R.A. and Lamb, C.J. (1988) Glutathione and fungal elicitor regulation of a plant defense gene promoter in electroporated protoplasts. *Proc. Natl. Acad. Sci. USA* **85**, 6738-6742.

Lamb, C.J., Lawton, M.A., Dron, M. and Dixon, R.A. (1989) Signals and transduction mechanisms for activation of plant defenses against microbial attack. *Cell* **56**, 215-224.

Schmid, J., Doerner, P.W., Clouse, S.D., Dixon, R.A. and Lamb, C.J. (1990) Developmental and environmental regulation of a bean chalcone synthase promoter in transgenic tobacco. *Plant Cell* **2**, 619-631.

Stermer, B.A., Schmid, J., Lamb, C.J. and Dixon, R.A. (1990) Infection and stress activation of bean chalcone synthase promoters in transgenic tobacco. *Mol. Plant-Microbe Int.,* in press.

SIGNALS IN PLANT DEFENSE GENE ACTIVATION

D. SCHEEL, C. COLLING, R. HEDRICH[1], P. KAWALLECK,
J. E. PARKER[2], W. R. SACKS, I. E. SOMSSICH, and K. HAHLBROCK
Max-Planck-Institut fuer Zuechtungsforschung
Department of Biochemistry
Carl-von-Linné-Weg 10
5000 Koeln 30, Federal Republic of Germany

[1]*University of Goettingen*
Department of Plant Physiology
Untere Karspuele 2
3400 Goettingen, Federal Republic of Germany

[2]*Sainsbury Laboratory*
John Innes Institute
Colney Lane
Norwich NR4 7UH, England

ABSTRACT. Plants respond to pathogen attack by rapid activation of defense genes. In selected experimental systems, the transcription of these genes is also induced when cultured plant cells or protoplasts are treated with elicitors. We have purified an extracellular glycoprotein of 42 kDa from the culture filtrate of the soybean pathogen *Phytophthora megasperma* f. sp. *glycinea* (Pmg), which is a potent elicitor of phytoalexin accumulation and defense gene activation in parsley (*Petroselinum crispum*) cells and protoplasts. Reversible binding of the ^{125}I-labelled glycoprotein to microsomes and protoplasts indicates the presence of specific binding sites on the parsley plasma membrane. The transduction of the elicitor signal from the cell surface to the nucleus was found to involve a rapid and transient uptake of Ca^{2+}, alkalization of the culture medium and effluxes of K^+ and Cl^-. These ion fluxes probably result from opening of elicitor-responsive ion channels, as indicated by preliminary results from patch-clamp analysis of parsley protoplasts in the presence of pure glycoprotein elicitor. Omission of Ca^{2+} from the culture medium substantially reduced the elicitor-induced transcription rates of plant defense genes. Treatment of cultured parsley cells or protoplasts with amphotericin B resulted in ion fluxes and defense gene activation similar to those elicited by the crude Pmg elicitor. The callose elicitors chitosan and digitonin, however, induced ion fluxes of different intensities, activated only some defense genes, and did not stimulate the synthesis of phytoalexins at concentrations optimal for elicitation of callose formation.

373

H. Hennecke and D. P. S. Verma (eds.),
Advances in Molecular Genetics of Plant-Microbe Interactions, Vol. 1, 373–380.
© 1991 *Kluwer Academic Publishers. Printed in the Netherlands.*

1. Introduction

Plants respond to attack by potential pathogens with the rapid activation of a variety of defense responses which are similar in non-host and host-incompatible interactions (Table 1). The induction of these reactions appears to follow a predetermined program with respect to timing and location, and many are initiated by activation of specific genes. The accumulation of callose, on the other hand, appears to be the result of allosteric enzyme activation [1], while the molecular basis of others, such as the hypersensitive response, is not well understood.

TABLE 1. Typical pathogen defense responses of plants.

Defense response	Mechanism of activation
Hypersensitive response	unknown
Phytoalexin accumulation	specific gene activation
Cell wall modifications	
impregnation with phenolics	specific gene activation
incorporation of proteins	specific gene activation
callose apposition	allosteric activation of $1,3-\beta$-glucan synthase
Accumulation of lytic enzymes	specific gene activation
Accumulation of pathogenesis-related proteins	specific gene activation

The initiation of defense responses requires the perception of appropriate signals by the plant cell and transmission to each specific site of activation. Since most components of a successful resistance reaction are restricted to localized areas of the infected plant tissue, it is difficult to study the underlying molecular mechanisms in the intact plant. For a number of plants, systems of reduced complexity have been developed consisting of cultured plant cells and elicitors derived from the cell wall or from the culture filtrate of a pathogen. We have used this type of system to study the non-host resistance of parsley to *Phytophthora megasperma* f. sp. *glycinea* (Pmg), a fungal pathogen of soybean.

2. Elicitor Recognition

Treatment of cultured parsley cells or protoplasts with crude elicitor derived from the *Pmg* cell wall induces the synthesis and secretion of furanocoumarin phytoalexins [2] and activates a characteristic set of defense genes [3]. An elicitor preparation from the fungal culture filtrate was found to have identical effects. The induced reactions, which appear to be elicited almost synchronously in all cells, were similar if not identical to the localized responses of leaves to inoculation with zoospores from *Pmg* [4, 5]. A 42 kDa glycoprotein elicitor was purified from culture filtrate by screening for phytoalexin accumulation in parsley protoplasts [6]. Elicitor activity was shown to reside in the protein moiety (Table 2). The pure glycoprotein activated the same set of genes as were activated by crude elicitor [6].

TABLE 2. Phytoalexin synthesis in parsley protoplasts upon treatment with various elicitor preparations from *Pmg*.

Elicitor preparation	Phytoalexins [% of maximum]
Cell wall	100
Culture filtrate	84
42 kDa glycoprotein	88
pronase-treated	0
trypsin-treated	0
deglycosylated	96

The glycoprotein elicitor was labelled radioactively without detectable loss of biological activity, by coupling ^{125}I-Bolton-Hunter reagent to free amino groups. Direct iodination of tyrosine residues, however, significantly reduced elicitor activity. Preliminary results from binding studies with intact parsley protoplasts and microsomal preparations have indicated the presence of specific binding sites for the 42 kDa glycoprotein on the plasma membrane. In representative examples of such experiments, 27 and 67 % of the total binding to protoplasts and microsomes, respectively, were found to be specific.

In contrast to parsley cells, soybean cells were found to be responsive, not to the 42 kDa glycoprotein, but rather to a heptaglucan from the *Pmg* cell wall, which is itself inactive in parsley [7]. Consistent with these results, the 42 kDa glycoprotein did not compete for binding of the heptaglucan to its binding sites on the soybean plasma membrane [Eric Cosio and Jürgen Ebel, personal communication]. The two plant species, therefore, not only recognize different elicitors from the cell wall of *Pmg*, but are, in addition, unable to respond to the compound recognized by the other plant.

3. Signal Transduction

Binding of elicitor to its target site has been proposed to initiate a signal transduction chain, leading to activation of the various defense responses. The earliest events that were observed upon elicitor treatment of cultured parsley cells were ion fluxes through the plasma membrane, all of which began simultaneously within two to four minutes of addition of elicitor and persisted for three to four hours. Ion-selective electrodes detected increases in extracellular levels of K$^+$ and

Cl⁻ (8.8 and 1.4 μmol·30 min^{-1}·g^{-1} fresh weight, respectively) and decreases in the concentration of H$^+$ in the culture medium (1.5 μmol·30 min^{-1}·g^{-1} fresh weight). In addition, a Ca^{2+} influx was measured directly with ^{45}Ca^{2+} (0.016 μmol·30 min^{-1}·g^{-1} fresh weight).

CELL–ATTACHED   WHOLE CELL   CELL–FREE PATCH

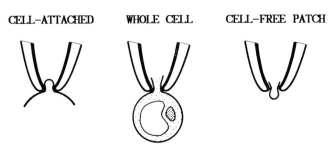

Figure 1. Schematic diagram of possible cell configurations during patch-clamping.

These ion fluxes may be caused by transient opening of specific ion channels, which can be detected by the patch-clamp technique [8]. Initial experiments demonstrated that freshly prepared parsley protoplasts are suitable for the analysis of ion channels by this method. Patch-clamping of protoplasts in the cell-attached configuration (Figure 1) indicated increased channel activity upon addition of the pure glycoprotein elicitor. Further characterization of the ion channel properties, such as mode of interaction with the elicitor, ion selectivity, voltage dependence, and channel number, will be the topic of future experiments.

TABLE 4. Effects of different effector molecules on ion fluxes in cultured parsley cells [% of flux caused by *Pmg* elicitor].

Effector molecule	Ca^{2+}influx	H$^+$influx	Cl$^-$efflux	K$^+$efflux
Pmg elicitor	100	100	100	100
Amphotericin B	48	44	44	32
Chitosan	855	130	747	274
Digitonin	40	33	31	17
None	0	0	0	0

These results, although demonstrating correlations between ion fluxes and defense responses, do not prove a causal relationship between the two processes. We therefore initiated a series of experiments, where we measured phytoalexin production after either inhibition of specific ion fluxes in the presence of *Pmg* elicitor or after stimulation of ion fluxes in the absence of elicitor. Omission of Ca^{2+} from the medium of elicitor-treated parsley cells reduced phytoalexin accumulation by 30 to 50 % and run-on transcription rates of elicitor-responsive genes by 30 to 80 % [9, 10]. Transcription of constitutively expressed genes remained unaffected. The Ca^{2+} channel blockers, $LaCl_3$ and flunarizine, inhibited both the elicitor-stimulated Ca^{2+} influx and phytoalexin accumulation. However, the Ca^{2+} ionophores A23187 and ionomycin, the K^+ ionophores valinomycin and nigericin, as well as combinations of A23187 and valinomycin, all failed to induce furanocoumarin synthesis in parsley cells and protoplasts in the absence of elicitor. These results indicate that influx of Ca^{2+} and/or efflux of K^+ are necessary but not sufficient to stimulate phytoalexin production in parsley.

TABLE 5. Effects of different effector molecules on activation of defense responses in cultured parsley cells.

Effector molecule	Phytoalexins[1]	Callose[2]	Activated defense genes[3]
Pmg elicitor	100	20	ELI:3,5,6,7,8,9,10,11 12,14,16,17,18,19 PAL,4CL,PR1,PR2
Amphotericin B	63	23	ELI:3,5,6,7,8,9,10,11 12,14,16,17,18,19 PAL,4CL,PR1,PR2
Chitosan	0	183	ELI:3,-,-,-,-,-,10,11 --,--,16,--,--,-- - , - ,PR1, -
Digitonin	0	152	ELI:-,-,-,-,-,-,--,11 --,14,16,17,18,19 - , - , - , -
None	0	15	none

[1]% of the amount of total coumarins produced by parsley cells 24 hours after addition of 50 µg·ml^{-1} Pmg elicitor.
[2]relative units derived from a calibration curve with pachyman [1], measured after 24 hours.
[3]as determined in run-on transcription experiments with nuclei isolated three hours after elicitor application [3].

The fungal cell wall constituent chitosan, the polyene antibiotic amphotericin B and the saponin digitonin are efficient elicitors of callose accumulation in plants [1]. In spite of their structural diversity, all three compounds stimulated qualitatively similar ion fluxes in cultured parsley cells as the Pmg elicitor (Table 4). The relative intensities of the individual fluxes, however, varied greatly. In order to analyze the ability of amphotericin B, chitosan and digitonin to elicit typical defense responses of parsley, we measured the accumulation of callose and phytoalexins as well as the transcriptional activation of a number of defense genes in cultured parsley cells [3] upon treatment with these compounds (Table 5). Amphotericin B was a strong elicitor of furanocoumarin synthesis and activated the same set of defense genes as the *Pmg* elicitor. Both compounds were equally poor inducers of callose accumulation. In contrast, chitosan and digitonin stimulated the accumulation of high levels of callose, but did not induce phytoalexin synthesis when added to parsley cells at concentrations optimal for elicitation of callose production. Within a narrow range of concentrations, chitosan caused furanocoumarin synthesis in cultured cells but not in protoplasts. Chitosan and digitonin, moreover, activated different subsets of defense genes.

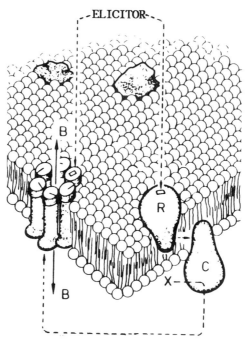

Figure 2. Three-dimensional diagram of the fluid mosaic model showing alternative mechanisms of elicitor-mediated opening of ion channels (B = ion channel, R = elicitor binding protein, C = signal coupling protein, X = intracellular signal).

These results indicate that the initiation of an appropriate pattern of ion fluxes in cultured parsley cells is sufficient to activate the complete set of defense genes. Changes in flux intensity of one or more ions, however, may cause qualitative changes in the gene activation pattern, thereby translating quantitative signal variations into distinct phenotypes.

Phosphorylation of proteins may also contribute to the signal transduction process, since a number of proteins were found to be specifically phosphorylated *in vivo* in a Ca^{2+}-dependent manner upon treatment of cultured parsley cells with *Pmg* elicitor [11]. These transient protein phosphorylations fit well into the time frame between induction of ion fluxes and activation of defense genes.

4. Conclusions

We have used plant cell cultures in conjunction with fungal elicitor as a model system to examine non-self recognition and early signalling events in defense gene activation. A glycoprotein elicitor purified from *Pmg* is active only in parsley, whereas soybean recognizes a glucan component of the *Pmg* cell wall. Therefore, no structural consensus seems to exist among the determinants of elicitor activity.

The parsley plasma membrane appears to possess a specific receptor-like target site for the glycoprotein elicitor. Binding of elicitor seems to cause transient opening of ion channels, resulting in increased uptake of Ca^{2+} and H^+ as well as efflux of K^+ and Cl^-. It is not known yet if the elicitor binds directly to an ion channel or to a distinct site, which then indirectly leads to opening of ion channels (Figure 2). Small changes in the intensities of specific ion fluxes dramatically alter the pattern of defense gene expression, which may provide an elegant and simple way to modulate responses to changing environmental stimuli.

5. Acknowledgements

Our work was supported by the Max-Planck-Gesellschaft, Fonds der Chemischen Industrie and an EMBO postdoctoral fellowship to Jane E. Parker during her stay in Köln.

6. References

[1] Kauss, H., Waldmann, T., Jeblick, W., Euler, G., Ranjewa, R., and Domard, A. (1989) "Ca^{2+} is an important but not the only signal in callose synthesis induced by chitosan, saponins and polyene antibiotics", in B. J. J. Lugtenberg (ed.), Signal Molecules in Plants and Plant-Microbe Interactions, Springer-Verlag, Berlin, pp. 107-116.

[2] Dangl, J. L., Hauffe, K. D., Lipphardt, S., Hahlbrock, K., and Scheel, D. (1987) "Parsley protoplasts retain differential responsiveness to u.v. light and fungal elicitor", EMBO J. 6, 2551-2556.

[3] Somssich, I. E., Bollmann, J., Hahlbrock, K., Kombrink, E., and Schulz, W. (1989) "Differential early activation of defense-related genes in elicitor-treated parsley cells", Plant Mol. Biol. 12, 227-234.

[4] Scheel, D., Hauffe, K. D., Jahnen, W., and Hahlbrock, K. (1986) "Stimulation of phytoalexin formation in fungus-infected plants and elicitor-treated cell cultures of parsley", in B. Lugtenberg (ed.), Recognition in Microbe-Plant Symbiotic and Pathogenic In-

380

teractions, Springer-Verlag, Berlin, pp. 325-331.

[5] Schmelzer, E., Krüger-Lebus, S., and Hahlbrock, K. (1989) "Temporal and spatial patterns of gene expression around sites of attempted fungal infection in parsley leaves", Plant Cell 1, 993-1001.

[6] Parker, J. E., Schulte, W., Hahlbrock, K., and Scheel, D. (1990) "An extracellular glycoprotein from *Phytophthora megasperma* f. sp. *glycinea* elicits phytoalexin synthesis in cultured parsley cells and protoplasts", Mol. Plant-Microbe Interact., in press.

[7] Parker, J. E., Hahlbrock, K., and Scheel, D. (1988) "Different cell-wall components from *Phytophthora megasperma* f. sp. *glycinea* elicit phytoalexin production in soybean and parsley", Planta 176, 75-82.

[8] Hedrich, R. and Schroeder, J. I. (1989) "The physiology of ion channels and electrogenic pumps in higher plants", Annu. Rev. Plant Physiol. Plant Mol. Biol. 40, 539-569.

[9] Scheel, D., Colling, C., Keller, H., Parker, J., Schulte, W., and Hahlbrock, K. (1989) "Studies on elicitor recognition and signal transduction in host and non-host plant/fungus pathogenic interactions", in B. J. J. Lugtenberg (ed.), Signal Molecules in Plants and Plant-Microbe Interactions, Springer-Verlag, Berlin, pp. 211-218.

[10] Ebel, J. and Scheel, D. (1990) "Elicitor recognition and signal transduction", in T. Boller and F. Meins (eds.), Genes Involved in Plant Defense, Plant Gene Research, Vol. 8, Springer-Verlag, Wien, in press.

[11] Dietrich, A., Mayer, J. E., and Hahlbrock, K. (1990) "Fungal elicitor triggers rapid, transient, and specific protein phosphorylation in parsley cell suspension cultures", J. Biol. Chem. 265, 6360-6368.

A SEARCH FOR RESISTANCE GENE-SPECIFIC RECEPTOR PROTEINS IN LETTUCE PLASMA MEMBRANE

Alison Woods-Tor, Peter Dodds & John Mansfield
Department of Biochemistry and Biological Sciences,
Wye College, University of London,
Ashford,
Kent. TN25 5AH
United Kingdom

ABSTRACT

We describe a novel approach towards the detection of receptor proteins in the plasma membrane of lettuce which are responsible for the recognition of avirulent isolates of the downy mildew fungus Bremia lactucae. Heat-shock (HS) causes the gene-specific suppression (for up to 3 days) of the hypersensitive reaction controlled by the resistance gene Dm5/8. This effect is consistent with the loss of a Dm5/8-specific receptor from the plasma membrane (PM) and its slow recovery. In order to detect PM proteins conforming to the criteria defined by HS experiments, PM-enriched fractions were isolated by 2-phase partitioning from heat-shocked and unheated cotyledons and their polypeptide composition successfully analysed by 2D IEF/SDS-PAGE. Two polypeptides of 52.5 kD (putative receptors) were significantly reduced in concentration after HS in cultivars containing Dm5/8 but were unaffected by HS in cultivars which lack this gene. Future analysis of the putative receptors is discussed.

1. Introduction

Race-specific resistance of lettuce to the obligate biotroph Bremia lactucae Regel involves a gene-for-gene relationship between host and parasite. In Lactuca sativa L. 16 different resistance (Dm) factors are now recognised of which 13 have been characterized as single dominant alleles and have been mapped to four linkage groups [3]. Resistance controlled by Dm genes and their complementary avirulence (A) genes in the fungus is associated with the hypersensitive reaction (HR) of penetrated cells [7,10] but neither the fungal elicitors of the HR [1,2] nor the host's receptors have been identified. We have found that heat-shock causes gene-specific suppression of resistance controlled by Dm5/8 and have used this phenomenon as a tool to identify potential receptor proteins in the lettuce plasma membrane.

381

H. Hennecke and D. P. S. Verma (eds.),
Advances in Molecular Genetics of Plant-Microbe Interactions, Vol. 1, 381–386.
© 1991 *Kluwer Academic Publishers. Printed in the Netherlands.*

2. Results and Discussion

2.1 TIMING OF IRREVERSIBLE MEMBRANE DAMAGE DURING THE HR

Irreversible membrane damage (IMD) to the plasma membrane of penetrated cells occurs during the early stages of the HR and may closely follow critical recognition events between host and pathogen. IMD can be detected by the failure of cells to plasmolyse in a hypertonic solution. Microscopical observations were made of different Dm/A gene interactions for the first 12h after inoculation; the type and size of fungal infection structures (Fig. 1) and the occurrence of IMD were noted. It was found that the stage of infection at which IMD occurred depended on the Dm/A genes being expressed and in some cases (e.g. Dm5/8:A5/8 interaction) IMD occurred within 1.5h of penetration [10] preceding the accumulation of autofluorescent phenolics by several hours.

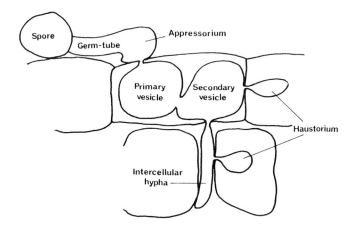

Fig. 1.   Infection structures of B. lactucae in L. sativa

2.2 EFFECTS OF HEAT-SHOCK ON IMD

Lettuce cotyledons were subjected to various treatments to suppress protein synthesis in order to observe effects on IMD after inoculation with B. lactucae [11]. Heat-shock (given by incubating cotyledons for 45s on a Petri dish floated on a water bath at 55°C) suppressed the HR in cvs. Valmaine (Dm5/8), Dandie (Dm3), Blondine (Dm1) or Diana (Dm1+Dm3+Dm7+Dm5/8) inoculated 0.5h after treatment. If inoculation was delayed for 17h after heat-shock, IMD determined by Dm1 or Dm3 was not affected, but the plant's response continued to be suppressed in interactions where only Dm5/8 was expressed.

The gene-specific effect of heat-shock was clearly demonstrated in the responses of cv. Diana (Dm1+Dm3+Dm7+Dm5/8) to isolate V0/11 (A at all loci). In unheated cotyledons IMD characteristic of the action of Dm5/8 was observed, i.e. during the development of the primary vesicle. Heat-shock 0.5h before inoculation suppressed IMD but if cotyledons were left 17h before inoculation IMD occurred at the stage of fungal development characteristic of Dm1/A1 or Dm3/A3 interactions. i.e. during growth of the secondary vesicle (see Fig. 1).

The prolonged effect of heat-shock on the expression of Dm5/8 was further demonstrated by the ability of isolates with A5/8 to sporulate on treated cotyledons. Resistance was not recovered until 3 days after heat-shock [11].

2.3 RECEPTOR PROTEINS AND THE HEAT-SHOCK RESPONSE

The transient suppression of IMD in cvs. Dandie (Dm3) and Blondine (Dm1) suggests that heat-shock temporarily affects protein synthesis or other metabolic processes required for HR. It is well known that heat-shock inhibits translation of mRNAs other than those newly synthesized to code for heat-shock proteins [9]. However, the heat-shock response diminishes after several hours as, for example, in the suppression of resistance of peas to Fusarium solani f.sp. phaseoli following heat treatment [4]. The more permanent suppression of IMD determined by Dm5/8 indicates alteration of a structural component with a slow turnover rate rather than an inducible protein whose rapid synthesis would be restored within several hours of heat-shock. The half-lives of plasma membrane proteins have been found to vary from 2 to 100h [8]. Therefore, the gene-specific effect is consistent with the loss from the plasma membrane and the slow recovery of a receptor to gene-specific determinants of avirulence in B. lactucae.

The heat-shock experiments have defined the following criteria to be fulfilled by the putative receptor. 1) By 17h after heat-shock the protein should be absent from or considerably reduced in all cultivars containing Dm5/8. 2) The protein should re-appear in membranes of these cultivars as they recover the ability to express resistance following heat-shock. 3) The protein should be absent from or unaffected by heat-shock in cultivars which do not possess Dm5/8.

Therefore, plasma membranes from heated and unheated cotyledons were isolated and analysed for proteins conforming to these criteria.

2.4 HEAT TREATMENT OF COTYLEDONS AND ISOLATION OF PLASMA MEMBRANES

In order to heat-shock the large number of cotyledons needed for membrane isolation a change in method was required. It was found that the suppression of resistance varied according to the temperature regime and two treatments were selected for subsequent experiments. Cotyledons were immersed in a water bath at either 55°C for 10s or 50°C for 30s. Effects on the expression of Dm5/8 were as follows.

1) Heating at 55°C for 10s suppressed resistance until 3 days after treatment (as in previous experiments) but senescence of some cotyledons was observed after prolonged incubation.

2) Heating at 50°C for 30s suppressed resistance until at most 2 days after treatment but no senescence of cotyledons was observed.

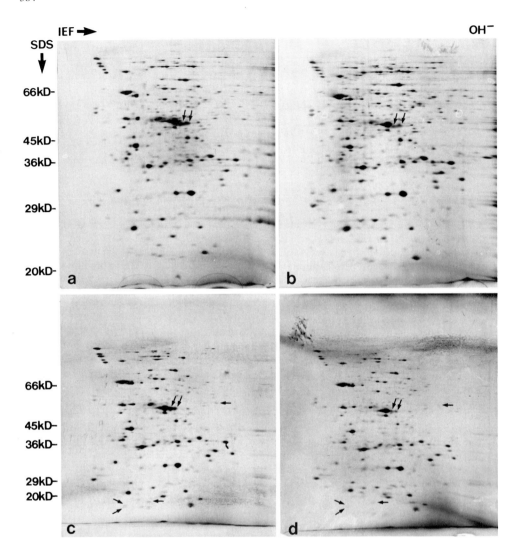

Fig. 2. Two-dimensional IEF/SDS-PAGE of plasma membrane proteins isolated 17h after treatment from a) and c) unheated cotyledons of cv. Valmaine (Dm5/8), b) cotyledons heated at 50°C for 30s and d) cotyledons heated at 55°C for 10s. Polypeptides considerably reduced after heat-shock (arrowed) include two of 52.5 kD which are affected by both heat treatments.

Membrane fractions enriched with plasmalemma were isolated from cotyledons by the method of two-phase partitioning adapted from Larsson [5] in which plasma membrane is recovered from the upper phase of PEG/Dextran partitions. Both the upper and lower phases were assayed for membrane marker enzyme activity using K^+-stimulated ATPase and vanadate-sensitive ATPase for plasmalemma, IDPase for Golgi, Cytochrome C oxidase for mitochondria and antimycin A-resistant NADPH Cytochrome C reductase for ER. The only major contaminant of the plasmalemma-rich upper phase we obtained appeared to be ER, but even this may not have been in large amounts since antimycin A-resistant NAD(P)H Cytochrome C reductase activity has been found in wheat plasma membrane [6]. Our plasmalemma preparations were therefore considered to be sufficiently pure for analysis.

2.5 TWO-DIMENSIONAL ELECTROPHORESIS OF PLASMA MEMBRANE PROTEINS

Proteins were separated by denaturing IEF (pH3-10) in the first dimension and SDS-PAGE (T=10%, C=2.7%) in the second dimension.

Comparison of 2D gels of plasma membrane proteins from heated (50°C, 30s) and unheated cotyledons of cv. Valmaine (Dm5/8) 17h after treatment revealed encouragingly few differences. Two polypeptides of molecular weight 52.5 kD were considerably reduced in concentration but this was not clearly observed in replicate experiments (Fig. 2a,b).

The more severe heat treatment of 55°C for 10s gave rise to more differences, several of which were consistently observed (Fig. 2c,d). Again, the two polypeptides of molecular weight 52.5 kD were particularly reduced in staining intensity. Could these be receptor proteins? Further investigations have supported this possibility. 1) Analysis of plasma membrane proteins from cv. Diana which also possesses Dm5/8 has shown a similar reduction in concentration of these polypeptides following heat-shock. 2) Although these polypeptides were present in cvs. Mildura (Dm1+Dm3) and Blondine (Dm1) they appeared unaffected by heat-shock.

3. Conclusions and Future Research

The gene-specific effects of heat-shock have been shown to provide a useful tool in the search for a receptor protein in the plasma membrane of lettuce responsible for the recognition of avirulent isolates of B. lactucae. The most successful method for analysis of plasma membrane proteins was found to be 2-dimensional IEF/SDS-PAGE. Our analyses have identified putative receptors.

Confirmation that the proteins fulfil the criteria defined for the receptor by heat-shock experiments will require analysis of protein turnover by radiolabelling and immunological identification of the putative receptors in cultivars carrying Dm5/8. Successful completion of these experiments should allow the gene for the Dm5/8 receptor protein to be cloned. The receptor may prove to be the product of the Dm5/8 gene for resistance to lettuce downy mildew disease.

We wish to acknowledge support from the Agricultural and Food Research Council.

4. References

1. Crucefix, D.N., Mansfield, J.W. and Wade, M. (1984) 'Evidence that determinants of race specificity in lettuce downy mildew disease are highly localized', Physiological Plant Pathology 24, 93-106.
2. Crucefix, D.N., Rowell, P.M., Street, P.F.S. and Mansfield, J.W. (1987) 'A search for elicitors of the hypersensitive reaction in lettuce downy mildew disease', Physiological and Molecular Plant Pathology 30, 39-54.
3. Farrara, B.F., Ilott, T.W. and Michelmore, R.W. (1987) 'Genetic analysis of factors for resistance to downy mildew (Bremia lactucae) in species of lettuce (Lactuca sativa and L. serriola)', Plant Pathology 36, 499-514.
4. Hadwiger, L.A. and Wagoner, W. (1983) 'Effect of heatshock on the mRNA-directed disease resistance response of peas' Plant Physiology 72, 553-556.
5. Larsson, C. (1985) 'Plasma membranes', in H.F. Linskens and J.F. Jackson (eds.), Modern methods of plant analysis, Springer-Verlag, Berlin, pp. 85-104.
6. Lundborg, T., Widell, S. and Larsson, C. (1981) 'Distribution of ATPases in wheat root membranes separated by phase partition, Physiologia Plantarum 52, 89-95.
7. Maclean, D.J. and Tommerup, I.C. (1979) 'Histology and physiology of compatibility and incompatibility between lettuce and the downy mildew fungus Bremia lactucae Regel', Physiological Plant Pathology 14, 291-312.
8. Morré, D.J., Kartenbeck, J. and Franke, W.W. (1979) 'Membrane flow and interconversions among endomembranes', Biochimica et Biophysica Acta (MR) 559, 71-152.
9. Nagao, R.T., Kimpel, J.A., Vierling, E. and Key, J.L. (1986) 'The heat-shock response: a comparative analysis', in B.J. Miflin (ed.), Oxford Surveys of Plant Molecular and Cell Biology Vol. 3, Oxford University Press, Oxford, pp. 384-438.
10. Woods, A.M., Fagg, J. and Mansfield, J.W. (1988) 'Fungal development and irreversible membrane damage in cells of Lactuca sativa undergoing the hypersensitive reaction to the downy mildew fungus Bremia lactucae', Physiological and Molecular Plant Pathology 32, 483-497.
11. Woods, A.M., Fagg, J. and Mansfield, J.W. (1989) 'Effects of heatshock and inhibitors of protein synthesis on irreversible membrane damage occurring during the hypersensitive reaction of Lactuca sativa L. to Bremia lactucae Regel', Physiological and Molecular Plant Pathology 34, 531-544.

PATHOGENESIS-RELATED PROTEINS EXHIBIT BOTH PATHOGEN-INDUCED AND DEVELOPMENTAL REGULATION

R. Fluhr, G. Sessa, A. Sharon, N. Ori and T. Lotan
Department of Plant Genetics
P. O. Box 26
Weizmann Institute of Science
Rehovot, Israel 76100

ABSTRACT. Antisera to acidic isoforms of pathogenesis-related (PR) proteins were used to measure the activity of these genes in tobacco plants. A novel endo-(1-4)-β-xylanase purified from fungal filtrates of *Trichoderma viride* was found to be a strong activator of PR proteins synthesis in tobacco leaves. The induction was not inhibited by blockers of either ethylene biosynthesis or ethylene action highlighting a novel ethylene independent pathway for PR proteins. Concomitant with the induction of PR proteins phytoalexins are induced. The regulation of the phytoalexin capsidiol showed identical ethylene dependent and independent pathways described for PR proteins.

In addition to the pathogen-induced regulation observed in leaves, PR proteins accumulate in developing flower organs in a unique spatial and developmental pattern. Antiserum raised against the leaf pathogen induced (1-3)-β-glucanases cross reacts with a stylar specific protein of apparent molecular weight of 41kD (sp41). Sp41 polypeptide was purified and found to have (1-3)-β-glucanase activity. cDNA clones corresponding to sp41 mRNA were isolated and sequenced. The cDNA clones show 52%-82% homology with the different acidic secreted (1-3)-β-glucanases from leaves, and represent distinct genes. The differential appearance of PR proteins during flower development, their *in situ* localization and post-translational processing point to alternate biological functions.

Materials and Methods

Nicotiana tabacum cv. 'Samsun NN' plants were grown in the greenhouse, in 18 h day, 26°C and 6 h night, 22°C diurnal cycles. All experiments were conducted in the greenhouse on young potted plants with three to four leaves at least 10 cm long. Immunoblots and ethylene measurements were as described (Lotan and Fluhr, 1990). Capsidiol was isolated according to

H. Hennecke and D. P. S. Verma (eds.),
Advances in Molecular Genetics of Plant-Microbe Interactions, Vol. 1, 387–394.
© 1991 *Kluwer Academic Publishers. Printed in the Netherlands.*

388

Stoessel at al. (1972), and quantitative analysis was carried out with GLC. (1-3)-β-glucanase was isolated according to Ori et al. (1990). Digests of glucans was carried out in 100 µl buffer which contained 5 mg laminarin, 100 ng purified enzyme, 20 CaCl₂ and 2 mM sodium acetate ph 4.8. Incubations were carried out at 37°C.

Introduction

Pathogenesis-related (PR) proteins are plant proteins that accumulate during pathogen attack (Bol et al. 1990). In tobacco 5 groups of acidic PR proteins were isolated and were classified according to serological relations. The groups are: 1. PR-1a, 1b, 1c; 2. PR-2,N,O; 3. PR-P,Q; 4. PR-r,s and PR-R,S (Bol et al.1990). The enzymatic functions of two groups of PR proteins have been described. One group, consisting of acidic and basic PR proteins, has (1-3)-β-glucanase activity the acidic polypeptides of this group are PR-2, N and O. A second group consisting of two basic and two acidic proteins has endochitinase activity; the acidic polypeptides of this group are PR-P and Q . Both hydrolases can digest the main cell wall components of some bacterial and fungal pathogens (Mauch et al. 1988).

Ethylene is thought to be the natural mediator of PR proteins accumulation as has been found for other defense functions. Physiological stresses that are associated with ethylene, such as flowering and aberrant hormonal levels, induce synthesis of PR proteins. Exogenous application of ethephon has the same effect. A plethora of abiotic elicitors can induce PR proteins synthesis; among them are polyacrylic acid, salicylic acid and some amino acid derivatives. Since many of these chemicals (except salicylic acid) trigger ethylene biosynthesis, chemical elicitation of PR proteins accumulation may be via ethylene. Recently we have shown that an endoxylanase purified from culture filtrates of *Trichoderma viride* induces PR proteins synthesis when applied directly to tobacco leaves. In contrast to other treatments, induction by xylanase was found to be independent of ethylene (Lotan and Fluhr, 1990).

The PR proteins also accumulate in leaves of flowering plants and in roots and flowers of healthy plants. We have recently shown that proteins that are cross-reactive with antibodies raised against three of the acidic PR protein classes are detected in flower parts of *N. tabacum* (Lotan et al. 1989). Their appearance was not related to pathogenesis. In infected leaves the PR proteins appear in concert, whereas in flowers their appearance is strictly compartmentalized. Protein cross-reactive with PR-1 antisera is present in sepal tissue. Chitinase antisera cross react with a protein in pedicels, sepals, anthers and ovaries. A glycosylated protein serologically related to the leaf (1-3)-β-glucanases but of higher apparent molecular weight is located in the style and stigma (Lotan et al. 1989). We have recently characterized this protein and find it to be a (1-3)-β-glucanase that is closely related, but distinct, from the pathogen induced (1-3)-β-glucanase (Ori et al. 1990).

Results

The accumulation of PR proteins was induced in intact leaves using both biotic and abiotic elicitors. The accumulation of PR proteins ethylene and capsidiol were monitored after 24 h with or without the ethylene biosynthesis inhibitor AVG (Table 1).

Table 1. Relative Ethylene, Chitinase and Capsidiol accumulation by intact leaves in the presence and absence of AVG.
Accumulation is expressed as percent of maximal response. Measurements of ethylene and chitinase were adapted from Lotan and Fluhr, (1990).
Maximal levels of ethylene were 69 µl/22 h/leaf and of capsidiol were 13.6 µg/g fresh weight. Chitinase was measured by scanning autoradiographs of immunoblots treated with antisera to PR-P,Q.
b.d.,below detection; n.d., not done

Treatment (AVG)	Ethylene		Chitinase		Capsidiol	
	-	+	-	+	-	+
Control	4	b.d.	b.d.	b.d.	n.d.	n.d.
Buffer phosphate	7	b.d.	b.d.	b.d.	b.d.	b.d.
Xylanase	26	7	40	40	100	72
P. syringae	33	1	100	100	n.d.	n.d.
α-Aminobutyric acid	100	8	60	10	75	n.d.

Table 1 shows that all elicitors of chitinase accumulation that were tested induced significant amounts of capsidiol. The amount of capsidiol induced are well above that generally used in bioassay examinations. Ethylene production was induced as well and in all cases was reduced to control levels by the application of AVG. However, in the case of xylanase treatment both chitinase and capsidiol accumulated in the presence of AVG. In the case of α-aminobutyric acid application of AVG severely reduced chitinase accumulation and completely abrogated capsidiol accumulation. These experiments suggest the existence of an ethylene independent pathway used by xylanase and also emphasize commonality in the regulation of two independent plant defenses.

We have previously reported that PR-proteins accumulate in a developmentally regulated and spatially unique manner in normal flowers, but that flowers still retain the ability to react to elicitors of pathogenesis (Lotan et al. 1989). A 41 kD glycoprotein was detected in styles that immuno-crossreacted with antisera to the leaf (1-3)-β-glucanases, PR-2,N,O. We purified this protein to homogeneity as shown in Fig. 1. The major steps in

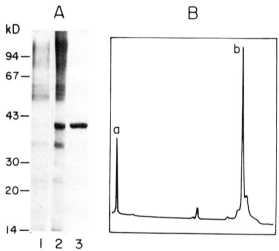

Fig. 1. Purification of sp41. A. Commasie stained polyacrylamide SDS gel of fractions from the different purification steps. The molecular weight is indicated in the left. Lane 1, the original stylar extract; lane 2, enzymatically active fractions from mono Q column; lane 3, enzymatically active fractions from Con A column. B. OD_{280} profile of the material from lane 3 in A, fractionated on a zorbax protein plus column.

purification took advantage of the polypeptides charge by fractionation on a mono-Q column (lane 2, Fig. 1A) and its glycosylation by fractionation on a ConA column (lane 3, Fig. Fig. 1A). The fractions in lane 3 were examined for purity in a reverse-phase column and judged to be over 95% pure (Fig. 1B).

To confirm that the immunoreactive polypeptide was indeed a (1-3)-β-glucanase, we tested its activity on the (1-3)-β-glucan substrate, laminarin as shown in Fig. 2. Limit digests of this polymer yielded low molecular weight oligosaccharides. The major size unit that accumulated was laminaritriose and the smallest size was laminaribiose. No free glucose was detected. When digest products were examined kinetically laminaritriose appeared immediately as the major product (data not shown). Taken together the data show that the 41 kD polypeptide is a endoglucanase (E.C. 3.2.1.39) and that the enzyme has a preference for cleavage of 3 residue-size portions from the end of the sugar polymer.

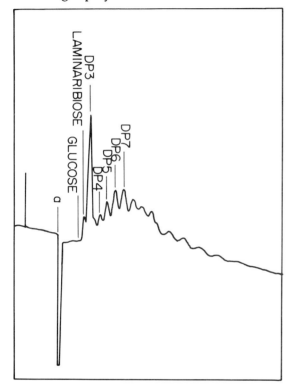

Fig. 2. Refractometric analysis of laminarin digests fractionated on amino phase HPLC. 0.5 mg Laminarin,was digested with (1-3)-β-glucanase as described in the Methods. The column was calibrated with oligosaccharide standards of glucose, sucrose, maltotriose, maltotetraose, pentaose, hexaose and heptaose, which are DP1-DP7 respectively, and developed in 65% acetonitrile.

Sequence comparison between sp41 and other (1-3)-β-glucanases are provided in Figure 3. The sp41 sequence shows a 47% identity with the monocot (1-3, 1-4)-β-glucanase, 47% with the monocot (1-3)-β-glucanase and 49% identity with the dicot basic, vacuolar, (1-3)-β-glucanase. Comparison with the dicot acidic extracellular glucanases shows two levels of homology: 40% to PR(35) and a higher homology of approximately 82% to PR(36) and PR (37). The strong similarity to the latter PR proteins agrees with the serological cross-reactivity of sp41 with the acidic pathogen-induced glucanases (Lotan et al. 1989).

392

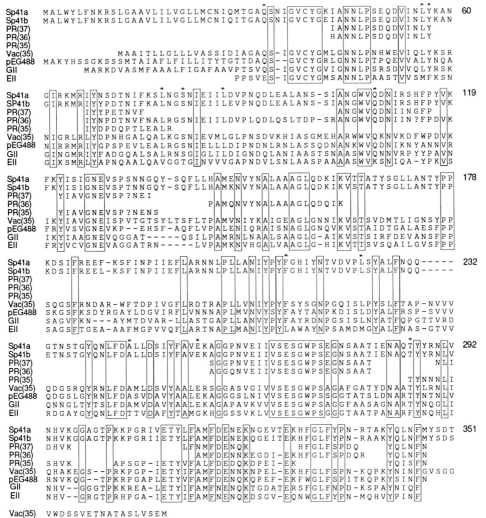

Fig. 3. Amino acid sequence comparison of sp41a and sp41b and other plant glucanases. Putative preprotein cleavage site is marked by an arrow. Gaps have been introduced to facilitate amino acid allignment. Boxed regions indicate positions of amino acids that are conserved in all the glucanases. Asterisks have been placed above rows of amino acid sequence that are conserved in all (1-3)-β-glucanase but not in the (1-3, 1-4)-β-glucanase. PR(37), PR(36) and PR(35), tobacco acidic (1-3)-β-glucanases (Van den Bulcke et al. 1989); Vac(35), tobacco basic-type (1-3)-β-glucanase, (Shinshi et al. 1988); pEG488, soybean (1-3)-β-glucanase, (Takeuchi et al. 1990); GII, barley (1-3)-β-glucanase (Høj et al. 1989); EII, (1-3, 1-4)-β-glucanase from barley (Fincher et al. 1986).

Discussion

Xylanase and certain pathogens have been shown to induce the accumulation of PR proteins in an ethylene independent manner. The elicitation does not require light (Lotan and Fluhr, 1990). Together with the induction of PR proteins by xylanase, the sesquiterpenoid, capsidiol, also accumulates in treated intact leaves. Thus, it appears that in tobacco, two defense-related plant responses, PR proteins induction and phytoalexin accumulation, share common features of regulation. It is interesting to speculate if additional defense mechanisms are coregulated.

The elucidation of novel developmentally regulated PR genes has considerably broadened the biological significance of these proteins. We have shown that a 41 kD glycosylated polypeptide belongs to the (1-3)-β-glucanase superclass (Ori et al. 1990). It is an endoglucanase indistinguishable enzymatically from its pathogen-regulated counterparts. The two types of glucanase classes in Figure 3 show substrate specficity for either (1-3) linked or (1-3, 1-4) linked β-glucan. In a few locations marked by asterisks in Figure 3, we observe that the (1-3, 1-4)-β-glucanase differs in residues found to be conserved in all (1-3)-β-glucanases. These positions should provide clues as to the evolution of the distinct substrate specificities of these closely related polypeptides.

The roles of (1-3)-β-glucanases in plants are unclear. As mentioned, it was proposed to act in the defence mechanism of plants, either by direct attack of the pathogens (Mauch et al.1988), or by releasing elicitor molecules (Darvill and Albersheim, 1984; Keen and Yoshikawa, 1983) from bacterial or fungal cell walls. Additional roles were proposed in cell growth and differentiation and in metabolism of callose, a possible substrate of (1-3)-β-glucanase. Based on the evidence that the 41 kD polypeptide accumulates massively in the extracellular matrix of the transmitting tract we have suggested that it influences the physiochemical properties of the matrix and may play a role in facilitating phase matrix driven pollen translocation (Ori et al. 1990).

Acknowledgements

We gratefully acknowledge support of the Israel National Council for Research and Development, the Gesellschaft fuer Biotechnologische Forschung MBH, Braunschweig and the Leo and Julia Forcheimer Center for Molecular Genetics at the Weizmann Institute of Science. R. Fluhr is a recipient of the Jack and Florence Goodman Career Development Chair. T. Lotan is supported by the L. Eshkol grant of the Ministry of Science, National Council for Research and Development.

394

References

Bol, J. F. and Linthorst, H. J. M. (1990) 'Plant Pathogenesis-related proteins induced by virus infection', Annu. Rev. Phytopathol. 28, 113-138.

Darvill, G.G., and Albersheim, P. (1984) 'Phytoalexins and their elicitors - a defense against microbial infection in plants', Annu. Rev. Plant Physiol. 35, 243-275.

Fincher, G.B., Lock, P.A., Morgan, M.M., Lingelbach, K., Wettenhall, R.E.H., Bercer, J.F.B., Brandt, A. and Thomsen, K.K. (1986) 'Primary structure of the (1->3,1->4) –β–D-glucan 4-glucohydrolase from barley aleurone', Proc. Natl. Acad. Sci. USA. 83, 2081-2085.

Hoj, P.B, Hartman, D.J., Morrice, N. A., Doan, D.N.P. and Fincher, G. B. (1989) 'Purification of (1-3)–β–glucan endohydrolase isoenzyme II from germinated barley and determination of its primary structure from a cDNA clone', Plant Mol. Biol. 13, 31-42.

Keen, N.T., and Yoshikawa, M. (1983) 'β-1,3-endoglucanase from soybean releases elicitor-active carbohydrates from fungus cell walls', Plant Physiol. 71, 460-465.

Lotan, T., Ori, N., and Fluhr, R. (1989) 'Pathogenesis-related proteins are developmentally regulated in tobacco flowers', Plant Cell 1, 881-887.

Lotan, T. and Fluhr, R. (1990) 'Xylanase, a novel elicitor of pathogenesis-related proteins in tobacco, uses a non-ethylene pathway for induction', Plant Physiol. 93, 811-817.

Mauch, F., Mauch-Mani, B., and Boller, T. (1988) 'Antifungal hydrolases in pea tissue. II Inhibition of fungal growth by combinations of chitinase and β-1,3-glucanase', Plant Physiol. 76, 607-611.

Ori, N., Sessa, G., Lotan, T., Himmelhoch, S., and Fluhr, R. (1990) 'A major stylar matrix polypeptide (sp41) is a member of the pathogenesis-related proteins superclass', EMBO J. 9, (in press)

Shinshi, H., Wenzler, H., Neuhaus, J.M., Felix, G., and Hofsteenge, J. (1988) 'Evidence for N- and C-terminal processing of a plant defense-related enzyme primary structure of tobacco prepro-β-1,3-glucanase', Proc. Natl. Acad. Sci. USA 85, 5541-5545.

Stoessel, A., Unwin, C., and Ward, E. (1972) 'Postinfectional inhibitors from plants. I. Capsidiol, an antifungal compound from Capsicum frutescens', Phytopathology Z. 74, 1414-152.

Takeuchi, Y., Yoshikawa, M., Takeba, G., Tanaka, K., Shibata, D., and Horino, O. (1990) 'Molecular cloning and ethylene induction of mRNA encoding a phytoalexin elicitor-releasing factor, β–1,3-endoglucanase, in soybean', Plant Physiol. 93, 673-682.

Van den Bulcke, M., Bauw, G., Castresna, C., Van Montagu, M., and Vandekerckhove, J. (1989) 'Characterization of vacuolar and extracellular β-(1,3)-glucanases of tobacco: evidence for strictly compartmentalized plant defense system', Proc. Natl. Acad. Sci USA 76, 4340-4354.

PATHOGEN-INDUCED GENES IN WHEAT

R. DUDLER, C. HERTIG, G. REBMANN, J. BULL, AND F. MAUCH
Institute for Plant Biology
University of Zurich
Zollikerstrasse 107
8008 Zurich
Switzerland

ABSTRACT. Resistance of wheat (*Triticum aestivum* L.) to *Erysiphe graminis* f. sp. *tritici* (powdery mildew) is increased following exposure to the incompatible *E. g.* f. sp. *hordei*. This resistance response correlates with the activation of certain genes. We have obtained genomic clones of some such genes using induced cDNAs as probes. Sequence analysis indicated that one of our clones encodes a glutathione-S-transferase (GST), an enzyme known to be involved in herbicide detoxification, but which so far has not been associated with response to pathogen attack. A second class of genes encodes a small protein of unknown function. Based on its deduced primary sequence and its high glycine and proline content we speculate that it is an integral membrane protein possibly interacting with the cell wall.

1. Introduction

The infection of wheat (*Triticum aestivum* L.) by *Erysiphe graminis* f. sp. *tritici* provokes a host response which includes the induction of new mRNA species. Some of these pathogen-induced mRNAs have been cloned as cDNAs by differential screening of cDNA libraries (WIR1-WIR6; Schweizer *et al.*, 1989). The cloned mRNA species are also associated with the local induced resistance against this pathogen observed following a primary infection with the non-host pathogen *E. g.* f. sp. *hordei* (Schweizer *et al.*, 1989). The identification and manipulation of the expression of such genes may allow elucidation of a possible causal connection between their activation and the increased resistance.

We cloned pathogen-induced wheat genes to obtain clues about the function of the encoded products and to study their regulation by promoter analysis. Below we summarize our results concerning two novel classes of pathogen-induced genes in wheat.

2. Results and Discussion

2.1. A PATHOGEN-INDUCED GENE ENCODES A PROTEIN HOMOLOGOUS TO GLUTATHIONE-S-TRANSFERASE

We used one of the induced cDNA clones, WIR5, to screen a wheat genomic λEMBL3 library and obtained 3 positive clones. Analysis of one of these clones, λWIR56, indicated that it contains a gene with 3 exons encoding a protein of 229 amino acids (Dudler *et al.*, 1990). The

395

H. Hennecke and D. P. S. Verma (eds.),
Advances in Molecular Genetics of Plant-Microbe Interactions, Vol. 1, 395–398.
© 1991 *Kluwer Academic Publishers. Printed in the Netherlands.*

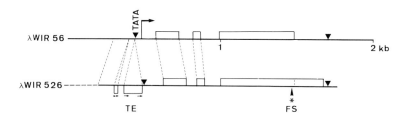

Fig. 1. Structure of λWIR56 and λWir526. Open bars above the line denote exons. The positions of TATA boxes and polyadenylation signals are indicated by filled triangles. The horizontal arrow marks the start of transcription. Compared to λWIR56, λWIR526 contains a frame shift (FS, marked with an asterisk) and two putative defective transposable elements (TE, open bars below the line) with terminal direct repeats (small arrows). The broken line in λWIR526 represents sequences not homologous to λWIR56.

organisation of this gene is shown in Figure 1. S1-mapping experiments showed that mRNA corresponding to the cloned gene is at least 20 times more abundant in leaves inoculated with *E. g. f. sp. hordei* than in leaves from control plants. In contrast, no induction was observed in leaves wounded by abrasion with carborundum.

As evident from the dot matrix comparison shown in Figure 2, the encoded protein exhibits significant homology over essentially the entire sequence to the glutathione-S-transferases (GST; EC. 2.5.1.18) of maize (Moore *et al.*, 1986; Shah *et al.*, 1986; Grove *et al.*, 1988). Sixty four percent of the amino acids of the wheat protein and maize GST I are identical or conserved. Therefore, the λWIR56 clone is proposed to contain a glutathione-S-transferase gene, named gstA1. Both the maize GST I (Shah *et al.*, 1986) and the wheat gstA1 genes contain two introns at exactly conserved positions.

Fig. 2. Matrix comparison of the sequences of the maize GST I (verticl axis) and the wheat gstA1 gene product (horizontal axis) using the DIAGON program of Staden-Plus software (Amersham, U.K.) with a window of 11 and the MDM 78 matrix.

Measuring GST activity with CDNB (1-chloro-2,4-dinitrobenzene) as a substrate, a 2 fold increase was found in leaves 48 h after inoculation with *E. g. f. sp. hordei*. Although this result confirms GST induction, the increase is much lower than that observed measuring gstA1-derived mRNA by S1-mapping. This might be due do the possibility of CDNB being a poor substrate for the induced GST isozymes (Timmerman, 1989). We are currently expressing the gstA1 encoded protein in *Escherichia coli* to (1) demonstrate enzyme activity and elucidate substrate specificity and (2) use it as an antigen for antibody production.

Assuming that the gstA1 gene product is a GST, what might be the function, if any, of this enzyme in relation to pathogen defense? GSTs are ubiquitous multifunctional enzymes that catalyze the conjugation of glutathione (γ-L-glutamyl-L-cysteinyl-glycine) to electrophilic centers of lipophilic compounds, leading to the detoxification of such compounds (Sies and Ketterer, 1988; for a review). Although known to be involved in herbicide detoxification (Timmerman, 1989; for a review), GST induction by pathogens has so far not been reported in plants. In animal systems, the highly toxic products of membrane lipid peroxidation are substrates of GST, which thus may contribute to protect tissue against oxidative damage (Pickett and Lu, 1989; for a review). Since formation of active oxygen species and membrane lipid peroxidation occur in plants as a result of pathogen attack, elicitor treatment and tissue damage (Croft *et al.*, 1990; Rogers *et al.*, 1988; Thompson *et al.*, 1987), GSTs may have a similar protective function in plants as well.

The gene contained in the second positive clone we sequenced (λWIR526; diagrammed in Figure 1) is very similar to gstA1 (95% identity at the amino acid level). However, it seems to be inactivated by mutations: a 1 bp deletion in the second to last codon results in a frame shift (asterisk, Figure 1) leading to an extension of the putative protein by 62 amino acids. In addition, the promoter region contains two insertions of 18 bp and 110 bp in the promoter region (Figure 1). Since they are flanked by a duplicated sequence of 6 bp and 10 bp, respectively, which is present only once in the gstA1 gene, they likely represent deletion remnants of transposons which duplicated the target site upon insertion. S1-mapping with a gene specific probe revealed a weak pathogen-inducible signal possibly, but not necessarily, originating from this gene (not shown). Thus, it seems likely that this gene is expressed at a greatly reduced level, if at all, in comparison to gstA1.

2.2. THE WIR1 GENE-PRODUCT: AN UNUSUAL MEMBRANE PROTEIN?

Using the WIR1 cDNA clone as a probe, 2 positive genomic clones containing identical genes were obtained. Sequence data suggested that they encode a protein of 88 amino acids. S1-mapping showed that the mRNA corresponding to WIR1 is greatly induced in infected leaves, whereas no such mRNA is detected in untreated or wounded plants. We do not know the function of the encoded protein. However, analysis of its primary sequence allow some testable speculations. The protein contains a very hydrophobic N-terminal domain which probably spans a membrane. The C-terminal part contains 2 cysteine residues and consists of 25% glycine and 20% proline, pointing to a structural role. Due to the charge distribution surrounding the putative transmembrane domain (Hartmann *et al.*, 1989), the N-terminus of the protein is hypothesized to face the cytoplasmic side, the C-terminal part being extracellular and possibly interacting with the cell wall. We are currently raising antibodies against synthetic peptides and fusion proteins which will be used to localize the WIR1 gene product.

2.3. CONCLUDING REMARKS

At present it is unclear whether the correlation between the activation of pathogen-induced wheat genes and the onset of induced resistance is based on a causal relationship. Experimental manipulation of the expression (e.g. by mutation or overexpression) of putative defense genes might help to answer this question. Hexaploid wheat is not a very suitable system for such experiments due to its large genome size and the difficulty of genetic transformation. However, the recent advent of transient regulated gene expression in intact tissues of rice and other cereals (Dekeyser *et al.*, 1990) might ameliorate this situation considerably. In spite of potential pitfalls, we have also initiated experiments in a heterologous system, namely transgenic tobacco, where work is in progress to assess the effect of overexpressing the putative wheat defense genes and to analyze their promoters.

3. Acknowledgements

P. Schweizer and collegues are thanked for giving the pathogen-induced cDNA clones to us. This work was supported by the KWF and by DR. R. MAAG AG, Dielsdorf.

4. References

Croft, K. P. C., Voisey, C. R., and Slusarenko, A. J. (1990) 'Mechanism of hypersensitive cell collapse: correlation of increased lipoxygenase activity with membrane damage in leaves of *Phaseolus vulgaris* (L.) inoculated with an avirulent race of *Pseudomonas syringae* pv. *phaseolicola*', Physiol. Mol. Plant Pathol. 36, 49–62.

Dekeyser, R. A., Claes, B., De Rycke, R. M. U., Habets, M. E., Van Montagu, M. C., and Caplan, A. B. (1990) 'Transient gene expression in intact and organized rice tissue', Plant Cell 2, 591–602.

Dudler, R., Hertig, C., Rebmann, G., Bull, J., and Mauch, F. (1990) 'A pathogen-induced wheat gene encodes a protein homologous to glutathione-S-transferase', Mol. Plant-Microbe Interact., in press.

Grove, G., Zarlengo, R. P., Timmerman, K. P., Li, N., Tam, M. F., and Tu, C. D. (1988) 'Characterization and heterospecific expression of cDNA clones of genes in the maize GSH S-transferase multigene family', Nucleic Acids Res. 16, 425–438.

Hartmann, E., Rapoport, T. A., and Lodish, H. F. (1989) 'Predicting the orientation of eukaryotic membrane-spanning proteins', Proc. Natl. Acad. Sci. USA 86, 5786–5790.

Moore, R. E., Davies, M. S., O'Connel, K. M., Harding, E. I., Wiegand, R. C., and Tiemeier, D.C. (1986) 'Cloning and expression of a cDNA encoding a maize glutathione-S-transferase in *E. coli*', Nucleic Acids Res. 18, 7227–7235.

Pickett, C. B., and Lu, A. Y. H. (1989) 'Glutathione-S-transferases: Gene structure, regulation and biological function', Annu. Rev. Biochem. 58, 743–764.

Rogers, K. R., Albert, F., and Anderson, A. J. (1988) 'Lipid peroxidation is a consequence of elicitor activity', Plant Physiol. 86, 547–553.

Schweizer, P., Hunziker, W., and Mösinger, E. (1989) 'cDNA cloning, *in vitro* transcription and partial sequence analysis of mRNAs from winter wheat (*Triticum aestivum* L.) with induced resistance to *Erysiphe graminis* f. sp. *tritici*', Plant Mol. Biol. 12, 643–654.

Shah, D. M., Hironaka, C. M., Wiegand, R. C., Harding, E. I., Krivi G. G., and Tiemeier, D. C. (1986) 'Structural analysis of a maize gene coding for glutathione-S-transferase involved in herbicide detoxification', Plant Mol. Biol. 6, 203–211.

Sies, H. and Ketterer, B., eds. (1988) Glutathione Conjugation, Academic Press, London.

Thompson, J. E., Legge, R. L., and Barber, R. F. (1987) 'The role of free radicals in senescence and wounding', New Phytol. 105, 317–344.

Timmerman, K. P. (1989) 'Molecular characterization of corn glutathione-S-transferase isozymes involved in herbicide detoxification', Physiol. Plant. 77, 465–471.

BIOLOGICAL ACTIVITY OF PR-PROTEINS FROM TOBACCO; CHARACTERIZATION OF A PROTEINASE INHIBITOR

M. LEGRAND, A. STINTZI, T. HEITZ, P. GEOFFROY, S. KAUFFMANN, and B. FRITIG
Institut de Biologie Moléculaire des Plantes
12 rue du Général Zimmer
67084 Strasbourg Cédex
France

ABSTRACT. Many metabolic changes that have been reported as associated to active defense in incompatible plant-fungi and plant-bacteria interactions have also been found during the hypersensitive reaction of plants to viruses. This is particularly the case for the Pathogenesis-Related (PR) proteins which represent major protein changes in tobacco hypersensitively reacting to tobacco mosaic virus (TMV). Many of these PR-proteins have been shown to have glycanhydrolase activities and to be responsible, at least in part, for the strong increase in ß-1,3-glucanase, chitinase and lysozyme activities which is associated to the hypersensitive reaction. For each enzyme several distinct isoforms are produced and have been characterized. In addition a proteinase inhibitor has been isolated and shown to be induced in parallel in the same material. The inhibitor is very effective against serine endoproteinases of microbial origin but inhibits poorly trypsin and chymotrypsin. It could represent a plant defense complementary to the potential lytic activity of hydrolases.

1 . Introduction

The hypersensitive response of plants to pathogens is one of the most efficient natural mechanism of defense. It leads ultimately to the necrotization of a few cells and to the localization of the parasite in a limited area close to the infection site. The hypersensitivity reaction is associated with the induction of intense metabolic alterations believed to contribute to the resistance observed. Among these, the accumulation of phytoalexins, the production of mechanical barriers and the synthesis of defense-related proteins and enzymes are well documented (Lamb et al. (1989), Fritig et al. (1990)).

Pathogenesis-related (PR) proteins account for the major quantitative changes in soluble proteins in the infected plants. They were first shown to accumulate during the hypersensitive reaction of tobacco to tobacco mosaic virus (TMV). Such PR-proteins have now been described in many plant species, including dicots and monocots and shown to be inducible by pathogens, by treatment with chemicals or under various physiological situations (for reviews see Van Loon (1985), Carr and Klessig (1989)). They have characteristic properties : they are selectively extractable at low pH, highly resistant to proteolytic enzymes and localized predominantly in the intercellular spaces. In tobacco 10 major acidic PR-proteins are identified according to their relative mobility in native polyacrylamide gel electrophoresis and are referred to as PR-1a, -1b, -1c, -2, -N, -O, -P, -Q, -R and -S in order of decreasing mobility. More recently other less abundant acidic PR-proteins and basic PR-like proteins have been characterized (Fritig et al. (1990)). The biological activities of some of them have been unravelled and this gives a clue to the possible role of PR-proteins in resistance against fungi or bacteria.

399

H. Hennecke and D. P. S. Verma (eds.),
Advances in Molecular Genetics of Plant-Microbe Interactions, Vol. 1, 399–402.
© 1991 *Kluwer Academic Publishers. Printed in the Netherlands.*

2 . Materials and Methods

2.1. PLANT MATERIAL

The three first fully expanded leaves at the top of three-month-old tobacco plants (*N. tabacum* cv. Samsun NN) were inoculated with a suspension of purified TMV. Leaves were harvested, frozen in liquid nitrogen and stored at -80°C.

2.2. EXTRACTION AND PURIFICATION OF PROTEINS

The procedures used for extraction and purification of the different proteins have been published previously (Legrand et al. (1987), Kauffmann et al. (1987), Kauffmann et al. (1990), Geoffroy et al. (1990)).

2.3. BIOLOGICAL ACTIVITIES

The ß-1,3-glucan laminarin was used as substrate for glucanase assay and enzyme activity was estimated by measuring the reducing power of the reaction products (Kauffmann et al. (1987)). Chitinase activity was assayed with chitin as substrate. The method used has been published previously (Legrand et al. (1987)). Lysozyme activity was tested at 540 nm with *Micrococcus* cells as substrate. Proteinases inhibitory activity was assayed by preincubating 1 μg of a commercial proteinase with the inhibitor solution for 3 min at 37°C and measuring the residual proteinase activity with Azocoll (Geoffroy et al. (1990)).

3 . Results

We have isolated up to now 25 defense-related proteins from tobacco reacting hyper-sensitively to TMV. These include the 10 major acidic PR-proteins which have been shown to belong to 4 distinct serological groups : the PR-1 group with -1a, 1b, 1c whose biological function remains unknown ; PR-2, N and O which are 1,3-ß-glucanases (Kauffmann et al. (1987)) serologically related to another acidic glucanase (named Q' because it migrates as Q on native gel) and to a basic glucanase. A sixth glucanase (O') which is not serologically related to the others is also induced by infection. In fact, the 6 different isoforms account for the increase of ß-1,3-glucanase activity in infected leaves (Figure 1). The increase in chitinase activity arises from the production of 4 major enzymes (Legrand et al. (1987)). Two of these are PR-proteins -P and -Q which are serologically related to the 2 other isoforms with higher isoelectric points. The latter basic chitinases have also lysozyme activity and are also responsible, at least in part, for the marked increase in lysozyme activity found in TMV-infected leaves (Figure 1). Two other enzymatic proteins with lysozyme activity (but no chitinase activity) have also been characterized (A. Stintzi, unpublished results). These 2 lysozymes are not serologically related to the 4 chitinases.

Proteins R and S are serologically related and are also related to a basic counterpart which has been demonstrated to be in fact osmotin, a protein known to be induced in tobacco cells during salt adaptation. Finally a fifth group of proteins have been characterized (Kauffmann et al. (1990)) but their biological activities as well as those of R, S and osmotin are still unknown.

In addition to the synthesis of numerous enzymatic proteins, TMV infection triggers the production of a proteinase inhibitor (Geoffroy et al. (1990)). The time course curve of inhibitory activity towards subtilisin, a bacterial enzyme, is presented in figure 1. This inhibitory activity is due to the accumulation of a small proteinaceous inhibitor which is highly active against 4 different serine endoproteinases of fungal and bacterial origin but inhibits poorly trypsin and chymotrypsin, 2 serine endoproteinases of animal origin. Its amino acid composition and its NH$_2$-terminal amino acid sequence indicate that the inhibitor belongs to the potato inhibitor I family.

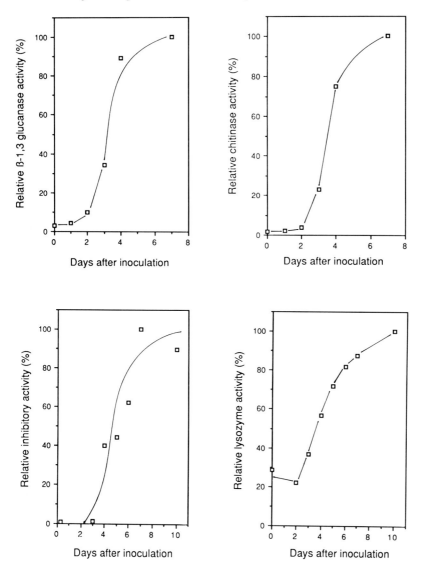

Figure 1. Time course curves of enzyme activities and proteinase inhibitory activity in tobacco leaves infected with TMV. Enzymatic activities were measured as described in Materials and Methods. Inhibitory activity was tested towards subtilisin.

4. Discussion

The tobacco PR-proteins which were first detected in viral infection were thought to be related to antiviral defense. It is now well established that a significant proportion of these proteins are glycanhydrolases which may represent very efficient "direct" defense enzymes against fungi and bacteria. The combination of chitinases and ß-1,3-glucanases has been shown to increase the rate of hydrolysis of fungal cell walls *in vitro* (Mauch et al. (1988)) and a similar synergistic effect may be hypothesized *in vivo* where the different enzymes are induced in parallel (Figure 1). But glycanhydrolases might also be involved in the production of chemical signals of pathogen and/or plant origin. Such regulatory molecules, called "elicitors", are usually carbohydrates or contain carbohydrates (Darvill and Albersheim (1984)).

The production of a proteinase inhibitor which is highly specific for serine endoproteinase of microbial origin suggests that it is involved in plant defense and may neutralize some of the pathogen's weapons. In fact, infection with TMV induces the synthesis of a whole set of antimicrobial proteins, whose role in the resistance to virus is not clear.

5. References

Carr, J. P. and Klessig, D. F. (1989) 'The pathogenesis-related proteins of plants', in J. K. Setlow (ed.), Genetic Engineering ; Principles and Methods, vol. 11, Plenum Press, New York and London, pp. 65-109.

Darvill, A. G. and Albersheim, P. (1984) 'Phytoalexins and their elicitors- a defense against microbial infection in plants', Annu. Rev. Plant Physiol. 35, 243-275.

Fritig, B., Kauffmann, S., Rouster, J., Dumas, B., Geoffroy, P., Kopp, M. and Legrand, M. (1990) 'Defence proteins, glycanhydrolases and oligosaccharide signals in plant-virus interactions', in R. S. S. Fraser (ed.), Recognition and Response in Plant-Virus Interactions, NATO ASI Series, Cell Biology, vol. 41, Springer-Verlag, Berlin, pp. 375-394.

Geoffroy, P., Legrand, M. and Fritig, B. (1990 'Isolation and characterization of a proteinaceous inhibitor of microbial proteinases induced during the hypersensitive reaction of tobacco to tobacco mosaic virus', Mol. Plant-Microbe Interac. 3 (in the press).

Kauffmann, S., Legrand, M., Geoffroy, P. and Fritig, B. (1987) 'Biological function of 'pathogenesis-related' proteins : four PR proteins of tobacco have 1,3-ß-glucanase activity', EMBO J. 6, 3209-3212.

Kauffmann, S., Legrand, M. and Fritig, B. (1990) 'Isolation and characterization of six pathogenesis-related (PR) proteins of Samsun NN tobacco', Plant Mol. Biol. 14, 381-390.

Lamb, C. J., Lawton, M. A., Dron, M. and Dixon, R. A. (1989) 'Signals and transduction mechanisms for activation of plant defenses against microbial attack', Cell 56, 215-224.

Legrand, M., Kauffmann, S., Geoffroy, P. and Fritig, B. (1987 'Biological function of "pathogenesis-related" proteins : four tobacco PR-proteins are chitinases', Proc. Natl. Acad. Sci. USA 84, 6750-6754.

Mauch, F., Mauch-Mani, B. and Boller, T. (1988) 'Antifungal hydrolases in pea tissue.II.Inhibition of gungal growth by combinations of chitinase and ß-1,3-glucanase', Plant Physiol. 88, 936-942.

Van Loon, L. C. (1985) 'Pathogenesis-related proteins', Plant Mol. Biol. 4, 111-116.

MOLECULAR RECOGNITION IN PLANTS: IDENTIFICATION OF A SPECIFIC BINDING SITE FOR OLIGOGLUCOSIDE ELICITORS OF PHYTOALEXIN ACCUMULATION

MICHAEL G. HAHN AND JONG-JOO CHEONG
The University of Georgia
Complex Carbohydrate Research Center
220 Riverbend Road
Athens, GA 30602
USA

ABSTRACT. A brief overview of previous studies on the molecular mechanisms by which plant cells perceive and respond to extracellular signals is presented. A relatively small number of signal molecules, including plant hormones and oligosaccharins, have been characterized in plants. Binding sites (putative receptors) for several of these signals have been identified, though proof of their function as receptors is still required. Some progress has also been made in identifying the cellular mechanisms (*e.g.*, second messengers, protein phosphorylation, ion flux) by which signal perception in plants is translated into changes in metabolism and gene expression.

The results of studies in our laboratory on the induction of phytoalexin biosynthesis in soybean cells by fungal cell wall-derived oligoglucosides are summarized. The smallest active fungal cell wall-derived elicitor is a hepta-β-glucoside having a β-(1→6)-linked backbone with two terminal β-(1→3)-glucosyl side-chains. This hepta-β-glucoside elicitor induces half-maximum accumulation of phyto-alexins in soybean cotyledons at concentrations of ~10^{-8} M. Structurally related oligoglucosides were only active at concentrations 10^3 to 10^4-fold higher. A binding site with properties expected of a physiological receptor for the hepta-β-glucoside elicitor has been identified in soybean microsomal membranes. Binding of a radio-iodinated tyramine conjugate of the elicitor-active hepta-β-glucoside was specific, reversible, saturable, and of high affinity (K_d = 7.5 x 10^{-10} M). Competitive displacement of the radiolabeled hepta-β-glucoside elicitor with a number of elicitor-active and inactive oligoglucosides demonstrated a direct correlation between the ability to displace the labeled elicitor and the elicitor activity of these oligoglucosides.

1. Introduction

Biochemical analysis of the interactions between plants and microbes has contributed to a greater understanding of the molecular basis for signal perception and transduction in plant cells (for recent reviews, see [30,73,102]). Research in this area has led to the discovery of new classes of signal molecules and provided useful model systems for molecular studies on signal perception, signal transduction, and gene regulation in plants. The purpose of this article is to give an overview of molecular signals and signal transduction in plants, emphasizing contributions from studies of plant-microbe interactions, and to summarize results from our research on one plant signal transduction system, the induction of phytoalexin accumulation by oligoglucoside elicitors.

H. Hennecke and D. P. S. Verma (eds.),
Advances in Molecular Genetics of Plant-Microbe Interactions, Vol. 1, 403–420.
© 1991 *Kluwer Academic Publishers. Printed in the Netherlands.*

2. Signalling in Plants

The process of cellular signalling can be divided into four parts: the signal molecules that trigger specific signal transduction pathways; signal perception, usually by cellular receptors that specifically recognize the signal molecules; signal transduction, that is, transmission of the signal to the site(s) of action within the cell, either directly (*e.g.*, steroid receptors in mammalian cells [38]) or indirectly (*e.g.*, via second messengers, changes in protein phosphorylation, G proteins, etc.); and signal translation, that is, conversion of the signal into a specific cellular response such as activation of specific genes. The following overview will focus on the first three steps. The activation of specific genes in response to particular stimuli is discussed elsewhere in this volume (see contributions by Hahlbrock and Lamb). One signal that exerts a powerful influence on cellular processes in plants, namely light, will not be discussed here.

2.1 PLANT SIGNALS

Several classes of signal molecules have been identified in plants. These include the well-known plant growth regulators (hormones), oligosaccharins, poly- and oligopeptides, and microbial toxins. Each of these classes of signal molecules have unique features that are discussed in the following sections. It is to be expected that additional signal molecules will be discovered as plant biochemists gain a better understanding of the molecular mechanisms underlying plant development and responses to external stimuli.

2.1.1 Plant Growth Regulators. Plant growth regulators (plant hormones) constitute a large group of natural compounds in plants that have the ability to affect a variety of plant developmental processes (for a review of these compounds and their activities, see the book by Davies [23]). Examples of compounds belonging to different groups of plant

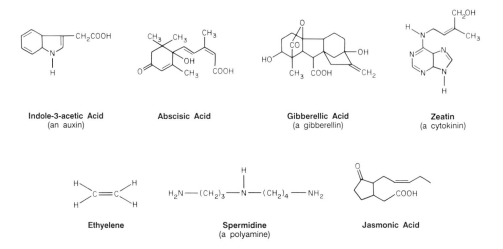

Indole-3-acetic Acid
(an auxin)

Abscisic Acid

Gibberellic Acid
(a gibberellin)

Zeatin
(a cytokinin)

Ethyelene

Spermidine
(a polyamine)

Jasmonic Acid

Figure 1. Examples of plant growth regulators (hormones).

hormones are shown in Figure 1. I have included in this class of signal molecules the familiar plant hormones (auxins, cytokinins, gibberellins, abscisic acid, and ethylene) and two additional signal molecules, polyamines and jasmonic acid, that have aroused more recent interest.

Plant hormones characteristically are pleiotropic in their action, that is, a given plant hormone will affect a wide variety of developmental and growth processes in plants depending on the tissue being studied, the concentration of the hormone, and the presence (or absence) of other plant hormones [24]. For example, the effects of auxins include induction of cell division and elongation, initiation of roots, and determination of apical dominance; these effects are often mediated by the presence of cytokinins or ethylene. Activities of other plant hormones include the induction of cell division and initiation of shoots by cytokinins, the induction of stem elongation upon bolting of long-day plants and the production of α-amylases in germinating seeds by gibberellins, the closure of leaf stomata and the stimulation of seed storage protein biosynthesis by abscisic acid, and fruit ripening by ethylene. Polyamines appear to play a role in plant growth and development, though their identification as plant hormones remains controversial [37,44]. Jasmonic acid and its methyl ester have recently been shown to induce vegetative storage proteins [3,80] and proteinase inhibitors [41]. Methyl jasmonate has been reported to act as a volatile signal molecule that induces proteinase inhibitors [41].

2.1.2 Oligosaccharins. Oligosaccharins (oligosaccharides with biological regulatory properties [1,2]) are a fairly recent addition to the family of plant signal molecules. A number of oligosaccharins have been identified and structurally characterized (Figure 2). They include oligosaccharides derived from fungal [7,67,109], plant [52,82,84,90,119], and most recently, bacterial [75] glycoconjugates. Evidence for additional biologically active oligosaccharides has been obtained, although the oligosaccharins have not yet been

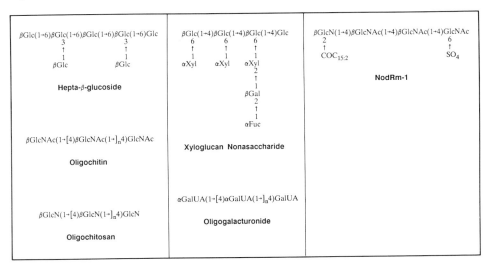

Figure 2. Structures of oligosaccharins. The oligosaccharins in the left hand panel were isolated from fungi, those in the middle panel from plants, and the one in the right hand panel from bacteria.

purified or their structures completely determined [31,45,118].

Biologically active oligosaccharides were first identified in studies on the activation of plant defense responses during microbial infection or upon treatment with elicitors (reviewed in [22,50,100]). Subsequently, evidence has accumulated that oligosaccharins also regulate plant hormone responses [13,82,83,119] and plant development [35,75,114]. At least some oligosaccharins are pleiotropic in their effects on plants. Oligogalacturonides, for example, have been shown to induce the biosynthesis and accumulation of phytoalexins [52,63,90,116], lignification of plant cell walls [97,98], and accumulation of proteinase inhibitors [8,9], as well as altering morphogenesis in tissue culture explants [35]. In addition to being individually active, different oligosaccharins in combination have been shown to synergistically activate plant defense responses [25,26]. The range of activities of many oligosaccharins has not been fully determined, primarily due to the small amounts of these compounds that have been purified to date.

One distinguishing feature of oligosaccharins when compared with the plant growth regulators described in the previous section is that oligosaccharins tend to show greater plant specificity. For example, the fungal wall derived hepta-β-glucoside is active as an elicitor of phytoalexin accumulation in soybean [109], but is not active in parsley [92]. Similarly, the rhizobial oligosaccharin NodRm-1 induces root-hair deformation on alfalfa but not on vetch [75].

2.1.3 Peptides. Poly- and oligopeptides constitute an additional class of signal molecules that are active in plants. Polypeptide signals include glycosylhydrolases that induce plant defense responses by releasing oligosaccharins from either plant or microbial cell surface polymers. Examples of such enzymes are β-glucanases of plant origin that release elicitor-active oligosaccharides from fungal cell walls [66], and microbial pectic-degrading enzymes (endopolygalacturonase, endo-pectate lyase) that release active oligogalacturonides from plant cell walls [14-16,27,98]. Other polypeptides that have no demonstrable enzymatic activity have also been shown to induce plant defense responses such as phytoalexin accumulation [21,39,69,101] and necrosis [96].

Recently, evidence that oligopeptides also function as signal molecules in plants has been obtained in two systems. De Wit and colleagues have isolated a 27 amino acid peptide from apoplastic fluids of a compatible plant pathogen interaction [105]. This peptide, which appears to be a fungal product, induces necrosis in resistant plant cultivars. Ryan and his colleagues have purified an 18 amino acid peptide that, when applied to the base of cut tomato seedlings, induces the accumulation of proteinase inhibitors in the leaves of the plants (C.A. Ryan, personal communication). The peptide, which is synthesized by the plant, is very active, requiring only femtomolar to picomolar amounts per plant to induce inhibitor accumulation. Chemical synthesis of the active peptide confirmed its structure. This oligopeptide is the first example of a peptide hormone in plants.

2.1.4 Microbial toxins. A number of phytopathogenic microorganisms secrete toxins that appear to play a major role in the disease process (see a recent collection of papers [47]. These toxins are generally divided into non-specific [99] and host-selective [103] types. The non-specific toxins affect a wide variety of plant cells, while the host-selective toxins affect only susceptible varieties of a particular plant. Structures of two microbial toxins are shown in Figure 3.

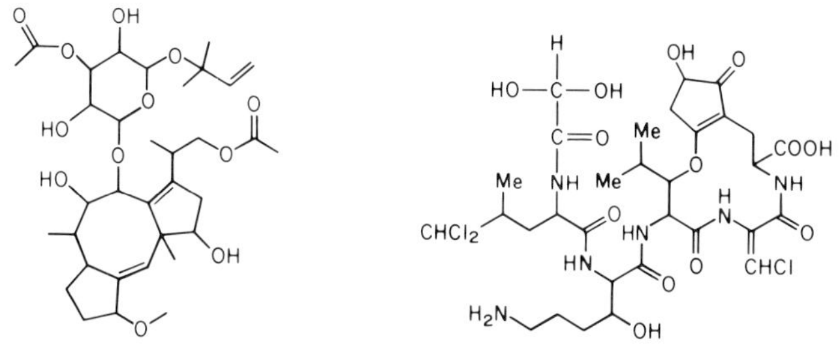

Fusicoccin   Victorin C

Figure 3.   Structures of two fungal toxins: a non-specific toxin, fusicoccin (produced by *Fusicoccum amygdali*) and a host-selective toxin, victorin C (produced by *Cochliobolus victoriae*).

The microbial toxins, as their name implies, are damaging to plant cells and tissues, although their mechanism of action can vary, depending on their precise site of action. Fusicoccin, for example, appears to induce major changes in the flow of ions across the plant cell membrane, leading to physiological changes such as acidification of the extracellular medium, uptake of K^+, and cell enlargement [79].

2.2 SIGNAL PERCEPTION

Research aimed at elucidating the molecular basis for specific recognition of signal molecules in plants has only recently begun to bear fruit. Evidence for the presence of specific binding sites for several of the signal molecules discussed in the previous section has been obtained (see Table 1). These studies utilized derivatives of the signal molecules radiolabeled to a high specific radioactivity. Photo-affinity labeling and/or affinity chromatography have been used to identify binding proteins for auxin, abscisic acid, and fusicoccin, while covalent attachment of labeled victorin C upon binding identified a binding protein for this signal molecule. The binding site for the hepta-β-glucoside elicitor will be discussed in greater detail below (also see contribution by Ebel *et al.* in this volume).

The binding sites identified thus far are membrane-localized, and have high affinity and specificity for their respective ligands (K_d = 10^{-8} to 10^{-10} M). They thus fulfill criteria expected of *bona fide* receptors for these signal molecules. However, proof that these binding proteins function as part of the signal transduction pathway, *i.e.*, are physiological receptors, must await reconstitution of a ligand responsive system using the purified binding protein or the gene encoding the binding protein. These types of experiments have been carried out for several receptors in mammalian systems (for examples, see [49]), but have yet to be accomplished for any plant receptor.

Purification of the auxin-binding protein made possible the isolation of cDNA clones by probing cDNA libraries with either antibodies raised against the purified binding protein

[113] or synthetic oligonucleotides based upon partial amino acid sequences of the binding protein [58,62]. Analysis of the deduced amino acid sequences for the auxin-binding protein revealed no identifiable transmembrane domain and the presence of an animal consensus sequence (KDEL) at the C-terminus for targeting of proteins to the endoplasmic reticulum (ER) [86]. The apparent dichotomy between the apparent ER targeting of the auxin-binding protein and its apparent functional localization to the plasma membrane is a subject of ongoing discussions and research [68].

TABLE 1. Specific binding sites in plants for signal molecules.

Ligand	Plant	Reference
Auxin	Maize	[64,77,110]
	Tomato	[60]
	Zucchini	[59]
Abscisic Acid	*Vicia faba*	[61]
Fusicoccin	Arabidopsis	[85]
	Maize	[93,94]
	Oats	[28]
	Spinach	[6]
	Vicia faba	[42]
Victorin C	Oats	[117]
Hepta-β-glucoside Elicitor	Soybean	[18-20,104]

The strategy of domain homology cloning [54] has recently been used to identify a cDNA encoding a putative membrane-localized receptor kinase in maize [115]. Mixtures of degenerate oligonucleotides corresponding to highly conserved amino acid sequences in serine/threonine-specific protein kinases were used to isolate a cDNA clone for a protein containing a protein kinase domain, a transmembrane domain, and an extracellular domain with similarity to the S-locus glycoproteins in *Brassica* [87,88]. The ligand and function of this putative receptor kinase remain unknown.

2.3 SIGNAL TRANSDUCTION

Our understanding of signal transduction mechanisms within plant cells remains fragmentary (for a recent collection of papers on this subject, see [12]). For the most part, plant signal transduction mechanisms have been examined with an eye toward animal paradigms.

Second messengers that have been identified in animal cells, such as cAMP, phosphoinositides and calcium, have been found to be present in plants [11,65,78,89]. However, the evidence that cAMP (*e.g.*, [53,71]) and phosphoinositides (*e.g.*, [70,111]) play roles in plant signal transduction remains ambiguous. The data is more convincing for a role of calcium in plant signal transduction, particularly in the activation of plant defense responses. Elicitor-induced phytoalexin accumulation [71,112] and callose synthesis [72] in suspension cultured cells were inhibited if the extra-cellular calcium concentration was reduced. A calcium ionophore was also shown to induce phytoalexin accumulation [71,112], further suggesting that ion fluxes are involved in the elicitor-stimulated signal cascade. There is still a need to demonstrate that the observed ion fluxes are directly involved in the signal transduction pathway, rather than an indirect response. Patch clamp techniques, which allow the direct measurement of ion flow in cells and membranes, have recently been applied to plant cells [106] and may prove a valuable tool in these analyses. There is also a need to perform these experiments with homogeneous preparations of signal molecules (elicitors), rather than with the heterogeneous preparations used in most studies to date, in order to unambiguously correlate a cellular response with a specific signal transduction system.

Signal transduction cascades in animal systems often involve guanine nucleotide-binding proteins (G proteins) and rapid changes in phosphorylation of specific proteins catalyzed by specific protein kinases. Guanine nucleotide-binding proteins [10,34,56,57] and analogs of mammalian *ras* oncogenes (also GTP-binding proteins) [81,91] have been identified in plants. Homology probing [54] using oligonucleotides corresponding to the highly conserved ATP-binding site of protein kinases [55] has resulted in the identification in plants of putative protein kinases [74,115]. However, a role for the G proteins and protein kinases in signal transduction pathways in plants has not yet been demonstrated.

Changes in protein phosphorylation during plant growth and development have been observed, though the physiological significance of these changes have, in most cases, not been determined (see review [95]). Recently, several groups have demonstrated specific changes in the pattern of phosphorylated proteins after treatment of soybean [46], parsley [29], or tomato [40] with various elicitor preparations. Again, the physiological significance of these changes in protein phosphorylation remains to be determined, as do the properties and functions of the phosphorylated proteins. However, the rapidity and specificity of the response (1-5 min) does suggest a connection with signal transduction.

Another rapid change observed in infected or elicitor-treated plant tissues is the generation of superoxide anion [32,33] or hydroxyl radicals [4,36,76]. These redox perturbations precede the activation of plant defense mechanisms such as the hypersensitive response or accumulation of phytoalexins. The rapidity of the response (< 1 min in elicitor-treated suspension-cultured soybean cells) and the potential involvement of the hydroxyl radicals in membrane lipid peroxidation and cell wall lignification (possible plant defense mechanisms) have led to the suggestion that this oxidative activity is an early step in the elicitor-stimulated signal cascade [4].

3. Specific Recognition of Oligoglucoside Elicitors by Soybean

We are studying signal transduction mechanisms in plants using the induction of phytoalexin accumulation in soybean by oligoglucoside elicitors as a model system. This system has the advantage that a considerable amount of information has been obtained at the molecular level about both ends of the signal transduction pathway. The structures of signal molecules (elicitors) that induce phytoalexin synthesis, in particular certain oligosaccharides derived from fungal and plant cell walls, have been determined [22], and the biosynthetic pathway for soybean phytoalexins (cellular response to the signal) has been largely elucidated [48].

The elicitor-active oligoglucoside being utilized in our studies is a branched hepta-β-glucoside (compound **1**, Figure 4) that was originally purified from a mixture of oligosaccharides released by partial acid hydrolysis of the mycelial walls of the soybean pathogen *Phytophthora megasperma* f. sp. *glycinea* [108,109]. This hepta-β-glucoside elicitor induces phytoalexin accumulation in soybean cotyledons at concentrations between 10^{-7} and 10^{-9} M [107,109]. The following sections summarize our recent results which identify the structural elements of the hepta-β-glucoside elicitor that are most important for its biological activity and demonstrate that a specific, high-affinity binding site for the hepta-β-glucoside elicitor exists in soybean membranes.

3.1 STRUCTURE-FUNCTION STUDIES

The approach chosen to identify structural elements of the hepta-β-glucoside elicitor that are essential for its biological activity was to examine the biological activity of nine chemically synthesized oligo-β-glucosides (Figure 4) structurally related to the elicitor [17]. In addition, several derivatives of the hepta-β-glucoside elicitor, modified at the reducing glucosyl residue, were prepared to determine whether the structure of the elicitor could be modified at that position without significantly altering its ability to induce phytoalexin accumulation.

The results of biological assays of the nine synthetic oligoglucosides demonstrated that oligoglucosides must have a specific conformation and structure in order to trigger the signal transduction pathway which ultimately leads to the *de novo* synthesis of phytoalexins. Four oligoglucosides (compounds **1-4**) were equally active as elicitors, each having an EC$_{50}$ $\approx$ 10 nM (concentration of oligosaccharide required to achieve half-maximum phytoalexin

Figure 4. Structures of synthetic oligoglucosides. Hydroxyl groups not involved in glycosidic linkages or at the reducing end of the oligosaccharides have been omitted from the structures for clarity.

accumulation). These four oligoglucosides have, as part of their structures, the branching pattern found in the elicitor-active hepta-β-glucoside identified previously [109]. Oligoglucosides 5-8 were not effective inducers of phytoalexin accumulation (Figure 5), demonstrating the structural specificity of recognition of elicitor-active oligoglucosides by soybean tissue. Substitution, in hexaglucoside 3, of the side chain glucosyl residue closest to the non-reducing terminus of the oligosaccharide (compound 7) or of the non-reducing terminal backbone glucosyl residue (compound 8) with N-acetylglucosaminyl residues reduced the elicitor activity ~1000- and 10,000-fold. The importance of the β-(1→6)-linked backbone residues in the recognition of elicitor-active oligoglucosides was also demonstrated by the fact that gentiobiose (β-(1→6)-linked diglucoside) had elicitor activity at ~10^{-3} M, while laminaribiose (β-(1→3)-linked diglucoside) was inactive at that concentration. The importance of the side-chain glucosyl residues in the recognition of elicitor-active oligoglucosides was shown by the lack of elicitor activity of a linear β-(1→6)-linked heptaglucoside. Furthermore, the different arrangement of side-chains in hepta-β-glucoside 5 substantially reduced (~1000-fold) the activity of the oligoglucoside, confirming previous findings [109].

Bioassays of reducing end derivatives of hepta-β-glucoside 1 demonstrated that attachment of bulky alkyl or aromatic groups (e.g., compound 10) to the oligosaccharide had no significant effect on the EC_{50} of the hepta-β-glucoside derivative. Reduction of the terminal glucosyl residue to its corresponding glucitol had previously been shown not to significantly affect the elicitor activity of hepta-β-glucoside 1 [107,109]. A tyramine conjugate of hepta-β-glucoside 5 was ~2.5 times as active as underivatized hepta-β-glucoside 5. Aromatic conjugates of the structurally unrelated hepta-α-glucoside, maltoheptaose, had no detectable elicitor activity. Thus, conjugation of the phenoxyl

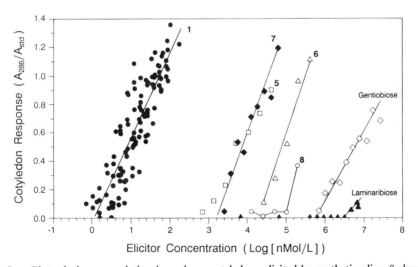

Figure 5. Phytoalexin accumulation in soybean cotyledons elicited by synthetic oligo-β-glucosides. The response of the cotyledons (A/A_{std}) to the oligoglucosides was determined using the cotyledon bioassay [5,51]. The activity curves are labeled with numbers identifying the oligoglucosides (see Figure 4).

group to biologically inactive oligosaccharides does not appear to lead to an activation of those oligosaccharides.

The results of the bioassays suggest that hexa-β-glucoside 3 (Figure 4) encompasses the minimum structural element required for an oligoglucoside to be a maximally effective elicitor. These results also demonstrate that reducing end derivatives of the hepta-β-glucoside elicitor can be prepared for use in the biochemical analysis of the recognition process. One such derivative, a radio-iodinated tyramine conjugate of hepta-β-glucoside 1, was used as a labeled ligand to demonstrate the presence of specific, high-affinity binding sites for the elicitor in soybean membranes (see following section).

3.2 LIGAND-BINDING STUDIES

The specificity of the response of soybean tissue to oligoglucoside elicitors of phytoalexin accumulation [17,109] suggests that a specific receptor for the hepta-β-glucoside elicitor exists in soybean cells. Ligand-binding assays using a radiolabeled hepta-β-glucoside derivative were carried out to demonstrate the existence of a specific binding site for the hepta-β-glucoside elicitor in soybean membranes [18]. This binding site has properties characteristic of a physiologically relevant receptor.

Radio-iodination of a tyramine conjugate of the hepta-β-glucoside elicitor (compound 10, Figure 4), using Iodogen [43] as the oxidizing agent, readily yielded radiolabeled hepta-β-glucoside with specific radioactivities of ~100 Ci/mmol. The high specific radioactivity permitted sensitive detection of binding of the hepta-β-glucoside to microsomal

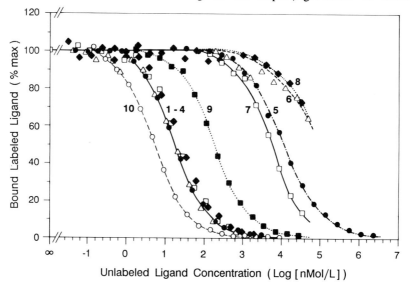

Figure 6. Competitive inhibition of binding of ^{125}I-hepta-β-glucoside-tyramine conjugate to soybean root membranes by unlabeled oligoglucosides. Radiolabeled hepta-β-glucoside 1 was incubated with soybean root microsomal membranes in the presence of increasing amounts of unlabeled synthetic oligoglucosides. Each inhibition curve is labeled with numbers identifying the oligoglucoside (see Figure 4). Binding assays were carried out as described [18].

membranes prepared from soybean leaf, hypocotyl, cotyledon, and root tissues. Membranes prepared from soybean roots bound the greatest amount of hepta-β-glucoside elicitor per mg of membrane protein.

Binding of radiolabeled hepta-β-glucoside to its binding site in root membranes was saturable over a ligand concentration range of 0.1 to 5 nM, which is comparable to, albeit somewhat lower than the range of hepta-β-glucoside concentrations (6 to 200 nM) required to saturate the cotyledon bioassay for phytoalexin accumulation ([17,107,109]; see previous section). A single class of binding sites with a high affinity for the hepta-β-glucoside elicitor (apparent K_d = 7.5 x 10^{-10} M) was identified in root membranes using saturation binding assays. Binding of the hepta-β-glucoside elicitor to its binding site was reversible; half-maximum binding was achieved at 0°C within 20 minutes after adding the ligand, while dissociation of the labeled hepta-β-glucoside, in the presence of an excess of unlabeled ligand, was half complete after 2.5 hours.

The hepta-β-glucoside binding site exhibited strong selectivity in terms of the oligoglucosides that were recognized efficiently (Figure 6). Oligoglucosides 1, 2, 3, and 4, in which the arrangement of glucosyl residues is identical to that found in the radiolabeled hepta-β-glucoside-tyramine conjugate (Figure 4), were equally effective inhibitors of ligand binding, requiring concentrations of 10-20 nM to achieve 50% inhibition of binding. Other oligoglucosides that either differed in the arrangement of glucosyl residues (5 and 9) or were substituted with N-acetyl groups (7 and 8) required higher concentrations (10- to

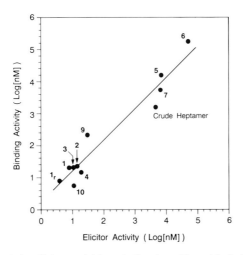

Figure 7. Correlation of the elicitor activities of oligoglucosides with their affinities for the hepta-β-glucoside binding site in soybean root membranes. The relative elicitor activity is defined as the concentration of an oligosaccharide required to give half-maximum induction of phytoalexin accumulation (A/A$_{std}$=0.5) in the cotyledon bioassay corrected to the standard curve for hepta-β-glucoside 1 (see Figure 5). The binding activity is defined as the concentration of oligosaccharide required to give 50% inhibition of the binding of radiolabeled hepta-β-glucoside 1 to its binding site in soybean root membranes (see Figure 6). Points are identified with numbers identifying the oligoglucosides (see Figure 4); 1$_r$ - reduced hepta-β-glucoside 1; **Crude Heptamer** - mixture of heptaglucosides prepared from mycelial wall hydrolyzates of *P. megasperma* [109].

10,000-fold) to achieve 50% inhibition of binding of the radiolabeled ligand. Interestingly, the tyramine conjugate of hepta-β-glucoside 1 was somewhat more efficient (3-fold) as a competitive inhibitor in the binding assays than the underivatized hepta-β-glucoside, suggesting that the tyramine moiety may contribute to the binding of the derivatized hepta-β-glucoside.

The effectiveness of oligosaccharides as competitors in the ligand binding assay correlated very well with the ability of the oligosaccharides to induce phytoalexin accumulation (Figure 7). Those oligoglucosides having a high elicitor activity were efficient competitors of the radiolabeled ligand, while biologically less active oligoglucosides were less efficient. Specifically, oligoglucosides 1-4 (Figure 4), ranging in size from hexa- to deca-glucoside, which were indistinguishable on the basis of their abilities to induce phytoalexin accumulation [17], were equally effective competitive inhibitors of binding of the radiolabeled hepta-β-glucoside to its binding site (Figure 6). Two other structurally related oligoglucosides, hexaglucoside 7 and heptaglucoside 5, that had several hundred-fold lower biological activity (Figure 5) were also several hundredfold less efficient as competitive inhibitors in the binding studies (Figure 6). Oligoglucosides 6 and 8, which were even less effective elicitors were also less efficient inhibitors of binding of the labeled hepta-β-glucoside.

The hepta-β-glucoside elicitor binding site identified in soybean membranes is the first putative receptor for an oligosaccharin. Purification and characterization of physiologically important oligosaccharin receptors is a first step toward elucidating the molecular mechanisms by which oligosaccharins induce changes in plant cellular metabolism.

4. Acknowledgements

We are grateful to P. Garegg (University of Stockholm) and T. Ogawa (RIKEN, Japan) and their colleagues for their generous gifts of synthetic oligoglucosides. We also thank C. L. Gubbins Hahn for drawing the figures. This work is supported by a grant from the National Science Foundation (DCB-8904574). Also supported in part by the USDA/DOE/NSF Plant Science Centers Program; this project was funded by Department of Energy grant DE-FG09-87ER13810.

5. References

1. Albersheim P, Darvill AG (1985) Oligosaccharins. *Sci Am* 253:58-64
2. Albersheim P, Darvill AG, McNeil M, Valent BS, Sharp JK, Nothnagel EA, Davis KR, Yamazaki N, Gollin DJ, York WS, Dudman WF, Darvill JE, Dell A (1983) Oligosaccharins: Naturally occuring carbohydrates with biological regulatory functions. In: Ciferri O, Dure L III (eds) Structure and Function of Plant Genomes. Plenum Publishing Corp., New York, NY, p 293-312
3. Anderson JM, Spilatro SR, Klauer SF, Franceschi VR (1989) Jasmonic acid-dependent increase in the level of vegetative storage proteins in soybean. *Plant Sci* 62:45-52
4. Apostol I, Heinstein PF, Low PS (1989) Rapid stimulation of an oxidative burst during elicitation of cultured plant cells. Role in defense and signal transduction. *Plant Physiol* 90:109-116

5. Ayers AR, Ebel J, Finelli F, Berger N, Albersheim P (1976) Host-pathogen interactions. IX. Quantitative assays of elicitor activity and characterization of the elicitor present in the extracellular medium of cultures of *Phytophthora megasperma* var. *sojae*. *Plant Physiol* 57:751-759

6. Ballio A, Federico R, Pessi A, Scalorbi D (1980) Fusicoccin binding sites in subcellular preparations of spinach leaves. *Plant Sci Lett* 18:39-44

7. Barber MS, Bertram RE, Ride JP (1989) Chitin oligosaccharides elicit lignification in wounded wheat leaves. *Physiol Mol Plant Pathol* 34:3-12

8. Bishop PD, Makus DJ, Pearce G, Ryan CA (1981) Proteinase inhibitor-inducing factor activity in tomato leaves resides oligosaccharides enzymically released from cell walls. *Proc Natl Acad Sci USA* 78:3536-3540

9. Bishop PD, Pearce G, Bryant JE, Ryan CA (1984) Isolation and characterization of the proteinase inhibitor-inducing factor from tomato leaves. Identity and activity of poly- and oligogalacturonide fragments. *J Biol Chem* 259:13172-13177

10. Blum W, Hinsch K-D, Schultz G, Weiler EW (1988) Identification of GTP-binding proteins in the plasma membrane of higher plants. *Biochem Biophys Res Commun* 156:954-959

11. Boss WF (1989) Phosphoinositide metabolism: Its relation to signal transduction in plants. In: Boss WF, Morré DJ (eds) Second Messengers in Plant Growth and Development. Alan R. Liss, Inc., New York, NY, p 29-56

12. Boss WF, Morré DJ (1989) Second Messengers in Plant Growth and Development. Alan R. Liss, Inc., New York, NY

13. Branca C, De Lorenzo G, Cervone F (1988) Competitive inhibition of the auxin-induced elongation by α-D-oligogalacturonides in pea stem segments. *Physiol Plant* 72:499-504

14. Bruce RJ, West CA (1982) Elicitation of casbene synthetase activity in castor bean. The role of pectic fragments of the plant cell wall in elicitation by a fungal endopolygalacturonase. *Plant Physiol* 69:1181-1188

15. Cervone F, De Lorenzo G, Degrà L, Salvi G (1987) Elicitation of necrosis in *Vigna unguiculata* Walp. by homogeneous *Aspergillus niger* endo-polygalacturonase and by α-D-galacturonate oligomers. *Plant Physiol* 85:626-630

16. Cervone F, Hahn MG, De Lorenzo G, Darvill A, Albersheim P (1989) Host-pathogen interactions. XXXIII. A plant protein converts a fungal pathogenesis factor into an elicitor of plant defense responses. *Plant Physiol* 90:542-548

17. Cheong J-J, Birberg W, Fügedi P, Pilotti Å, Garegg PJ, Hong N, Ogawa T, Hahn MG (1990) Structure-function relationships of oligo-β-glucoside elicitors of phytoalexin accumulation in plants. *The Plant Cell*, in press

18. Cheong J-J, Hahn MG (1990) A specific high-affinity binding site for the hepta-β-glucoside elicitor exists in soybean membranes. *The Plant Cell*, in press

19. Cosio EG, Frey T, Ebel J (1990) Solubilization of soybean membrane binding sites for fungal β-glucans that elicit phytoalexin accumulation. *FEBS Lett* 264:235-238

20. Cosio EG, Pöpperl H, Schmidt WE, Ebel J (1988) High-affinity binding of fungal β-glucan fragments to soybean (*Glycine max* L.) microsomal fractions and protoplasts. *Eur J Biochem* 175:309-315

21. Cruickshank IAM, Perrin DR (1968) The isolation and partial characterization of monilicolin A, a polypeptide with phaseollin-inducing activity from *Monilinia fructicola*. *Life Sci* 7:449-458

22. Darvill AG, Albersheim P (1984) Phytoalexins and their elicitors - A defense against microbial infection in plants. *Annu Rev Plant Physiol* 35:243-275

23. Davies PJ (1987) Plant Hormones and their Role in Plant Growth and Development. Martinus Nijhoff Publishers, Dordrecht, The Netherlands

24. Davies PJ (1987) The plant hormones: Their nature, occurence, and functions. In: Davies PJ (ed) Plant Hormones and their Role in Plant Growth and Development. Martinus Nijhoff Publishers, Dordrecht, The Netherlands, p 1-11

416

25. Davis KR, Darvill AG, Albersheim P (1986) Host-pathogen interactions. XXXI. Several biotic and abiotic elicitors act synergistically in the induction of phytoalexin accumulation in soybean. *Plant Mol Biol* 6:23-32

26. Davis KR, Hahlbrock K (1987) Induction of defense responses in cultured parsley cells by plant cell wall fragments. *Plant Physiol* 85:1286-1290

27. Davis KR, Lyon GD, Darvill AG, Albersheim P (1984) Host-pathogen interactions. XXV. Endopolygalacturonic acid lyase from *Erwinia carotovora* elicits phytoalexin accumulation by releasing plant cell wall fragments. *Plant Physiol* 74:52-60

28. de Boer AH, Watson BA, Cleland RE (1989) Purification and identification of the fusicoccin binding protein from oat root plasma membrane. *Plant Physiol* 89:250-259

29. Dietrich A, Mayer JE, Hahlbrock K (1990) Fungal elicitor triggers rapid, transient, and specific protein phosphorylation in parsley cell suspension cultures. *J Biol Chem* 265:6360-6368

30. Dixon RA (1986) The phytoalexin response: elicitation, signalling and control of host gene expression. *Biol Rev* 61:239-291

31. Doares SH, Bucheli P, Albersheim P, Darvill AG (1989) Host-pathogen interactions XXXIV. A heat-labile activity secreted by a fungal phytopathogen releases fragments of plant cell walls that kill plant cells. *Mol Plant-Microbe Interac* 2:346-353

32. Doke N (1983) Involvement of superoxide anion generation in the hypersensitive response of potato tuber tissues to infection with an incompatible race of *Phytophthora infestans* and to the hyphal wall components. *Physiol Plant Pathol* 23:345-357

33. Doke N, Ohashi Y (1988) Involvement of an O_2^- generating system in the induction of necrotic lesions on tobacco leaves infected with tobacco mosaic virus. *Physiol Mol Plant Pathol* 32:163-175

34. Drobak BK, Allan EF, Comerford JG, Roberts K, Dawson AP (1988) Presence of guanine nucleotide-binding proteins in a plant hypocotyl microsomal fraction. *Biochem Biophys Res Commun* 150:899-903

35. Eberhard S, Doubrava N, Marfà V, Mohnen D, Southwick A, Darvill A, Albersheim P (1989) Pectic cell wall fragments regulate tobacco thin-cell-layer explant morphogenesis. *The Plant Cell* 1:747-755

36. Epperlein MM, Noronha-Dutra AA, Strange RN (1986) Involvement of the hydroxyl radical in the abiotic elicitation of phytoalexins in legumes. *Physiol Mol Plant Pathol* 28:67-77

37. Evans PT, Malmberg RL (1989) Do polyamines have roles in plant development? *Annu Rev Plant Physiol* 40:235-269

38. Evans RM (1988) The steroid and thyroid hormone receptor superfamily. *Science* 240:889-895

39. Farmer EE, Helgeson JP (1987) An extracellular protein from *Phytophthora parasitica* var. *nicotianae* is associated with stress metabolite accumulation in tobacco callus. *Plant Physiol* 85:733-740

40. Farmer EE, Pearce G, Ryan CA (1989) *In vitro* phosphorylation of plant plasma membrane proteins in response to the proteinase inhibitor inducing factor. *Proc Natl Acad Sci USA* 86:1539-1542

41. Farmer EE, Ryan CA (1990) Interplant communication: Airborne methyl jasmonate induces expression of protease inhibitor genes in plant leaves. *Proc Natl Acad Sci USA*, in press

42. Feyerabend M, Weiler EW (1989) Photoaffinity labeling and partial purification of the putative plant receptor for the fungal wilt-inducing toxin, fusicoccin. *Planta* 178:282-290

43. Fraker PJ, Speck JC Jr (1978) Protein and cell membrane iodinations with a sparingly soluble chloramide, 1,3,4,6-tetrachloro-3a,6a-diphenylglycoluril. *Biochem Biophys Res Commun* 80:849-857

44. Galston AW, Kaur-Sawhney R (1987) Polyamines as endogenous growth regulators. In: Davies PJ (ed) Plant Hormones and their Role in Plant Growth and Development. Martinus Nijhoff Publishers, Dordrecht, The Netherlands, p 280-295

45. Gollin DJ, Darvill AG, Albersheim P (1984) Plant cell wall fragments inhibit flowering and promote vegetative growth in *Lemna gibba* G3. *Biol Cell* 51:275-280

46. Grab D, Feger M, Ebel J (1989) An endogenous factor from soybean (*Glycine max* L.) cell cultures activates phosphorylation of a protein which is dephosphorylated in vivo in elicitor-challenged cells. *Planta* 179:340-348

47. Graniti A, Durbin RD, Ballio A (1989) Phytotoxins and Plant Pathogenesis. NATO ASI Series, Volume H27. Springer Verlag, Berlin, FRG

48. Grisebach H, Edelmann L, Fischer D, Kochs G, Welle R (1989) Biosynthesis of phytoalexins and nod-gene inducing isoflavones in soybean. In: Lugtenberg BJJ (ed) Signal Molecules in Plants and Plant-Microbe Interactions. NATO ASI Series, Volume H36. Springer Verlag, Berlin, p 57-64

49. Hahn MG (1989) Animal receptors - Examples of cellular signal perception molecules. In: Lugtenberg BJJ (ed) Signal Molecules in Plants and Plant-Microbe Interactions. NATO ASI Series, Volume H36. Springer Verlag, Heidelberg, FRG, p 1-26

50. Hahn MG, Bucheli P, Cervone F, Doares SH, O'Neill RA, Darvill A, Albersheim P (1989) The roles of cell wall constituents in plant-pathogen interactions. In: Kosuge T, Nester EW (eds) Plant-Microbe Interactions. Molecular and Genetic Perspectives, Vol. 3. McGraw Hill Publishing Co., New York, NY, p 131-181

51. Hahn MG, Darvill A, Albersheim P, Bergmann C, Cheong J-J, Koller A, Lò V-M (1990) Preparation and characterization of oligosaccharide elicitors of phytoalexin accumulation. In: Bowles DJ (ed) Molecular Plant Pathology: A Practical Approach. IRL Press, Inc., Oxford, UK, p In press

52. Hahn MG, Darvill AG, Albersheim P (1981) Host-pathogen interactions. XIX. The endogenous elicitor, a fragment of a plant cell wall polysaccharide that elicits phytoalexin accumulation in soybeans. *Plant Physiol* 68:1161-1169

53. Hahn MG, Grisebach H (1983) Cyclic AMP is not involved as a second messenger in the response of soybean to infection by *Phytophthora megasperma* f. sp. *glycinea. Z Naturforsch* 38c:578-582

54. Hanks SK (1987) Homology probing: Identification of cDNA clones encoding members of the protein-serine kinase family. *Proc Natl Acad Sci USA* 84:388-392

55. Hanks SK, Quinn AM, Hunter T (1988) The protein kinase family: conserved features and deduced phylogeny of the catalytic domains. *Science* 241:42-52

56. Hasunuma K, Funadera K (1987) GTP-binding protein(s) in green plant, *Lemna paucicostata. Biochem Biophys Res Commun* 143:908-912

57. Hasunuma K, Furukawa K, Tomita K, Mukai C, Nakamura T (1987) GTP-binding proteins in etiolated epicotyls of *Pisum sativum* (Alaska) seedlings. *Biochem Biophys Res Commun* 148:133-139

58. Hesse T, Feldwisch J, Balshüsemann D, Bauw G, Puype M, Vandekerckhove J, Löbler M, Klämbt D, Schell J, Palme K (1989) Molecular cloning and structural analysis of a gene from Zea mays (L.) coding for a putative receptor for the plant hormone auxin. *EMBO J* 8:2453-2461

59. Hicks GR, Rayle DL, Jones AM, Lomax TL (1989) Specific photoaffinity labeling of two plasma membrane polypeptides with an azido auxin. *Proc Natl Acad Sci USA* 86:4948-4952

60. Hicks GR, Rayle DL, Lomax TL (1989) The *Diageotropica* mutant of tomato lacks high specific activity auxin binding sites. *Science* 245:52-54

61. Hornberg C, Weiler EW (1984) High-affinity binding sites for abscisic acid on the plasmalemma of *Vicia faba* guard cells. *Nature* 310:321-324

62. Inohara N, Shimomura S, Fukui T, Futai M (1989) Auxin-binding protein located in the endoplasmic reticulum of maize shoots: Molecular cloning and complete primary structure. *Proc Natl Acad Sci USA* 86:3564-3568

63. Jin DF, West CA (1984) Characteristics of galacturonic acid oligomers as elicitors of casbene synthetase activity in castor bean seedlings. *Plant Physiol* 74:989-992

64. Jones AM, Venis MA (1989) Photoaffinity labeling of indole-3-acetic acid-binding proteins in maize. *Proc Natl Acad Sci USA* 86:6153-6156

65. Kauss H (1987) Some aspects of calcium-dependent regulation in plant metabolism. *Annu Rev Plant Physiol* 38:47-72

66. Keen NT, Yoshikawa M (1983) β-1,3-endoglucanase from soybean releases elicitor-active carbohydrates from fungus cell walls. *Plant Physiol* 71:460-465

67. Kendra DF, Hadwiger LA (1984) Characterization of the smallest chitosan oligomer that is maximally antifungal to *Fusarium solani* and elicits pisatin formation in *Pisum sativum*. *Exp Mycol* 8:276-281

68. Klämbt D (1990) A view about the function of auxin-binding proteins at plasma membranes. *Plant Mol Biol* 14:1045-1050

69. Kogel G, Beissman B, Reisener HJ, Kogel KH (1988) A single glycoprotein from *Puccinia graminis* f. sp. *tritici* cell walls elicits the hypersensitive lignification response in wheat. *Physiol Mol Plant Pathol* 33:173-185

70. Kurosaki F, Tsurusawa Y, Nishi A (1987) Breakdown of phosphatidylinositol during the elicitation of phytoalexin production in cultured carrot cells. *Plant Physiol* 85:601-604

71. Kurosaki F, Tsurusawa Y, Nishi A (1987) The elicitation of phytoalexins by Ca^{2+} and cyclic AMP in carrot cells. *Phytochemistry* 26:1919-1923

72. Köhle H, Jeblick W, Poten F, Blashek W, Kauss H (1985) Chitosan-elicited callose synthesis in soybean cells as a Ca^{2+}-dependent process. *Plant Physiol* 77:544-551

73. Lamb CJ, Lawton MA, Dron M, Dixon RA (1989) Signals and transduction mechanisms for activation of plant defenses against microbial attack. *Cell* 56:215-224

74. Lawton MA, Yamamoto RT, Hanks SK, Lamb CJ (1989) Molecular cloning of plant transcripts encoding protein kinase homologs. *Proc Natl Acad Sci USA* 86:3140-3144

75. Lerouge P, Roche P, Faucher C, Maillet F, Truchet G, Promé JC, Dénarié J (1990) Symbiotic host-specificity of *Rhizobium meliloti* is determined by a sulphated and acylated glucosamine oligosaccharide signal. *Nature* 344:781-784

76. Lindner WA, Hoffmann C, Grisebach H (1988) Rapid elicitor-induced chemiluminescence in soybean cell suspension cultures. *Phytochemistry* 27:2501-1503

77. Löbler M, Klämbt D (1985) Auxin-binding protein from coleoptile membranes of corn (*Zea mays* L.). I. Purification by immunological methods and characterization. *J Biol Chem* 260:9848-9853

78. Marmé D (1989) The role of calcium and calmodulin in signal transduction. In: Boss WF, Morré DJ (eds) Second Messengers in Plant Growth and Development. Alan R. Liss, Inc., New York, NY, p 57-80

79. Marrè E (1979) Fusicoccin: A tool in plant physiology. *Annu Rev Plant Physiol* 30:273-288

80. Mason HS, Mullet JE (1990) Expression of two soybean vegetative storage protein genes during development and in response to water deficit, wounding, and jasmonic acid. *The Plant Cell* 2:569-579

81. Matsui M, Sasamoto S, Kunieda T, Nomura N, Ishizaki R (1989) Cloning of *ara*, a putative *Arabidopsis thaliana* gene homologous to the *ras*-related gene family. *Gene* 76:313-319

82. McDougall GJ, Fry SC (1988) Inhibition of auxin-stimulated growth of pea stem segments by a specific nonasaccharide of xyloglucan. *Planta* 175:412-416

83. McDougall GJ, Fry SC (1989) Structure-activity relationships for xyloglucan oligosaccharides with antiauxin activity. *Plant Physiol* 89:883-887

84. McDougall GJ, Fry SC (1989) Anti-auxin activity of xyloglucan oligosaccharides: the role of groups other than the terminal α-L-fucose residue. *J Exp Bot* 40:233-238

85. Meyer C, Feyerabend M, Weiler EW (1989) Fusicoccin-binding proteins in *Arabidopsis thaliana* (L.) Heynh. *Plant Physiol* 89:692-699

86. Munro S, Pelham HRB (1987) A C-terminal signal prevents secretion of luminal ER proteins. *EMBO J* 48:899-907

87. Nasrallah JB, Kao T-H, Chen C-H, Goldberg ML, Nasrallah ME (1987) Amino-acid sequence of glycoproteins encoded by three alleles of the *S* locus of *Brassica oleracea*. *Nature* 326:617-619

88. Nasrallah JB, Kao T-H, Goldberg ML, Nasrallah ME (1985) A cDNA clone encoding an *S*-locus-specific glycoprotein from *Brassica oleracea*. *Nature* 318:263-267

89. Newton RP, Brown EG (1986) The biochemistry and physiology of cyclic AMP in higher plants. In: Chadwick CM, Garrod DR (eds) Hormones, Receptors and Cellular Interactions in Plants. Cambridge University Press, Cambridge, UK, p 115-153

90. Nothnagel EA, McNeil M, Albersheim P, Dell A (1983) Host-pathogen interactions. XXII. A galacturonic acid oligosaccharide from plant cell walls elicits phytoalexins. *Plant Physiol* 71:916-926

91. Palme K, Diefenthal T, Hesse T, Nitsche K, Campos N, Feldwisch J, Garbers C, Hesse F, Schwonke S, Schell J (1989) Signalling elements in higher plants: Identification and molecular analysis of an auxin-binding protein, GTP-binding regulatory proteins and calcium sensitive proteins. In: Lugtenberg BJJ (ed) Signal Molecules in Plants and Plant-Microbe Interactions. NATO ASI Series, Volume H36. Springer Verlag, Berlin, FRG, p 71-83

92. Parker JE, Hahlbrock K, Scheel D (1988) Different cell-wall components from *Phytophthora megasperma* f. sp. *glycinea* elicit phytoalexin production in soybean and parsley. *Planta* 176:75-82

93. Pesci P, Cocucci SM, Randazzo G (1979) Characterization of fusicoccin binding to receptor sites on cell membranes of maize coleoptile tissue. *Plant Cell Environ* 2:205-209

94. Pesci P, Tognoli L, Beffagna N, Marrè E (1979) Solubilization and partial purification of a fusicoccin-receptor complex from maize microsomes. *Plant Sci Lett* 15:313-322

95. Ranjeva R, Boudet AM (1987) Phosphorylation of proteins in plants: Regulatory effects and potential involvement in stimulus/reponse coupling. *Annu Rev Plant Physiol* 38:73-93

96. Ricci P, Bonnet P, Huet J-C, Sallantin M, Beauvais-Cante F, Bruneteau M, Billard V, Michel G, Pernollet J-C (1989) Structure and activity of proteins from pathogenic fungi *Phytophthora* eliciting necrosis and acquired resistance in tobacco. *Eur J Biochem* 183:555-563

97. Robertsen B (1986) Elicitors of the production of lignin-like compounds in cucumber hypocotyls. *Physiol Mol Plant Pathol* 28:137-148

98. Robertsen B (1987) Endo-polygalacturonase from Cladosporium cucumerinum elicits lignification in cucumber hypocotyls. *Physiol Mol Plant Pathol* 31:361-374

99. Rudolph K (1976) Non-Specific Toxins. In: Heitefuss R, Williams PH (eds) Encyclopedia of Plant Physiology, New Series, Vol. 4, Physiological Plant Pathology. Springer Verlag, Berlin, FRG, p 270-315

100. Ryan CA (1987) Oligosaccharide signaling in plants. *Annu Rev Cell Biol* 3:295-317

101. Scheel D, Colling C, Keller H, Parker J, Schulte W, Hahlbrock K (1989) Studies on elicitor recognition and signal transduction in host and non-host plant/fungus pathogenic interactions. In: Lugtenberg BJJ (ed) Signal Molecules in Plants and Plant-Microbe Interactions. NATO ASI Series, Volume H36. Springer Verlag, Berlin, FRG, p 211-218

102. Scheel D, Parker JE (1990) Elicitor recognition and signal transduction in plant defense gene activation. *Z Naturforsch* 45c:569-575

103. Scheffer RP, Livingston RS (1984) Host-selective toxins and their role in plant diseases. *Science* 223:17-21

104. Schmidt WE, Ebel J (1987) Specific binding of a fungal glucan phytoalexin elicitor to membrane fractions from soybean *Glicine max*. *Proc Natl Acad Sci USA* 84:4117-4121

105. Scholtens-Toma IMJ, de Wit PJGM (1988) Purification and primary structure of a necrosis-inducing peptide from the apoplastic fluids of tomato infected with *Cladosporium fulvum* (syn. *Fulvia fulva*). *Physiol Mol Plant Pathol* 33:59-67

106. Schroeder JI, Hedrich R (1989) Involvement of ion channels and active transport in osmoregulation and signaling of higher plant cells. *Trends Biochem Sci* 14:187-192

420

107. Sharp JK, Albersheim P, Ossowski P, Pilotti Å, Garegg PJ, Lindberg B (1984) Comparison of the structures and elicitor activities of a synthetic and a mycelial-wall-derived hexa(β-D-glucopyranosyl)-D-glucitol. *J Biol Chem* 259:11341-11345

108. Sharp JK, McNeil M, Albersheim P (1984) The primary structures of one elicitor-active and seven elicitor-inactive hexa(β-D-glucopyranosyl)-D-glucitols isolated from the mycelial walls of *Phytophthora megasperma* f. sp. *glycinea*. *J Biol Chem* 259:11321-11336

109. Sharp JK, Valent B, Albersheim P (1984) Purification and partial characterisation of a β-glucan fragment that elicits phytoalexin accumulation in soybean. *J Biol Chem* 259:11312-11320

110. Shimomura S, Sotobaya T, Futai M, Fukui T (1986) Purification and properties of an auxin-binding protein from maize shoot membranes. *J Biochem* 99:1513-1524

111. Strasser H, Hoffmann C, Grisebach H, Matern U (1986) Are polyphosphoinositides involved in signal transduction of elicitor-induced phytoalexin synthesis in cultured plant cells? *Z Naturforsch* 41c:717-724

112. Stäb MR, Ebel J (1987) Effects of Ca^{2+} on phytoalexin induction by fungal elicitor in soybean cells. *Arch Biochem Biophys* 257:416-423

113. Tillmann U, Viola G, Kayser B, Siemeister G, Hesse T, Palme K, Löbler M, Klämbt D (1989) cDNA clones of the auxin-binding protein from corn coleoptiles (*Zea mays* L.): isolation and characterization by immunological methods. *EMBO J* 8:2463-2467

114. Tran Thanh Van K, Toubart P, Cousson A, Darvill AG, Gollin DJ, Chelf P, Albersheim P (1985) Manipulation of the morphogenetic pathways of tobacco explants by oligosaccharins. *Nature* 314:615-617

115. Walker JC, Zhang R (1990) Relationship of a putative receptor protein kinase from maize to the S-locus glycoproteins of *Brassica*. *Nature* 345:743-746

116. Walker-Simmons M, Hadwiger L, Ryan CA (1983) Chitosans and pectic polysaccharides both induce the accumulation of the antifungal phytoalexin pisatin in pea pods and antinutrient proteinase inhibitors in tomato leaves. *Biochem Biophys Res Commun* 110:194-199

117. Wolpert TJ, Macko V (1989) Specific binding of victorin to a 100-kDa protein from oats. *Proc Natl Acad Sci USA* 86:4092-4096

118. Yamazaki N, Fry SC, Darvill AG, Albersheim P (1983) Host-pathogen interactions. XXIV. Fragments isolated from suspension-cultured sycamore cell walls inhibit the ability of the cells to incorporate [^{14}C]leucine into proteins. *Plant Physiol* 72:864-869

119. York WS, Darvill AG, Albersheim P (1984) Inhibition of 2,4-dichlorophenoxyacetic acid-stimulated elongation of pea stem segments by a xyloglucan oligosaccharide. *Plant Physiol* 75:295-297

PERCEPTION OF PATHOGEN-DERIVED ELICITOR AND SIGNAL TRANSDUCTION IN HOST DEFENSES

J. EBEL, E.G. COSIO, AND T. FREY
Biologisches Institut II
Universität Freiburg
Schänzlestr. 1
D-7800 Freiburg
Federal Republic of Germany

1. INTRODUCTION

The biochemical mechanisms of resistance of plants against attack by microorganisms include a wide range of inducible defense responses. One inducible defense mechanism is the production of phytoalexins at the site of attempted infection. A host-parasite system in which phytoalexins have been implicated as one defense factor is soybean (Glycine max L.) and Phytophthora megasperma f.sp. glycinea, the fungus causing stem and root rot in this plant [1, 2]. The phytoalexin defense response in soybean can be activated, not only upon challenge of different tissues by the fungus, but also upon exposure to elicitors. Elicitors are now widely used in experimental systems of reduced complexity, such as soybean cell cultures, to study certain common features of inducible plant resistance reactions. These include: (i) perception of pathogen-derived signal(s); (ii) intracellular signal transduction and (iii) initiation of defense response [2].

2. SIGNAL PERCEPTION

The ß-glucan elicitors of P. megasperma are components of prominent structural polysaccharides of the cell walls of oomycetes. Investigations on the structure, distribution, and biological activity of these glucans have shown that the configuration required for elicitor activity is contained in oligosaccharide fractions with 3-, 6-, and 3,6-linked glucosyl residues [3, 4]. Structure-activity relationships have also been investigated with a series of chemically synthesized oligoglucosides [5], including a hexa-ß-glucosyl glucitol and its corresponding hepta-ß-glucoside.

H. Hennecke and D. P. S. Verma (eds.),
Advances in Molecular Genetics of Plant-Microbe Interactions, Vol. 1, 421–427.
© 1991 Kluwer Academic Publishers. Printed in the Netherlands.

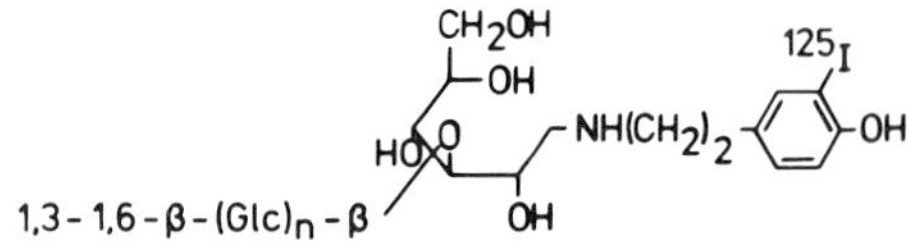

Figure 1. Chemical structures of ß-glucan derivatives used for binding studies (compounds 1 and 2) and for affinity chromatography (compound 3). Derivatives were synthesized by reductive amination using either a fungal ß-glucan fraction (1, 3; average n = 17) or a synthetic hepta-ß-glucoside (2) and different amines. For affinity chromatography, fungal glucan was bound to an aminopropyl support. For binding studies, aryl alkyl-amine conjugates were iodinated using ^{125}I [7, 8].

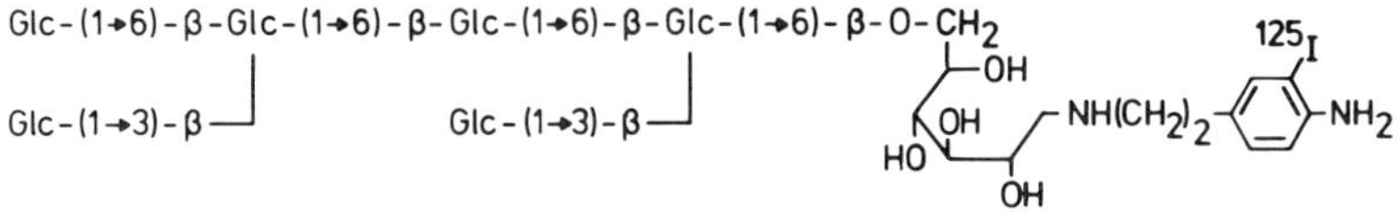

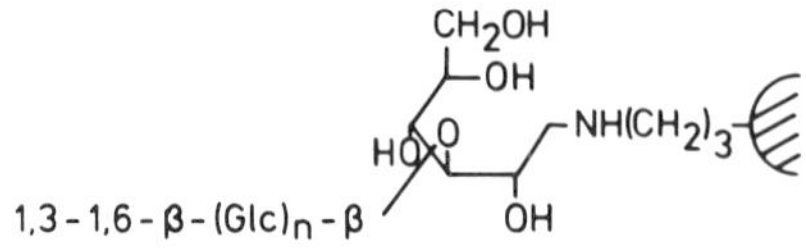

The initial event in the activation by the fungal ß-glucan of genes involved in phytoalexin synthesis is thought to be its interaction with a specific primary target site or receptor on the host cell surface [1, 2]. The identification of ß-glucan-binding sites in soybean tissues has been achieved by using elicitor-active glucan fragments which had been obtained by size fractionation of P. megasperma cell wall hydrolysates [6]. The initial binding studies have thus employed fungal glucan preparations consisting of a mixture of structural isomers of large degree of polymerization (DP) within a narrow molecular size range (average DP of 18) [6, 7]. The use of larger glucans with a variety of branching patterns has assured a broad enough approach for the identification of glucan elicitor-binding sites, something a single structure might have excluded. Glucan binding to soybean membranes has fulfilled the criteria of saturability, reversibility, localization on the plasma membrane, and high affinity (K_ds between 10 and 40 nM) [7, E.G. Cosio and T. Frey, unpublished results]. Specific ß-glucan binding has been detected in membrane preparations from soybean roots, hypocotyls, cotyledons, and cell suspension cultures, but not in soluble protein extracts from these tissues.

A series of radioactively labelled glucan ligands has been prepared and used in our binding studies (Fig. 1) [6-8]. Apart from a ^{3}H-labelled ligand [6] all others have involved modification of the oligo-ß-glucans by reductive amination and radioiodination of the conjugates to yield specific radioactivities of at least 100 Ci/mmol [7, 8]. Recently, an aminophenylethylamine conjugate has been synthesized [8]. This type of derivative (Fig. 1) has been chosen because the aryl amino group offers a greater flexibility for further chemical modifications than the aryl hydroxy group of tyramine used earlier [7]. In all cases studied, including even biotinylation and complex formation with avidin (E. G. Cosio, unpublished results), it has been apparent that covalent modification at the reducing end of the glucan had no major effect on its binding ability. The elicitor activity of the derivatives and of unmodified glucan has proved to be similar [7].

Ligand competition studies and elicitor activity assays with a variety of poly- and oligosaccharides have been carried out to investigate the specificity of ß-glucan binding to a soybean membrane fraction and the relative biological activity of the ligands [6-8]. A strong positive correlation has been found between the ability of the different competitors to displace the labelled glucan ligand and to elicit phytoalexin accumulation in a bioassay (Table 1) [6-8]. The most likely explanation for the pronounced dependence of ligand affinity on the apparent DP of the fungal glucan fractions (Table 1) [8] would be that the soybean binding sites recognize one or more "basic" structural motifs within the glucans. The probability for a certain glucan of containing such structures increases with its size.

Most interestingly, the ß-glucan-binding sites have displayed the highest affinity of all glucan fractions tested (K_d about 3 nM) for a hepta-ß-glucoside (Table 1) which was synthesized according to a published method [9]. The heptaglucoside has also shown the highest phytoalexin elicitor activity in this study. Displacement of the P. megasperma ^{125}I-glucan from soybean membranes by the syn-

TABLE 1. Correlation between binding affinity (IC_{50}) of soybean binding sites for branched ß-glucan fractions and corresponding activity of the glucans as phytoalexin elicitors (EC_{50}) in a bioassay

ß-Glucan fraction	IC_{50} (nM)[a]	EC_{50} (µM)[b]
App. DP = 5[c]	12,000	200
App. DP = 10	410	12
App. DP 16 to 20	100	0.6
Synthetic heptaglucoside	16	0.22

[a]IC_{50}, concentration required for half-maximal displacement of the ^{125}I-labelled derivative of the synthetic heptaglucoside; [b]EC_{50}, concentration resulting in half-maximal phytoalexin accumulation in a bioassay; [c]App. DP, apparent degree of polymerization.

thetic ligand has been complete and followed a uniform sigmoidal pattern indicating access, with similar affinity, of the heptaglucoside to all sites available to the ^{125}I-labelled fungal glucan fraction. The results suggest that all soybean ß-glucan-binding sites which have been identified with the help of the ^{125}I-labelled fungal glucan mixture display maximal affinity for a single glucan structure. This structure may possibly be that of the heptaglucoside identified by Sharp et al. [4].

Further elucidation of the role of the ß-glucan-binding sites in mediating the elicitor stimulus in soybean tissues requires their isolation and characterization. Recently, glucan-binding sites have been solubilized using detergents [10]. Partial purification of the putative binding protein(s) has been achieved by a procedure including chromatography on Q Sepharose and on a glucan affinity matrix (Table 2). This matrix has been obtained by coupling of a fungal ß-glucan fraction to an inert support (Fig. 1).

3. SIGNAL TRANSDUCTION AND INITIATION OF DEFENSE RESPONSE

Defense gene activation, as involved in the phytoalexin response of soybean, requires the transduction of (an) elicitor signal(s) from the site of primary perception at the cell surface to the nucleus where transcription of specific genes is initiated [2]. In several plant-elicitor systems, increasing evidence has been obtained for the involvement of Ca^{2+} in defense gene activation [2]. In soybean cell cultures, a requirement of external Ca^{2+} for the activation of the phytoalexin response by the glucan elicitor from P. megasperma has been demonstrated [11]. The observations [11] suggest the possibility that an early response following the interaction of the elicitor with the surface of soybean cells is a number of metabolite (ion) flux changes across the plasma membrane. These fluxes might include or affect Ca^{2+} fluxes [1, 2].

TABLE 2. Partial purification of detergent-solubilized ß-glucan-binding sites from soybean roots

Step	Protein	Specific activity	Total activity	Purification	Recovery
	mg	pmol/mg	pmol	-fold	%
Detergent solubilizate	110	0.9	100	1	100
Q Sepharose chromatography	19	1.8	34	2	34
Glucan affinity chromatography	0.05	40	2	44	2

Little is known at present about the integration of Ca^{2+} in the transduction pathway linking elicitor perception to the initiation of the phytoalexin response [2]. Our results have demonstrated that elicitor treatment of soybean cells rapidly affected phosphorylation of several proteins [12]. The results are compatible with the hypothesis that protein phosphorylation is involved in the transduction pathway following elicitor perception at the surface of soybean cells.

Induction of phytoalexin accumulation in soybean tissues comprises a whole complement of enzymes of the biosynthetic pathway [1] and is very likely regulated by tem-

426

porary gene activation. Soybean cultivars are differentially resistant to several P. megasperma races. Only resistant soybean plants rapidly accumulate the phytoalexins at
infection sites and fungal growth is arrested early in the
immediate vicinity of the inoculation site. Race:cultivar-
specific temporal differences have also been observed for
the increases in the catalytic activities of several
enzymes involved in phytoalexin biosynthesis and in the
activities (amounts) of the mRNAs of some of the enzymes
during early stages of infection (until 11 h) [1, 13]. The
kinetics of the changes in mRNA activities and amounts, as
observed in resistant plants, has been remarkably similar
to that measured after stimulation of soybean cell cultures
by glucan elicitor from P. megasperma [13]. The rapidity of
mRNA induction in both the resistant plant-fungus and the
elicitor-cell culture interactions might imply that at
least some of the components involved in the perception and
the transduction of the signal(s) in the two challenged
tissues are similar and are available within a short period
after treatment.

4. ACKNOWLEDGEMENT

This work was supported by the Deutsche Forschungsgemeinschaft (SFB 206) and Fonds der Chemischen Industrie. We
thank Sandoz A.G., Basel, for providing the synthetic
hepta-ß-glucoside.

5. REFERENCES

1.   Ebel, J. and Grisebach, H. (1988) Defense strategies
      of soybean against the fungus Phytophthora megasperma
      f.sp. glycinea: a molecular analysis, Trends Biochem.
      Sci. 13, 23-27.
2.   Ebel, J. and Scheel, D. (1990) Elicitor recognition
      and signal transduction, in T. Boller and F. Meins,
      Jr. (eds.), Plant Gene Research, Vol. 8, Springer,
      Wien, in press.
3.   Darvill, A.G. and Albersheim, P. (1984) Phytoalexins
      and their elicitors - A defense against microbial in-
      fections in plants, Ann. Rev. Plant Physiol. 35, 243-
      275.
4.   Sharp, J.K., McNeil, M. and Albersheim, P. (1984) The
      primary structures of one elicitor-active and seven
      elicitor-inactive hexa (ß-D-glucopyranosyl)-D-glu-
      citols isolated from the mycelial walls of Phy-
      tophthora megasperma f.sp. glycinea, J. Biol. Chem.
      259, 11321-11336.

5. Hahn, M.G., Cheong, J.-J., Birberg, W., Fügedi, P., Piloti, Å, Garegg, P., Hong, N., Nakahara, Y. and Ogawa, T. (1989) Elicitation of phytoalexins by synthetic oligoglucosides, synthetic oligogalacturonides, and their derivatives, in B.J.J. Lugtenberg (ed.), Signal Molecules in Plants and Plant-Microbe Interactions, NATO ASI Ser., Vol. H36, Springer, Berlin, pp. 91-97.

6. Schmidt, W.E. and Ebel, J. (1987) Specific binding of a fungal glucan phytoalexin elicitor to membrane fractions from soybean Glycine max, Proc. Natl. Acad. Sci. USA 84, 4117-4121.

7. Cosio, E.G., Pöpperl, H., Schmidt, W.E. and Ebel, J. (1988) High-affinity binding of fungal ß-glucan fragments to soybean (Glycine max L.) microsomal fractions and protoplasts, Eur. J. Biochem. 175, 309-315.

8. Cosio, E.G., Frey, T., Verduyn, R., van Boom, J. and Ebel, J. (1990) High-affinity binding of a synthetic heptaglucoside and fungal glucan phytoalexin elicitors to soybean membranes, FEBS Letters, in press.

9. Fügedi, P., Birberg, W., Garegg, P.J. and Pilotti, Å (1987) Syntheses of a branched heptasaccharide having phytoalexin-elicitor activity, Carbohydr. Res. 164, 297-312.

10. Cosio, E.G., Frey, T. and Ebel, J. (1990) Solubilization of soybean membrane binding sites for fungal ß-glucans that elicit phytoalexin accumulation, FEBS Letters 264, 235-238.

11. Stäb, M.R. and Ebel, J. (1987) Effects of Ca^{2+} on phytoalexin induction by a fungal elicitor in soybean cells, Arch. Biochem. Biophys. 257, 416-423.

12. Grab, D., Feger, M. and Ebel, J. (1989) An endogenous factor from soybean (Glycine max L.) cell cultures activates phosphorylation of a protein which is dephosphorylated in vivo in elicitor-challenged cells, Planta 179, 340-348.

13. Habereder, H., Schröder, G. and Ebel, J. (1989) Rapid induction of phenylalanine ammonia-lyase and chalcone synthase mRNAs during fungus infection of soybean (Glycine max L.) roots or elicitor treatment of soybean cell cultures at the onset of phytoalexin synthesis, Planta 177, 58-65.

PHOSPHOPROTEIN-CONTROLLED CHANGES IN ION TRANSPORT ARE COMMON EVENTS IN SIGNAL TRANSDUCTION FOR CALLOSE AND PHYTOALEXIN INDUCTION

H. Kauss
FB Biologie, Universität Kaiserslautern
Postfach 3049
D-6750 Kaiserslautern
BRD

ABSTRACT. Synthesis of the 1,3-ß-glucan callose in suspension-cultured plant cells can be elicited by chemically different substances. Some callose elicitors can also be used to induce coumarin synthesis in parsley cells, similar to a fungal glycoprotein. Early events in the induction of both defense reactions are Ca^{2+} uptake, external alkalinization and K^+ leakage. The protein kinase inhibitor K252a and the protein phosphatase inhibitor okadaic acid influence the above responses, indicating the involvement of phosphoproteins in the signal transduction chain.

Plants make use of several metabolic reactions to defend themselves against microbes [1-4]. The cell physiology and biochemistry of two induced defense reactions, callose and phytoalexin synthesis, have been subject to intensive research in the past years. Rapid callose deposition onto the cell wall, e.g. in papillae basipetal to the sites of fungal penetration, appears to operate by enzyme activation, while phytoalexin synthesis involves long-term gene activation resulting in *de novo* synthesis of enzymes required for the formation of respective toxic secondary metabolites (phytoalexins). Nevertheless, callose and phytoalexin synthesis appear to utilize some common initial reactions related to signal transduction.

Fig. 1 lists some examples (substances not boxed in) of chemically very different compounds which have been empirically found to be elicitors of callose formation. Chitosan, a constituent of the cell walls of many phytopathogenic fungi, appears to bind ionically to the plasma membrane over a broad surface area, presumably to phospholipid head groups [4,5] whereas other callose elicitors have been suggested to bind to sterols or intercalate into the lipid phase of the membrane [3]. Thus the callose elicitors appear not to bind initially to complementary receptor proteins in the classical sense but to various general membrane constituents. In any case, a rapid K^+ efflux, extracellular alkalinization and a Ca^{2+} uptake into the cells is associated with callose formation. Although considerable quantitative differences between the various callose elicitors exist with regard to these effects, the signal resulting from binding of the elicitors to membrane constituents appears to be transmitted by a common but unknown signal transmitting system (ST in Fig. 1). It first appeared possible that the callose elicitors facilitated a nonspecific permeation of Ca^{2+}, but the inhibition of digitonin-induced Ca^{2+} uptake and callose synthesis by nifedipine [6] suggests the involvement of specific ion transport sites.

428

H. Hennecke and D. P. S. Verma (eds.),
Advances in Molecular Genetics of Plant-Microbe Interactions, Vol. 1, 428–431.
© 1991 *Kluwer Academic Publishers. Printed in the Netherlands.*

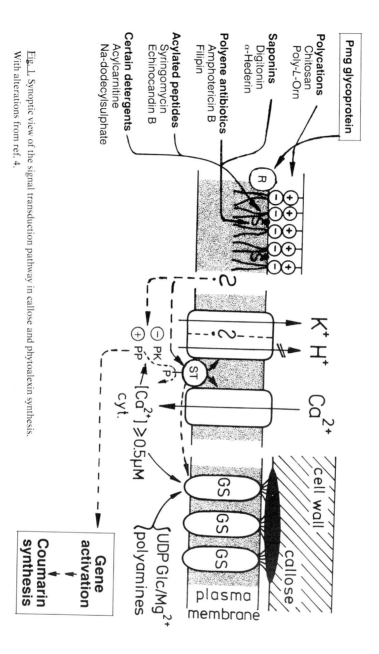

Pmg glycoprotein

Polycations
Chitosan
Poly-L-Orn

Saponins
Digitonin
α-Hederin

Polyene antibiotics
Amphotericin B
Filipin

Acylated peptides
Syringomycin
Echinocandin B

Certain detergents
Acylcarnitine
Na-dodecylsulphate

K⁺ H⁺ Ca²⁺

PK P
PP

[Ca²⁺] ≥ 0.5μM
cyt.

cell wall

callose

GS GS GS {UDP Glc/Mg²⁺ {polyamines

plasma membrane

Gene activation

Coumarin synthesis

Fig. 1. Synoptic view of the signal transduction pathway in callose and phytoalexin synthesis. With alterations from ref. 4.

It was originally thought that the major signal which couples changes in ion permeability to callose synthesis was an increase in $[Ca^{2+}]$ at the cytoplasmic side of the plasma membrane. This transient increase in $[Ca^{2+}]$ could directly activate the Ca^{2+}-dependent, plasma membrane-localized 1,3-ß-glucan synthase [3]. This enzyme complex [7] is also stimulated *in vitro* by polyamines and certain lipids [3]. However, dose-response curves with amphotericin B [4] of the $K^+/H^+/Ca^{2+}$ response and of callose synthesis, as well as experiments with the nonselective bacterial phytotoxin syringomycin [8] suggested that Ca^{2+} influx was an important but not the only signal triggering the 1,3-ß-glucan synthase. A similar conclusion could be drawn from experiments with the Ca^{2+} ionophore A-23187 [6]. Recent experiments using digitonin, amphotericin B and syringomycin as callose elicitors have shown that a potent inhibitor of protein phosphatases, okadaic acid [9], can be used to inhibit induced Ca^{2+} uptake and, to a lesser extent, also callose synthesis. These results again suggest that a gated Ca^{2+} transport is involved, which is regulated by a more general signal transduction system involving the control phosphoproteins (H. Kauss, W. Jeblick, unpublished results). The signal that in addition to Ca^{2+} is involved in the activation of the 1,3-ß-glucan synthase might result either directly from the activated ST (Fig. 1) in analogy to the action of activated G-proteins on target enzymes in animal plasma membranes, or indirectly from changes in the membrane lipid composition [3].

Performing the above experiments on callose synthesis we observed that some callose elicitors can also be used to induce phytoalexins. Chitosan and digitonin, for instance, induce glyceollin synthesis in soybean cells [3], while with parsley cells, conditions have not yet been found where digitonin acts as an elicitor. However, chitosan [10] and amphotericin B (U. Conrath, W. Jeblick, H. Kauss, unpublished results) elicit coumarin synthesis, and their efficiency is comparable to the glycoprotein contained in extracts from cell walls of *Phytophthora megasperma* f.sp. *glycinea* (Pmg elicitor). This glycoprotein is assumed [1,2] to initially bind to a still hypothetical plasma membrane-localized receptor protein (R in Fig. 1). Obviously a generalization in regard to signal perception is not yet possible. However, the three chemically different coumarin elicitors found for parsley cells all appear to deliver information to a joint signal transduction system (ST in Fig. 1), which might have components in common with the signal transduction pathway presumed for callose induction. The latter suggestion is not only based on the fact that some elicitors are identical but also on some early responses in common. It was first shown that the Pmg-elicitor when added to parsley cells suspended in diluted growth medium also leads within a few min to K^+ efflux, external alkalization and a concommitant Ca^{2+} uptake into the cells, all effects lasting for about 20 min (C. Colling, K. Hahlbrock, H. Kauss, D. Scheel, unpublished results). The transient nature of the Ca^{2+} uptake induced by Pmg elicitor indicates an adaptation process and is in contrast to the effect of callose elicitors such as chitosan or digitonin which cause a Ca^{2+} uptake lasting far longer periods. A rapid Pmg elicitor-induced increase of some phosphoproteins has been recently observed [11]. Accordingly, other features pointing to the involvement of a common and more general signal transducing system are just becoming evident from experiments with the protein kinase inhibitor K252a [12]. This antibiotic greatly increases coumarin synthesis induced by low concentrations of Pmg elicitor, indicating that a decrease in the degree of phosphorylation of a critical phosphoprotein might render the signal transducing system more sensitive (U. Conrath, W. Jeblick, H. Kauss, unpublished). Interestingly, K252a given alone causes a marked Ca^{2+} uptake but no coumarin synthesis. Given together with Pmg elicitor, K252a can partially inhibit the $K^+/H^+/Ca^{2+}$ response. These results indicate that Ca^{2+} is not the only signal involved in the activation of genes of the coumarin pathway. The protein phosphatase inhibitor okadaic acid [9], decreases Pmg-induced Ca^{2+} uptake and coumarin synthesis, the dose-response indicating that protein dephosphorylation is not only involved in early signal transduction but also at later steps in the signal chain. Taken together these inhibitor studies strongly suggest that the stimuli of various phytoalexin elicitors are transmitted by protein phosphorylation/dephosphorylation which in turn affects rapid changes in ion transport.

References

1. Hahlbrock, K. and Scheel, D. (1989) 'Physiology and molecular biology of phenylpropanoid metabolism', Annu. Rev. Plant Physiol. Plant Mol. Biol. 40, 347-369.
2. Dixon, R.A. and Lamb, C.J. (1990) 'Molecular communication in interactions between plants and microbial pathogens', Annu. Rev. Plant Physiol. Plant Mol. Biol. 41, 339-367.
3. Kauss, H. (1990) 'Role of the plasma membrane in host-pathogen interactions', in C. Larsson and I.M. Møller (eds.), The Plant Plasma Membrane, Springer-Verlag, Berlin Heidelberg, pp. 320-350.
4. Kauss, H., Waldmann, T. and Quader, H. (1990) 'Ca^{2+} as a signal in the induction of callose synthesis', in R. Ranjeva and A. Boudet (eds.), Signal Perception and Transduction in Higher Plants, Springer-Verlag, Heidelberg, in press.
5. Kauss, H., Jeblick, W. and Domard, A. (1989) 'The degrees of polymerization and N-acetylation of chitosan determine its ability to elicit callose formation in suspension cells and protoplasts of Catharanthus roseus', Planta 178, 385-392.
6. Waldmann, T., Jeblick, W. and Kauss, H. (1988) 'Induced net Ca^{2+} uptake and callose biosynthesis in suspension-cultured plant cells', Planta 173, 88-95.
7. Fink, J., Jeblick, W. and Kauss, H. (1990) 'Partial purification and immunological characterization of 1,3-ß-glucan synthase cells of Glycine max', Planta 171, 130-135.
8. Kauss, H., Waldmann, T., Jeblick, W. and Takemoto T.Y. (1990) 'Syringomycin-induced callose synthesis and Ca^{2+} uptake in suspension-cultured cells of Catharanthus roseus. Physiol. Plant., in press.
9. Cohen, P., Holmes, C.F.B. and Tsukitani, Y. (1990) 'Okadaic acid: a new probe for the study of cellular regulation', TIBS 15, 98-102.
10. Conrath, U., Domard, A. and Kauss, H. (1989) 'Chitosan-elicited synthesis of callose and of coumarin derivatives in parsley cell suspension cultures', Plant Cell Reports 8, 152-155.
11. Dietrich, A., Mayer, J.E. and Hahlbrock, K. (1990) 'Fungal elicitor triggers rapid, transient, and specific protein phosphorylation in parsley cell suspension cultures', Journal of Biol. Chem. 265, 6360-6368.
12. Kase, H., Iwahashi, K., Nakanishi, S., Matsuda, Y., Yamada, K., Takahashi, M., Muraka, C., Sato, A. and Kaneko, M. (1987) 'K-252 compounds, novel and potent inhibitors of protein kinase C and cyclic nucleotide-dependent protein kinases', Biochem. Biophys. Res. Commun. 142, 436-440.

INDUCED SYSTEMIC RESISTANCE IN CUCUMBER IN RESPONSE TO 2,6-DICHLORO-ISONICOTINIC ACID AND PATHOGENS

J.P. METRAUX, P. AHL GOY, TH. STAUB, J. SPEICH,
A. STEINEMANN, J. RYALS*) AND E. WARD*)
Agricultural Division, CIBA-GEIGY AG, 4002 Basel, Switzerland and *)
Agricultural Biotechnology Research Unit, CIBA-GEIGY Ltd, Research
Triangle Park, NC 27709, U.S.A

ABSTRACT. 2,6-dichloro-isonicotinic acid (CGA 41396) and its ester derivative (CGA 41397) induce local and systemic resistance in cucumber against *C. lagenarium* as well as other pathogens. These compounds have no direct fungicidal effet comparable to standard fungicides and extracts from treated plants do not show the presence of fungitoxic metabolites. In addition both compounds modify the physiology of the host plant by inducing chitinase. Thus CGA 41396 and 41397 could be considered as novel inducers of resistance.

1. Introduction

Localized infections with necrotrophic pathogens often result in the development of local and systemic disease resistance. Since this protection operates against a wide variety of pathogens such as fungi, bacteria or viruses (*Kuc 1982, Ross 1966*), it is likely to be the result of a combination of resistance mechanisms. Evidence from studies in cucumber and tobacco indicate that lignification of the cell walls (*Hammerschmidt et al, 1982*) or the induction of hydrolytic enzymes (*Boller 1987*) and other PR-proteins *(van Loon 1985)* are part of the induced defence mechanisms. Resistance has been proposed to be induced systemically by an endogenous signal produced in the infected leaf and translocated to other parts of the plant (*Kuc 1982, McIntyre et al 1981, Ross 1966*). Induced systemic resistance has been extensively studied in tobacco and cucumber, but has also been observed in other plants such as tomato or rice. The nature of the resistance mechanisms and the identity of the putative signal for induced resistance are among the major questions currently under investigation by many laboratories.

Pretreatments of crop plants to improve resistance against pathogens might be an attractive disease control practice, especially if protection covers a broad spectrum of pathogens. Such an approach could readily complement conventional chemical control, and be a valuable tool for plant breeders.

We report here that application to cucumber plants of 2,6-dichloro-isonicotinic acid (CGA 41396), or of its ester derivative (CGA 41397)(Fig.1), results in local and systemic protection. These compounds are likely to act indirectly by stimulating the mechanisms of resistance in the host plant.

432

H. Hennecke and D. P. S. Verma (eds.),
Advances in Molecular Genetics of Plant-Microbe Interactions, Vol. 1, 432–439.

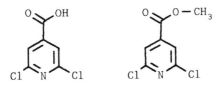

Figure 1. Chemical structures of CGA 41396 (left) and CGA 41397 (right).

2. The effect of CGA 41396 and CGA 41397 against pathogens in vivo.

Fig. 2 shows results from a glasshouse trial on the effect of CGA 41396, CGA 41397 and pretreatments with tobacco necrosis virus (TNV) or *Colletotrichum lagenarium* on the protection against *C. lagenarium* in cucumber plants. Protection was obtained after either foliar or drench application. When compared at optimal concentrations, protection was in the same range as that obtained after preinoculation of the lower leaf with necrotizing pathogens. Treatments with the chemicals are accompanied by slight yellowing of the leaves at the highest concentrations. Protection could also be observed against *Pseudomonas lachrymans* (Fig. 3), either after foliar or drench application. Thus treatments with CGA 41396 and CGA 41397 result in a spectrum of protection which is not limited to fungal diseases only, which is similar to the situation observed after pretreatments with pathogens (*Kuc 1982*). These compounds were also evaluated under field conditions against a variety of pathogens (Table 1). Rice is another example where in the same crop activity was observed as well against fungal as bacterial diseases.

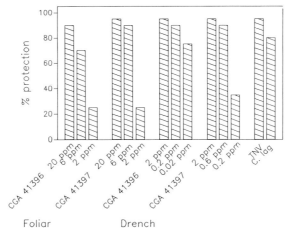

Figure 2. Glasshouse trial of CGA 41396 and CGA 41397 on protection of cucumber (2nd leaf stage) against *C. lagenarium*. Plants were treated 2 days before challenge with the pathogen. Protection obtained after preinoculation on leaf 1 with TNV or with *C. lagenarium* 7 days before challenge is also shown. Disease was rated visually as % protection relative to untreated controls. The diseased leaf area in untreated controls was 60 % (n=3).

434

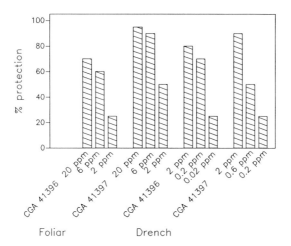

Foliar   Drench

Figure 3. Glasshouse trial of CGA 41396 and CGA 41397 on protection of cucumber (2nd leaf stage) against *P. lachrymans*. Plants were treated 2 days before challenge with the pathogen. Ratings are as described under Fig. 2 (n=3).

Table 1. Effect of CGA 41396 and CGA 41397 on various host-pathogen systems under field conditions. Treatments were by foliar application for pear and pepper. For rice the compounds were applied into the water.

Host/pathogen	Rate	% infected leaf area				
		Check	41396	41397	Standard	
Pear/*Erwinia amylovora*	25 g/hl	45	18	14	4	1)*
Pepper/*Xanthomonas vesicatoria*	25 g/ha	66	10	29	65	2)*
Rice/*Pyricularia oryzae*	2 kg/ha	2.2	0.1	0.1	0.3	.3)*
Rice/*Xanthomonas oryzae*	2 kg/ha	8.2	0.2	1	0.1	4)*
Tobacco/*Peronospora tabacina*	20 g/hl	29	12		17	5)*

* 1) Streptomycin 25 g/hl; 2) Copper+Mancozeb 200+100 g/hl; 3) Beam 2 kg/ha;
4) TF-130 (non-commercial standard) 2 kg/ha; 5) Ridomil+Mancozeb 24+196 g/hl

3. The effect of CGA 41396 and 41397 against pathogens in vitro

Table 2 shows the effect of CGA 41396 and CGA 41397 on the growth of various pathogenic fungi and bacteria. In most of the cases the EC-50 values were higher than those of commercial standard products. In the case of tests on fungi, the surface expansion of a mycelium plug is determined and interference of a test compound on spore germination would possibly be overlooked. However the EC-50 values of CGA 41396 and CGA 41397 on spore germination were largely above those of the standard benomyl. CGA 41396 and CGA 41397 do not appear to be strongly antibiotic neither against fungi nor bacteria.

Table 2. Effect of CGA 41396 and CGA 41397 on fungal and bacterial growth (n=4).

Pathogen used	41396	41397	Standard	Standard
Colletotrichum lagenarium a)	378	559	0.2	Benomyl
Colletotrichum lagenarium b)	>100	>100	<0.2	Benomyl
Pyricularia oryzae a)	>60	>60	<0.2	Benomyl
Xanthomonas oryzae	>100	>100	0.5	Streptomycin
Pseudomonas lachrymans	>100	25	2.5	Streptomycin

a) spore germination tests, b) mycelial growth test *values in ppm*

Plants treated with CGA 41396 and CGA 41397 did not contain fungitoxic metabolites as evidenced by thin-layer chromatography bioassays of various leaf extracts using *C. lagenarium* for the bioassay (Table 3). Fungitoxic metabolites were detected in water or organic extracts of benomyl-treated plants only. These metabolites cochromatographed with the position of the parent compound as well in the water as in the organic extracts. In organic extractions benomyl typically broke down into 2 active metabolites. Similar results were obtained using *C. cucumerinum* as another test fungus for the bioassay (data not shown).

Table 3. Rf's of fungitoxic metabolites in cucumber leaves extracted with various solvents 2 days after treatment with CGA 41396 and CGA 41397 (200 ppm, foliar). The extracts were separated by TLC (silica gel; dichloromethane/H_2O, 1/1, v/v) and bioassayed by spraying a spore suspension of *C. lagenarium* on the plates.

		Treatment	
Extract tested	41396	41397	Benomyl
Water	nd	nd	0.1
Methanol	nd	nd	0.1; 0.63
Acetone	nd	nd	0.1; 0.63
Dichloromethane	nd	nd	0.1; 0.63
Hexane	nd	nd	0.1; 0.63

nd = not detected

4. The effect of CGA 41396 and 41397 on the host plant

The time-course of resistance development was compared between cucumber plants preinoculated with TNV or pretreated with CGA 41396 on the lower leaf (leaf 1, Fig. 4). Both types of treatments were systemic: resistance and induction of chitinase activity could be observed in the upper untreated leaf (leaf 2). Thus CGA 41396 induces similar changes in the host plant as those which typically accompany induced resistance in cucumber (*Métraux & Boller 1986*). Resistance and induction of chitinase in leaf 2 appeared already 1 day after treatment with CGA 41396. This is faster than the changes observed after TNV pretreatments, where resistance and chitinase induction increased after 2 to 3 days in the upper leaf. Chitinase activity induced by

CGA 41396 and by TNV both resulted from an accumulation of protein and chitinase m-RNA in the tissue (Figs 5 and 6).

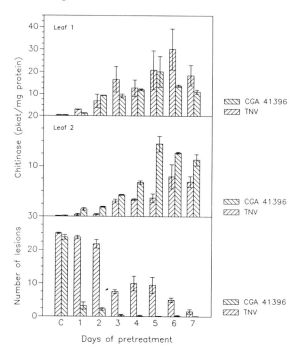

Figure 4. Time-course of resistance development and chitinase induction in cucumber plants treated on leaf 1 with either TNV or CGA 41396 (20 ppm). Both treatments were applied by gently rubbing leaf 1 with a brush. Plants were all challenged at the same time (age of plants: 3.5 weeks). Only controls pretreated 7 days before challenge are indicated. No effect on controls pretreated 1 to 6 days were observed (data not shown). (n=2; upper and lower values are indicated)

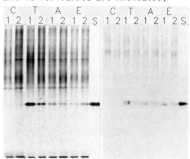

Figure   5. Non-denaturing PAGE electrophoresis (left) and western blot (right) of soluble leaf proteins from cucumber. Lane 1: proteins from the first leaf; lane 2: proteins from the second leaf. C: mock-inoculated control; T: TNV; A: CGA 41396 and E: CGA 41397 (both 20 ppm applied on leaf 1); S: purified cucumber chitinase.

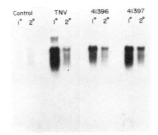

Figure 6. Northern blots of RNA extracted of cucumber leaves hybridized with a chitinase/lysozyme probe *(Métraux et al 1989)*. Lane 1°: RNA from the first leaf; lane 2°: RNA from the second leaf. CGA 41396 and 41397 were used at 20 ppm, foliar application. TNV was preinoculated on leaf 1 and controls were mock-inoculated on leaf 1.

5. Translocation of CGA 41396

We have followed the fate of ^{14}C-labelled CGA 41396 after leaf injection (80 nCi, 12 injection sites) of the lower leaf of a cucumber plant at the 2 leaf stage. Autoradiography of the plant 1 day after treatment showed radioactive material distributed in the roots and shoot of the plant (Fig 7). The labelled material consisted at least to a large part of CGA 41396 and to a minor degree of an uncharacterized metabolite (Table 5). Thus CGA 41396 can be rapidly translocated in the entire plant after application by leaf injection. Other types of applications such as foliar or drench led to the same findings (data not shown).

Table 5. Distribution of radioactivity in the residue and in the soluble fractions of leaf 2 of cucumber plants. Leaf 1 was injected with ^{14}C-CGA 41396 (30 µl in 12 small leaf panels) and the measurements were carried out after 24 h incubation. Leaves were homogenized in 80 % methanol and the soluble material was dried and suspended in H2O (aqueous phase) which was then extracted with dichloromethane (organic phase). Separation was by two-dimensional TLC (insert right); spot 1 and 2 cochromatographed with CGA 41396. TLC plates were viewed with a radio-chromatogram camera, the radioactive spots were then scraped off and counted.

Fraction	nCI	%	Aqueous phase	Organic phase
Initial injection	87.9	100		
Homogenate	16.7	19		
Aqueous phase spot 1	15.1	17		
Organic phase spot 2	0.3	0.3		
Organic phase spot 3	0.03	0.03		
Insoluble residue	0.7	0.8		

Figure 7. *Cucumber plant (left) and its autoradiograph (right) exposed 24 h after injection of CGA ^{14}C-41396 into leaf 1. Sites of injection are visible on leaf 1 (left leaf of the plant).*

6. Discussion

Systemic induced resistance is characterized by a broad resistance against a wide variety of pathogens and can be induced by preinoculations with various pathogens. We have used cucumber which has proven previously to be a useful experimental system, to examine the effect of chemical pretreatments with 2,6-dichloro-isonicotinic acid on systemic induced resistance. Foliar and drench treatments with CGA 41396 and CGA 41397 protected cucumber and other plants against fungal and bacterial diseases. This effect is unlikely to result from a direct antibiotic effect of these compounds. Alternatively the chemicals could be converted into antimicrobial metabolites or could induce the formation of antimicrobial metabolites in the plants. We did not find evidence which supports this hypothesis.

We have also evaluated the possibility that treatments with CGA 41396 and CGA 41397 result in the induction of resistance. Chitinase, an antifungal enzyme (*Mauch et al 1988, Schlumbaum et al 1986*) which accompanies the onset of induced resistance in cucumber (*Métraux and Boller 1986*), was strongly induced by pretreatments with CGA 41396 and CGA 41397, indicating an effect on the physiology of the host plant. The possibility that these compounds might affect other defence reactions has not been checked but might be likely since the protection can be effective against fungi and bacteria. It is assumed that pretreatments with pathogens result in the production of a signal for the systemic induction of resistance (*Kuc 1982, McIntyre et al 1981, Ross 1966*). The onset of resistance and chitinase induction by CGA 41396 and CGA 41397 was however rapid and although we cannot exclude that CGA 41396 induces an endogenous signal, we hypothesize that after a rapid uptake and translocation to other parts of the plant CGA 41396 interferes directly with defence reactions, and not through the intermediate of a systemic signal. The nature of this interaction will be of great interest to understand the molecular events unfolded during induced resistance.

In conclusion, the present results show that CGA 41396 and 41397 can protect plants against fungal and bacterial diseases and suggest that these compounds function indirectly by increasing the resistance of the plant against pathogens.

7. References

Boller, T. (1987) In *Plant microbe interactions, molecular and genetic perspectives Vol 2,* T. Kosuge and E.W. Nester (eds.), Macmillan, New York, pp 385-413.

Hammerschmidt, R., and Kuc, J. (1982) 'Lignification as a mechanism for induced systemic resistance in cucumber', Physiol. Plant Pathol. 20, 61-71.

Hammerschmidt, R., Nuckles, E., and Kuc J. (1982) 'Association of enhanced peroxidase activity with induced systemic resistance of cucumber to *C. lagenarium*', Physiol. Plant Pathol. 20, 73-82.

Kuc, J. (1982) 'Induced immunity to plant disease', Bioscience 32, 854-860.

Mauch, F, Mauch-Mani, B, and Boller, T. (1988) 'Antifungal hydrolases in pea tissue. II. Inhibition of fungal growth by combinations of chitinase and ß-1,3 glucanase', Plant Physiol 88, 936-942.

Mc Intyre, J.L., Dodds, J.A., and Hare, J.D. (1981) 'Effects of localized infections of *N. tabacum* by brome mosaic virus on systemic resistance against diverse pathogens and insects', Phytopathology 71, 297-301.

Métraux, J.P., and Boller, T. (1986) 'Local and systemic induction of chitinase in cucumber plants in response to viral, bacterial and fungal infections', Physiol. Molec. Plant Pathol. 56, 161-169.

Métraux, J.P., Burkhart, W., Moyer, M., Dincher, S., Middlesteadt, W., Williams, S., Payne, G., Carnes, M., and Ryals, J. (1989) 'Isolation of a complementary DNA encoding a chitinase with a structural homology to a bifunctional lysozyme/chitinase', Proc. Natl. Acad. Sci. USA 86, 896-900.

Schlumbaum, A., Mauch, F., Vögeli, U., and Boller, T. (1986) 'Plant chitinases are potent inhibitors of fungal growth', Nature 324, 365-367.

Ross, A.F. (1966) In *Viruses of plants*, A.B.R. Beemster and J. Dijkstra (eds.), North Holland Publishing Company, Amsterdam, pp. 127-150.

Van Loon, L.C. (1985) 'Pathogenesis-related proteins', Plant Molec. Biol. 4, 111-116.

Section   V

BIOCONTROL AND RHIZOSPHERE ASSOCIATIONS

GENETIC ASPECTS OF PHENAZINE ANTIBIOTIC PRODUCTION BY FLUORESCENT PSEUDOMONADS THAT SUPPRESS TAKE-ALL DISEASE OF WHEAT

L. S. Thomashow and L. S. Pierson III
USDA-ARS, 362 Johnson Hall
Washington State University
Pullman, Washington 99164-6430 USA

Introduction

Microorganisms isolated from the rhizosphere of plants have potential value as supplements or alternatives to disease controls of soilborne pathogens that rely on chemical pesticides and cultural practices. Our research has focused on the use of fluorescent pseudomonads for biological control of take-all of wheat, caused by the fungus *Gaeumannomyces graminis* var. *tritici* (*Ggt*), and the mechanisms responsible for suppression.

For *Pseudomonas fluorescens* 2-79 and *P. aureofaciens* 30-84, disease suppression depends largely on the production of phenazine antibiotics. Phenazines are pigmented compounds with broad-spectrum activity against bacteria and fungi. For both strains, transposon Tn5 mutants defective in phenazine production (Phz⁻) fail to inhibit *Ggt* in vitro and are greatly reduced in their ability to suppress the disease on wheat seedlings (3; L. Pierson and L. Thomashow, unpublished). Phenazine-1-carboxylate (PCA) has been isolated from the roots of wheat colonized by either of these two phenazine-producing strains, and disease symptoms were significantly reduced when the antibiotic was present (4). Phenazine synthesis in situ also contributes to the ecological competence of strains 2-79 and 30-84. When Phz⁺ strains or Phz⁻ mutant derivatives were added to raw soil and 5 successive crops of wheat were grown for 20 days, the wildtype strains persisted in significantly larger populations than the mutant strains both in the soil and in the rhizosphere of wheat (M. Mazzola and R. J. Cook, unpublished).

Mechanisms such as the production of fluorescent pyoverdin siderophores (2) or cyanide (6) that have been implicated in the control of soilborne pathogens, including *Ggt* (7, and references cited therein), by other fluorescent pseudomonads appear to be of little or no significance in the suppression of take-all by these strains (H. Hamdan and L. Thomashow, unpublished; L. Pierson and L. Thomashow, unpublished). However, a nonphenazine antifungal factor (Aff; see below) produced by strain 2-79 may have a minor role in disease suppression. We report here on two genetic loci required for the production of phenazine antibiotics by strains 2-79 and 30-84. The first of these is involved in the synthesis of both Aff and PCA in 2-79 and

443

H. Hennecke and D. P. S. Verma (eds.),
Advances in Molecular Genetics of Plant-Microbe Interactions, Vol. 1, 443–449.
© 1991 *Kluwer Academic Publishers. Printed in the Netherlands.*

may have a regulatory function, whereas the second encodes structural genes for phenazine biosynthesis in both strains.

Results and Discussion

THE MUTATIONS IN STRAINS 2-79-B46 AND 2-79-782 ARE PLEIOTROPIC

The Phz⁻ mutants 2-79-B46 and 2-79-782, which contain Tn5 insertions in contiguous 0.6 and 12.2 kb EcoRI fragments, respectively (3), did not inhibit Ggt on the low iron medium King B agar despite production of a fluorescent siderophore. However, on the same medium both 2-79-892B, a Phz⁻ mutant generated by treatment of 2-79 with nitrosoguanidine (7), and its nonfluorescent (Flu⁻) Tn5 derivative 2.79-892.224 were inhibitory (H. Hamdan and L. Thomashow, unpublished). Fungal inhibition under iron limitation was attributed to a previously undetected antifungal factor (Aff) that is functionally distinct from both PCA and the pyoverdin siderophore. Synthesis of Aff is linked genetically to PCA production because neither substance was produced in the mutants 2-79-B46 and 2-79-892. Aff was not detected in bioassays on media supplemented with iron, suggesting that either its antifungal activity or synthesis is iron-regulated.

In iron-limited media, and particularly when tryptophan was added (L. Thomashow, unpublished), Flu⁻Aff⁺ strains such as 2-79-892.224 or Flu⁻ mutants of 2-79 fluoresced violet under long-wave UV illumination, but no fluorescence was visible in iron-supplemented cultures or in cultures of 2-79-59.34.24, a Flu⁻ Tn5 derivative of the Aff⁻Phz⁻ strain 2-79-5.12 in which the 0.6 kb EcoRI fragment required for PCA production (see above) is deleted (H. Hamdan, and L. Thomashow, unpublished). After extraction into acidified benzene the fluorescent material from Aff⁺ cultures co-chromatographed with, and was indistinguishable by UV-visible spectroscopy from anthranilic acid. Furthermore, anthranilic acid itself inhibited Ggt at concentrations comparable to those produced in culture.

Anthranilic acid, a tryptophan precursor, is a probable intermediate in the synthesis of pyocyanin, a blue phenazine pigment produced by P. aeruginosa. P. aeruginosa has two anthranilate synthase gene pairs, one of which (trpE and trpG) participates mainly in tryptophan synthesis and the other (phnAB), mainly in pyocyanin production (1). The plasmid pPHZ49-61 contains cloned sequences homologous to the mutated locus in strains 2-79-B46 and 2-79-782 and complements these mutants to Phz⁺Aff⁺. However, this plasmid does not rescue trpE mutants of P. putida from tryptophan auxotrophy, nor is it homologous to the P. aeruginosa genes trpE, trpG, and phnAB as determined by Southern hybridization at low stringency. In fact, genomic blots of strain 2-79 appear to lack sequences homologous to phnAB (L. Thomashow and D. Essar, unpublished). These observations suggest that the phenazine biosynthetic pathway in 2-79 may differ somewhat from that in P. aeruginosa. Further, because 2-79-B46 and 2-79-782 are not auxotrophic for aromatic compounds including tryptophan, the complementary gene(s) encoded on pPHZ49-61 are likely to be regulatory in nature or only indirectly involved in phenazine synthesis.

Phz⁻Aff⁻ strains also differ from the wildtype in several phenotypic characteristics that are consistent with alterations at the cell surface. Cells of strain 2-79 aggregate into clumps in broth cultures, especially in defined media, whereas cultures of 2-79-B46 and 2-79-782 are uniformly dispersed. On solid media colonies of strain 2-79 appear slightly granular after several days and are somewhat larger in diameter than colonies of the mutants, which are more elevated and glossy. Finally, the mutants grow as discrete colonies on motility agar whereas the wildtype forms a thin, spreading film indicative of active motility (B. Ownley, unpublished). In broth culture, however, the mutants also are motile as determined by microscopy. Membrane proteins and lipopolysaccharide currently are being examined to determine whether phenotypic differences between the wildtype and the mutants are correlated with structural changes in the cell envelope.

Mutagenesis of pPHZ49-61 with the transposon Tn3HoHo1 indicated that sequences required to complement 2-79-B46 and 2-79-782 to Aff⁺Phz⁺ are located within a region encompassing approximately 2.6 kb. Marker exchange of Tn3HoHo1 insertions within this region into the genome of strain 2-79 also resulted in an Aff⁻ Phz⁻ phenotype. Preliminary results with two marker-exchanged derivatives that contain Tn3HoHo1 in the same orientation permissive for β-galactosidase expression indicate that in a glucose-based minimal medium supportive of phenazine production by wildtype 2-79, the locus is transcribed constitutively but only at a low level (20-30 Miller units) suggestive of a possible regulatory, rather than structural function. Expression began during logarithmic growth and was unaffected by exogenous iron, anthranilate or tryptophan.

A PHENAZINE BIOSYNTHETIC LOCUS

PCA is the second phenazine in the proposed biosynthetic pathway and the only phenazine produced by strain 2-79. However, other *Pseudomonas* and *Streptomyces* spp. further modify this compound to yield more highly derivatized antibiotics (5). In addition to PCA, *P. aureofaciens* strain 30-84 produces lesser quantities of 2-hydroxy-phenazine-1-carboxylate and 2-hydroxy-phenazine. These hydroxylated compounds, which give cultures of 30-84 their characteristic orange-red color and were identified by their spectral, chemical and physical properties, also inhibit *Ggt* in vitro and contribute to disease suppression on wheat seedlings (L. Pierson and L. Thomashow, unpublished).

Localization of Phenazine Structural Genes and Expression in Escherichia coli. A single 20.4 kb genomic clone from a library of wildtype strain 30-84 complemented several Tn5 Phz⁻ mutants of both 30-84 and 2-79 to Phz⁺. A mutant of 30-84 that produced only PCA also was restored for production of both hydroxyphenazines, suggesting that the cloned DNA encoded at least part of a phenazine biosynthetic pathway. The complementary region was contained on contiguous *Eco*RI fragments of 11.2 and 9.2 kb, and was further localized to a segment of approximately 5 kb by deletion mapping and Tn5 mutagenesis. Most of this 5 kb segment was present on the 9.2 kb EcoRI fragment, which was subcloned into the unique EcoRI site in the chloramphenicol resistance determinant of pBR325. Chloramphenicol-sensitive E. coli transformants were recovered that produced all three phenazine antibiotics as determined by thin layer chromatography and HPLC. Transformants with the 9.2 kb fragment present in the reverse orientation did not produce phenazines, indicating the transcriptional orientation of the phenazine structural genes and that transcription was initiated from within the vector, probably from the chloramphenicol promoter. E. coli clones containing either the entire 20.4 kb segment from strain 30-84 or a homologous 35 kb fragment from strain 2-79 that complemented Phz⁻ mutants also failed to produce phenazines, suggesting either that native phenazine promoters are not recognized in E. coli or that an additional factor(s) required for phenazine gene expression is lacking.

Construction of a PCA::lacZ Gene Fusion. Inconsistent performance from one field to another is an obstacle to the commercialization of these biocontrol agents for take-all. Part of the inconsistency may be due to variability in the levels of phenazine produced, or in antibiotic activity, as a function of different environmental conditions that occur among soils. The relative importance of soil chemical and physical factors on take-all suppression by phenazine producers was evaluated on wheat treated with 2-79 or various Phz⁺ and Phz⁻ mutant derivatives and sown in ten different soils infested with *Ggt* (B. Ownley and D. Weller, unpublished). Disease suppression was positively correlated with pH, sodium, sulfate-sulfur, zinc, ammonium nitrogen and sand, and negatively correlated with iron, manganese, cation exchange capacity, silt, clay and organic matter. Some of these factors may have a greater impact on phenazine production whereas others (eg. clay content) are more likely to affect antibiotic activity.

Two approaches are available to determine the effect of environmental factors specifically on phenazine production. PCA is readily measured in vitro and also can be monitored directly on the roots of wheat (4). However, the extractions are laborious and multiple replications are not feasible. Gene fusions offer a simpler experimental approach, especially in situ. A modified version of *P. aureofaciens* 30-84 therefore was generated in which the structural region required for PCA production was fused to a *lacZ* reporter gene.

The fusion initially was constructed in pLSP18-6H3del3, a plasmid in which the 9.2 kb EcoIRI fragment containing the phenazine structural region is positioned downstream of the E. coli lac promoter. By using this construct, insertion and expression of the *lacZ* cartridge could be screened directly in E. coli, thereby avoiding the need to mobilize and monitor each construct in strain 30-84. The 9.2 kb EcoRI fragment contains a unique SalI restriction site within the region required for PCA synthesis. This site is flanked on either side by several kb of 30-84 DNA. Introduction of a promoterless lacZ gene cartridge into this SalI site in the correct reading frame resulted in expression of β-galactosidase activity in only one of the two possible orientations. A fusion expressing β-galactosidase was transferred to the broad host range vector pLAFR3, and mobilized into wildtype strain 30-84. Plasmid pLAFR3 confers resistance to tetracycline but is unstable in *P. aureofaciens* without antibiotic selection. Exconjugants were grown in the absence of tetracycline and individual blue colonies on media containing the chromogenic indicator X-gal were screened for loss of the vector as indicated by sensitivity to tetracycline. Recombinants in which the PCA::*lac*Z had marker-exchanged into the chromosome occurred at low frequency, and did not produce any phenazines due to insertion of *lac*Z into sequences required for production of PCA, the presumptive hydroxyphenazine precursor. The resulting modified strain 30-84Z contained a single, stably integrated chromosomal copy of the *lac*Z gene, avoiding potential gene dosage artifacts in subsequent studies of phenazine gene regulation.

Production of β-galactosidase. Expression of β-galactosidase by strain 30-84Z paralleled synthesis of PCA by wildtype 30-84 both temporally and spatially. On agar plates containing X-gal, colonies appeared blue (30-84Z) or orange (30-84) after about 8 hours, indicating production of β-galactosidase and phenazine antibiotics, respectively. Colonies near the edge of the plate were the first to show color, followed by those in the center. For studies in broth culture, strain 30-84Z was grown overnight, centrifuged, washed and diluted 100-fold into fresh medium. Cultures were shaken at $28°C$, growth was monitored spectrophotometrically, and β-galactosidase activity was assayed as a function of time. For strain 30-84Z in a glycerol-based medium, β-galactosidase production began approximately 8 hours after inoculation and then increased, reaching a maximum of approximately 700 Miller units after 24 hours. In control experiments strain 30-84 did not produce significant levels of β-galactosidase.

The pH of the medium affected both the growth of strain 30-84Z and the production of β-galactosidase (L. Pierson and L. Thomashow, unpublished). Although bacterial growth was essentially the same in AB minimal broth between pH 6 and pH 8, β-galactosidase production was optimum at pH 7. At pH 8, production of β-galactosidase was somewhat reduced relative to that at pH 7, and at pH 6 the onset of expression was delayed 1-2 hr but eventually neared the level obtained at pH 7. At pH 9 strain 30-84Z had an extended lag phase but in exponential growth its doubling time was the same as that in cultures at pH 6 to 8. However, at pH 9 production of β-galactosidase never exceeded 68 Miller units. Strain 30-84Z grew poorly at pH 5, although by 26 hours the density was the same as at pH 9. β-galactosidase production at pH 5 occurred rapidly once growth began, and reached a level similar to that obtained at pH 6 to 8. PCA was produced by cultures of wildtype 30-84 between pH 5 and 8, but very little was produced above pH 8. In soil, suppression of take-all occurs over a range from pH 5 to pH 8.5, and is optimal between pH 6 and pH 7.6 (B. Ownley and D. Weller, unpublished).

Iron as either ferric ammonium citrate or $FeSO_4$ stimulated β-galactosidase production 5- to 8-fold in strain 30-84Z (L. Pierson and L. Thomashow, unpublished). This observation, which contrasts with the negative correlation between root disease and iron availability in soil (see above), suggests either that phenazine production responds differently to iron in strains 2-79 and 30-84, or that soil iron impacts mainly on parameters other than phenazine production. Results from other studies support the latter hypothesis. Iron stimulates PCA production in strain 2-79 (7), and may not be limiting to synthesis even in soils of low iron availability (4; H. Hamdan and L. Thomashow, unpublished).

Glucose initially inhibited β-galactosidase production, but when it was exhausted due to bacterial growth, production increased rapidly to a level comparable to that present in iron-supplemented media. Addition of glucose periodically over the course of the experiment repressed β-galactosidase production, consistent with a possible catabolite repression-like effect. Preliminary experiments with other carbon sources suggest that, in general, those that are optimal for bacterial growth (gluconate, glucose, glycerol) do not necessarily support the highest levels of β-galactosidase production (L. Pierson and L. Thomashow, unpublished).

It also is interesting that nitrogen as NO_3 stimulated β-galactosidase production whereas NH_4 was inhibitory, particularly in view of the positive correlation between take-all suppression and ammonium nitrogen in soils. It is important to remember that many of the soil factors with which disease suppression is correlated (see above) are interrelated and may impact on phenazine activity as well as production. Thus, the apparent discrepancy in the effect of ammonium nitrogen in situ as compared to in vitro may be due to the influence of ammonium on rhizosphere pH or other parameters that can affect phenazine activity or disease suppression in a more general way, rather than to a specific effect on phenazine production *per se*. There results illustrate the importance of monitoring the expression of phenazine genes directly in the rhizosphere of wheat. Studies with this objective currently are in progress.

LITERATURE CITED

1. Essar, D. W., Eberly, L., Hadero, A. and Crawford, I.P. 1990. Identification and characterization of genes for a second anthranilate synthase in *Pseudomonas aeruginosa*: interchangeability of the two anthranilate synthases and evolutionary implications. J. Bacteriol. **172**:884-900.
2. Leong, J. 1986. Siderophores: Their biochemistry and possible role in the biocontrol of plant pathogens. Annu. Rev. Phytopathol. **24**:187-209.
3. Thomashow, L. S. and Weller, D. M. 1988. Role of a phenazine antibiotic from *Pseudomonas fluorescens* in biological control of *Gaeumannomyces graminis* var. *tritici*. J. Bacteriol. **170**:3499-3508.
4. Thomashow, L. S., Weller, D. M., Bonsall, R. F. and Pierson, L. S. 1990. Production of the antibiotic phenazine-1-carboxylic acid by fluorescent *Pseudomonas* species in the rhizosphere of wheat. Appl. Environ. Microbiol. **56**:908-912.

5. Turner, J. M., and Messenger, A. J. 1986. Occurrence, biochemistry and physiology of phenazine pigment production. Adv. Microbial Physiol. 27:211-275.

6. Voisard, C., Keel, C., Haas, D. and Defago, G. 1989. Cyanide production by *Pseudomonas fluorescens* helps suppress black root rot of tobacco under gnotobiotic conditions. EMBO J. **8**:351-358.

7. Weller, D. M., Howie, W. J. and Cook, R. J. 1988. Relationship between *in vitro* inhibition of *Gauemannomyces graminis* var. *tritici* and suppression of take-all of wheat by fluorescent pseudomonads. Phytopathology **78**:1094-1100.

SECONDARY METABOLITES OF *PSEUDOMONAS FLUORESCENS* STRAIN CHA0 INVOLVED IN THE SUPPRESSION OF ROOT DISEASES

D. HAAS[1], C. KEEL[2], J. LAVILLE[1], M. MAURHOFER[2],
T. OBERHÄNSLI[2], U. SCHNIDER[1], C. VOISARD[1],
B. WÜTHRICH[2] & G. DEFAGO[2]

Departments of Microbiology[1] and Plant Sciences / Phytomedicine[2]

Eidgenössische Technische Hochschule

CH-8092 Zürich

Switzerland

ABSTRACT. *Pseudomonas fluorescens* strain CHA0 protects plants from diseases caused by various soil-borne pathogenic fungi. Strain CHA0 produces and releases several metabolites: pyoverdine (pseudobactin), indoleacetate, HCN, pyoluteorin (Plt), and 2,4-diacetylphloroglucinol (Phl). We have begun to assess the importance of these compounds in disease suppression by testing non-producing CHA mutants in gnotobiotic systems and by cloning relevant genomic fragments from strain CHA0. No major role was found for pyoverdine and indoleacetate. HCN helped suppress black root rot caused by *Thielaviopsis basicola* (*Tb*) on tobacco. In contrast, an Hcn⁻ mutant did not differ significantly from the wildtype in the suppression of wheat disease due to *Gaeumannomyces graminis* var. *tritici* (*Ggt*). The antifungal and herbicidal compound Phl was found to be produced in the rhizosphere of wheat by the wildtype CHA0. A Phl⁻ mutant gave reduced suppression of diseases caused by *Tb* and *Ggt*; a recombinant cosmid restored both Phl production in the rhizosphere and disease suppression. Another cosmid carrying a 22 kb fragment from the genome of strain CHA0 enhanced the production of Phl and Plt *in vitro* and improved the protection of cucumber from *Pythium ultimum*, a Plt-sensitive fungus. Interestingly, the same 22 kb amplification rendered strain CHA0 deleterious to cress, presumably because of a herbicidal effect of Phl and Plt on cress. Thus, the production of antifungal compounds contributes importantly to the suppressive properties of strain CHA0, but overproduction of such compounds may have deleterious effects on plants.

1. Introduction

Fields near Payerne (Switzerland) are naturally suppressive to black root rot of tobacco, a disease caused by the fungus *Thielaviopsis basicola. Pseudomonas fluorescens* strain CHA0 was isolated from this suppressive soil by incubation of tobacco roots on King's B medium, which favors growth of fluorescent pseudomonads (Stutz *et al.*, 1986). Tests carried out in both non-sterile natural soils and sterilized artificial soils have shown that strain CHA0 has the ability to suppress a variety of root diseases under growth chamber or greenhouse conditions

450

H. Hennecke and D. P. S. Verma (eds.),
Advances in Molecular Genetics of Plant-Microbe Interactions, Vol. 1, 450–456.
© 1991 *Kluwer Academic Publishers. Printed in the Netherlands.*

(Stutz *et al.*, 1986; Keel *et al.*, 1989; Stutz *et al.*, 1989; Défago *et al.*, 1990). Thus, strain CHA0 can be included in a growing list of fluorescent *Pseudomonas* strains having plant-beneficial and biocontrol properties (Schroth and Hancock, 1982; Schippers *et al.*, 1987; Kloepper *et al.*, 1988; Davison, 1988; Weller, 1988; Kloepper *et al.*, 1989; Défago and Haas, 1990).

Strain CHA0 has also been applied to field plots naturally infested with *Gaeumannomyces graminis* var. *tritici* (*Ggt*), which is the causative agent of take-all disease of wheat. Wheat health was improved by the application of strain CHA0 in several, but not all, field trials conducted over 4 years (Défago *et al.*, 1990; B. Wüthrich, unpublished results). These results confirm a general conclusion derived from many independent experiments, namely that bacterial inocula can significantly improve crop yields but that performance may be inconsistent (Weller, 1988; Kloepper *et al.*, 1989).

Our knowledge of the mechanisms by which fluorescent pseudomonads protect plants from soilborne pathogens is still rudimentary and the reasons for variable performance in the field are largely a matter of speculation. Disease suppression is determined by a number of biotic and abiotic factors, and depends on complex interactions between plants, pathogens, soil, and the biocontrol agent. Advances in the area of biocontrol with bacteria require a better understanding of the molecular mechanisms involved in disease suppression and of their regulation by environmental conditions.

It is generally assumed that effective root colonization by fluorescent pseudomonads is a prerequisite for biocontrol activity. A cucumber root surface glycoprotein causes agglutination of *P. fluorescens* and *P. putida*. Mutants of *P. putida* lacking agglutinability (Agg⁻) protect cucumber less effectively from *Fusarium* wilt and colonize the roots at lower levels than does the parental wildtype strain. Agg⁻ mutants show alterations of membrane proteins (Anderson *et al.*, 1988; Tari and Anderson, 1988). Pili on *P. fluorescens* appear to mediate the attachment of the bacterium to corn roots (Vesper, 1987). Motility and chemotaxis towards root exudates probably also contribute to the ability of pseudomonads to colonize roots (Weller, 1988).

Secondary metabolites produced by fluorescent pseudomonads seem to have a key role in disease suppression. The yellow-green, fluorescent siderophores of fluorescent *Pseudomonas* strains (pyoverdine, pseudobactin) are high-affinity Fe^{3+} chelators. Under conditions of limited iron availability in the rhizosphere, these siderophores are assumed to give the pseudomonads a selective growth advantage over pathogens and deleterious microorganisms having less potent siderophores. Evidence that this kind of competition for iron is an important mechanism in disease suppression comes from studies on siderophore-negative mutants and effects of purified siderophores (Schroth and Hancock, 1982; Leong, 1986; Weller, 1988; Kloepper *et al.*, 1988). Plant growth promoting compounds, such as indole-3-acetic acid (IAA), are produced by fluorescent pseudomonads but a causal role in plant growth promotion or disease control has not been demonstrated conclusively (Kloepper *et al.*, 1989). Another group of secondary metabolites, antibiotic compounds, play a major role in the suppression of several diseases. *P. fluorescens* 2-79 produces phenazine-1-carboxylic acid *in vitro* and in the rhizosphere of wheat (Thomashow *et al.*, 1990). Mutants defective in phenazine synthesis provide significantly less control of take-all due to *Ggt* than does the wildtype. Suppression of take-all can be restored in the mutants by complementation with recombinant cosmids (Thomashow and Weller, 1988). Further evidence for a participation of antibiotic compounds in disease control has been reviewed recently (Weller, 1988; Fravel, 1988; Défago and Haas, 1990; Keel *et al.*, 1990[a]).

We have initiated a research program to identify those metabolites produced and released by *P. fluorescens* strain CHA0 that are important for this strain's biocontrol activity. First, we have established a gnotobiotic system, as a means to reduce the complexity of interactions and to improve the reproducibility of plant protection tests. In the gnotobiotic system, the composition of the artificial sterilized soil (containing quartz grains of different sizes, clay minerals and modified Knop nutrient solution), the numbers of the pathogen and of the *P. fluorescens* cells, and the environmental conditions (humidity, temperature, light) can all be kept constant. In this way, the severity of plant diseases and the suppressive effects of strain CHA0 can be controlled reproducibly and quantitated in terms of plant weight (Keel *et al.*, 1989); moreover, an appropriate containment of genetically manipulated bacteria can be ensured. The second step was to develop genetic methods for strain CHA0: transposon mutagenesis using IncI1 suicide vectors, complementation by broad-host-range recombinant cosmids (which can be transferred from *Escherichia coli* to strain CHA0 by modified IncP helper plasmids), and transformation by plasmid DNA (Voisard *et al.*, 1988; Voisard *et al.*, 1989). With these tools we have been able to demonstrate that two metabolites of *P. fluorescens* strain CHA0, hydrogen cyanide (HCN) and 2,4-diacetylphloroglucinol, contribute to the suppression of tobacco black root rot, whereas no significant role could be attributed to the bacterial siderophore, pyoverdine (Keel *et al.*, 1989; Voisard *et al.*, 1989; Keel *et al.*, 1990[a]). In this communication we extend these investigations to other plants and pathogens, in particular to wheat and *Ggt*. We also examine the importance of IAA production by strain CHA0 in disease suppression. Finally, we present an example of a genetically engineered derivative of strain CHA0 which can be either beneficial or deleterious, depending on the plant colonized.

2. Materials and Methods

The organisms, genetic manipulations, and gnotobiotic systems have been described previously (Stutz *et al.*, 1986; Voisard *et al.*, 1988; Keel *et al.*, 1989; Voisard *et al.*, 1989; Keel *et al.*, 1990[a]).

3. Results and Discussion

3.1. SECONDARY METABOLITES OF STRAIN CHA0

P. fluorescens CHA0 produces and releases several metabolites (Défago *et al.*, 1990). Here we are concerned with the effects of five metabolites: pyoverdine, IAA, HCN, 2,4-diacetyl-phloroglucinol (Phl), and pyoluteorin (Plt). Phl and Plt have broad-spectrum antibiotic and herbicidal properties (Reddi *et al.*, 1969; Reddi and Borovkov, 1970; Broadbent *et al.*, 1976; Ohmori *et al.*, 1978; Howell and Stipanovic, 1980). Virtually nothing is known about the mode of action of these compounds but recently Yoneyama *et al.* (1990) showed that phloroglucinols with two electron-withdrawing substituents are potent photosystem II inhibitors.

3.2. ROLE OF PYOVERDINE

A non-fluorescent, pyoverdine-negative mutant obtained after Tn*1733* insertion mutagenesis, strain CHA400, protects tobacco from black root rot with wildtype efficiency under gnotobiotic conditions (Keel *et al.*, 1989). Similarly, it was found that wheat diseases caused by *Ggt* or *Pythium ultimum* were suppressed to the same extent by the wildtype CHA0 and the mutant CHA400. In one experiment, wheat was grown in an artificial illitic soil without added iron until the first symptoms of iron deficiency (chlorosis) appeared. Even under these conditions, there was no difference between strains CHA0 and CHA400 with respect to suppression of *P. ultimum* - induced disease. Although these experiments failed to demonstrate a significant effect of pyoverdine in disease suppression, they do not rule out competition for iron as a general mechanism. It may be that *P. fluorescens* CHA0 synthesizes iron chelator(s) in addition to pyoverdine. One such siderophore, salicylate, has indeed been detected in culture fluids of strain CHA0 (M.A. Abdallah, personal communication).

3.3. ROLE OF IAA

P. fluorescens strain CHA0 has two pathways leading from L-tryptophan to the growth hormone IAA (T. Oberhänsli, G. Défago and D. Haas, manuscript in preparation). The first pathway is initiated by tryptophan side chain oxidase (TSO), an enzyme forming indoleacetaldehyde and several other indole compounds from tryptophan (Narumiya *et al.*, 1979). Indoleacetaldehyde is converted to IAA by an aldehyde dehydrogenase (Narumiya *et al.*, 1979). In the second pathway, L-tryptophan is transaminated to indolepyruvate, whose decarboxylation yields indoleacetaldehyde as well. The IAA biosynthetic pathway of *P. savastanoi*, which involves indoleacetamide as an intermediate (Morris, 1986), could not be detected in strain CHA0. The amount of IAA produced by *P. fluorescens* is smaller by several orders of magnitude than that synthesized by *P. savastanoi* (T. Oberhänsli *et al.*, manuscript in preparation). In natural and artificial soils strain CHA0 does not stimulate plant growth significantly in the absence of pathogens. However, since strain CHA0 improves growth of plants when they are attacked by root pathogens, we decided to test the importance of bacterial IAA synthesis in disease suppression. A TSO-negative mutant, CHA750, was isolated after Tn5 mutagenesis. Resting cells of strain CHA750, at pH6, produced 20% of IAA; the amount synthesized by the wildtype is taken as 100%. The mutant CHA750 suppressed black root rot of tobacco and take-all of wheat with wildtype efficiency in gnotobiotic systems at pH6 and pH7. Therefore, bacterial IAA production does not appear to have a significant effect on the suppression of these diseases.

3.4. ROLE OF HCN

In a previous communication (Voisard *et al.*, 1989) we have shown that the HCN-negative mutant CHA5, which was constructed by insertion of a mercury resistance cassette into the chromosomal *hcn* genes, suppresses black root rot of tobacco with reduced efficiency under gnotobiotic conditions. The *hcn*+ recombinant plasmid pME3013 complements for HCN production and restores disease suppression (Voisard *et al.*, 1989). We have now repeated these experiments with the *hcn* deletion mutant CHA77; similar results were obtained although the loss of suppressive potential was somewhat less pronounced than in experiments with strain CHA5 (data not shown). In contrast, strain CHA77 protected wheat from *Ggt* at wildtype levels in a gnotobiotic system. This result raises the interesting question whether a protective effect of bacterial cyanogenesis depends on the plant, the pathogen, or both.

3.5. ROLE OF 2,4-DIACETYLPHLOROGLUCINOL

A Phl-negative mutant (CHA625) was obtained by Tn5 insertion. The recombinant cosmid pME3101, which carries a 22 kb insert of chromosomal DNA from strain CHA0, complements CHA625 for Phl synthesis both *in vitro* and in the rhizosphere of wheat (Table 1). As we have shown previously (Keel *et al.*, 1990[ab]), the *phl*::Tn5 mutation causes a marked reduction in the suppressive effect on black root rot of tobacco and take-all of wheat (Table 1). The cosmid pME3101 in strain CHA625 restored the suppressive effect, albeit not to the full extent (Table 1). These results suggest that Phl contributes importantly to the suppression of black root rot and take-all. The amount of Phl produced by the wildtype in the rhizosphere (Table 1) was clearly too low to give a visible herbicidal effect since the bacteria did not affect plant growth in the absence of the pathogens. However, it is conceivable that Phl, like other herbicides (Altman and Campbell, 1977; Cohen *et al.*, 1986), can induce plant defence mechanisms against the pathogens. An alternative hypothesis is that Phl synthesized in the rhizosphere locally inhibits growth of the pathogens. The fungitoxic effect of Phl has been demonstrated *in vitro* (Reddi *et al.*, 1969; Keel *et al.*, 1990[ab]).

TABLE 1. Role of 2,4-diacetylphloroglucinol (Phl) in disease suppression

Strain	Phl production		Suppressive effect (%)[b] on	
	on malt agar[a] (μg/ml)	in rhizosphere[b] of wheat (μg/g)	black root rot (tobacco)	take-all (wheat)
CHA0	2.5	0.03	100	100
CHA625 (phl::Tn5)	<0.02	ND	31	56
CHA625/pME3101 (phl+)	1.8	0.01	79	76

[a] At 27°C, after 7 d
[b] In gnotobiotic system; suppressive effect in terms of plant weight
ND, not detectable

3.6. BENEFICIAL OR DELETERIOUS ?

Does enhanced antibiotic production improve the suppressive properties of *P. fluorescens* ? This question was addressed in the following experiment. Recombinant cosmids were mobilized to strain CHA0 and tested for increased inhibition of *Pythium ultimum,* a fungus that is sensitive to pyoluteorin (Plt) (Howell and Stipanovic, 1980). A cosmid (pME3090) was identified causing a large zone of inhibition in King's B agar with *Pythium* as the indicator. This cosmid had a 22 kb insert of *P. fluorescens* DNA and, in a CHA0 background, enhanced the production of Plt 4-fold and that of Phl 2-fold on agar plates. The performance of strains CHA0 and CHA0/pME3090 was then tested on plants infected with *Pythium*. Cucumber plants treated with either strain, in the absence of the pathogen, had the same weight as had plants grown without bacteria. Under conditions of severe *Pythium*-induced disease in a gnotobiotic system, strain CHA0 gave 40% protection whereas strain CHA0/pME3090 afforded 80% protection (in terms of plant weight). Thus, enhanced antibiotic production was correlated with improved disease suppression. Cress, under similar conditions, was effectively protected from *P. ultimum* by strain CHA0 (60% protection). Strain CHA0/pME3090 did not show improved suppression but, on the contrary, suppressed disease poorly. Moreover, this strain was deleterious to cress in the absence of *Pythium* (50% reduction in plant weight), whereas the wildtype CHA0 had no adverse effect on cress. We believe that the production of Plt and Phl by strain CHA0/pME3090 was high enough to exert a herbicidal effect on cress. In conclusion, strain CHA0 carrying the recombinant plasmid pME3090 was both beneficial and deleterious - depending on the plant.

4. Acknowledgements

We thank Valeria Gaia und Maria Sie for their help with some experiments and Marie-Thérèse Lecomte for secretarial assistance. This work was supported by the Schweizerische Nationalfonds (project 3.100-25321).

5. References

Altman, J. and Campbell, C. L. (1977) Effect of herbicides on plant diseases, *Annu. Rev. Phytopathol.* **15**, 361-385.

Anderson, A. J., Habibzadegah-Tari, P., and Tepper, C. S. (1988) Molecular studies on the role of root surface agglutinin in adherence and colonization by *Pseudomonas putida*, *Appl. Environ. Microbiol.* **54**, 375-380.

Broadbent, D., Mabelis, R. P., and Spencer, H. (1976) C-Acetylphloroglucinols from *Pseudomonas fluorescens*, *Phytochemistry* **15**, 1785.

Cohen, R., Riov, J., Lisker, N., and Katan, J. (1986) Involvement of ethylene in herbicide-induced resistance to *Fusarium oxysporum* f. sp. *melonis*, *Phytopathology* **76**, 1281-1285.

Davison, J. (1988) Plant beneficial bacteria, *Bio/Technology* **6**, 282-286.

Défago, G., and Haas, D. (1990) Pseudomonads as antagonists of soilborne plant pathogens: modes of action and genetic analysis, in J.-M. Bollag and G. Stotzky (eds.), Soil Biochemistry, vol. 6, Marcel Dekker, Inc., New York, pp. 249-291.

Défago, G., Berling, C. H., Burger, U., Haas, D., Kahr, G., Keel, C., Voisard, C., Wirthner, P., and Wüthrich, B. (1990) Suppression of black root rot of tobacco and other root diseases by strains of *Pseudomonas fluorescens*: potential applications and mechanisms, in R. J. Cook, Y. Henis and D. Hornby (eds.) Biological Control of Soil-borne Plant Pathogens, CAB International, pp. 93-108.

Fravel, D. R. (1988) Role of antibiotics in the biocontrol of plant diseases, *Annu. Rev. Phytopathol.* **26**, 75-91.

Howell, C. R., and Stipanovic, R.D. (1980) Suppression of *Pythium ultimum*-induced damping-off of cotton seedlings by *Pseudomonas fluorescens* and its antibiotic, pyoluteorin, *Phytopathology* **70**, 712-715.

Keel, C., Voisard, C., Berling, C. H., Kahr, G., and Défago, G. (1989) Iron sufficiency, a prerequisite for the suppression of tobacco black root rot by *Pseudomonas fluorescens* strain CHA0 under gnotobiotic conditions, *Phytopathology* **79**, 584-589.

Keel, C., Wirthner, P., Oberhänsli, T., Voisard, C., Burger, U., Haas, D., and Défago, G. (1990[a]) Pseudomonads as antagonists of plant pathogens in the rhizosphere: role of the antibiotic 2,4-diacetylphloroglucinol in the suppression of black root rot of tobacco, *Symbiosis 9*, pp. 15 (in press).

Keel, C., Maurhofer M., Oberhänsli, Th., Voisard, C., Haas, D., and Défago, G. (1990[b]) Role of 2,4-diacetylphloroglucinol in the suppression of take-all of wheat by a strain of *Pseudomonas fluorescens* . In: Proceedings of the First Conference of the EFPP on Biotic Interactions and Soil-Borne Diseases, Wageningen, The Netherlands, in press.

Kloepper, J. W., Lifshitz, R., and Schroth, M. N. (1988) *Pseudomonas* inoculants to benefit plant production, *ISI Atlas of Science-Animal and Plant Sciences*, 60-64.

Kloepper, J. W., Lifshitz, R., and Zablotowicz, R. M. (1989) Free living bacterial inocula for enhancing crop productivity, *Trends Biotechnol.* **7**, 39-44.

Leong, J. (1986) Siderophores: their biochemistry and possible role in the biocontrol of plant pathogens, *Annu. Rev. Phytopathol.* **24**, 187-209.

Morris, R. O. (1986) Genes specifying auxin and cytokinin biosynthesis in phytopathogens, *Annu. Rev. Plant Physiol.* **37**, 509-538.

Narumiya, S., Takai, K., Tokuyama, T., Noda, Y., Ushiro, H., and Hayaishi, O. (1979) A new metabolic pathway of tryptophan initiated by tryptophan side chain oxidase, *J. Biol. Chem.* **254**, 7007-7015.

Ohmori, T., Hagiwara, S., Veda, A., Minoda, Y., and Yamada, K. (1978) Production of pyoluteorin and its derivatives from *n*-paraffin by *Pseudomonas aeruginosa* S10B2, *Agr. Biol. Chem.* **42**, 2031-2036.

Reddi, T. K. K., Khudyakov, Ya. P., and Borovkov, A. V. (1969) *Pseudomonas fluorescens* strain 26-o, a producer of phytotoxic substances, *Mikrobiologiya* **38**, 909-913.

Reddi, T. K. K. and Borovkov, A. V. (1970) Antibiotic properties of 2,4-diacetylphloroglucinol produced by *Pseudomonas fluorescens* strain 26-o, *Antibiotiki (Moscow)* **15**, 19-21.

Schippers, B., Bakker, A. W., and Bakker, P. A. H. M. (1987) Interactions of deleterious and beneficial rhizosphere microorganisms and the effect of cropping practices, *Annu. Rev. Phytopathol.* **25**, 339-358.

Schroth, M. N. and Hancock, J. G. (1982) Disease-suppressive soil and root-colonizing bacteria, *Science* **216**, 1376-1381.

Stutz, E. W., Défago, G., and Kern, H. (1986) Naturally occurring fluorescent pseudomonads involved in suppression of black root rot of tobacco, *Phytopathology* **76**, 181-185.

Stutz, E. W., Kahr, G., and Défago, G. (1989) Clays involved in suppression of tobacco black root rot by a strain of *Pseudomonas fluorescens*, *Soil Biol. Biochem.* **21**, 361-366.

Tari, P. H. and Anderson, A. J. (1988) *Fusarium* wilt suppression and agglutinability of *Pseudomonas putida*, *Appl. Environ. Microbiol.* **54**, 2037-2041.

Thomashow, L. S., and Weller, D. M. (1988) Role of phenazine antibiotic from *Pseudomonas fluorescens* in biological control of *Gaeumannomyces graminis* var. *tritici*, *J. Bacteriol.* **170**, 3499-3508.

Thomashow, L. S., Weller, D. M., Bonsall, R. F., and Pierson III, L. S. (1990) Production of the antibiotic phenazine-1-carboxylic acid by fluorescent *Pseudomonas* species in the rhizosphere of wheat, *Appl. Environ. Microbiol.* **56**, 908-912.

Vesper, S. J. (1987) Production of pili (fimbriae) by *Pseudomonas fluorescens* and correlation with attachment to corn roots, *Appl. Environ. Microbiol.* **53**, 1397-1405.

Voisard, C., Rella, M., and Haas, D. (1988) Conjugative transfer of plasmid RP1 to soil isolates of *Pseudomonas fluorescens* is facilitated by certain large RP1 deletions, *FEMS Microbiol. Lett.* **55**, 9-14.

Voisard, C., Keel, C., Haas, D., and Défago, G. (1989) Cyanide production by *Pseudomonas fluorescens* helps suppress black root rot of tobacco under gnotobiotic conditions, *EMBO J.* **8**, 351-358.

Weller, D. M. (1988) Biological control of soilborne plant pathogens in the rhizosphere with bacteria, *Annu. Rev. Phytopathol.* **26**, 379-407.

Yoneyama, K., Konnai, M., Honda, I., Yoshida, S., Takahashi, N., Koike, H., and Inoue, Y. (1990) Phloroglucinol derivatives as potent photosystem II inhibitors, *Z. Naturforsch.* **45c**, 317-321.

TESTS OF SPECIFICITY OF COMPETITION AMOUND *PSEUDOMONAS SYRINGAE* STRAINS ON PLANTS USING RECOMBINANT ICE-STRAINS AND USE OF ICE NUCLEATION GENES AS PROBES OF *IN SITU* TRANSCRIPTIONAL ACTIVITY

S. E. LINDOW
Department of Plant Pathology
University of California
147 Hilgard Hall
Berkeley, CA 94720

Abstract

Pseudomonas syringae is the most commonly observed and the most active ice nucleatins bacterium on the surfaces of healthy plants. This and other ice nucleation active (Ice$^+$) bacterial species can contribute to frost damage to plants by initiating ice formation in plants that would otherwise supercool and avoid damaging ice formation. The biological control of frost injury can be achieved by the application of non-ice nucleation active bacteria to plant surfaces before they become colonized by Ice$^+$ species. Frost injury to plants is reduced proportional to the reduction of the log population size of Ice$^+$ bacteria on plants, and non-ice nucleating bacteria reduce frost injury by competitive exclusion of Ice$^+$ bacterial species. A functional diversity exists between Ice$^+$ *P. syringae* strains; the resultant specificity in competition should limit the effectiveness of the competitive exclusion of all Ice$^+$ *P. syringae* strains by the application of a single Ice$^-$ strain. Tests of the functional diversity among *P. syringae* strains were made by producing by several isogenic Ice$^-$ recombinant mutants of *P. syringae* using cloned *ice* genes by a process of marker exchange mutagenesis. Recombinant Ice$^-$ strains of *P. syringae* that were applied to potato under field conditions colonized inoculated plants in high numbers (population sizes in excess of 10^6 cells/g fresh wt) for a period of one month or more. The population size of indigenous Ice$^+$ *P. syringae* strains were reduced from 30 to 180 fold on inoculated plants compared to untreated control plants. However, the population size of isogenic Ice$^+$ *P. syringae* strains co-inoculated with the respective Ice$^-$ mutant was reduced over 1,000 fold, indicating that specificity of competition does occur on leaf surfaces. Due to the high sensitivity by which ice nucleation activity can be measured, the more than linear increase in ice nucleation activity exhibited in strains with increases in ice protein content, and lack of endogenous ice nucleating agents active in temperatures of -5°C or higher, a new method of assessing the transcriptional activity of genes has been developed based on cloned ice nucleation genes. The transcriptional activity of an iron regulated genomic region required for the biosynthesis of siderophores in *P. syringae* was assessed *in situ* on leaves and roots by appropriate fusions of this region with a promotorless *ina*Z gene. Ice nuclei produced by *ina*Z in response to transcriptional activity of the fused siderophore gene were recovered on leaves and roots but only at levels of less than 10% that expected had the siderophore biosynthesis gene been fully transcriptionally active *in situ*.

H. Hennecke and D. P. S. Verma (eds.),
Advances in Molecular Genetics of Plant-Microbe Interactions, Vol. 1, 457–464.
© 1991 *Kluwer Academic Publishers. Printed in the Netherlands.*

Ecological Role of Ice Nucleation Active Bacteria

Several bacterial species including about half of the described pathovars of *Pseudomonas syringae*, some strains of *P. fluorescens, Erwinia herbicola, P. viridiflava*, and a few pathovars of *Xanthomonas campestris* can catalyze ice formation in water at temperatures only slightly below 0°C [1-6]. All of these species have the potential to colonize the surfaces of healthy plants in high numbers. Ice nucleation active (Ice+) strains of these bacteria, particularly *P. syringae*, occur in population sizes ranging from undetectable to an excess of 10^6 cells/g fresh wt on different plant species at different times of the year [2,7,8]. Frost injury to sensitive plant species can be incited by epiphytic populations of any Ice+ bacterium. The incidence of freezing injury to plants is proportional to the logarithm of the population size of ice nucleation active bacteria [9]. However, most plant species do not contain indigenous ice nucleation agents capable of catalyzing ice at temperatures warmer than about 5°C [9]. Thus in the absence of heterologous ice nucleating agents such as Ice+ bacteria, the water in plant tissues can supercool and the plant can avoid damaging ice formation. Many ice nucleation active bacteria including *P. syringae, P. viridiflava*, and *X. campestris* pathovars are potential pathogens on the plants in which they live. While many plant pathogens exist as saprophytes on the surfaces of healthy plants prior to infection, factors which lead to the transition from a saprophytic to a pathogenic phase are largely unknown [10]. It is possible that ice nucleation activity is a temperature conditional virulence factor that these bacteria employ to gain entrance into plant tissue. Bacteria may damage tissues directly by the initiation of ice crystals which create portals for the invasion of plants or the bacteria may be introduced into tissues by the freezing and rehydration process occurring in frost tolerant plant species.

Biological Control of Frost Injury

The incidence of freezing injury to frost sensitive plants has been reduced significantly by the application of naturally occurring non-ice nucleating bacteria to plants [8,11,12]. Naturally occurring bacterial strains differ greatly in their ability to reduce frost injury to plants inoculated with Ice+ *P. syringae* strains. The reduction of incidence in frost injury to plants is correlated with the reduction in population size of Ice+ bacteria on plants at the time of freezing stress [13]. The population size of Ice+ bacteria can be reduced from 10 to 1000 fold on plants inoculated with non-ice nucleation active bacteria compared to control plants. Populations of Ice+ bacteria are apparently reduced by a process of competitive exclusion. While the population size of established Ice+ bacteria are not reduced by the subsequent application of a non-nucleating bacterium, the prior establishment of a non-Ice+ bacterium can greatly reduce the potential multiplication of an Ice+ bacterial strain subsequently applied to plants [14]. Both non-ice nucleation active and Ice+ bacteria therefore appear to compete for limiting environmental resources that are contributed by the leaf. We expect however, that Ice+ *P. syringae* strains are ecologically diverse (utilize different limiting resources).

We have tested the specificity of competition among *P. syringae* strains on leaves by constructing several isogenic Ice- recombinant mutants of Ice+ strains. We have hypothesized that if different Ice+ bacteria utilize different resources on leaf surfaces, then isogenic bacterial strains which should be identical to Ice+ parental strains in their resources requirements, should be optimally competitive towards that strain [15]. In contrast, non-isogenic Ice- mutants derived from heterologous Ice+ *P. syringae* strains should not be as capable of the a competitive exclusion of an Ice+ strain as the isogenic Ice- derivative. Ice genes were cloned from three different *P. syringae* strains and deletions internal to the coding ice regions were made *in vitro* [16,17]. Isogenic Ice⁻ derivatives of each parental strains were made by a process of marker exchange mutagenesis using modified *ice* genes on the non-replicative plasmid pBR325 as a suicide delivery vector [17]. While ice genes from different *P. syringae* strains share considerable sequence homology, we observed that sufficient diversity existed among these sequences that marker exchange mutagenesis was not practical unless homologous *ice* genes were utilized in marker exchange mutagenesis.

While marker exchange mutants of *P. syringae* strains were functionally Ice⁻ they did not differ measurably in any other phenotype measured in culture, in the colonization of plant tissues, or in survival characteristics measured in laboratory and greenhouse experiments [17]. Such strains therefore were useful for our tests of the specificity of competition among *P. syringae* strains on leaves because their behavior on plants surfaces appeared not to be altered by the removal of the appropriate *ice* gene.

Recombinant Ice⁻ mutants of *P. syringae* were inoculated onto potato plants under field conditions in what constituted the first authorized release of recombinant microorganisms in the United States [18]. Ice⁻ mutants were applied to potato seed pieces at the time of planting and were subsequently sprayed onto young plants as they emerged from the soil. Ice⁻ mutants were detected in high populations (in excess of 10^5 cells/g fresh wt) for between 4-7 weeks following emergence of plants from the soil. Population size of indigenous Ice⁺ bacteria that had immigrated to inoculated plants was reduced approximately 100 fold on plants inoculated with Ice⁻ mutants compared to untreated control plants, indicating that competitive exclusion had occurred under field conditions as in greenhouse and growth chamber studies [19]. The incidence of frost injury to plants was reduced from about 50% to nearly 85% during natural frost events which occurred at the experimental sites during 1987 and 1988 [19]. The lowered incidence of frost injury to treated plants was close to that expected based on the reduction of population size of indigenous Ice⁺ bacteria on treated plants. It was noteworthy however, that the population size of isogenic Ice⁺ *P. syringae* strains that were co-inoculated in some treatments with a given recombinant Ice⁻ mutant was reduced up to 1000 fold. The larger reduction of isogenic bacterial strains compared to that of the reduction of a mixture of indigenous Ice⁺ *P. syringae* strains at the field site indicated that some specificity was involved in the competition among bacteria on plant surfaces [19]. These results strongly suggest that populations of Ice⁺ *P. syringae* strains may be ecologically diverse. More than one ecotype of a give bacterial species such as *P. syringae* may exist at a given field site. Therefore, any one Ice⁻ strain or naturally occurring bacterium may not be able to compete effectively with all other Ice⁺ bacteria present at that site (Figure 1).

Similarly, while ecological characteristics of Ice⁻ mutants which condition their growth and competition on plant surfaces probably coincide exactly with that of a subset of the indigenous Ice⁺ bacterial strains, a similar amount of ecological overlap may also exist among some other non-nucleating bacterial species. Until we better understand what resources the bacteria compete for on leaf surfaces, the production of antagonistic bacteria by the removal of deleterious traits such as ice nucleation genes provide a method by which at least a portion of a target bacterial population can be antagonized without elaborate screening methods to identify strains from among randomly collected bacteria.

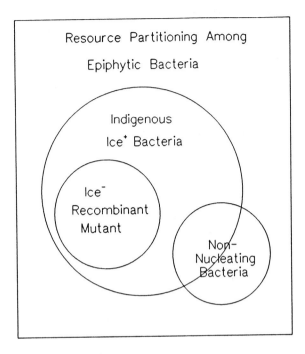

Figure 1

Ice Nucleation Activity as a Reporter of Transcriptional Activity

In both basic studies of the mechanisms by which bacteria interact with plants as well as in ecological studies of the association of bacteria with plants and with themselves on plants it is necessary to know both a) when a particular gene of interest is active and b) what biological or environmental factors regulate gene activity. In many instances of bacterial-plant interactions such as those involving pathogenesis, a gene whose product has not been fully characterized or, a gene product whose function is not known may be the sole basis for a particular investigation into that association. Alternatively, some bacterial components such as antibiotics or toxins thought to be important in plant-microbe interactions may be well characterized in culture but may be sufficiently labile that they cannot be recovered from natural habitats or they may not be recoverable because of their irreversible association with environmental constituents such as clay particles, etc. which render them non-extractable. It should be noted that while bacteria elaborate many interesting products or express a number of interesting traits that are easily measurable or observable in culture, these triats frequently cannot be detected in nature. It is unclear whether this represents general phenotypic plasticity by which a different collection of gene products are elaborated in natural environments than are expressed in culture conditions or whether methods by which gene products are assessed in natural environments are inadequate to measure the presence of these products in natural environments.

A powerful method to assess the activity of genes for which the product is unknown or difficult to measure is to place the transcriptional activity of a "reporter" gene, whose product is easily measured, under the regulatory control of the gene to be studied. Such genes as *lac*Z and *cat* have been useful as "reporter genes" in cultures since their products can be easily measured and quantified under the controlled conditions of many laboratory applications [20,21]. However, the products of some of these reporter genes such as ß-galactosidase encoded by *lacZ* cannot be readily quantified in natural habitats such as in or on plants. For example, plant enzymes having ß-galactosidase activity yield too high a background activity for the ease of detection of poorly transcribed bacterial genes fused to *lacZ* when placed in or on plants. Other reporter genes such as *gus* and *lux* while not having problems with specificity leading to high background levels because of the lack of similar enzymatic activities in plants [22] may have other limitations in some applications. For example, emission of significant amounts of light from luciferase encoded products of the *lux* genes is dependent on high levels of metabolic activity in bacterial cells in which gene activity is being assessed. High numbers of cells containing *lux* or *gus* gene fusions may be necessary for the detection of activity of genes at low levels of transcription when fused to these reporter genes.

We recently have developed a novel reporter gene system based on ice nucleation activity [23]. A promotorless *inaZ* gene has proven to be a versatile reporter of transcriptional activity of genes cloned 5' to this sequence because of the high sensitivity of this system. The expression of ice nucleation activity in a single bacterial cell containing *inaZ* gene fusions is easily measured if high levels of transcription are present. Even poorly transcribed genes can be detected in as few as 1000 cells. Ice nucleation activity was found to be more that 10^5 times more sensitive than ß-galactosidase as a measure of the transcription [23]. Small differences in levels of transcription of *inaZ* gene fusions are measurable because of the amplified nature of the ice nucleation activity produced in response to gene fusions. A non-linear relationship exists between the content of the *inaZ* gene product in a cell and ice nucleation activity expressed in that cell [23]. Ice nucleation activity increases by the relationship of X^2 to X^3 with each unit X of transcript conferred by the gene of interest. Because ice nucleation activity at temperatures such as -5°C is thought to require the proper assembly of several *inaZ* encoded proteins, the assembly of multimers of the ice protein is quite sensitive to the concentration of *inaZ* gene product present. Ice nucleation activity has proven to be a highly specific assay of gene transcription products since most substances including plant tissue, do not contain significant concentrations of ice nuclei active at temperatures as warm as -5°C, a typical assay temperature for bacterial ice nuclei. Ice nucleation activity can be used as a versatile "reporter gene" since all gram negative bacterial species efficiently express introduced ice nucleation genes [23]. Ice nucleation activity has also been expressed with modest efficiency when appropriately modified *ice* genes have been introduced into yeast or gram positive bacteria (D. Pridmore, personal communication). Ice nucleation activity is also expressed reasonably well in appropriate expression vectors in tobacco and *Solanum* species (D. Baertlein, S. E. Lindow, and T. Chen, and N. J. Panopoulos, unpublished).

A better understanding of the regulation of bacterial genes involved in plant associations have been ascertained utilizing ice nucleation as a reporter gene. Several *hrp* and *avr* genes have been fused with promotorless ice nucleation genes both by direct cloning 5' to the promotorless *inaZ* gene or via the introduction of the promotorless *inaZ* into cloned genes on the vector Tn3 spice [23,24]. The use of ice nucleation as a promotor probe has revealed rapid increases in transcriptional activity of *hrp* genes when introduced into the plant environment and of *avr* genes exposed to particular balances of carbon and nitrogen compounds which may be reflective of the plant environment.

Recent information on the transcriptional activity of the siderophore biosynthetic genes in *P. syringae in situ* has been obtained utilizing a promotorless *inaZ* gene. Siderophores, compounds with a high affinity for iron that are elaborated in culture conditions deficient in Fe^{+3} are considered important in the uptake of iron needed for cellular metabolism and have been speculated to be involved in the interactions of microorganisms with each other on plants via the sequestering of iron [25]. While siderophores are prominently produced in iron deficient culture media, they have not been detected on plant surfaces or in other natural environments. Their lack of detection may reflect low levels of production in natural environments or their binding to constituents of the natural environment such as plant components or minerals making their recovery difficult. Plasmid pJEL1701 was constricted by cloning an 8 kb DNA fragment from the cosmid pSF112 [26] which is essential for fluorescent siderophore production (Flu) of *P. syringae* upstream of a promotorless *inaZ* gene in the stable plasmid pVSP61. *P. syringae* and *P. fluorescens* strains harboring pJEL1701 expressed approximately one ice nucleus per cell when grown in an iron-deplete minimal medium and only about 10^{-5} ice nuclei per cell when the medium was supplemented with 10^{-4} molar or more $FeCl_3$ or more. In contrast, cells harboring pJEL1703 comprised of an ice nucleation gene (*iceC*) transcribed from its native iron constituent promotor cloned in pVSP61, expressed about 10^{-3} ice nuclei in both iron replete and iron deficient media [27]. On leaf surfaces, *P. syringae* and *P. fluorescens* cells harboring pJEL1701 expressed ice nucleation activity at a level about 3-fold less than those harboring pJEL1703. On root surfaces, *P. fluorescens* cells harboring pJEL1701 expressed ice nucleation activity at a level about 10-fold less than those harboring pJEL1703. *P. fluorescens* cells harboring pJEL1701 expressed less ice nucleation activity on root surfaces in soil amended with iron EDDHA or iron EDTA, than in unamended soil. Results suggest that the Flu gene is transcribed on leaf and root surfaces but only at about 5% the level observed in an iron deficient medium.

1. Lindow, S. E. (1983) 'The role of bacterial ice nucleation in frost injury to plants', Ann. Rev. Phytopathology, 21,363-384.
2. Lindow, S. E., Arny, D. C. and Upper, C. D. (1978) 'Distribution of ice nucleation active bacteria on plants in nature', Appl. Envir. Microbiol., 36,831-838.
3. Lindow, S. E., Arny, D. C. and Upper, C. D. (1978) '*Erwinia herbicola*: a bacterial ice nucleus active in increasing frost injury to corn', Phytopathology 68,523-527.
4. Maki, L. R., K. J. Willoughby, (1978) 'Bacteria as biogenic sources of freezing nuclei', J. Appl. Meterol., 17,1049-1053.
5. Maki, L. R., Galyon, E. L., Chang-Chien, M., and Caldwell, D. R. (1974) 'Ice nucleation induced by *Pseudomonas syringae*, Appl. Microbiol., 28,456-460.
6. Lim, H. K., Orser, C.S., Lindow, S. E. and Sands, D. C. (1987) '*Xanthomonas campestris* pv. *translucens* strains active in ice nucleation', Plant Dis., 71,994-997.
7. Gross, D. C. , Cody, Y. S., Proebsting, E. L., Radamaker, G. K., and Spotts, R. A. (1983) 'Distribution, population dynamics, and characteristics of ice nucleation active bacteria in deciduous fruit tree orchards', Appl. Environ. Microbiol., 46,1370-1379.
8. Lindow, S. E. (1983) 'Methods of preventing frost injury caused by epiphytic ice nucleation active bacteria', Plant Disease, 67,327-333.
9. Lindow, S. E., Arny, D. C. and Upper, C. D. (1982) 'Bacterial ice nucleation: a factor in frost injury to plants, Plant Physiol., 70,1084-1089.
10. Hirano, S. S., and Upper, C. D. (1983) 'Ecology and epidemiology of foliar plant pathogens', Annu. Rev. Phytopathol, 21,243-269.
11. Lindow, S. E., Arny, D. C. and Upper, C. D. (1983) 'Biological control of frost injury II: Establishment and effects of an antagonistic *Erwinia herbicola* isolate on corn in the field', Phytopathology 73,1102-1106.
12. Lindow, S. E. (1985) 'Strategies and practice of biological control of ice nucleation active bacteria on plants' in N. Fokkema (ed.), Microbiology of the Phyllosphere, Cambridge University Press, pp. 293-311.
13. Lindow, S. E. (1982) 'Population dynamics of epiphytic ice nucleation active bacteria on frost sensitive plants and frost control by means of antagonistic bacteria', in P. H. Li and A. Sakai (eds.), Plant Cold Hardiness, Academic Press, New York, pp. 395-416.
14. Lindow, S. E., Arny, D. C. and Upper, C. D. (1983) 'Biological control of frost injury I: An isolate of *Erwinia herbicola* antagonistic to ice nucleation-active bacteria', Phytopathology, 73,1097-1102.
15. Lindow, S. E. (1988) 'Construction of isogenic Ice⁻ strains of *Pseudomonas syringae* for evaluation of specificity of competition on leaf surfaces', in F. Megusar and M. Gantar (eds.), Microbial Ecology, Slovene Society for Microbiology, Ljuvljana, pp. 509-515.
16. Orser, C. S., Staskawicz, B. J., Panopoulos, N. J., Dahlbeck, D., and Lindow, S. E. (1985) 'Cloning and expression of bacterial ice nucleation genes in *Escherichia coli*', J. Bacteriol, 164,359-366.
17. Lindow, S. E. (1985) 'Ecology of *Pseudomonas syringae* relevant to the field use of Ice⁻ deletion mutants constructed *in vitro* for plant frost control' in Engineering organisms in the environment: Scientific issues, Am. Soc. Microbiology, Washington, D.C., pp. 23-25.

464

18. Lindow, S. E. and Panopoulos, N. J. (1988) 'Field tests of recombinant Ice⁻ *Pseudomonas syringae* for biological frost control in potato' in M. Sussman, C. H. Collins, and F. A. Skinner (eds.), Proc. First Internat'l. Conf. on Release of Genetically Engineered Microorganisms, Academic Press, London, pp. 121-138.

19. Lindow, S. E., Panopoulos, N. J., Pierce, C., Andersen, G. and Lim, G. (1988) 'Biological control of frost injury to potato with recombinant Ice⁻ strains of *Pseudomonas syringae* ant their survival and dispersal at a test site', Phytopathology, 78,1552.

20. Stachel, S. E., On, G., Flores, C. and Neiter, E. W. (1985) 'A Tn3*lacZ* transposon for the random generation of ß-galactosidase gene fusions: application to the analysis of gene expression in *Agrobacterium*', EMBO J., 4,891-898.

21. Osbourn, A. E., Barker, C. E. and Doniele, M. J. (1987) Identification of plant-induced genes of the bacterial pathogen *Xanthomonas campestris* pv. *campestris* using a promoter-probe plasmid', EMBO J., 6,23-28.

22. Jefferson, R. A. (1989) 'The GUS reporter gene system', Nature, 342,837-838.

23. Lindgren, P. B., Govindarajan, A. G., Frederick, R., Panopoulos, N. J., Staskawicz, B. J. and Lindow, S. E. (1989) 'An ice nucleation reporter gene system: identification of inducible pathogenicity genes in *Pseudomonas syringae* pv. *phaseolicola*', EMBO J., 8,1291-1301.

24. Huynh, T., Dahlbeck, D. and Staskawicz, B. J. (1989) 'Bacterial blight of soybean: regulation of pathogen gene determining host cultivar specificity', Science, 245,1374-1377.

25. Loper, J. E. (1988) 'Role of fluorescent siderophore production in biological control of *Pythium ultimum* by *Pseudomonas fluorescens* strain', Phytopathology, 78,166-172.

26. Loper, J. E., Orser, C. S., Panopoulos, N. J. and Schroth, M. N. (1984) 'Genetic analysis of fluorescent pigment production *Pseudomonas syringae* pv. *syringae*', J. Gen. Microbial., 130,1507-1515.

27. Lindow, S. E. and Loper, J. E. (1990) 'Transcriptional activity of fluorescent siderophore genes from *Pseudomonas syringae* in situ on leaf and root surfaces', Phytopathology, (in press).

REGULATION OF THE SYNTHESIS OF INDOLE-3-ACETIC ACID IN AZOSPIRILLUM

W. ZIMMER and C. ELMERICH
Unité de Physiologie Cellulaire, CNRS URA 1300
Institut Pasteur, Dept. des Biotechnologies
25 rue du Dr. Roux
75724 Paris Cedex 15
France

ABSTRACT. Bacteria of the genus *Azospirillum* are associated with the roots of *gramineae* and produce the phytohormone indole-3-acetic acid (IAA). In order to characterize genes involved in IAA biosynthesis, a cosmid library of *A. brasilense* Sp7 was transferred by conjugation into *A. irakense* KA3, a low producer of IAA. IAA production was increased in two transconjugants, both containing an identical 18.5 kb *Hind*III fragment of the *A. brasilense* genome. Tn5 mutagenesis was performed with one of these cosmids. Tn5 insertions at 36 different positions in the 18.5 kb fragment were transferred into strain KA3 and the IAA production of the recipient clones was screened by HPLC. The Tn5 insertions of 4 clones with decreased IAA production were mapped on a 2 kb *Sal*I/*Sph*I fragment. Recombination of two of these Tn5 insertions into the genome of strain Sp7 led to Trp auxotrophic mutants. Sequencing of the region revealed that it contains part of *trpG*, complete *trpD* and part of *trpC*, organized in a single operon. To explain that introducing *trpD* into strain KA3 enhances IAA production, a model is proposed in which anthranilate inhibits IAA synthesis. Several experimental results corroborate this model.

1. Introduction

Production of phytohormones by *Azospirillum* ,which lives in association with the roots of grasses, has been proposed to be responsible for the increase of the root surface of the host-plant (reviewed by Elmerich et al. 1990). It has been established that indole-3-acetic acid (IAA) production by *Azospirillum* is tryptophan (Trp)-dependent (Tien et al. 1979, Zimmer and Bothe 1988). Mutants of *A. lipoferum* obtained by random Tn5 mutagenesis (Abdel-Salam and Klingmüller 1987) showed a reduced IAA production rate. No mutant, completely unable to synthesize IAA was isolated, suggesting that *Azospirillum* possesses more than a single pathway for IAA synthesis. The biosynthetic pathway(s) has(have) not yet been elucidated. Moreover, a regulation mechanism probably controls IAA synthesis, since the release of this compound was only observed when the bacteria were supplied with Trp. Results reported here show that IAA synthesis in *Azospirillum* is inhibited by anthranilate.

2. Results and Discussion

2.1. SEARCH FOR THE PATHWAY OF INDOLE-3-ACETIC ACID SYNTHESIS IN AZOSPIRILLUM BRASILENSE

The best investigated pathway for IAA synthesis was first described in the phytopathogen *Pseudomonas syringae* (Magie et al. 1963). A Trp monoxygenase catalyses the conversion of Trp into indole-3-acetamide (IAM). The second step, the formation of IAA is catalysed

465

H. Hennecke and D. P. S. Verma (eds.),
Advances in Molecular Genetics of Plant-Microbe Interactions, Vol. 1, 465–468.
© 1991 *Kluwer Academic Publishers. Printed in the Netherlands.*

by an IAM hydrolase.The genes coding for Trp monoxygenase (*iaaM*) and for IAM hydrolase (*iaaH*) of *P. syringae*, and for the IAM hydrolase (*bam*) of *Bradyrhizobium* have been isolated and sequenced (Yamada et al. 1985; Sekine et al. 1989). No hybridization signal was detected between *A. brasilense* total DNA and intragenic probes from *iaaM* and *iaaH* of *P. syringae*, or *bam* of *Bradyrhizobium* (data not shown). In addition HPLC-analyses showed that *Azospirillum* was unable to convert IAM to IAA. It was therefore concluded that this pathway is not present in *Azospirillum*.

2.2. CLONING AND MAPPING OF AN *A. BRASILENSE* SP7 LOCUS INVOLVED IN THE SYNTHESIS OF INDOLE-3-ACETIC ACID

A cosmid library of strain Sp7 constructed by cloning *Hind*III fragments in the broad host range cosmid vector pVK100 (Fogher et al. 1985) was used to screen for genes involved in IAA synthesis. 400 cosmids were transferred by conjugation into *A. irakense* KA3, a recently described *Azospirillum* species (Khammas et al. 1989), which was found to produce only 1/10 of the IAA amount of *A. brasilense* after 3 days of incubation in a Trp (100 mg/l) containing medium. Two transconjugants, containing pAB1005 and pAB1289 showed a significantly enhanced Trp-dependent IAA production when compared to the control strain, which contained pVK100 without any insert (Fig. 1). Cosmids pAB1005 and pAB1289 carried an identical 18.5 kb *Hind*III fragment, which was verified to be a genuine fragment of strain Sp7 genome by hybridization with *Hind*III digested total DNA of strain Sp7 (data not shown). The region was mapped by restriction analysis (Fig. 1). To localize the locus responsible for the enhancement of the IAA production, Tn5 mutagenesis was performed with pAB1005. 36 different insertions were obtained in the 18.5 kb *Hind*III fragment. After transfer into *A. irakense* KA3, the IAA production of the recipient clones was assayed by reversed phase HPLC. Four transconjugants showed an IAA production level comparable to that of the strain containing the control cosmid pVK100. The others produced as much IAA as strain KA3 containing the wildtype cosmid pAB1005 (Fig. 1). The Tn5 insertions of the 4 clones with reduced IAA production were mapped on a 3 kb *Sal*I fragment. Two of the Tn5 containing fragments were subcloned in the suicide vector pSUP202 and recombined into the genome of wildtype strain Sp7. Surprisingly, the two recombinant strains were Trp auxotroph and their Trp-dependent IAA production was almost as high as that of the wildtype strain Sp7.

2.3. SEQUENCING OF THE LOCUS AND IDENTIFICATION OF A *trp*-OPERON

The nucleotide sequence of the 2 kb *Sal*I/*Sph*I fragment of pAB10053 (Fig. 1) was established. The 4 Tn5 insertions which abolished the enhanced IAA production in *A. irakense*, were mapped in this 2 kb fragment. One complete open reading frame (ORF) encoding a polypeptide of 355 amino acids was localized in the middle of the 2 kb fragment (Fig. 1). The amino acid sequence showed homology with TrpD of *E.coli* (Yanofsky et al. 1981). The highly conserved regions found in TrpD polypeptides (Crawford 1989) were present in the amino acid sequence coded by this ORF. Another incomplete ORF was localized upstream from the 5'end of *trpD*. The deduced amino acid sequence of this ORF showed homology with the C-terminal part of *E. coli* TrpG. A third incomplete ORF starts directly at the 3'end of *trpD* (Fig. 1). Homology between the amino acid sequence coded by this ORF and the N-terminal part of TrpC of *E. coli* was detected. As the three genes *trpG*, *trpD* and *trpC* were adjacent, with the stop codon of the previous gene overlapping the start codon of the next, it is likely that the 3 *trp* genes of *A. brasilense* Sp7 are organized in a single operon.

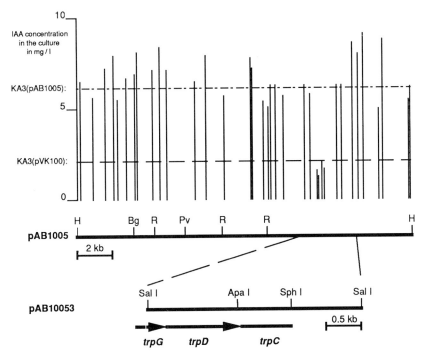

Figure 1. Physical map of pAB1005 and pAB10053 and IAA-production by *A. irakense* KA3 clones, carrying Tn*5* insertions in pAB1005. Restriction sites: H: *Hin*dIII, Bg: *Bgl*II, R: *Eco*RI, Pv: *Pvu*II. The vertical lines in the diagram represent the IAA amount, released in the culture medium by *A. irakense* KA3 clones, carrying a Tn*5* insertion in pAB1005 at the corresponding position. IAA amounts produced by strain KA3 carrying pVK100 and pAB1005 are shown by interrupted horizontal lines. Determination of IAA was performed by reversed phase HPLC (C_{18} column, detection at 278 nm, flow rate 1ml/min, mobile phase methanol : 1% H_3PO_4, 40/60 V/V) after 2 days of incubation in minimal medium supplemented with 100 mg/l Trp and 5 mg/l kanamycin. The arrows under the physical map of pAB10053 indicate the open reading frames established by sequencing.

2.4. MODEL FOR THE REGULATION OF INDOLE-3-ACETIC ACID SYNTHESIS IN *AZOSPIRILLUM*

The *trpD* gene encodes the phosphoribosylanthranilate transferase, which phosphoribosylates anthranilate in the second step of Trp synthesis. HPLC analysis of culture supernatants of *A. irakense* KA3 grown in the presence of Trp (100 mg/l) revealed that this strain consumed Trp faster than *A. brasilense* Sp7. Moreover it was found that strain KA3 released more anthranilate (16 mg/l in 2 days) into the medium than strain Sp 7 (less than 0.5 mg/l in 2 days), suggesting that strain KA3 was able to convert Trp into anthranilic acid. In contrast, strain KA3 released about 1.6 mg/l IAA whereas strain Sp7 released 16 mg/l under these conditions. The production of anthranilate by transconjugants of strain KA3 carrying the *trpD* gene of strain Sp7 on pAB1005 was reduced to 1/10 and the synthesis of IAA was enhanced more than 3 fold compared to the control strain containing pVK100.

These observations led us to propose a model for the regulation of IAA synthesis in *Azospirillum*. According to this model anthranilate inhibits the conversion of Trp to IAA.

In addition to the phenotypes of the KA3 mutants, this model accounts also for the observation that *Azospirillum brasilense* releases IAA only in the presence of Trp (Zimmer and Bothe 1988). In the absence of Trp, the genes for Trp synthesis are transcribed and translated. Consequently, the level of anthranilate, the first intermediate of the pathway, increases. According to the model, this inhibits IAA synthesis and prevents a loss of Trp from the cells. In the presence of external Trp, internal Trp synthesis is blocked and the intracellular level of anthranilate is low, enabling IAA synthesis.

In strain KA3, where the anthranilate concentration is high even in the presence of Trp, IAA synthesis is blocked. The anthranilate concentration in cultures of clones carrying *trpD* of strain Sp7 on pAB1005 was found to be 90% decreased suggesting that TrpD is functional in strain KA3. According to the model, the lower level of anthranilate enabled the observed enhanced IAA production.

The model is also in agreement with the behavior of the Tn*5* insertion TrpD⁻-mutants of strain Sp7 (mentioned above), whose Trp-dependent IAA production was decreased by adding anthranilate to the medium, without affecting growth. Other experiments are in progress to corroborate this model.

3. Acknowledgements

The authors wish to thank Prof. J.-P. Aubert for helpful discussions and Mrs Aparicio for skillful technical assistance. WZ was recipient of a postdoctoral fellowship from the Deutsche Forschungsgemeinschaft. This work was supported by research funds from the University of Paris 7.

4. References

Abdel-Salam, M.S. and Klingmüller, W. (1987), 'Transposon Tn*5* mutagenesis in *Azospirillum lipoferum*: isolation of indole acetic acid mutants', Mol. Gen. Genet. 210, 165-170.

Crawford, I.P. (1989), 'Evolution of a biosynthetic pathway: The tryptophan paradigm', Annu. Rev. Microbiol. 43, 567-600.

Elmerich, C., Zimmer, W. and Vieille, C. (1990), 'Associative nitrogen fixing bacteria', in H. Evans, R. Burris and G. Stacey (eds.), Biological nitrogen fixation, Chapman and Hall, New York, (submitted).

Fogher, C., Bozouklian, H., Bandhari, S.K. and Elmerich, C. (1985), 'Construction of a genomic library of *Azospirillum brasilense* and cloning of the glutamine synthetase gene', in W. Klingmüller (ed.), Azospirillum III: Genetics, Physiology, Ecology, Springer Verlag, Heidelberg, pp.41-52.

Khammas, K.M., Ageron, E., Grimont, P.A.D. and Kaiser, P. (1989), '*Azospirillum irakense* sp. nov., a nitrogen-fixing bacterium associated with rice roots and rhizosphere soil', Res. Microbiol. 140, 679-693.

Magie, A.R., Wilson, E.E. and Kosuge, T. (1963), 'Indoleacetamide as an intermediate in the synthesis of indoleacetic acid in *Pseudomonas savastanoi*', Science 141, 1281-1282.

Sekine, M., Watanabe, K. and Syono, K. (1989), 'Molecular cloning of a gene for indole-3-aceteamide hydrolase from *Bradyrhizobium japonicum*', J. Bacteriol. 171, 1718-1724.

Tien, T.M., Gaskins, M.H. and Hubbel, D.H. (1979), 'Plant growth substances produced by *Azospirillum brasilense* and their effect on the growth of pearl millet (*Pinnesetum americanum*)', Appl. Environm. Microbiol. 37, 219-226.

Yamada, T., Palm, C.J., Brooks, B. and Kosuge, T. (1985), 'Nucleotide sequences of *Pseudomonas savastanoi* indoleacetic acid genes show homology with *Agrobacterium tumefaciens* T-DNA', Proc. Natl. Acad. Sci. USA 82, 6522-6526.

Yanofsky, C., Platt, T., Crawford, I.P., Nichols, B.P., Christie, G.E., Horowitz, H., van Cleemput, M. and Wu, A.M. (1981), 'The complete nucleotide sequence of the tryptophan operon of *Escherichia coli*', Nucl. Acids Res. 9, 6647-6668.

Zimmer, W. and Bothe, H. (1988) 'The phytohormonal interactions between *Azospirillum* and wheat', Plant Soil 110, 239-247.

AUTHOR INDEX

Current Plant Science and Biotechnology in Agriculture

KLUWER ACADEMIC PUBLISHERS – DORDRECHT / BOSTON / LONDON